W9-CUK-314

Methods in X-Ray Crystallography

Methods in X-Ray Crystallography

J. W. JEFFERY

Department of Crystallography,
Birkbeck College,
University of London

1971

ACADEMIC PRESS

London and New York

CHEMISTRY

ACADEMIC PRESS INC. (LONDON) LTD.
24/28 Oval Road
London, NW1

U.S. Edition published by
ACADEMIC PRESS INC.
111 Fifth Avenue,
New York, New York 10003

Copyright © 1971 By ACADEMIC PRESS INC. (LONDON) LTD.

All Rights Reserved

No part of this book may be reproduced in any form by photostat, microfilm, or any other means, without written permission from the publishers

Library of Congress Catalog Card Number: 71–153523
ISBN: 0–12–382250–5

Printed in Great Britain by
ROYSTAN PRINTERS LIMITED
Spencer Court, 7 Chalcot Road
London N.W.1

QD945
J36
1971

CHEMISTRY
LIBRARY

Preface

I have a much clearer picture of the origin of this book than of most other 'happenings' of my life. As an undergraduate I had attended classes under J. D. Bernal at Cambridge. My practical work was to make a gas X-ray tube out of glass tubing, two end plates and some sealing wax, and to take a powder photograph of a material that 'Sage', as J. D. B. was called, wanted identifying. I was then sent, with his characteristically very vague directions, to the Crystallographic reference library to see if the resultant d-values tallied with the cell dimensions of some material whose name I had never heard of, but which I would find characterised in what sounded like the 'Structerberite'. It was not difficult to guess that this was a German publication, but I knew it would be no good to confess that I did not know a word of German, since 'the numbers and chemical symbols are the same'. At this point my memory becomes very vague, so maybe I never even found the library. But the exhilaration of actually taking an X-ray photograph with ones 'own' X-rays was proof against such frustrations, and when the opportunity offered, after the war, to join Bernal's research team at Birkbeck College, I jumped at it.

My first serious research project was to determine the structure of 'alite', the most important component of Portland cement. The crystals had no recognisable faces, but I had managed to get one of them nearly set about a direction which showed every sign of being the c_H axis of a rhombohedral lattice.

However, there were hardly any zero layer reflections and the standard setting method would not work. Sage was impatiently waiting for the rotation photograph which would show whether the crystal was really rhombohedral, and I was just looking miserably at my fourth failure, when he came along, saw the photograph and said, quite unnecessarily, 'But this isn't set yet'. Then, with very uncharacteristic acerbity and lack of modesty, he added, 'It must be possible to do this systematically. It's all in my 1926 paper. Go and look at it!' That remark was as near to the origin of this book as one is likely to get.

Of course, when one is generation 3, or at least $2\frac{1}{2}$, of X-ray crystallographers, one cannot expect to do much direct quarrying in that famous paper whose product, the Bernal chart, is known to every crystallographer. Others

v

9658

have been there first, and of these I owe a debt to no-one more than Martin Buerger. I have never found it possible to improve significantly on his treatment of Weissenberg geometry, and although it is twenty years since I studied 'X-ray Crystallography', I suspect that the flattery of imitation in the Weissenberg section of the present book will be very obvious.

I had not originally intended to include divergent beam geometry in the book, but during its preparation I came across a paper, 'The use of Miller indices and the reciprocal lattice concept in the interpretation of divergent-beam X-ray diffraction patterns', by K. J. H. Mackay. This was such a beautiful illustration of the further use of the reciprocal lattice as a tool for the interpretation of X-ray diffraction that I decided to add a chapter on divergent beam photographs. Ken Mackay not only gave me permission to reproduce some of his illustrations (Figs. 32, 35 and 36), but went to considerable trouble to make suggestions (most of which were incorporated) for improving the text of Chapter 5. However, he is not responsible for any errors which may remain, nor for the details of the method of presentation, which have been altered to fit in with the previous development of the reciprocal lattice in this book. I am very grateful for his collaboration.

Although in most of the book a knowledge of point and space group symmetry operations is assumed, in Appendix II on lattice theory, especially in the section on the effects of centered cells and translation symmetry elements, it was not possible to avoid dealing in some detail with symmetry considerations. Here considerable potential confusion was avoided by incorporating a number of suggestions from Kathleen Lonsdale made during her recent fatal illness. My gratitude for her help however, does not make her responsible for the result, but only for the improvement. It is difficult to realise that it is now no longer possible to rely on receiving her kindly but penetrating comments.

In a text book of this kind acknowledgements could run into pages. My colleagues on the teaching staff of the Crystallography Department at Birkbeck College have cheerfully taken over some of my normal commitments, and allowed me priorities which I would not have got away with under normal conditions! The technical staff of the Department have helped in many ways, in particular Arthur Benton, whose skill and patience have contributed so much to the preparation of the many X-ray photographs which have been specially taken to illustrate the book; Doug Parry, our photographer, who produced the prints for the half tone illustrations; our workshop staff under Len Stevens, who have been responsible for the many pieces of special apparatus, back to the first prototype fine focus X-ray tube in 1950, which have enabled the laboratory to develop effectively and produce the results shown in the illustrations; and John Painter, whose skill with H.T. apparatus and cables, control systems and cameras, wiring and plumbing, has kept us in operation under conditions of fire and flood in the old building and through all the difficulties of moving into the new one.

A whole series of Ph.D. students, especially Ken Rose, Alan Whitaker, Narayan Datta and Roy Baker have helped to develop the methods described in the Appendices, and generations of M.Sc. students have, by their comments and questions, their incomprehensions as well as their triumphant comprehensions, formed the approach to the elucidation of some of the difficult problems which face a newcomer to crystallography. In particular they have reinforced the view that, in the last analysis, you learn to do something by doing it. The examples therefore are an integral part of the book, and many of the finer practical points, as well as alternative methods of approach, are given in the worked answers.

A book must finally get to the printer in typescript, and it is a pleasure to acknowledge my indebtedness to Margaret Watts, and in the last hectic months, to Brenda Holman, who have been responsible for typing the many preliminary drafts, as well as the final copy.

As I am forbidden to mention my wife's part in the production of this book, I can only say that without my literary and technical assistant it could never have been prepared for the press.

Finally, the book itself pays tribute to Academic Press and Roystan Printers, but I would like to add my thanks for their cooperation in dealing with the many problems involved in getting the book into print.

I am grateful to the following for permission to reproduce diagrams or plates: J. M. Robertson, Fig. 1; M. F. Perutz, Fig. 28; E. G. Chirer, Fig. 151; C. A. Taylor, Figs. 65 and 66; G. Bell and Sons, Ltd., Fig. 29; Enraf–Nonius, Ltd., Fig. 138; Siemens (U.K.) Ltd., Fig. 153 (c); H. P. Stadler, Fig. 265.

Most of the diagrams are new, but some have been redrawn with amendments from existing diagrams, and acknowledgements are made in the caption if the origin is known. However a few derive from old slides whose origin could not be traced. They have certainly stood the test of time, and the lack of acknowledgement is solely due to my ignorance of their origin.

No-one writing a textbook in an established branch of science is likely to be able to claim much originality except in matters of detail. In the present book the crude energy analysis in Chapter 1 is as far as I know original, probably because the uncertainty in the starting point (the X-ray flux from a normal tube) is still so great that no-one else has thought the analysis to be worth doing. Nevertheless, although only an order of magnitude analysis, it shows quite dramatically the limitations of simple kinematic diffraction theory and the corresponding limitations of crystals to which it can be applied.

The analysis of precession movement and the derivation of the velocity factor for a precession camera on the basis of stereographic representation, was suggested to me by a slide from a paper on 'A Camera with Pure Precession Motion' by Y. Tomiie, K. Nakatsu and Y. Hukao, given at the VIIIth International Congress of Crystallography at Stonybrook, New York in 1968. The accidents of timetable irregularities allowed me to hear only the end of the paper, but the representation of precession motion on a stereogram suggested that the velocity factor might be derived in this way. I found that it did indeed give a particularly simple derivation, as well as greater physical insight into the origin of the irregularities in what was originally thought to be uniform motion about the precession axis.

Any other claims to originality are only in the details of presentation and are not important enough to list.

The plan of the book is outlined in the 'Students' Notes' on p. xxvi. These are essential reading before attempting to use the book.

J. W. Jeffery 1971

Contents

Part I. The Theory of X-ray Diffraction

Chapter 1. THE USES AND LIMITATIONS OF X-RAY DIFFRACTION

Chapter 2. THE GEOMETRY OF X-RAY DIFFRACTION FROM CRYSTALS   18

Part II.   Instrument Geometry and Interpretation

Appendices—Headings

List of symbols and abbreviations

Only the main uses are given. Where several symbols are closely linked they are given together under the first member of the set. Similar uses of a symbol are given under one entry. Different uses have separate entries. Where relevant a page reference is given in brackets.

(i) *Scalars etc. (Latin alphabet).*

a, b, c	lengths of the three axes $(\mathbf{a}, \mathbf{b}, \mathbf{c})$ of a crystal cell or lattice. (22, 324, 340)
a^*, b^*, c^*	lengths of the three axes $(\mathbf{a^*}, \mathbf{b^*}, \mathbf{c^*})$ of the reciprocal unit cell or lattice. (29, 30)
a	amplitude of wave scattered by a single electron. (6)
Å	Ångstrom unit $= 10^{-10}$m. (5)
A–Z	points on a diagram are indicated by italic or thin line capitals. matrices are represented by italic capitals.
A	amplitude of an electromagnetic wave. (6)
A, B	X and Y components of a Structure Factor, $\mathbf{F}$. (real part and imaginary part). (99)
A	absorption factor. (283)
A*	absorption correction factor $= 1/A$. (282)
$A(B, C)$	symbols for an $A(B$ or $C)$-face centred crystal cell or lattice. (127, 342)
B	temperature factor. (120)
c	velocity of light. (10)
d_{hkl} (d)	interplanar spacing of set of planes (hkl). (24)
d_{hkl}^* (d^*)	length of reciprocal lattice (RL) vector $\mathbf{d}_{hkl}^*$ $(\mathbf{d^*})$. (27)
D	optical density of blackening on a photographic film. (260)
D_0	observed density (mass/volume) of a crystal. (173)
D_c	density calculated from cell contents and parameters. (174)
e, m	charge and mass of the electron. (6)

E	energy. (113)
E′	partial energy (E = Σ E′) (113); power E′ = E/t. (118)
f	atomic scattering factor for an atom at rest, in terms of a, the amplitude scattered by one electron. (15, 124)
F	symbol for an all-face-centred crystal cell or lattice. (127, 342)
F_{hkl} (F)	absolute value (amplitude) of the Structure Factor, **F_{hkl}** (**F**). (99)
g	length of vector **g** in 'real' space. (340, 365)
H, K, L; h, k, l	triplets of integers—also used as subscripts; coordinates of a relp in terms of **a***, **b***, **c***, indices of a reflection. (25)
$\bar{h}, \bar{k}, \bar{l}$	−h, −k, −l.
(hkl)	indices of a set of planes or crystal face. (23)
{hkl}	a 'form' of sets of planes or faces, i.e. all those planes related by symmetry to (hkl)
[HKL]	a direction (H**a** + K**b** + L**c**) in the crystal lattice; a zone axis.
i	$\sqrt{-1}$; an operator which operates (by multiplication) on a 'real' vector, i.e. a vector in the OX direction, and transforms it by anti-clockwise rotation through 90°, into an 'imaginary' vector in the OY direction. (337)
I	intensity (energy area^{-1} s^{-1}) = A^2. (6, 112) integrated intensity ≡ relative total energy, E. (265)
I_0	incident intensity. (112)
I	symbol of a body centered cell or lattice. (127, 342)
J	**Integrated Reflection.** (115)
K	inner electron shell and spectra connected with it (Kα, Kβ). (6, 54)
K	constant.
l	general length.
L	'Lorentz' or velocity factor. (118, 381)
L	origin of the RL net (≡O for the zero layer). (179)
m, n, p	general integers
M	molecular weight. (174)
M	number of unit cells in a perfect crystal block. (111)
N	number of scattering units. (8)
O	origin.

O_n origin of construction for the nth layer of the RL. (179)

p, q, r a triplet of integers—also used as subscripts.

p   polarisation factor. (117)

p' multiplicity factor—rotation photographs. (118)

p'' multiplicity factor—powder photographs. (119)

p_{hkl} ratio of relp density in a given direction on a powder sphere to average relp density (i.e. that for a randomly oriented specimen). (252)

P REciprocal Lattice Point (RELP). (30, 176)

P' reflection corresponding to P. (176)

P symbol of a primitive cell or lattice. (127, 342)

P_l power (energy s^{-1}) in a length l of a powder ring. (388)

r radius of a cylindrical camera, or annular radius of an available layer screen for a precession camera. (226)

r' radius of a Weissenberg layer screen (197), or theoretical annular radius of a precession camera layer screen. (226)

r radial distance of a reflection on a flat plate photograph, from the direct beam. (78)

r radius of a spherical crystal. (265)

r length of a general position vector, **r**.

R resultant amplitude on the Argand diagram. (13, 335)

R   distance of crystal from intensity measuring device. (6, 112)

R_{Hex} symbol for a rhombohedral lattice referred to hexagonal axes, giving a doubly centered hexagonal cell. (127, 207)

R_n radius of the circle of reflection for the nth layer. (197)

symmetry symbols are standard [4, 5].

s distance from specimen to flat film. (78)

s position of an available layer screen. (226)

s' theoretical position of a layer screen. (226)

S position of a film cassette. (198)

t time.

T time of one revolution. (387)

$v(\delta v)$ volume of a crystal (crystallite). (114)

$V(kV)$ voltage (kilovolts).

V	volume of a unit cell in the crystal lattice. (361)
V^*	volume of a unit cell in the RL. (360)
x, y, z x', y', z'	coordinates of a point in 'real' space, in terms of $\mathbf{a}, \mathbf{b}, \mathbf{c}$ or $\mathbf{i}, \mathbf{j}, \mathbf{k}$, i.e. $\mathbf{r} = x\mathbf{a} + y\mathbf{b} + z\mathbf{c}$ or $x'\mathbf{i} + y'\mathbf{j} + z'\mathbf{k}$
$X; Y$	axes on the Argand diagram. (335)
X, Y, Z	coordinates of a point in reciprocal space, in terms of unit orthogonal axes.
X	distance between camera axes in mm ($\equiv 1$ R.U.). (219)
z	distance between equivalent axial reflections along the axial line on a zero layer Weissenberg photograph. (201)
Z	Atomic number. (15)
Z	Number of molecules/unit cell. (174)

(ii) *Vectors*

$\mathbf{a}, \mathbf{b}, \mathbf{c}$	axes of a crystal unit cell or lattice. (22, 324, 340)
$\mathbf{a}^*, \mathbf{b}^*, \mathbf{c}^*$	axes of a RL. (29, 30)
$\mathbf{A}$	resultant vector on the Argand diagram. (377)
$\mathbf{d}^*_{hkl} (\mathbf{d}^*)$	RL vector $= h\mathbf{a}^* + k\mathbf{b}^* + l\mathbf{c}^*$. (29)
$\mathbf{D}^*$	general position vector in reciprocal space. (94, 375)
$\mathbf{F}_{hkl} (\mathbf{F})$	Structure Factor—a vector on the Argand diagram. (375)
$\mathbf{g}_{pqr} (\mathbf{g})$	crystal lattice vector $= p\mathbf{a} + q\mathbf{b} + r\mathbf{c}$. (31)
$\mathbf{i}$	unit vector in the direction of the incident beam. (12)
$\mathbf{i}, \mathbf{j}, \mathbf{k}$	unit orthogonal R.H. set of coordinate vectors; Cartesian axes. (326)
$\mathbf{L}$	resultant vector on the Argand diagram. (377)
$\mathbf{n}$	vector (usually a unit vector) in a particular direction, e.g. normal to a plane, bisector of an angle, etc.
$\mathbf{r}$	general position vector in real space $= x\mathbf{a} + y\mathbf{b} + z\mathbf{c}$ or $x'\mathbf{i} + y'\mathbf{j} + z'\mathbf{k}$.
$\mathbf{r}_m$	position vector of the mth atom, etc.
$\mathbf{R}$	resultant vector on the Argand diagram. (13, 335)
$\mathbf{s}$	unit vector in the direction of the scattered beam. (12)
$\mathbf{V}$	resultant vector on the Argand diagram. (377)

(iii) *Greek alphabet*

α (alpha)	phase angle of the Structure Factor, $\mathbf{F}$. (17)

α	correction angle on an arc, or rotation angle of a preferred orientation specimen. General orientation angle.
α, β, γ	the angles between **a**, **b** and **c**; α opposite **a**, i.e. between **b** and **c**, etc. (330)
$\alpha^*, \beta^*, \gamma^*$	the corresponding angles between **a***, **b*** and **c***.
β (beta)	$v(I)/v(r)$—ratio of the coefficient of variation of intensity (I), to the coefficient of variation of crystal radius (r). (266)
γ (gamma)	the angle between a central RL plane and the X-ray beam in the back reflection Laue method. (167)
Δ (Delta)	differential operator; a small increment of $-$. This symbol is used where it is desirable to avoid confusion with the interplanar spacing, d.
δ (delta)	a small increment of $-$.
δ	general angle.
ε (epsilon)	strain. (83)
ε	orientation angle about the X-ray beam of the RL plane in the back reflection Laue method. (167)
ζ (zeta)	vertical distance in R.U. between horizontal RL planes, i.e. the interplanar spacing in the RL for such planes; the height of a point above the horizontal plane through the origin of the RL, i.e. one of the cylindrical coordinates of the point in relation to a vertical axis; the perpendicular distance of a RL layer from the zero layer. (176)
ζ	see ξ.
η (eta)	the angle between a RL vector and the X-ray beam in the back reflection Laue method. (167)
θ (theta)	the Bragg angle, or half the angle of deviation of a scattered beam.
λ (lambda)	wavelength.
μ (mu)	angle of inclination on a Weissenberg camera (199); precession angle. (219)
μ_i	angles between the incident beam and the lattice axes ($i = 1, 2, 3$).
μ	linear absorption coefficient. (118), [1(63); 7(157)]
v (nu)	angle of inclination of reflected beams (197) ($= \mu$ for equi-inclination Weissenberg camera). Cone angle of reflected rays in a precession camera. (222)
v_j	angles between scattered beam and lattice axes ($j = 1, 2, 3$).

v	frequency (c/s). (7)
v	see ξ.
$v(r)$	coefficient of variation of $r(=100\sigma(r)/r)$. (265)
ξ (xi)	distance of a point in R.U. from an axis in reciprocal space, i.e. one of the cylindrical coordinates of the point. (176)
ξ, ζ, v	small fractional coordinates in reciprocal space; $\Delta\mathbf{d}^* = \xi\mathbf{a}^* + \zeta\mathbf{b}^* + v\mathbf{c}^*$. (376)
ρ (rho)	the angle between a RL vector $\mathbf{d}^*_{hkl}$ and the rotation axis. One of the spherical coordinates of the relp hkl (the other two are ϕ and d^*_{hkl}). (56, 152)
ρ	the angle of tilt about an axis. (156)
$\rho(\mathbf{r})$ (ρ_{xyz})	the electron density at the point $\mathbf{r} = x\mathbf{a} + y\mathbf{b} + z\mathbf{c}$. (109)
$\sigma(x)$ (sigma)	the standard deviation of x.
ϕ (phi)	the angle between the plane containing the rotation axis and $\mathbf{d}^*_{hkl}$, and that containing the rotation axis and the direct beam. Both a spherical and a cylindrical coordinate of the relp hkl (the other two cylindrical coordinates are ζ and ξ). (58, 152)
ϕ	the phase angle (on the Argand diagram and in general). (11)
ψ (psi)	general geometrical angle.
ψ	the angle between a RL vector in a RL plane and the central line of that plane, in the back reflection Laue method. (167)
Ω (Omega)	solid angle.
Ω	angular velocity.
ω (omega)	resultant angular velocity.
ω	angle of rotation

(iv) *Abbreviations*

RL	Reciprocal Lattice.
relp	reciprocal lattice point.
R.U.	Reciprocal Unit (λ/d).
S.G.	Space Group
L.S.	Least Squares.
E.M.	Electro-magnetic.
L.H.	Left Hand.
R.H.	Right Hand.

6.352 ± 3 means ± 3 in the last figure quoted, i.e. between 6·349 and 6·355. Unless otherwise stated the figure represents about 3σ, or the maximum error arising from unavoidable errors in measurement, depending on the context.

(v) *Wavelengths*

Element, Atomic Number	Line	Wavelength Å
Cu, 29	Kα	1·542
	Kα_1	1·540 51
	Kα_2	1·544 33
	Kβ	1·392
Mo, 42	Kα	0·711
	Kα_1	0·709 26
	Kα_2	0·713 54
	Kβ	0·632

Notes for Students

This book makes no attempt to be self-contained. It assumes a knowledge of symmetry and stereographic projection which can be obtained from a parallel study of Phillips, 'An Introduction to Crystallography' [4].

It was originally planned to include a section on powder diffraction and an appendix on the production and properties of X-rays; but the recent publication of 'Interpretation of X-ray Powder Diffraction Patterns' by Lipson and Steeple [1] which has a chapter on the Physics of X-rays, has made these unnecessary. References to Lipson and Steeple will be given in connection with powder diffraction and a knowledge of the Physics of X-rays up to the level of their Chapter 3 will be assumed. However MoK and CuK wavelengths are given for easy reference after the table of symbols.

Access to the various volumes of the Interntaional Tables [5, 6, 7] is an essential requirement for a serious student of X-ray crystallography. Make full use of the index in Vol. III.

The theoretical development is as simple as possible, (and in places may well involve dangers of giving rise to over simplified conceptions, e.g. on the existence of pairs of individual electrons in an atom related by a centre of symmetry, or on the basis of 'direct' phase determinations). The aim has been to give an insight into the physical principles involved rather than to develop rigorous mathematical formulations. However it is impossible to deal with 3-dimensional diffraction without spherical geometry and vector algebra. The minimum requirement is facility in the use of vector algebra up to scalar products. Appendix I (1–9, 15, 16) gives an outline of the results required. For those whose vector algebra is rusty or who have only a nodding acquaintance with it, a short course with numerous examples is an absolute requirement before tackling this book. Weatherburn's Elementary Vector Analysis [16] Chapters I–III, covers the ground admirably with numerous examples, and those who wish to include vector multiplication (which is freely used in the Appendices) should go on to Chapter IV.

Two sets of examples on vector algebra are provided for the student to test the adequacy of his preparation in this respect. The second set is directly related to crystallographic problems.

No attempt is made to deal with details of the techniques of Structure

Analysis, although the physical bases of some of them are outlined in Part I. There are a growing number of texts on this subject starting with Vol. III of the Crystalline State, by Lipson and Cochrane [20]. The present book aims, in this context, at dealing with the problems involved in producing a corrected set of F_{hkl}^2 values, which are the starting material of a structure analysis.

The book is in five parts. Part I covers the theoretical aspects of X-ray diffraction and interpretation of X-ray photographs. Part II covers instrumentation and the detailed techniques of interpretation. The Examples form the third part, the Appendices the fourth and the Worked Answers the fifth. The Appendices contain detailed derivations or descriptions of techniques which would have cluttered up the main text. Fairly rigorous proofs requiring more than elementary vector algebra have also been removed to Appendices. Some parts of the Appendices have been included because no suitable reference was known, even although the subject matter e.g. the anisotropic temperature factor coefficients, would not otherwise have justified inclusion at length.

The Examples are an integral part of the book, and in a number of cases the Worked Answers include alternatives or extensions which are supplementary to the main text.

Since all the parts of the book are intimately related, considerable cross referencing is provided, by section and page number e.g. Section 2.3.5, p. 126, if the section referred to is in the same chapter. If it is in another chapter, the chapter number is given in bold type, e.g. Section **4**.2.3.5, p. 126. However in both cases the page number (either the first page or a page within the section if the reference is sufficiently specific) is the main means of locating the reference. Any sub-division beyond that indicated above (which does not often occur) is made with small letters or small Roman numerals.

Figures are numbered serially throughout the book, and reference is self-explanatory e.g. (Fig. 12, p. 20). If the Figure is on the same or a neighbouring page, the page number is omitted. Tables have been treated as Figures. In many cases the captions to Figures contain information which supplements the main text.

Formulae are numbered from (1) on each page, and referred to by a page number followed by the formula number e.g. 'From 283(5)...' refers to formula (5) on page 283. If no page number is given the reference is to a formula on the same page.

Appendices are referred to by upper case Roman numerals (followed by a section number if necessary) with the page reference at the end e.g. 'In Appendix VIII (4) (p. 397)', enables the reference to be turned up immediately.

Book (and occasional paper) references are in square brackets e.g. '[11]' refers to the book number 11 in the list on page 555. Where appropriate, a page number in brackets follows the book number, e.g. [11(34)] refers to page 34 in reference 11. Book references have been kept to a minimum in order not to confuse the system of cross referencing. If in doubt, consult the index.

Suggestions are made of the points at which Examples might best be done, by printing e.g. **Example 9** (p. 270) at the appropriate point. However this should in general be at the second or third reading, and starred examples should be taken last.

The early examples are necessarily somewhat artificial, but are designed to give facility in the use of new concepts such as the sphere of reflection and the reciprocal lattice. Solutions should be sought by geometrical construction, even though analytical methods might be quicker in such simple cases.

The content of the examples includes on occasion matter from books referred to in the text, as well as from the text itself.

The Figures required for measurement in Examples have been reproduced as nearly as possible the same size as the original. (Working stereograms are printed the same size as the Wulff net). Some of these Figures also serve as illustrations to the text; and when doing such Examples it is essential to work with a number of card markers for easy reference to the Figures and relevant sections of the text.

The alphabetical list of symbols and abbreviations on p. xviii should be consulted to clear up doubts about their meaning in the text, or to obtain a page reference to a more detailed definition.

Tracing paper for marking the relevant parts of a Figure without spoiling the original and for comparison between two Figures, is also essential in working the examples.

A number of charts are required for measurement of the photographs used in the Examples. These are (a) the Bernal chart (Fig. 107. PC 2); (b) the ρ, ϕ and θ chart (Fig. 94, PC 5); (c) the Weissenberg coordinate chart (Fig. 119, PC 32); (d) the Grenninger chart (Fig. 103, PC 4). In addition (e) the Wulff stereographic net ([4(32)], PC 18) is required for working with stereographic projections. The references (PC 2, etc.) specify charts on transparent paper. The addition of a P at the end of the reference (PC 2P, etc.) specifies a chart on clear stable plastic sheet.

These charts can be obtained post free from the Institute of Physics, 1, Lowther Gardens, Prince Consort Road, London, S.W.7, at $2\frac{1}{2}$p each for the paper charts and 75p each for the plastic ones. In addition PC 20 (a 20 cm diameter Wulff net on card) also costs $2\frac{1}{2}$p, but the clear plastic version, PC 20P, is £1.00. A 30 cm diameter Wulff net on card, PC 30, is also available at 10p per card.

If a Weissenberg camera (cassette diameter 57·3 mm) is used as a rotation camera, the Bernal chart for this diameter, PC 37 and PC 37P, is available on the same terms as PC 2 and PC 2P.

A magnifying glass is essential for use with the charts if the readings are to have the required accuracy, e.g. ± 0.0025 R.U. on the Bernal chart.

There are two general ways that should be noted, in which description has been simplified. In most cases a small crystal is attached to a rotation axis by a supporting fibre. In the simplest X-ray camera this axis is vertical, with the 'free' end of the crystal upwards. Looking at the free end of the crystal is referred to as 'Looking down on to the crystal'; and 'above', 'below' and 'equatorial', in relation to the rotation axis, have corresponding meanings, even when the axis is actually horizontal. Similarly, in order to facilitate comparison between diagrams, the direct X-ray beam is usually drawn coming from left to right on the page, with the rotation axis vertical, so that the diagram may require reorientation to correspond with the physical directions on an actual camera. 'Horizontal' and 'vertical' however always have their normal meanings.

Some concession has been made to S.I. units, mainly by expressing most lengths in mm, although where the existing tables are expressed in terms of cm e.g. absorption coefficients, this has been taken into account. Ångstrom units have, however, been retained throughout.

Lastly a few remarks on the method of using the book may be useful, especially for those approaching the study of X-ray crystallography for the first time.

A preliminary reading of at least the main sections of Appendix I is essential, if only to check the vector conventions which are used.

Then start with a rapid first reading of Part I, without bothering with cross-references, and omitting Chapter 5 and Sections 6.53 and 6.54. Note references to other books which you need to study before a second reading. Leaf through Part II to get a general idea of the contents. After studying the book references, start on a second reading, making use of the cross references and undertaking most of the examples. Reconcile your answers with the Worked Answers in the book, *after* your answers have been completed. Study such Appendices as you feel are necessary for a preliminary understanding of the subject. If in doubt draw a diagram.

Finally I hope the book may be a useful reference work in helping to solve the problems which arise in crystallographic research whether industrial or academic. In using the book in this way refer to the table of contents as well as the index.

Part II and many of the Appendices are designed to supplement laboratory work. If the book is used as part of an organised course of study the general suggestions above will obviously be replaced by the requirements of that course.

A printed book has the property of appearing far more complete and authoritative than it really is. It is legitimate to start acquiring knowledge of a new subject without being too critical of the text one is studying. As one progresses to applying the knowledge first gained, one acquires a much deeper and extended understanding at the same time. When one is mainly applying the knowledge to real problems and learning things not contained in the text, then a critical approach should be possible which reveals short-comings in the text. I could, at this point, merely quote a sentence from a Trinidadian airline travel guide which reads, 'In case you find any errors in this magazine, please remember they were put there for the benefit of the people who are always looking for mistakes!' There is however only one deliberate mistake in the book, and this is exposed at the right place. So the others, which are certainly there for you to find, will provide a test of your ability to develop that critical approach which is the final aim of any study. When you find them, please let me know.

A good deal of trouble has been taken in obtaining the X-ray photographs reproduced in the book. Although it is not possible to produce on a print the same effect as can be obtained by viewing a transparency, nevertheless the prints themselves are better than most run-of-the-mill films. The student should not, therefore, expect to achieve results as good as these illustrations in his first attempts, and for some purposes the time required to do so would not be justified.

Part I

The Theory of X-ray Diffraction

Chapter 1

The Uses and Limitations of X-ray Diffraction

1. The Uses of X-ray Diffraction

X-ray diffraction is a tool for the study of the arrangement of matter. It has much in common with the optical and electron microscopes, with one vital difference. The information contained in the radiation scattered by an object cannot, in the case of X-rays, be displayed directly in a magnified image, but can only be extracted by the use of monochromatic radiation and a complicated mathematical analysis. This is because no means is available for physically recombining the scattered beams of X-rays to form an image.

The analysis, moreover, faces a special difficulty. In the case of a microscope the scattered beams from an object have not only different amplitudes, but also different phases relative to the direct beam. The preservation of these relative phases by the optical train of the microscope is an essential part of image formation. Any alteration of these phase relationships alters the appearance of the image, as, for instance, in phase contrast microscopy. To construct an image mathematically from measurements of the scattered beams thus requires a knowledge of their relative phases. In the case of X-rays no means has yet been found of measuring the relative phases. The intensities (and therefore absolute amplitudes) of the scattered beams can be measured, but not their phases. Although the intensities and geometry of the scattered beams can be of great use by themselves, as in identification by X-ray powder diffraction, the great problem of X-ray analysis is to find the phases of the scattered beams by indirect means. Once they are found the image can be computed without difficulty. (Fig. 1)

The reason for using X-ray diffraction can be seen from consideration of the Abbé diffraction theory of the microscope.

Those readers to whom diffraction is a rather theoretical concept should take an opportunity to look through an umbrella at a distant lamp (an old-fashioned street lamp, or an incandescent bulb through an uncurtained window will be a sufficiently small source) and observe the diffraction pattern produced by the fine mesh of the fabric. If two umbrellas are available, the better one (with the finer mesh) will give the larger diffraction pattern. It is interesting to

1

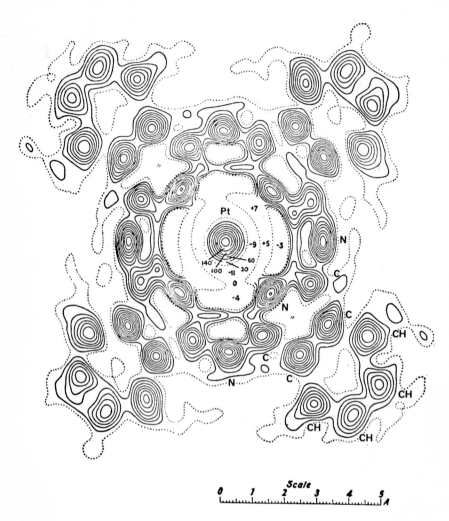

FIG. 1. The construction of the image of phthalocyanine ($N_8C_{32}H_{16}$) from the measured scattered beams of X-rays given by a crystal of the Pt complex. This is a rare (almost unique) case in which the heavy Pt atom at the centre of the molecule determined the phases of the scattered beams, and allowed the image to be constructed directly. The contours give the electron density of the atoms in the molecule, projected on to a plane about 45° inclined to the plane of the molecule. This accounts for the width of the molecule being apparently smaller than the length. Hydrogen atoms have such small scattering power that they do not show up in the image in this case.

note that the pattern can be seen with the fabric right up close to the eye, and that the pattern turns as the mesh of the fabric is turned.

For objects large enough to be seen by the naked eye, these diffracted beams are very weak except in directions parallel, or at very small angles, to the incident light. For macroscopic objects we therefore have a very good approximation to rectilinear propagation.

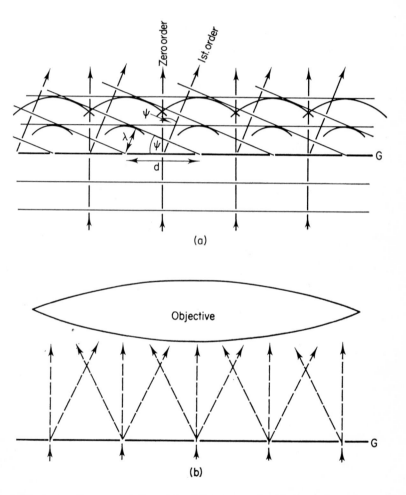

FIG. 2. Diagrammatic representation of the Abbé theory of resolution, using a line grating G, as object. Full lines—wave fronts; dashed lines—wave normals. (a) The grating law—$n\lambda = d \sin \psi$. (b) Zero and first orders enter objective. Lines will just be resolved in the image. Second order would not enter objective. (c) A finer grating. Only the zero order enters the objective, giving a uniform field in the image.

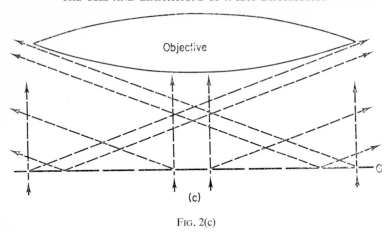

(c)

FIG. 2(c)

For microscopic objects however the diffracted beams are detectable at larger angles to the incident light; and it is the recombination of these beams by the optical train of a microscope which enables an enlarged image to be formed which the eye can resolve.

The lines on an optical grating cannot be seen by the naked eye; but if we look at them with a high power microscope they will be visible in the image, if the grating is not too fine. For such a grating the light entering the microscope consists of the direct beam going straight through, and the various orders of diffraction. The amount of detail seen in the image depends on the number of orders of diffracted beams taken in by the objective. (Fig. 2). The first order will show only the line spacing. To see any details of the line profile requires several more orders to be collected. But the finer the grating the greater the angle of deviation ψ of the diffracted beams. When the separation of the lines is equal to the wavelength of light used, the deviation becomes 90° and no image can be formed. (**Example 1,** p. 285).

Interatomic distances in the alkali halides (Å)

	Li	Na	K	Rb	Cs
F	2·01	2·31	2·66	2·82	3·00
Cl	2·57	2·81	3·14	3·27	3·57
Br	2·75	2·98	3·29	3·43	3·71
I	3·00	3·23	3·53	3·66	3·95

X-ray wavelengths (Å)

MoKα	CuKα	CoKα	CrKα
0·71	1·54	1·79	2·29

FIG. 3. Table of typical interatomic distances and X-ray wavelengths.

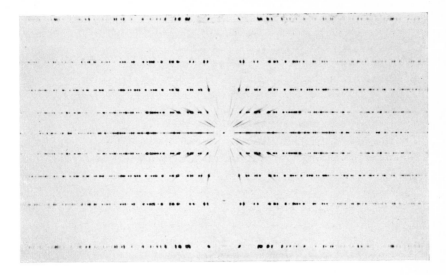

FIG. 4. The central portion of a single crystal rotation photograph. The film was bent into a cylinder of radius 30 mm, with the crystal on the axis, while the photograph was being taken. (Figs. 24, p. 33 and 87, p. 136). CuKα radiation, $\lambda = 1\cdot542$ Å.

If the wavelength is reduced finer detail can be seen, as in the ultra-violet microscope. To 'see' detail of atomic dimensions, radiation with a wavelength of the same order as atomic size is required. Atomic distances are of the order of 1 Å = 10^{-10} m or 10^{-7} mm, and monochromatic X-ray beams can be generated with similar wavelengths. (Fig. 3)

Theoretically the electron microscope, with even smaller wavelengths, should also be able to resolve atomic detail; but it has many limitations, the most important being the very small aperture of the objective lens which only allows a resolution of about 5–10 Å. In the case of X-rays all the diffracted beams which are produced can be collected, as shown in Fig. 4. The resolution obtained is therefore the maximum theoretically possible.

Electron and neutron diffraction give similar information to that obtained by X-rays, but tend to be limited, by practical difficulties, to specialised fields. Notable examples are diffraction of electrons by crystals too small to give measurable X-ray diffraction; and of neutrons by interaction with atomic magnetic fields (to which X-rays are completely insensitive).

2. The Limitations of X-ray Diffraction

The practical limitations of X-ray diffraction are fortunately very similar to the theoretical limitations of the extremely simplified diffraction theory to be presented here. It is as well, however, that these limitations should be appreciated from the start. A simple, order-of-magnitude, energy analysis brings the main limitations out very strikingly. The only theoretical results required are:

1. the intensity of a monochromatic beam I (energy, area^{-1}, s^{-1}) = A^2, where A is the absolute amplitude of the beam;
2. in the case of constructive interference, as for an optical grating, the resultant amplitude is the sum of the amplitudes of the scattered beams;†
3. in the case where the phase relationships between scattered beams are random, the resultant *intensity* is the sum of the *intensities* of the scattered beams.†

For simplicity we shall consider the X-rays to be vibrating perpendicular to the plane of scattering, since polarisation effects are small compared with the uncertainties of the analysis.

2.1 Intensity Available from a Standard X-ray Tube

The power dissipated in a normal tube is about 1 kW = 10^3 J s^{-1}. Of this some 99·9 per cent appears as heat in the target, and of the 0·1 per cent given off as X-rays, about one third appears outside the tube as the characteristic, monochromatic X-rays, [1(53)]‡ which we wish to use. The energy given off as the Kα characteristic X-rays is therefore taken as 0·03 per cent of the total. (The exact figures are still the subject of controversy, and this percentage may be in error by a factor of at least 2).

We assume that the specimen to be irradiated is at 100 mm from the tube target, and that the scattered X-rays are measured at a distance of 30 mm from the specimen. For the purposes of calculating the total energy, X-rays are assumed to be given off uniformly in the hemisphere above the plane of the target; and the foreshortening of the line focus [1(75, 83)] is assumed to give an intensity I in the desired direction, three times the average $\hat{I}$ over the hemisphere. Total energy over the hemisphere per sec. $= 2\pi 100^2\hat{I} = 10^3 \times 3 \times 10^{-4}$ J

$$\therefore \quad \hat{I} = 4\cdot8 \times 10^{-6} \text{ J mm}^{-2}\text{ s}^{-1}$$
$$\therefore \quad I = A^2 = 4\cdot8 \times 10^{-6} \times 3 = 1\cdot4 \times 10^{-5} \text{ J mm}^{-2}\text{ s}^{-1}.$$

2.2 Scattered Intensity from an Object Irradiated by the Primary Beam

2.2.1 An Electron

Let a be the amplitude of the radiation scattered from this primary beam by 1 electron, at a distance of 30 mm (for an electron this is independent of direction). From the classical Thomson formula

$$a = \frac{A}{R} \cdot \frac{e^2}{mc^2} = \frac{A}{30} \times 2\cdot8 \times 10^{-12}$$

$$\approx A \times 10^{-13} \tag{1}$$

$$\therefore \quad a^2 = 1\cdot4 \times 10^{-5} \times 10^{-26} = 1\cdot4 \times 10^{-31} \text{ J mm}^{-2}\text{ s}^{-1}.$$

† See Appendix I, (p. 336).

‡ See notes for students (p. xxvi).

The energy of a photon for CuKα radiation, $\lambda = 1.54$ Å, is

$$h\nu = 6.6 \times 10^{-34} \times \frac{3 \times 10^8}{1.5 \times 10^{-10}} = 1.3 \times 10^{-15} \text{ J}$$

$\therefore$ Flux of quanta from one electron $= \dfrac{1.4 \times 10^{-31}}{1.3 \times 10^{-15}}$

$$= 1.1 \times 10^{-16} \text{ mm}^{-2} \text{ s}^{-1}.$$

2.2.2. A Single Molecule

Consider a single molecule composed of the equivalent of 27 oxygen atoms. This might give a resultant scattering amplitude (depending on the arrangement of the atoms) equivalent to 10 electrons scattering in phase, and therefore an intensity 100 times that for one electron. For such a molecule:—

$$\text{Flux of quanta} = 10^2 \times 1.1 \times 10^{-16}$$

$$= 1.1 \times 10^{-14} \text{ mm}^{-2} \text{ s}^{-1} \tag{1}$$

$\therefore$ 1 quantum in 3,000,000 years!

This can hardly be considered statistically significant, and demonstrates why we cannot hope to 'see' a single molecule by means of X-rays.

2.2.3. A Gas

If we take 1 mm^3 of gas at S.T.P. containing 3×10^{16} such molecules, then the flux of quanta becomes $1.1 \times 10^{-14} \times 3 \times 10^{16} = 3 \times 10^2$ mm^{-2} s^{-1}. Here the *energies* are added because the phase relationships are random.

This is certainly a measurable quantity; and even if the average scattering of a gas molecule is smaller than that assumed, and the volume irradiated smaller than 1 mm^3, it would still be appreciable.

In fact, valuable information has been gained from the investigation of X-ray scattering from gases (especially the *accurate* determination of bond lengths in molecules of known shape); but its major limitation is that scattering takes place from the molecules in all possible orientations so that information is only obtained about a spherically symmetrical 'averaged' molecule.

2.2.4. Liquids and Glasses

In the case of liquids and glasses the scattering is more intense, because of the greater number of molecules per unit volume; but here short range order has an effect on the distribution of scattered intensity and makes the results even more difficult to interpret.

2.2.5. *Crystals*

In all cases discussed so far, scattering is continuous in direction. When we come to scattering from crystalline solids, we are dealing with a three-dimensional diffraction grating. Scattering is discontinuous in direction; and when it does take place constructive interference occurs and the intensity is N^2 times the intensity for a single 'molecule', where N is the number of scattering units. Here 'molecule' refers to the contents of a unit cell—the building block from which the crystal is constructed—and may have little relation to a normal chemical molecule.

The great advantage of a crystal is that all the 'molecules' are oriented in the same way, so that the scattering is an amplified version of that obtained from a single 'molecule'. It is true we can only sample the scattering for a single 'molecule' in the directions in which constructive interference occurs for the crystal as a whole; but these samples are sufficient to enable us to calculate the image of the 'molecule' when we have been able to determine the phases of the diffracted beams.

It remains to calculate the scattering which would be obtained in a number of typical cases. In each case a cubic unit cell of 10 Å side, containing the scattering unit ('molecule') defined in 7(1), and a crystal of cubic shape will be assumed.

(a) A crystal of side $0 \cdot 1$ mm $= 10^6$ Å. Number of unit cells (scattering units) $= (10^5)^3 = 10^{15} = N$.

$\therefore$ Scattering flux $= N^2 \times 1 \cdot 1 \times 10^{-14} = 1 \cdot 1 \times 10^{16}$ qu. $mm^{-2} s^{-1}$.

But the flux in the incident beam

$$= \frac{1 \cdot 4 \times 10^{-5}}{1 \cdot 3 \times 10^{-15}} = 1 \cdot 1 \times 10^{10} \text{ qu. } mm^{-2} s^{-1}. \tag{1}$$

That is to say, the intensity of the scattered beam is 10^6 times that of the incident beam! Even if we allow for the smaller cross section of the diffracted beam (approximately equal to the crystal cross section), the energy in the diffracted beam is still 10^4 times that of an incident beam of 1 mm^2 cross section, most of which does not hit the crystal anyway!

This illustrates the first limitation of simple diffraction theory. It can only be applied to very small crystals. The reason for this can easily be seen by considering scattering points deep in a 'large' crystal. The energy scattered by points in the upper part will reduce the intensity incident on those below. The simple theory assumes that the incident intensity is everywhere the same.

This is only a sufficiently good approximation for very small crystals. The full theory for larger crystals is very complicated, and shows that for a 'large' perfect crystal, with negligible absorption, total 'reflection' of the incident beam takes place over a very small angular range.

(b) If we repeat the calculation for a crystal of side $0 \cdot 1 \mu m$ ($= 10^3$ Å), $N = 10^6$ and scattered flux $= 1 \cdot 1 \times 10^{-2}$ qu. $mm^{-2} s^{-1}$. This is certainly smaller than the incident beam intensity; and if we take account of the area of cross section of the diffracted beam, the actual number of quanta is reduced to $1 \cdot 1 \times 10^{-10}$ qu. s^{-1}. Apart from the difficulty of manipulating a crystal of this size, the flux of quanta is obviously much too small to measure.

However, a larger crystal can be imagined as made up of such small blocks arranged parallel to one another but with micro-cracks, or disordered layers, between them, so that the phase relations between blocks are random. We then have a model of a 'mosaic' crystal. Most real crystals approximate to this model. For a crystal of $0 \cdot 1$ mm side there will be 10^9 blocks, and the total diffracted energy will be $1 \cdot 1 \times 10^{-10} \times 10^9 = 1 \cdot 1 \times 10^{-1}$ qu. s^{-1}. This would seem to be a satisfactory, if rather small, result until we try the effect of increasing the size of the mosaic blocks to $1 \mu m$, keeping the total size $0 \cdot 1$ mm. The diffracted energy would then be $1 \cdot 1 \times 10^4$ qu. s^{-1}, i.e. 100,000 times that for the previous case. This illustrates a second limitation—even if the mosaic blocks are smaller than the maximum size set by the first limitation (about $1 \mu m$), the intensity still varies with the size of the blocks. It is partly for this reason that the maximum intensity cannot be used as a measure of diffracting power. However, this limitation can be overcome by measuring the 'integrated reflection' as defined in Section 9.2.1 (p. 113).

The above calculations have assumed that all the blocks are exactly parallel. This would not be the case, and in fact must not occur, or shielding effects will reduce the intensity incident on the lower blocks. Only a proportion of the blocks will be in the correct orientation for diffraction, and this proportion may very well be different for different crystals, and even for different directions in the same crystal, thus introducing a further variable into the calculation. Finally the size of a block determines the 'spread' of the reflection on either side of the maximum. For a given angular distribution of blocks this determines the number of blocks making a contribution (and the size of the contribution, which will, in general, not be the maximum) in any given direction. This effect counteracts considerably the effect of the size calculated for parallel blocks above, but introduces more unknown variables into the calculation. Fortunately all these variables can be dealt with by the same method as the first.

With these limitations in mind, the following simple diffraction theory will be found sufficient for most purposes.

3. Elementary Diffraction Theory

3.1. THE SCATTERING PROBLEM AND THE ARGAND DIAGRAM

The essential problem is shown in Fig. 5. If an electromagnetic wave originating from a source S, is scattered at two points A and B, what is the combined effect of the two scattered waves at C? In the cases we shall consider, SA and CA are both large compared to AB, so that both incident and scattered beams can be considered as parallel. This simplifies the geometry, but the physical problem remains the same. The following simplified outline of the basic theory is given as an introduction to its application to X-ray diffraction. It is assumed at this stage that the waves are scattered from point scatterers such as A and B, with either no phase change or the same change for each. Since we can only deal with relative phases there is no difference between the two cases. (A successful X-ray laser however might enable the direct beam to be used as a phase reference in the future.)

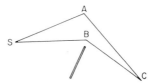

FIG. 5. The basic diffraction problem. S is a source of radiation scattered at points A and B. The combined effect of the scattered waves at C has to be calculated.

Consider a plane wave originating from a distant oscillator of frequency v and travelling with a velocity c mm s^{-1}. Its wavelength $\lambda = c/v$ mm. The oscillator can be represented by a vector rotating at v revolutions per second, and its *phase* is the angle ϕ between the vector and an arbitrary fixed origin line (Fig. 6). The phase of any wave front is the phase of the oscillator at the instant the wave front left it.

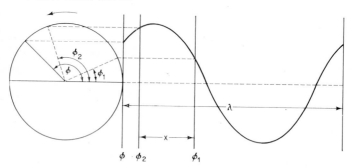

FIG. 6. Diagrammatic representation of a source of radiation by a rotating vector.

$$\phi_2 - \phi_1 = 2\pi \frac{x}{\lambda}.$$

The phase change per second $= 2\pi v$. Therefore phase change per milli-metre $= 2\pi v/c = 2\pi/\lambda$. The difference of phase between two wave fronts x mm apart $= 2\pi x/\lambda$. The wave front nearer the source has the larger phase angle, since it left the source later. A shorter path length therefore corresponds to a phase advance (or a positive phase) relative to a wave front with a longer path.

In physical terms, the ordinates of Fig. 6 represent the electric field strength in the wave fronts. As the wave fronts sweep through a point P, the electric field at P is changing; and since a complete oscillation occurs in an outward movement of λ mm, this oscillation occurs $c/\lambda = v$ times a second. The field at P can therefore also be represented by the ordinate of a rotating vector, the position of the vector at any instant being that 'imprinted' on the wavefront passing through P at that instant.

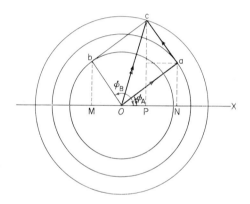

Fig. 7. The combination of rotating vectors representing the two waves scattered from A and B in Fig. 5.

When two or more wavefronts recombine, the resultant field is the algebraic sum of the ordinates of the various wavefronts. In the case of Fig. 5, let the path lengths from S to C be l_A and l_B. At a given instant let the phase angles of the two wave fronts passing through C be ϕ_A and ϕ_B. Then $\phi_B - \phi_A = 2\pi(l_A - l_B)/\lambda$. Let the amplitudes (i.e. maximum value of the ordinates or the lengths of the rotating vectors) be Oa and Ob (Fig. 7). Then the electric fields at this instant, due to the waves scattered from A and B, are aN and bM, and the resultant is their algebraic sum. If we complete the parallelogram $bOac$ and draw the diagonal Oc, then it is clear that $cP = aN + bM$, i.e. the ordinate of the vector resultant† of Oa and Ob, represents the combined field at C.

† The elementary vector algebra required in this book is set out in Appendix I.

The resultant of three or more waves will clearly be given by the vector sum of the individual rotating vectors, i.e. by the closing side of the vector polygon. Thus at any given point the variation with time of the resultant field will be given by the rotating resultant vector. The vector polygon which gives the resultant rotating vector for a number of waves (of the same frequency) is called the Argand diagram.

Let a plane wave (i.e. a wave whose distance from the source is large compared with the size of the scatterer) be scattered by a small distribution of equal scattering points P_m, whose positions are given by the vectors $\mathbf{r}_m$, relative to an arbitrary origin, O (Fig. 8). Assume that the wave is scattered without change of phase or with a change which is the same for all points and can be ignored in considering relative phases.† For convenience, O is taken within the array of scattering points.

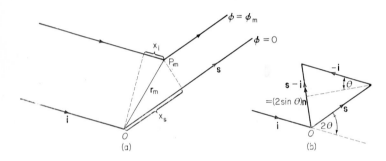

FIG. 8. (a) Scattering of a parallel beam by a set of scattering points, P_m. If $\mathbf{i}$ and $\mathbf{s}$ are in the plane of the paper, $\mathbf{r}_m$, in general, is not. (b) Relation between $\mathbf{i}$, $\mathbf{s}$, $\mathbf{n}$ and θ. $\mathbf{s}-\mathbf{i}=(2 \sin \theta)\mathbf{n}$ is necessarily in the same plane as $\mathbf{i}$ and $\mathbf{s}$.

Let $\mathbf{i}$ be a unit vector in the direction of the normal to the incident wave front, and $\mathbf{s}$ a unit vector in the direction in which the scattered wave is to be measured at a distant point, N.

The phase ϕ_m of the wave scattered from P_m is $2\pi(x_s - x_i)/\lambda$ relative to the wave scattered from O.

$$x_s = \mathbf{r}_m \cdot \mathbf{s}; \quad x_i = -\mathbf{r}_m \cdot -\mathbf{i} = \mathbf{r}_m \cdot \mathbf{i}$$

Therefore

$$x_s - x_i = \mathbf{r}_m \cdot (\mathbf{s} - \mathbf{i}) = \mathbf{r}_m \cdot \mathbf{n} \, (2 \sin \theta) \tag{1}$$

† There must not, of course, be any detectable change in frequency on scattering. Compton scattering, in which there is both change in frequency and random change of phase, is excluded from this discussion.

where **n** is a unit vector in the direction of **s** − **i**, that is bisecting the angle between −**i** and **s**, and θ is half the angle of deviation of **s** relative to **i**. (Fig. 8b). Therefore

$$\phi_m = 2\pi \frac{\mathbf{r}_m \cdot \mathbf{n}\,(2\sin\theta)}{\lambda} \tag{1}$$

The resultant amplitude of the wave scattered by all the points is the vector sum of the individual amplitudes, i.e. the closing side of the polygon on the Argand diagram (Fig. 9).

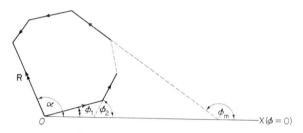

FIG. 9. The Argand polygon, giving the resultant amplitude. The phase of the resultant, relative to the wave scattered from the origin (O, Fig. 8(a)) is α. All the vectors are rotating at about 10^{18} revolutions s^{-1}. The diagram is drawn for the instant when the origin vector is along the X-axis. In this case there is no scattering point at the origin and the amplitude of scattering is therefore zero for $\phi_0(=0)$.

3.2. SCATTERING WHEN r_m IS SMALL COMPARED WITH λ

The greatest value of ϕ_m is $4\pi r_m/\lambda$. If all the values of r_m are small compared to λ, all the phase angles will be small and the sides of the Argand polygon will be all nearly parallel to OX. Thus the resultant for all values of θ will be nearly equal to the algebraic sum of the amplitudes, and the scattering intensity will be nearly the same in all directions. Optically this corresponds to a very small pinhole, and in the case of X-rays to an electron, which is relatively so small that, apart from polarisation effects which are considered later, its scattering is taken as independent of angle.

3.3. SCATTERING WHEN r_m IS LARGE COMPARED WITH λ

Only the case where there is a large number of scattering points need be considered. For points with large r_m, ϕ_m will increase rapidly as θ increases from zero (except for a small number where $\mathbf{r}_m$ is nearly perpendicular to **n**). For random distributions of scattering points the polygon will curl up into an irregular spiral, and the resultant scattering will be small except for very small θ, i.e. very close to the incident beam. This gives rise to the rectilinear propagation of light in relation to large objects, and is the basis of radiographs in the case of X-rays.

An exception occurs in the case of large distributions which contain regularities comparable to λ. Here the spiral, in certain directions of scattering, straightens out to give constructive interference. $\phi_m = 2\pi n$ (where n is an integer) in all cases, and the resultant amplitude is the algebraic sum of the individual scattered amplitudes. The optical diffraction grating is a well known example of this effect; and crystals act in the same way in relation to X-rays, with the added complication of three dimensions.

3.4. SCATTERING WHEN r_m IS OF THE SAME ORDER AS λ

This, in conjunction with the second case of Section 3.3, is the region of main interest, because the only general statement which can be made is that the polygon tends to curl up with increasing θ, so that the average intensity of scattering falls off with increasing deviation of the scattered beam. But for a particular orientation of the scattering points, relative to the incident and scattered directions, the phase angles ϕ_m depend entirely on the relative arrangement of the scattering points, so that the Argand polygon may be curled up or nearly straightened out, giving small or large intensity depending on this relative arrangement. By investigating the scattering in various directions it is possible to deduce the relative arrangement of the scattering points, or, as in the case of the optical microscope, to display them on an enlarged image. The optical microscope can investigate a single array of scattering points of a size corresponding to its maximum resolution. We have seen that this is not possible in the case of X-rays; but by combining the scattering from a large number of similar arrays, taking advantage of constructive interference from the regular three-dimensional arrangement found in crystals, we can achieve a similar result.

Information of a much more limited character can be obtained from scattering by gases and liquids. Only the case of crystals will be considered in detail.

4. Scattering of X-rays by Atoms

4.1. ATOMIC SCATTERING FACTORS

Since the scattering units, the electrons, are arranged in atomic groups, the amplitude vectors for each group can be summed separately on the Argand diagram, and the resultants for each group then added to give the final polygon, with its closing side as the resultant amplitude for all the atoms in the group.

To a first approximation the electron distribution in an atom is spherically symmetrical. If we take the origin at the centre of the atom (Fig. 10a) then for any electron position vector $\mathbf{r}_m$, there will be another, $-\mathbf{r}_m$. Thus for any angle of deviation 2θ, the amplitude vectors (all equal in length, since the scattering is from equal electrons and is also independent of θ) can be collected in pairs with phase angle $\pm\phi_m$. The Argand polygon for this arrangement is shown in Fig. 10b, and the resultant is obviously along OX, i.e. the phase angle of the resultant is either 0 or π. For the electron distribution in atoms it is always 0 and therefore the same as that of a wave scattered from the centre of the atom. We can thus replace the atom by a single scattering point at O, with a scattering amplitude equal to R. The final resultant for a group of such point atoms will then be exactly the same as if the summation had been made at random over all the electrons in a group of real atoms.

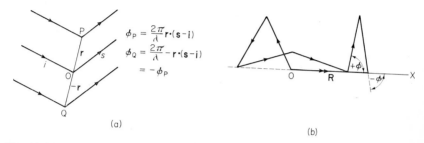

$$\phi_P = \frac{2\pi}{\lambda}\mathbf{r}\cdot(\mathbf{s}-\mathbf{i})$$

$$\phi_Q = \frac{2\pi}{\lambda}-\mathbf{r}\cdot(\mathbf{s}-\mathbf{i})$$

$$= -\phi_P$$

(a)   (b)

FIG. 10. (a) Scattering from two electrons at P and Q, related by a centre of symmetry at the origin O. (b) The phase of the resultant scattering for an atom is the same as that of the scattering from the origin point at the centre. This is because the electrons can be taken in pairs with phases $\pm\phi$, and because the electron density distribution in an atom is such that it always gives a resultant on the positive side of O (i.e. a resultant phase of 0 and not π).

R will vary with θ. The magnitude a, of the electron scattering amplitude, is independent of λ. Therefore R is a function only of the phase angles ϕ_m. Since the electronic configuration of a given atom or ion is constant and spherically symmetrical, the expression 13(1) shows that ϕ_m is proportional to $(\sin\theta)/\lambda$, i.e. $\lambda = 2$ and $\sin\theta = 0.5$ will give the same Argand diagram and resultant as $\lambda = 1$ and $\sin\theta = 0.25$. R can therefore be plotted against $(\sin\theta)/\lambda$. Actually a quantity f_m, called the atomic scattering factor and defined by $R = af_m$ (a is the amplitude scattered by one electron) is plotted. The values of f_m have been calculated and tabulated against $(\sin\theta)/\lambda$ for all atoms, and verified in many cases from gas scattering and diffraction from crystals of simple structure. $f = Z \pm p$ for $\theta = 0$ since all the electrons scatter in phase in the forward direction (Z is the atomic number of the atom concerned, and p is the increase or decrease of the number of electrons due to ionisation). A selection of curves of f versus $(\sin\theta)/\lambda$ is shown in Fig. 11.

The varying configuration of the valence electrons must have a perturbing effect on f, but in most cases this is negligible.

Resonance effects (Section **9**.5.5, p. 123) will affect both the value and phase of f especially when λ is near the absorption edge of an atom [1(65)].

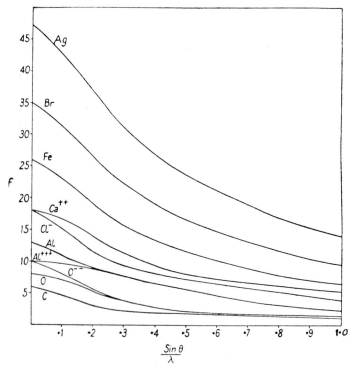

FIG. 11. Atomic scattering factor (f) curves (after Bunn). The influence of the outer electrons is small and ions can only be distinguished from neutral atoms at low values of $\sin \theta/\lambda$.

4.2. RESULTANT SCATTERING FROM A GROUP OF ATOMS

Let $\mathbf{r}_m$ (Fig. 8, p. 12) now be the position vectors of the atomic centres from an arbitrary origin, O. We have to sum a set of amplitude vectors of length f_m a and phase

$$\phi_m = 2\pi \frac{(2 \sin \theta) \, \mathbf{r}_m \cdot \mathbf{n}}{\lambda}. \tag{1}$$

From the Argand diagram (Fig. 9, p. 13):

$$X \text{ component} \quad = \Sigma f_m \, a \cos \phi_m = a \, \Sigma f_m \cos \phi_m = aA \tag{2}$$

Y component $= \Sigma f_m \, a \sin \phi_m = a \, \Sigma f_m \sin \phi_m = aB$ (1)

$$R = (X^2 + Y^2)^{\frac{1}{2}} = a(A^2 + B^2)^{\frac{1}{2}} = aF; \quad \alpha = \tan^{-1} \frac{B}{A}.$$ (2)

Scattering from a group of atoms is therefore equivalent to scattering from an arbitrary point O, with the scattering power of F electrons and a phase shift α, compared to the wave directly scattered from O. F is known as the Structure Factor for the arrangement of atoms concerned.

If this group of atoms is repeated regularly by translations given by three non-coplanar vectors, **a**, **b**, and **c**, it forms a crystal, and the arbitrary origin point chosen for the first group repeats to form a three-dimensional net or 'crystal lattice'. Scattering from such a crystal is therefore equivalent to scattering from the lattice points, with amplitude and phase as given above. For a given direction of scattering the 'phase shift' is the same for all points and can therefore be ignored, although it must be considered in comparing scattering in different directions. Scattering from a crystal therefore reduces to scattering from a regular array of points.

Chapter 2

The Geometry of X-ray Diffraction from Crystals

1. Scattering from a 3-Dimensional Point Lattice

1.1. Scattering from a Line of Regularly Spaced, Equally Weighted Points

This is the three-dimensional equivalent of an optical line grating. Let the repeat translation be given by $\mathbf{a}$ (Fig. 12a). Destructive interference occurs for all directions, except that for which the path difference for waves scattered from adjacent points is $n\lambda$, when the amplitudes are all 'in phase'. Taking the origin at one of the scattering points, the phase of the scattered wave from the adjacent point will be $2n\pi$, and the phases are all of the form $\phi_m = mn2\pi$, where $m\mathbf{a}$ is the position vector of a scattering point and therefore m is a positive or negative integer. The amplitude vectors of the Argand diagram all lie along OX, and the polygon has become a straight line, with $R = NA$, where N is the number of points and A the absolute scattering amplitude from any point.

From Fig. 12a, this condition is fulfilled when:

$$\mathbf{a} \cdot \mathbf{s} - (-\mathbf{a} \cdot -\mathbf{i}) = \mathbf{a} \cdot (\mathbf{s} - \mathbf{i}) = H\lambda$$

where H is an integer. In terms of angles:

$$a \cos v_1 - a \cos \mu_1 = H\lambda \tag{1}$$

$$\cos v_1 = \cos \mu_1 + H\frac{\lambda}{a}. \tag{2}$$

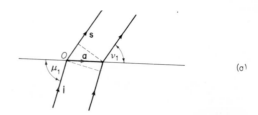

(a)

Fig. 12(a). Scattering from a one-dimensional lattice.

18

If **a** and **i** remain fixed, **s** can be in any direction which makes an angle v_1 (given by 18(2)) with **a**. The diffracted beams will therefore lie in the surfaces of cones whose common axis is the line of points, and whose semi-vertical angles are $v_1(H)$ where $H = 0, 1, 2, 3$ etc. (Fig. 12b).

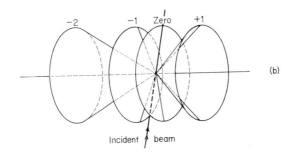

FIG. 12(b). Cones of diffracted rays from a one-dimensional lattice.

1.2. SCATTERING FROM A 2-DIMENSIONAL NET

Let the second vector defining the net be **b** (Fig. 13). A second condition:

$$\mathbf{b} \cdot \mathbf{s} - \mathbf{b} \cdot \mathbf{i} = \mathbf{b} \cdot (\mathbf{s} - \mathbf{i}) = K\lambda \qquad (1)$$

is imposed, and diffracted beams will only occur along the lines of intersection of two sets of cones on **a** and **b** as axes, (Fig. 14).

In the case of a net, **a** and **b** are chosen arbitrarily. Other vectors, such as **a'** and **b'** (Fig. 13) could be chosen to define the same net. These two sets are obviously not independent, and it is left as an exercise to the student to show that the conditions for **a'** and **b'** are equivalent to those for **a** and **b**.

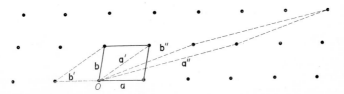

FIG. 13. A two-dimensional lattice, showing various ways in which axes can be chosen.

1.3. SCATTERING FROM A 3-DIMENSIONAL NETWORK OR SPACE LATTICE

If **c** is the third vector defining the lattice, then the condition $\mathbf{c} \cdot (\mathbf{s} - \mathbf{i}) = L\lambda$ must also be satisfied. Diffracted beams can only occur when cones on **a**, **b** and **c** all intersect in one line (Fig. 15). In general there is no such direction; and if a parallel beam of monochromatic X-rays impinges on a perfect crystal it is unlikely to be in a position to give rise to a diffracted beam.

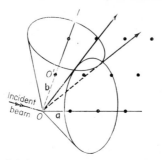

FIG. 14. Intersecting cones defining possible scattering directions for a two-dimensional lattice. Each parallel row with origin at O, O' etc. will give rise to the same diffraction condition, and this will apply for both **a** and **b** directions. Since the repeat distance of the net is a few Å, and the cones are observed at about 30 mm from the origin, the shift from O to O' etc. will have negligible effect.

The angles of the cones v_i can be altered by varying the wavelength or moving the crystal so that three cones intersect in one line, and all three diffraction conditions ($\cos v_1 = \cos \mu_1 + H\lambda/a$; $\cos v_2 = \cos \mu_2 + K\lambda/b$; $\cos v_3 = \cos \mu_3 + L\lambda/c$) are satisfied simultaneously.

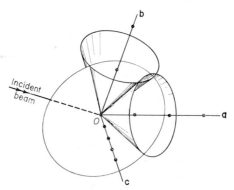

FIG. 15. Cones on three axes (**a**, **b**, **c**) defining possible diffraction directions (where three cones intersect in one line) from a three-dimensional lattice. In general, no diffracted beam is possible even if all the possible cones on each axis are taken into account, instead of just one.

1.3.1. *Varying* λ, μ_j *constant.* (*Stationary crystal*).

Actually 'white radiation' [1(51)], with a continuous range of wavelengths, is used, and the crystal picks out the wavelengths which satisfy the three

'Laue conditions' above for various values of the integers H, K and L. Historically this was the method used by von Laue in discovering the diffraction of X-rays by crystals, but because the wavelengths of the diffracted beams are all different (and any one reflection contains a whole series of wavelengths (Section **6**.2, p. 53)) it has only a restricted use in modern conditions. Figure 43 (pp. 59–62) shows 'Laue photographs' taken under these conditions.

1.3.2. *Varying μ_j, λ constant (moving crystal)*

Varying the directions of **a**, **b** and **c** relative to **i** is most simply achieved by rotating the crystal. The angles of the cones will vary continuously. When three of them intersect the diffraction conditions will be satisfied for the monochromatic radiation employed, and a diffracted beam will flash out. The problem is to relate the directions in which these diffracted beams occur to the crystal lattice vectors, **a**, **b** and **c**. As it stands at present, this is a formidable problem, and it is necessary to develop geometrical tools which can reduce it to reasonable proportions.

1.3.3. *Rotation photograph and the Laue condition for a lattice axis along the rotation axis.*

There is one special problem however which can be solved almost as simply from the Laue conditions, as it can by means of the elaborate armoury of geometrical aids which have been developed for solving the general problem.

If a crystal is rotated about one of the three lattice axes, say **c**, and a beam of monochromatic X-rays is incident on it perpendicular to the rotation axis, then the third Laue condition becomes $\cos v_3 = L\lambda/c$. Since this condition does not vary during the rotation, all diffracted beams must lie somewhere on the cones determined by this condition. If a flat film is rolled into a cylinder coaxial with the (vertical) rotation axis of the crystal, these cones will intersect it in horizontal circles (Fig. 16); and when the film is unrolled the spots produced by the diffracted beams during the rotation will lie on straight 'layer lines' (Fig. 4, p. 5). The central line, containing the direct beam, comes from the flat, zero cone, when $L = 0$. The first lines above and below correspond to $L = \pm 1$ and so on. If y_L mm is the height of the Lth layer line above the zero one (Fig. 16) and r mm the radius of the film, then $\tan v_3 = r/y_L$ and the length of the lattice vector **c** can be found from $c = L\lambda/\cos v_3$. (**Example 2**, p. 285).

There are slightly easier ways of finding c from such a rotation photograph than even this simple procedure, and it is not suggested that the method should normally be used in practice; but the example serves to link the Laue conditions with normal experimental problems. It can also be used, as in **Example 3**, (p. 285) to gain an insight into the sort of angular spread of diffracted beams to be expected in X-ray work. But before attempting any further interpretation of such photographs it is necessary to develop the geometrical tools required for the job.

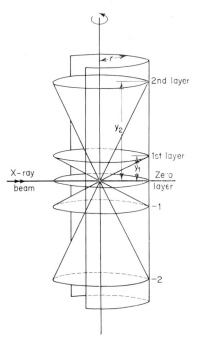

FIG. 16. Cones on crystal axis (c, set along the rotation axis) showing intersection with a co-axial cylindrical film.

2. X-ray Diffraction as a Form of Reflection from Lattice Planes

2.1 SOME PROPERTIES OF THE CRYSTAL LATTICE

The following three results of lattice theory are required here. The proofs are given in Appendix II (p. 340).

(a) Any three non-colinear lattice points define a lattice plane. A series of parallel lattice planes drawn through all lattice points will give a set of equidistant, similar planes. (Fig. 17).

(b) The nearest plane to the origin will cut the three lattices axes, **a**, **b** and **c** at points distant from the origin a/h, b/k, c/l where h, k and l are integers having no common factor. These integers $h, k, l,$ are called the indices of the set of planes. (A given lattice can be defined in an infinite number of ways (see Fig. 13, p. 19) and the indices of a given set of planes will depend on the choice of the defining vectors **a, b, c**. The indices of a set of planes or a crystal face are always given in parentheses (*hkl*)) (**Example 4**, p. 286).

(c) Any three integers having no common factor are the indices of a set of planes all of which contain lattice points, and which pass through all the points of the lattice. If the triplet contains a common factor *p*, then only $1/p$ of the set of planes defined by it contains lattice points. The others will lie equally spaced between these planes of points.

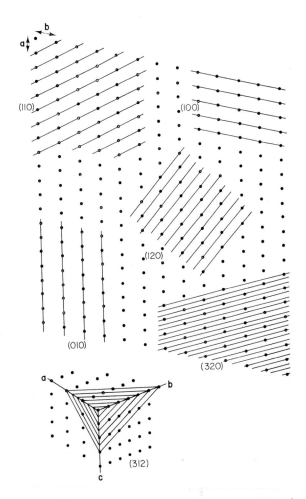

Fɪɢ. 17. Traces of various sets of lattice planes on (001) and a three-dimensional representation of the set (312).

2.2. The Development of the Laue Conditions in Terms of Lattice Planes

Since $\mathbf{s} - \mathbf{i} = \mathbf{n}\,(2 \sin \theta)$ (Fig. 8, p. 12) the Laue conditions are

$$\left. \begin{array}{l} \mathbf{a} \cdot \mathbf{n}\,(2 \sin \theta) = H\lambda = ph\lambda \\[4pt] \mathbf{b} \cdot \mathbf{n}\,(2 \sin \theta) = K\lambda = pk\lambda \\[4pt] \mathbf{c} \cdot \mathbf{n}\,(2 \sin \theta) = L\lambda = pl\lambda \end{array} \right\} \tag{1}$$

where h, k and l contain no common factor. These equations can be written

$$\frac{\mathbf{a}}{h} \cdot \mathbf{n} = \frac{\mathbf{b}}{k} \cdot \mathbf{n} = \frac{\mathbf{c}}{l} \cdot \mathbf{n} = \frac{p\lambda}{2 \sin \theta} \tag{2}$$

These are the equations of a plane cutting the axes at a/h, b/k, c/l (i.e. the nearest plane to the origin of the set (hkl)) whose normal is $\mathbf{n}$ and whose distance from the origin is $p\lambda/(2 \sin \theta)$ (Appendix I, p. 329). Since the next plane of the set passes through the origin, $p\lambda/(2 \sin \theta) = d_{hkl}$, the interplanar spacing. The incident and diffracted beams make equal angles $(90 - \theta)$ with the normal $\mathbf{n}$ to the set of planes, and $\mathbf{n}$ is necessarily in the same plane as $\mathbf{s}$ and $\mathbf{i}$. (Fig. 8, p. 12). The incident and diffracted beams obey the ordinary laws of reflection from the set of planes (hkl), but in addition reflection only occurs when the condition $p\lambda = 2d \sin \theta$ is satisfied.

2.3. Direct Proof of Bragg's Law

The above relation was originally deduced directly by W. L. Bragg without reference to the Laue conditions. If diffraction of X-rays is treated as reflection from lattice planes, then the reflection has to be considered as occurring not just at the surface, but from all the planes of the set. Constructive interference will only occur if the path difference between waves reflected from adjacent planes is a whole number of wavelengths. In Fig. 18

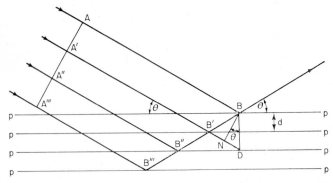

Fig. 18. Direct derivation of Bragg's Law. pp—planes; AA'''—wavefront.

consider the rays reflected at B and B'. Produce $A'B'$ to meet the third plane at D. Because of the equality of the angles of incidence and reflection $BB'D$ is isosceles and BD is perpendicular to the planes. Draw BN normal to $A'B'D$. Then

$$AB = A'N.$$

Path difference $= A'B' + B'B - AB = A'B' + B'D - A'N = ND = n\lambda,$

for constructive interference. But

$$ND = BD \sin \theta = 2d \sin \theta$$

where d is the interplanar spacing.

$$\therefore \qquad n\lambda = 2d \sin \theta. \qquad (1)$$

2.4. THE FURTHER REDUCTION OF BRAGG'S RELATION

If we write the relation as $(d_{hkl}/n) \, 2 \sin \theta = \lambda$ we can use Section 2.1(c) (p. 23) to substitute

$$d_{nh, \, nk, \, nl} \text{ for } \frac{d_{hkl}}{n} \, .$$

The relation then takes the more general form

$$2d_{HKL} \sin \theta = \lambda \qquad (2)$$

where HKL are no longer restricted to a triplet without a common factor. If a common factor is present it indicates the order of diffraction.

The first task in interpreting X-ray diffraction (henceforth called reflection) from single crystals is to determine $\mathbf{a}$, $\mathbf{b}$ and $\mathbf{c}$. The second is to find the indices (HKL) of the set of planes which give rise to a particular reflection, i.e. *to index the reflection*. The lower case letters hkl (without parentheses) will henceforth be used without restriction when refering to the indices of a reflection.

2.5. LIMITING VALUES FOR THE INDICES hkl

All triplets of integers are possible indices of an X-ray reflection, so long as the corresponding value of d_{hkl} is not so small that $\sin \theta$ would become greater than unity. The limiting value of d_{hkl} is given by

$$d_{hkl} \geqslant \frac{\lambda}{2} \, . \qquad (3)$$

Chapter 3

The Reciprocal Lattice – a Geometrical Tool for the Interpretation of 3-Dimensional Diffraction

1. The Geometry of Planes Represented by the Geometry of Points

The geometry of the external faces of a crystal has long been represented by points on the surface of a sphere where the face normals cut it (Fig. 19). Since such faces are parallel to sets of lattice planes, it is natural to try an extension of this method in their case. Some means must be found of representing the interplanar spacing d_{hkl}. One method, which turns out to have outstanding advantages, is to place the representative point, not at a constant distance from the origin, but at a distance K/d_{hkl}. proportional to the reciprocal of the interplanar spacing. For some theoretical purposes the constant K is taken as unity, but for practical use in interpretation it is taken as the value of the wavelength λ, in Å. The distance of the representative point from the origin will then be said to be in Reciprocal Units (R.U.). For calculations where the result is independent of the wavelength, λ can always be put equal to 1·0 Å.

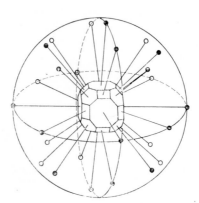

Fig. 19. Representation of crystal faces by points where the face normals cut the projection sphere (the "poles" of the faces).

26

A set of planes (hkl) is thus represented by a point P, given by the position vector

$$\mathbf{d}^*_{hkl} = \frac{\lambda}{d_{hkl}} \mathbf{n} \qquad (1)$$

where $\mathbf{n}$ is the unit vector perpendicular to the planes. Such a point can be immediately related to reflection from the set of planes by the construction shown in Fig. 20. O is the origin for the position vector $\mathbf{d}^*_{hkl}$, and is taken on the X-ray beam, XO. A sphere of unit radius (in R.U.) is constructed to a convenient scale (e.g. 100 mm = 1 unit) on OX as diameter. C is the centre of the sphere. $OP = \mathbf{d}^*_{hkl}$ is drawn to the same scale, and P is shown just as it is moving through the surface of the sphere, as a result of rotation about an axis or any other form of angular movement. The trace of the origin plane of the set (hkl) and the glancing angle of incidence θ (both in the plane XOP) are shown. From the geometry of the sphere, OPX is a right angle and, since $POX = 90 - \theta$, $OXP = \theta$.

$$\therefore OP/2 = \sin\theta.$$

But

$$OP = d^*_{hkl} = \lambda/d_{hkl}$$

$$\therefore \lambda/2d_{hkl} = \sin\theta$$

or

$$\lambda = 2d_{hkl}\sin\theta.$$

If P moves off the surface of the sphere the angle of incidence θ will alter and Bragg's law will no longer be obeyed. Thus the set of planes (hkl) is in the reflecting position if, and only if, the reciprocal point P_{hkl} is on this sphere,

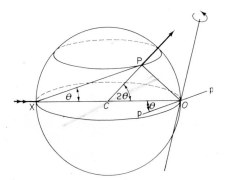

FIG. 20. The relation between the Sphere of Reflection, Reciprocal Point P_{hkl}, and the reflection from the set of lattice planes (hkl). pp is the trace of the plane through the origin in the plane XPO. The (instantaneous) rotation axis can be in any direction.

which is therefore called the '*Sphere of Reflection*' (or, in honour of its originator, the Ewald Sphere). (**Example 5**, p. 286).

Since CP makes an angle of 2θ with the X-ray beam and lies in the plane of the beam and normal, it must be the direction of the reflected beam. For many purposes it is convenient to take the crystal on its real rotation axis at C; but the origin for reciprocal points must always be at O, through which a conceptual parallel and synchronous rotation axis lies.

From this simple construction the position of the crystal for reflection from a given set of planes can be found and also the direction of the reflected beam. But before the Sphere of Reflection construction can be used we must find the relation between the various reciprocal points. (**Example 6**, p. 287).

2. The Reciprocal Lattice

In Fig. 21, Ox, Oy, Oz are the directions of the crystal lattice axes **a, b** and **c**. ABC is the nearest plane of the set (hkl) to the origin. By definition therefore, it cuts off intercepts $OA = a/h$, $OB = b/k$, $OC = c/l$ on the three axes. ON is drawn perpendicular to ABC and produced to P, so that $OP = \lambda/ON$. Since another plane of the set (hkl) must pass through the origin, ON is the interplanar spacing, d_{hkl}. Therefore $OP = \lambda/d_{hkl}$, and P is the representative point for the set of planes (hkl), henceforth called the reciprocal point, hkl. Draw PD perpendicular to Oz and join NC. Let $\widehat{CON} = \psi_3$.

Then
$$\cos \psi_3 = \frac{ON}{OC} = \frac{d_{hkl}}{c/l} = \frac{l d_{hkl}}{c}$$

$$OD = OP \cos \psi_3 = \frac{\lambda}{d_{hkl}} \cdot \frac{l d_{hkl}}{c} = l\frac{\lambda}{c}.$$

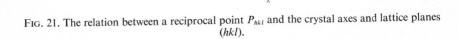

FIG. 21. The relation between a reciprocal point P_{hkl} and the crystal axes and lattice planes (hkl).

This means that all reciprocal points for a particular value of l lie in a plane (or *layer*) perpendicular to OZ. For $l = 0$ the layer passes through O; for $l = 1$ it cuts OZ at λ/c; for $l = 2$, at $2\lambda/c$, etc., i.e. the layers are equally spaced. Similarly all reciprocal points with a particular value of k lie on a layer perpendicular to Oy, and for h, perpendicular to Ox.

The intersections of the 3 layers through the origin give the directions of three axes, $\mathbf{a}^*, \mathbf{b}^*, \mathbf{c}^*$ on which lie the reciprocal points $h00, 0k0, 00l$ respectively. (Only for orthogonal crystal lattices will these lie in the same directions as $\mathbf{a}$, $\mathbf{b}$ and $\mathbf{c}$).

$a^* = d^*_{100} = \lambda/d_{100} = (\lambda bc \sin \alpha)/V$ where V is the volume of the crystal cell (the 'real' cell). (Appendix II, p. 360).

$\mathbf{a}^*$ lies on the intersection of planes perpendicular to Oy and Oz, and is therefore perpendicular to yOz. This is otherwise obvious, since yOz is one of the set of planes (100) and $\mathbf{d}^*_{100} = \mathbf{a}^*$ must, by definition, be normal to (100). Similar relations are obviously true for $\mathbf{b}^*$ and $\mathbf{c}^*$. (Fig. 22).

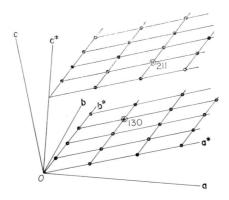

FIG. 22. The relation between the crystal and reciprocal lattices.

Thus the reciprocal points lie on a lattice similar to the crystal lattice, and the indices of a point are its coordinates in the lattice, i.e.

$$\mathbf{d}^*_{hkl} = h\mathbf{a}^* + k\mathbf{b}^* + l\mathbf{c}^*. \tag{1}$$

If an X-ray reflection can be shown to derive from a particular reciprocal point, it is at once possible to index it by inspection of the coordinates of the point, as in the case of the ringed points in Fig. 22. The coordinates of a reciprocal point hkl are given without parentheses, as in the case of indices of a reflection.

This network of points, each one of which is derived from and represents a set of crystal planes, is called the 'Reciprocal Lattice', and, for convenience, the geometry of this lattice is described as occurring in 'reciprocal space' to distinguish it from the crystal lattice in 'real space'.

A 'REciprocal Lattice Point' will be abbreviated to 'relp', and 'Reciprocal Lattice' to 'RL'.

As will be shown shortly, X-ray photographs can be considered as more or less distorted pictures of the RL. It will therefore be as well to summarise its properties by means of the following vector development.

3. The Vector Development of the Reciprocal Lattice (RL)

We start by defining the RL axes $\mathbf{a}^*$, $\mathbf{b}^*$ and $\mathbf{c}^*$ in terms of the crystal lattice axes, $\mathbf{a}$, $\mathbf{b}$ and $\mathbf{c}$ and then deduce the representative character of the relps. The defining relations are:

$$\mathbf{a}^* \cdot \mathbf{a} = \mathbf{b}^* \cdot \mathbf{b} = \mathbf{c}^* \cdot \mathbf{c} = \lambda \tag{1}$$

$$\mathbf{a}^* \cdot \mathbf{b} = \mathbf{a}^* \cdot \mathbf{c} = \mathbf{b}^* \cdot \mathbf{c} = \mathbf{b}^* \cdot \mathbf{a} = \mathbf{c}^* \cdot \mathbf{a} = \mathbf{c}^* \cdot \mathbf{b} = 0. \tag{2}$$

Relations (2) give the direction of the RL axes i.e. $\mathbf{a}^*$ perpendicular to $\mathbf{b}$ and $\mathbf{c}$, etc., and relations (1) define the lengths a^*, b^* and c^*.

Consider any RL vector $\mathbf{d}^*_{hkl}$, defined by:

$$\mathbf{d}^*_{hkl} = h\mathbf{a}^* + k\mathbf{b}^* + l\mathbf{c}^* \tag{3}$$

where hk and l are integers.

From (1) and (2)

$$\mathbf{d}^* \cdot \mathbf{a}/h = \mathbf{d}^* \cdot \mathbf{b}/k = \mathbf{d}^* \cdot \mathbf{c}/l = \lambda. \tag{4}$$

Let the unit vector $\mathbf{d}^*/d^* = \mathbf{n}$. Divide (4) by d^*. Then

$$\mathbf{n} \cdot \mathbf{a}/h = \mathbf{n} \cdot \mathbf{b}/k = \mathbf{n} \cdot \mathbf{c}/l = \lambda/d^*.$$

These are the equations of a plane whose normal is $\mathbf{n}$ and such that it cuts off intercepts on the three axes of a/h, b/k, c/l. But this is the nearest plane to the origin of the set (hkl), and the distance from the origin along the normal, which is the interplanar spacing, d_{hkl}, is given by:

$$d_{hkl} = \lambda/d^*_{hkl} \quad \text{or} \quad d^*_{hkl} = \lambda/d_{hkl}. \tag{5}$$

Thus the RL vector $\mathbf{d}^*_{hkl}$ is normal to the set of planes (hkl) and its length is given by the reciprocal relationship (5). The point defined by the position vector $\mathbf{d}^*_{hkl}$ is thus the representative point for the set of planes (hkl) as defined previously, and therefore has all the properties of such a point. (**Example 7**, p. 287).

From the complete symmetry of the vector definition of the RL it follows that any vector $\mathbf{g}_{pqr}$ in the crystal lattice, defined by:

$$\mathbf{g}_{pqr} = p\mathbf{a} + q\mathbf{b} + r\mathbf{c} \tag{1}$$

is perpendicular to a set of planes in the RL whose interplanar spacing is λ/g_{pqr}. In practice, $\mathbf{g}$ is set along the (normally vertical) rotation axis of an X-ray camera; and the interplanar spacing in the RL, measured from an X-ray photograph, has the symbol ζ.

$$\therefore \zeta = \lambda/g \quad \text{and} \quad g = \lambda/\zeta. \tag{2}$$

If p, q and r have no common factor, then g is the distance of the first crystal lattice point from the origin, i.e. it is the distance between points along the *central lattice line†* in the direction $\mathbf{g}$, and all RL planes of the set (pqr) will therefore contain relps and will register on an X-ray photograph. The interplanar spacing ζ, measured from such a photograph, enables the distance between points on the central lattice line perpendicular to the planes to be obtained from (2). In particular, if

$$\mathbf{g} = \mathbf{a}, \quad a = \lambda/\zeta \tag{3}$$

and the length of the crystal lattice axes can be obtained from rotation photographs about these axes. This is equivalent to the previous deduction from the corresponding Laue relationship; but, as will be seen later (Section 14.1.2, p. 175), charts are available from which ζ can be obtained directly, and so a is usually calculated from (3). (**Example 8**, p. 288).

---

† Any straight line of lattice points passing through the origin is a *central lattice line.*

Chapter 4

The Photography of the Reciprocal Lattice (RL)

1. General

All X-ray photographs, especially those taken with monochromatic radiation, can be considered as more or less distorted pictures of the RL. If the distortions can be sorted out the reflections can be immediately indexed. Since the photograph is taken on a two-dimensional film the RL can only be photographed a layer at a time, if the distortion is not to include a reduction in the number of dimensions. In the case of a rotation photograph this reduction does take place. The two-dimensional RL nets perpendicular to the rotation axis are reduced to one-dimensional 'layer lines' on the photograph (Fig. 4, p. 5). Only the distances of the relps from the rotation axis are preserved, and this makes it difficult to index the spots on these layer lines. (The difficulty can be partially overcome by taking a series of 15° oscillations instead of a complete rotation). However, the third dimension, the distance apart of the RL

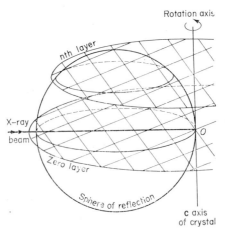

FIG. 23. Reciprocal lattice layers rotating in their own planes through the Sphere of Reflection.

layers, is presented all the more clearly because of the reduction of the layers to lines. The production of these layer lines is simply explained in terms of the RL. In Fig. 22, (p. 29) let **c** lie along the rotation axis. Then the **a*b*** RL planes will be perpendicular to the rotation axis and will rotate in their own planes. The relps on these planes will cut the sphere of reflection on circles (Fig. 23), and the reflected beams for all the points of one RL layer will lie on a cone defined by this circle and the centre C. If a cylindrical film is placed with its axis along the crystal rotation axis, the cone will intersect it in a circle perpendicular to the axis (Fig. 24). When the cylindrical film is unrolled the reflections all appear along straight lines called *layer lines* (Fig. 4, p. 5). In Fig. 24, $y/r = \tan v$; $\zeta/1 = \sin v$ and $c = \lambda/\zeta$. This is the equivalent calculation in terms of the RL to that using the Laue condition (Section 2.1.3.3, p. 21). ζ is, however, normally read direct from a chart (Section 14.1.2, p. 175).

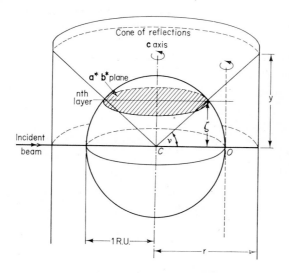

Fɪɢ. 24. Rotating reciprocal lattice layer giving rise to reflections on a cone, which intersects a cylindrical film in a 'layer line'.

If the RL is photographed one layer at a time with specially designed cameras, it is possible to obtain a completely undistorted 'contact print' of a layer of the RL (Fig. 25) (Section 16, p. 213); but even with a simple rotation camera and cylindrical film it is possible to obtain an almost undistorted picture of part of a layer, using a limited oscillation of the crystal.

2. An almost Undistorted Picture of part of a Vertical Zero RL Layer

The oscillation starts with **c** along the X-ray beam and the (vertical) **a*b*** zero layer tangential to the sphere at *AO* (Fig. 26). The crystal is oscillated through 10–15° about an axis perpendicular to the beam. The part of the layer shown in plan and elevation in Fig. 26 passes through the sphere, and reflected beams occur as the points pass through the sphere. If these reflections are registered on a cylindrical film they give an almost undistorted picture of the relps within the circle on the **a*b*** layer. The **a*b*** axes can be chosen or recognised and the reflections indexed by inspection. Figure 27a shows such a photograph. In Fig. 27b the circle at the centre has been outlined, as has the crescent shaped area arising from the first non-zero layer parallel to **a* b*** (*ED*. Fig. 26). Figure 28 shows these effects very clearly in the case of a crystal with a very small *a*b** mesh. (**Example 9**, p. 289).

3. An Optical Analogue of the Reciprocal Lattice

Optical diffraction can give a pattern which is essentially that of a zero layer of the RL (Section 9.1.2, p. 96). Figure 29 shows the relation between

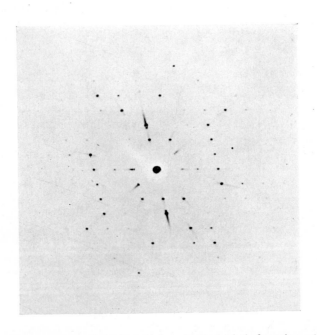

FIG. 25. The central portion of a 'retigraph', i.e. a 'contact print' of a reciprocal lattice layer (the **a*c*** zero layer of sucrose). Large thermal vibrations, due to weak intermolecular bonds, are responsible for the rapid fall-off in intensity towards the edges of the photograph.

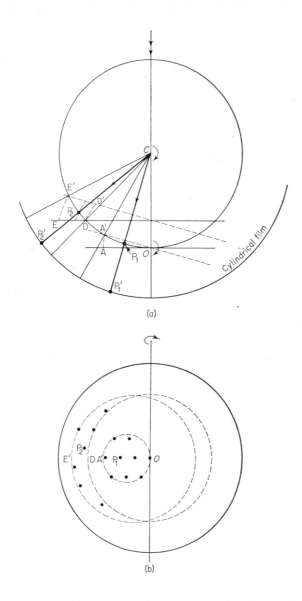

FIG. 26. The production of an almost undistorted picture of a vertical zero reciprocal lattice layer by means of a rotation camera. The reflections on the L.H.S. from the first layer back along the beam are also shown. A non-orthogonal centred lattice is pictured, so that P_2 and P_2' in (a) are not in the plane of the diagram. In (a) the vertical rotation axis, and in (b) the X-ray beam is coming out of the diagram at O. In (b) the cylindrical film is not shown, but the pattern on the sphere is projected forwards from C on to the film.

the two-dimensional repeat patterns of holes in the mesh and the corresponding diffraction patterns.

4. The 'Tore of Reflection' and the 'Limiting Sphere'

It is of interest to consider which relps can come into the reflecting position during a complete rotation, and which remain outside the sphere of reflection and therefore have to be investigated by rotation about another axis. This can best be done by an artifice which is often useful. If we consider the X-ray tube and camera to be fixed we can give the crystal a *clockwise* rotation, looked at from above. We can however consider the crystal to be fixed, and the X-ray tube and camera to be rotating *anti-clockwise*. If we do this for a complete rotation the sphere of reflection will trace out a 'tore' (a solid figure like an O-ring with zero internal diameter, Fig. 30) and all relps inside this tore will have passed through the sphere and given rise to reflections. However, since

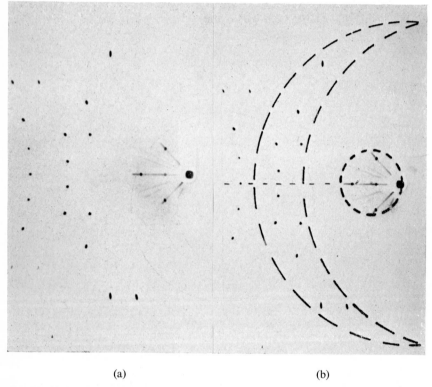

(a)   (b)

FIG. 27. (a) A nearly undistorted picture of the zero and first layers of the RL taken as in Fig. 26.   (b) As for (a), with the RL zero and first layers outlined. The layers are 'staggered' by lattice centering, so that the sequence of layers shows up particularly clearly.

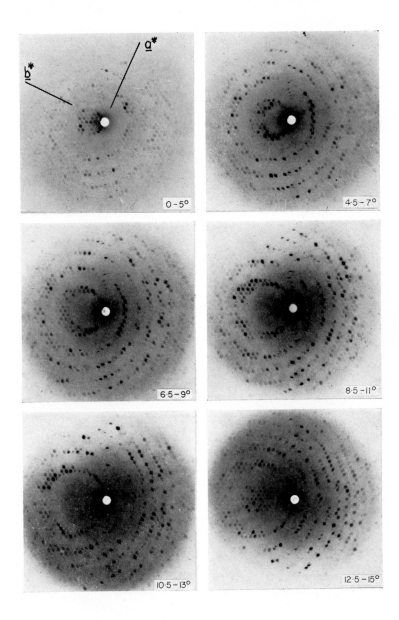

F<small>IG</small>. 28. Oscillation photographs of wet horse methaemoglobin, showing **a*b*** layers. Flat films, 50 mm from the crystal. Filtered CuK radiation. Oscillation about [110]. **c** along beam at 0°. In the final oscillation the −1 layer has just come into the sphere of reflection, giving a patch of reflections in the centre of the 'hole' left by the zero layer.

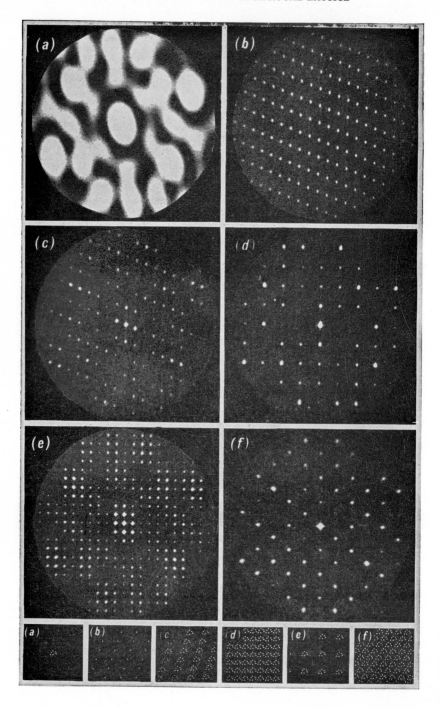

any relp within a distance of 2 R.U. of the origin can be brought into the surface of the sphere, there is a considerable number of relps remaining outside the tore. The sphere of radius 2 R.U. is called the 'Limiting Sphere', since no point outside it can give rise to a reflection for the wavelength being used. The limiting condition is thus $d^* \leqslant 2$, i.e. $\lambda/d \leqslant 2$ or $d \geqslant \lambda/2$, as deduced previously from Bragg's equation.

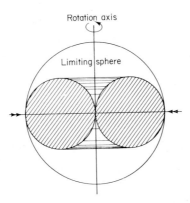

Fig. 30. A section through the 'tore' of reflection swept out by a complete rotation of the sphere of reflection relative to a fixed crystal and reciprocal lattice. The 'limiting sphere' is also shown.

Fig. 29. The optical anlogue of the zero layer of a RL, illustrating the effect of the size and shape of the unit cell. The top, large diagrams (a) to (f) are photographs ($\times 50$) of the optical diffraction from the masks, shown ($\times \frac{1}{4}$) in the bottom, small diagrams (a) to (f). (Portions only of the masks are shown in (b) to (f)).
    (a) Diffraction from a single, asymmetric 'molecule'.
    (b) Diffraction from a two-dimensional lattice of single holes.
    (c) Diffraction from a pattern of 'molecules' based on the same lattice. (d), (e) and (f) show the effects of variations in the size and shape of the lattice. Notice how lines of points in one lattice are perpendicular to 'layers' of points in the other, and that the sizes of the two lattices have a reciprocal relationship. The fact that the uniform points of (b) have, in (c), (d), (e) and (f) the intensity of the corresponding points in the diffraction diagram from the mask of a single 'molecule', is referred to in Section 9.1.3 (p. 97). (From *Optical Transforms* by Lipson and Taylor.)

Chapter 5

The Reciprocal Lattice and Divergent Beam Photographs

1. The Production of Divergent Beam Photographs

Very early in the history of X-ray diffraction W. H. Bragg noticed that when a large single crystal was rotated through the reflecting position, the transmitted beam was reduced in intensity as the Bragg reflection occurred. This is not surprising since the direct beam is not only attenuated by the normal processes of absorption, but is also having energy reflected away as well.†

This effect is well illustrated by the type of photograph which is obtained using a divergent beam of X-rays. In Fig. 31 a beam of monochromatic X-rays of wavelength λ, diverging from a point source S, strikes one set of lattice planes (hkl) in a crystal. Reflection will occur for those rays which are incident on the (hkl) planes at the Bragg angle θ.

For neighbouring rays, at angles smaller or greater than θ, no reflection will take place. A photographic plate placed behind the crystal will be more or less uniformly blackened, except in those directions corresponding to rays incident at θ. There, the transmitted intensity will have been reduced by an amount depending on the intensity of the reflection.

From Fig. 31 it is clear that such directions lie on a cone of semi-angle $(90 - \theta)$, with apex at S and axis perpendicular to the planes (hkl), i.e. the axis is parallel to $\mathbf{d}_{hkl}^*$. Such cones of diminished intensity (deficiency cones) will be produced by all the different sets of lattice planes for which Bragg reflection is possible. A flat photographic plate placed on the far side of the

---

† The implicit assumption that normal processes of absorption are unaffected by Bragg reflection is almost always a sufficiently good approximation. But with very perfect crystals a system of standing waves is set up in the crystal, and if the nodes are at the main absorption centres (e.g. heavy atoms) energy can get through largely unabsorbed. This accounts for the Borrmann effect, when 'anomalous transmission' can be obtained through normally completely opaque crystals, for X-rays incident at the Bragg angle. It was first observed in calcite by Borrmann and has since been observed in Si, Ge, Cu and various III-V semi-conductors.

The production of large very perfect crystals for semi-conductor work has made anomalous transmission a commonplace occurrence where such crystals are available, but otherwise it is very rarely observed.

40

crystal from S will be intersected by a whole series of 'deficiency cones', with axes in different directions. The more or less uniform background blackening of the film will be crossed by the 'deficiency conics' arising from the intersections of these cones with the flat plate. Figure 32 is a positive print from such an exposure, with the 'deficiency cones' showing as dark lines.

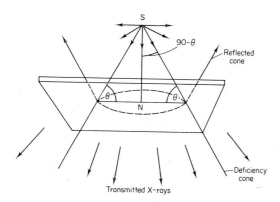

FIG. 31. Diagram showing the production of a deficiency and a reflection cone from one set of lattice planes.

This kind of effect was observed as early as 1915 by Rutherford and Andrade, using radiation from radium emanation in a fine glass tube (presumably emitting the internal conversion Kα lines of elements 82, 83 and 84, with an average wavelength of 0·163 Å.) The crystal used was rock salt ($d_{200} = 2·81$ Å), and the pattern was described by Bragg as 'casting shadows' of the reflecting planes. The pattern consisted of two pairs of 'straight lines' intersecting at right angles, the angular separation of each pair under the conditions of the experiment being 'about 3°'. (In fact, these lines are sections of very flat cones, of semi-angle 88·3°).

The full value of this aspect of X-ray diffraction has only recently been appreciated, largely through the development of very fine X-ray sources relying on the micro-focus techniques of electron-microscopy. Earlier work, particularly by Kossel (the deficiency conics are often called 'Kossel lines') and by Lonsdale, used less elaborate point-sources. Techniques have been evolved for determining the orientation of crystals, for measuring lattice parameters with high precision, and for examining the degree and axes of strain in suitable crystals.

The basic experimental arrangement required to obtain divergent beam photographs is extremely simple, compared with other techniques. Given a fine source of X-rays, the divergent beam method requires no motion either of specimen or of photographic plates. Results which can be quoted to the highest accuracy (1 in 100,000 or better) can be obtained without any measurements whatsoever of the physical dimensions of the experimental set-up. Useful results can also be obtained, without ambiguity, from the simultaneous use of several characteristic lines from the X-ray source (e.g. Cu and Zn radiations from a brass target).

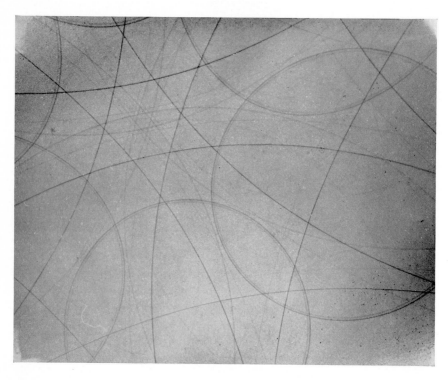

FIG. 32. The hexad region of a divergent beam deficiency photograph from a beryllium single crystal, using FeK radiation. The reproduction has reversed the intensity. (After K. J. H. Mackay).

The crystal, and its relation to the source, must satisfy certain simple conditions. Its size, coupled with its distance from the source, must be such that the crystal subtends an angle of about 30° at the source.

The crystal must not be so thin that the radiation reflected out is too small a proportion of that getting through; nor so thick that short wave length white radiation forms the major part of the background, with the Kα component so small that the deficiency is lost. In practice however, the thickness of the crystal is not found to be very critical (a few μm of Au and 10 mm of Be have both produced usable patterns); and provided an indexing pattern (e.g.

from a similar crystal in known orientation, with the same radiation) has been obtained (or constructed graphically) beforehand, it is also possible to make good use of crystals subtending smaller angles than 30°, as, for example, the grains in a metal foil specimen. Overlapping patterns from superimposed crystallites can be confusing, so it is helpful—for the purposes of inter-pretation—if the specimen is only one crystallite layer thick.

Crystal perfection, in theory, is a factor which affects the possibility of obtaining a pattern. Too perfect a crystal will result in less energy being re-flected out of the incident beam. In addition the Borrmann effect may largely compensate for the energy which is reflected out, by reducing normal absorption so that there is very little or no difference between the Bragg direction and the background. In practice however it is rare for this condition to limit contrast, although cases have been quoted where the contrast has, in fact, been reversed, due to 'anomalous transmission' at the reflection angle.

If the crystal is imperfect, the divergent beam pattern may still be useful. The imperfection may consist of a small number of relatively large mis-alignments of perhaps up to 5°. These give rise to a series of discrete over-lapping patterns, which can still be used satisfactorily, and moreover convey information about the relative displacement in the crystal. Another type of imperfection is a series of very small misalignments which will cause line-broadening, generally preferentially to certain deficiency traces. This again conveys information, though of less accuracy, regarding the axes and extent of the misorientation. Only if the crystallites are too small, too numerous and too randomly arranged, will no pattern be observed.

Additional information can be obtained from a divergent beam photograph using a very small source of X-rays, because it is also a transmission radio-graph of the crystal on a magnified scale. This radiograph displays the internal structure and inclusions within the specimen. (A simple procedure makes it possible to produce *stereo*-microradiographs by this means; two exposures are made before and after a controlled lateral or rotational displacement, the resultant prints being viewed in a stereoscope).

Reflection cones are also produced and can sometimes be seen as dark, broader lines on a lighter background. Usually they are too broad (the width being normally comparable with the size of the crystal) to be visible. Where they occur their geometry is less precisely defined but otherwise very similar to that of the narrow deficiency cones.

2. The Geometrical Interpretation of Divergent Beam Photographs

Faced with a pattern consisting of a complex tracery of intersecting conic sections (e.g. Fig. 32, p. 42) how can one start to make use of it? Fortunately the RL provides the tool required for interpretation of this aspect of X-ray diffraction, as it does for the more usual cases.

In Fig. 33, S is the point source of X-rays and also the origin of the RL. AB is the trace of one of a set of planes (hkl), and the section is taken so that the normal to the planes, DN, (parallel to $\mathbf{d}^*_{hkl}$) is in the plane of the paper. The ray SD is incident on the plane at the Bragg angle, and some of the energy is diffracted away in the direction DR. The diminished direct beam goes on in the direction DQ. Draw the RL vector $\mathbf{d}^*_{\bar{h}\bar{k}\bar{l}}$ from S, defining the relp $\bar{h}\bar{k}\bar{l}$ at P. Draw the plane through P perpendicular to $\mathbf{d}^*_{\bar{h}\bar{k}\bar{l}}$ cutting SQ at Q. We may call this the 'indexing plane'. $PSQ = (90-\theta)°$ and all rays which make this angle with $\mathbf{d}^*_{\bar{h}\bar{k}\bar{l}}$ will be incident at the Bragg angle θ on the planes (hkl) and will be reflected. The incident directions will lie on a cone (as in Fig. 31, p. 41 and defined by the circle round P in Fig. 33) with semi-vertical angle $(90-\theta)°$, and $\mathbf{d}^*_{\bar{h}\bar{k}\bar{l}}$ as axis. This cone will intersect the indexing plane through P in a circle and the flat photographic film ME in the corresponding conic.

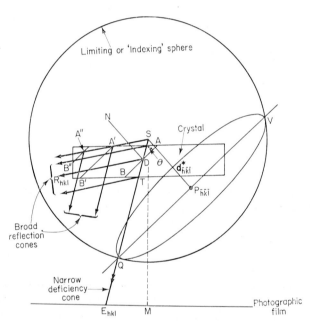

FIG. 33. The geometry of the production of deficiency and reflection cones in divergent beam photographs.

Rays parallel to DR, reflected from planes parallel to AB, will arise all the way along the part of the line ST which is within the crystal. These rays will constitute a broad reflection, which will also occur along the other generating lines of a cone with a semi-vertical angle of $(90-\theta)°$, but inverted with respect to the direct beam cone.

$SA'A''$ is incident on the reverse side of the set of planes (hkl) and gives rise to the reflection $\bar{h}\bar{k}\bar{l}$. This reflection occurs all the way along the part of the line $A'A''$ which is within the crystal, and also gives rise to a broad band to one side of the narrow deficiency direction SDQ. Neither this band nor the hkl reflection band will normally be visible above the background.

The incident beam SA' is parallel to the reflected beam DR, so that when DR is parallel to the crystal plate, this is the limit beyond which no reflected line parallel to the deficiency line can be formed.

$SA'A''$ will also be a generating line of a deficiency cone. In general, deficiency cones and reflection cones go in pairs, but the reflection cone is usually too broad to be visible.

2.1 INDEXING DIVERGENT BEAM PHOTOGRAPHS

The deficiency conic must be labelled with the indices hkl of the reflection. In most cases it is necessary not only to know the approximate cell parameters and the directions of the RL axes, but to have oriented the crystal so that a RL axis is along the normal to the flat film, and the direction of the other two axes relative to the photograph are known. (If the first condition is fulfilled it will generally be possible to locate the positions of the other axes).

Indexing can be carried out by noticing that in Fig. 33:

$$SQ = \frac{SP}{\sin \theta} = \frac{d^*}{d^*/2} = 2 \text{ R.U.}$$

Thus all the circles, in which the cones of direct beams corresponding to the various Bragg angles intersect the indexing planes, lie on a sphere of radius 2 R.U. about S as centre. This is the limiting sphere (Section **4**.4, p. 36), and there can be no direct beam cone corresponding to a relp outside it, as would be expected.

However, the limiting sphere has here an additional function as 'indexing sphere'. Suppose that $\mathbf{c}^*$ is along the normal to the film SM, and it is required to index the $h0l$ conics. These will all be symmetrical about the trace of the $\mathbf{a}^*\mathbf{c}^*$ plane on the photograph (a line through the centre point, M), i.e. they will meet this line at right angles. Construct the $\mathbf{a}^*\mathbf{c}^*$ section of the RL and draw in the indexing sphere (Fig. 34). Mark in the extreme positions SE, SF of the incident rays received on the film. Draw those traces of indexing planes through the relps which have at least one intersection with the circle within the limiting angle. These traces are the diameters of the 'deficiency cones' of direct beams where they intersect the indexing sphere (corresponding to QV in Fig. 33). Produce the direct beams from S to the plate at M, making SM equal to the experimental distance from source to film (allowing for any photographic enlargement in reproduction). The line of intersection of the

a*c* plane on the photograph will have conics crossing it at right angles at the points marked on *EF*, and by comparison these conics can be given the indices marked against the points. If there is any ambiguity in the indexing, the direction of concavity and curvature of each conic can be used to distinguish between close pairs.

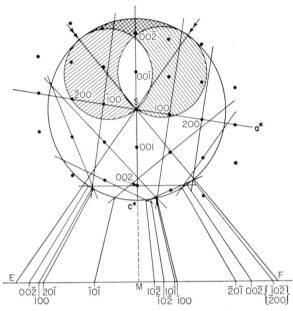

FIG. 34. The use of the RL and 'indexing' sphere in defining the position of the deficiency conic of a given reflection, *hkl*.

If the direction of **a*** is not known, the various lines through *M*, which have a number of conics intersecting them at right angles, must be compared with the points on *EF*. For high symmetry crystals these lines will be obvious symmetry lines. The **b*c*** conics can be identified and indexed in a similar way.

Lines corresponding to $K\alpha_1$ and $K\alpha_2$ are close doublets such as can be seen in Fig. 32 (p. 42) and present no indexing problems. Lines due to $K\beta$ radiation can be interpreted using the same RL net, and drawing a β indexing sphere of radius $2\lambda_{K\alpha}/\lambda_{K\beta}$ to the original scale i.e. if the RL should have been smaller, the sphere is made larger in proportion instead.

Figure 34 can also be used to find which relps will give rise to conics within the limiting angle of the divergent beam. For this section of the RL the divergent beam corresponds to a parallel beam oscillating from one limiting direction to the other. If a reflection occurs during this 'oscillation' a part of the

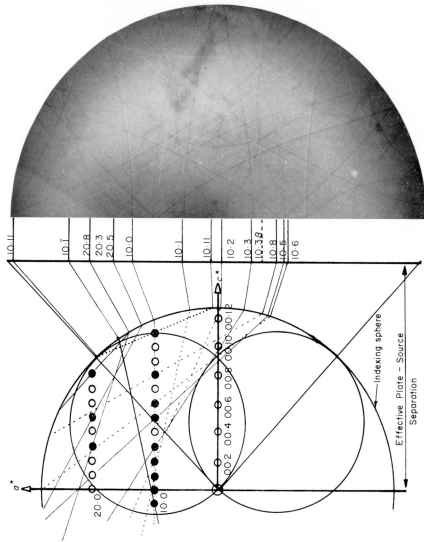

FIG. 35. The indexing of the deficiency conics given by MoS₂ with FeKα radiation. Only the relps corresponding to known strong reflections have been used in the indexing of the conics, which cut the line of section normally and are concave upwards. Others are indexed by considerations of symmetry. (After K. J. H. Mackay.)

The reproduction has reversed the intensities—notice the occurrence of some sharp reflection conics, which appear as *light* lines on the reproduction.

The reasons for the appearance of these lines are not well understood; but since their geometry is not nearly as perfect as in the case of deficiency conics, it is possible that the crystal morphology plays some part in their production.

corresponding conic will appear on the film, cutting the trace of the **a*c*** plane at right angles. Whether or not a reflection occurs can be found by the normal technique for interpreting oscillation photographs (Section **14**.1.2, p. 175). Draw the two limiting directions of the X-ray beam coming *towards* the origin of the RL at *S* (i.e. project the X-ray directions back beyond the source) and draw the two circles of reflection. The two shaded lunes contain the relps which will have passed through the sphere during the 'oscillation'. The relps in the double shaded area will have passed through twice, which means that the whole conic (a circle or ellipse) will appear on the photograph. If a relp *h0l* is in one of the lunes, the plane to determine the direction of the direct beam conic will be drawn through the inverse relp $\bar{h}0\bar{l}$. Since the zero layer of a RL necessarily has two-fold rotor symmetry, the circles can be constructed on the other side of the origin to determine directly the inverse relps through which the planes have to be drawn as in Fig. 35.

If some reflections in the *h0l* zone are known to be strong, these can be specially marked, since in general only strong reflections will give visible deficiency conics. Planes need then only be drawn through these relps. This can be a considerable help in identification, especially if the RL has been drawn from approximate cell dimensions.

In most cases there is no need to distinguish between positive and negative directions of the RL axes, and therefore no necessity to invert the RL vectors to get the correct indices. Only high symmetry crystals are normally investigated by divergent beam methods, although this may not always be the case in the future.

An actual example of indexing is shown in Fig. 35.

3. The Use of Divergent Beam Photographs in Accurate Lattice Parameter Determination

The rather remarkable claim was made earlier that this technique is capable of giving lattice parameter measurements to an accuracy of 1 in 100,000 without any physical measurements on the apparatus whatsoever. In fact, all that is required is to find three deficiency conics which intersect exactly or very nearly, in a point.

3.1. The use of Triple Intersection Points

If three indexed conics intersect at a point, the corresponding indexing planes through the relps must intersect in a point on the indexing sphere. The equations of the indexing planes are:

$$\mathbf{r_1} \cdot \mathbf{d_1^*} = d_1^{*2}; \qquad \mathbf{r_2} \cdot \mathbf{d_2^*} = d_2^{*2}; \qquad \mathbf{r_3} \cdot \mathbf{d_3^*} = d_3^{*2} \qquad (1)$$

where $\mathbf{d}_i^* = \mathbf{d}_{h_i k_i l_i}^*$ and $\mathbf{r}_i$ is a position vector of a point on the plane. The point of intersection of the three planes is the vector,

$$\mathbf{r} = x\mathbf{a} + y\mathbf{b} + z\mathbf{c} \tag{1}$$

which satisfies all three equations.

Expanding 48(1) in components for $\mathbf{r} = \mathbf{r}_1 = \mathbf{r}_2 = \mathbf{r}_3$

$$(x\mathbf{a} + y\mathbf{b} + z\mathbf{c}) \cdot (h_1\mathbf{a}^* + k_1\mathbf{b}^* + l_1\mathbf{c}^*) = d_1^{*2}$$

$$h_1 x\lambda + k_1 y\lambda + l_1 z\lambda = d_1^{*2}$$

$$\therefore h_1 x + k_1 y + l_1 z = d_1^{*2}/\lambda$$

similarly
$$h_2 x + k_2 y + l_2 z = d_2^{*2}/\lambda \tag{2}$$

$$h_3 x + k_3 y + l_3 z = d_3^{*2}/\lambda.$$

The general expression for d^{*2} is given in Appendix II (p. 363) and can be calculated in the three cases. The equations are then solved for the values of x, y, z, which are the co-ordinates of the point of intersection of the three indexing planes, in terms of the crystal lattice axes. If, in addition, this point is on the sphere of radius 2 units, then:

$$r^2 = 4 = x^2 a^2 + y^2 b^2 + z^2 c^2 + 2\,xyab \cos\gamma + 2\,yzbc \cos\alpha$$
$$+ 2\,zxca \cos\beta. \tag{3}$$

In the general case six triple intersection points would be required to solve for the six 'unknowns' (the *accurate* parameters are unknown) as functions of x, y and z, which are themselves proportional to $1/\lambda$. Even in the orthorhombic case, three equations and therefore three triple intersections are required. In the light of these requirements it is not surprising that most work has been done on cubic or hexagonal crystals. In the cubic case the equations reduce to:

$$h_1 x + k_1 y + l_1 z = \frac{a^{*2}}{\lambda}\,({h_1}^2 + {k_1}^2 + {l_1}^2)$$

$$h_2 x + k_2 y + l_2 z = \frac{a^{*2}}{\lambda}\,({h_2}^2 + {k_2}^2 + {l_2}^2) \tag{4}$$

$$h_3 x + k_3 y + l_3 z = \frac{a^{*2}}{\lambda}\,({h_3}^2 + {k_3}^2 + {l_3}^2)$$

$$a^2(x^2 + y^2 + z^2) = 4. \tag{5}$$

The first application of this technique (using spherical trigonometry for the interpretation) was by K. Lonsdale in 1947. For diamond, the conics 220, 313 and 133 intersect for CuKα_1 radiation. Putting these indices in the equations 49(4) for cubic crystals gives:

$$2x + 2y \quad\quad = 8a^{*2}/\lambda$$

$$3x + y + 3z = 19a^{*2}/\lambda$$

$$x + 3y + 3z = 19a^{*2}/\lambda$$

These are easily solved to give $x = 2a^{*2}/\lambda;\quad y = 2a^{*2}/\lambda;\quad z = \tfrac{11}{3}a^{*2}/\lambda$

$$\therefore a^2 \cdot \frac{a^{*4}}{\lambda^2}\left(4 + 4 + \tfrac{121}{9}\right) = 4 \quad \text{from 49(5)}$$

$$\frac{\lambda^2}{a^2}\left(\tfrac{193}{9}\right) = 4$$

$$\therefore a = \lambda\frac{\sqrt{193}}{6}$$

$$= 3 \cdot 55974 \text{ kX, for } \lambda_{K\alpha_1} = 1 \cdot 537416 \text{ kX}$$

The value obtained by Lonsdale was $3 \cdot 55974 \pm 5$ kX.†

3.2. INTERPOLATION AND THE USE OF CONICS WITH A COMMON TANGENT

It is possible to use the separation of the α_1, α_2 conics to interpolate in the case of near coincidence to obtain the wavelength that would give exact coincidence.

Figure 36 shows a case of near tangency of two conics (400 and 222). In case of exact tangency, the tangent point would lie on the indexing sphere and

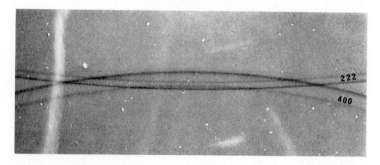

FIG. 36. Enlargement of a portion of a divergent beam photograph of nickel using NiK radiation, showing near tangency of the 400 and 222 deficiency conics. (After J. K. H. Mackay.)

† For the relation of kX to Å see [1(60)].

on the line of intersection of the two indexing planes through the corresponding relps. It would also lie on the central plane containing the two RL vectors, since the cones are symmetrical about this plane. The equation of this third plane is:

$$\mathbf{r} \cdot \mathbf{d}_1^* \times \mathbf{d}_2^* = 0. \tag{1}$$

For orthogonal lattices this reduces to

$$xa^2(k_1l_2 - k_2l_1) + yb^2(l_1h_2 - l_2h_1) + zc^2(h_1k_2 - h_2k_1) = 0.$$

and for a cubic lattice

$$x(k_1l_2 - k_2l_1) + y(l_1h_2 - l_2h_1) + z(h_1k_2 - h_2k_1) = 0. \tag{2}$$

For the two planes 400, 222, (2) gives:

$$-8y + 8z = 0 \qquad \text{or} \qquad y = z. \tag{3}$$

The equations 49(4) of the two indexing planes are:

$$4x = \frac{a^{*2}}{\lambda}(4^2); \quad \therefore x = 4a^{*2}/\lambda, \text{ and} \tag{4}$$

$$2x + 2y + 2z = \frac{a^{*2}}{\lambda}(2^2 + 2^2 + 2^2); \quad \therefore x + y + z = 6a^{*2}/\lambda. \tag{5}$$

From (3), (4) and (5):

$$x + 2y = 6a^{*2}/\lambda$$

$$2y = 2a^{*2}/\lambda$$

$$y = a^{*2}/\lambda$$

$$z = a^{*2}/\lambda$$

$$\therefore a^2 \cdot \frac{a^{*4}}{\lambda^2}(16 + 1 + 1) = 4 \quad \text{from } 49(5)$$

$$18 \, \lambda^2/a^2 = 4$$

$$a = 3 \, \lambda/\sqrt{2}.$$

Interpolation showed that exact tangency would occur for a wavelength which differs from $NiK\alpha_2$ by one-sixth of the doublet separation, i.e. for $\lambda = 1 \cdot 66106$ Å. This gives $a = 3 \cdot 5238 \pm 1$. The higher error is due to the slight spread of the lines due to some imperfection in the specimen.

3.3. Non-Significant Triple Inersections

If three indexing planes intersect in a line, this line will cut the indexing sphere in a point of triple intersection, whatever the wavelength, and will therefore be non-significant as far as parameter determination is concerned. This will occur if, in addition to the previous criteria for a triple intersection, the three RL vectors are co-planar, i.e.

$$\mathbf{d}_1^* \cdot \mathbf{d}_2^* \times \mathbf{d}_3^* = 0.$$

Expansion in components gives

$$V^* \cdot \begin{vmatrix} h_1 \, k_1 \, l_1 \\ h_2 \, k_2 \, l_2 \\ h_3 \, k_3 \, l_3 \end{vmatrix} = 0 \qquad (1)$$

so that if the determinant of the indices of the planes involved in a triple intersection is zero, the intersection is non-significant.

Chapter 6

The Reciprocal Lattice and White Radiation

1. Radial Lines in Reciprocal Space

A RL point is necessarily on the normal to its set of planes, and the distance of the point from the origin is proportional to λ. Therefore for a continuous distribution of λ, there will be a continuous series of points, i.e. a weighted radial line in reciprocal space. For white radiation therefore the RL becomes a set of weighted radial lines through the points corresponding to a particular wavelength.

2. The Length and Composition of the Radial Lines

Figure 37 shows the distribution of intensity in the white radiation spectrum. This will also represent the distribution of weight along the radial line for one particular reflection. From the origin to the point corresponding to the short wavelength cut-off [1(52)] ($\lambda_{min} = 12\cdot4/V$ Å, where V is the potential on the tube in kilovolts) the line will have zero weight, and beyond 2–3 Å the weight will become negligible. However, along any radial line in reciprocal space, the weight is due to the superposition of the weights for many orders of reflection. This is shown schematically in Fig. 38.

It will be noticed that the weighted line for mh, mk, ml is m times as long as that for hkl and the weight is spread out more and more thinly as we go to higher orders. This explains why the Laue streak on a photograph, arising from such a rod (Fig. 39) is most noticeable at low angles, although polarisation and velocity factors (Section **9.4**, p. 117) enhance the effect considerably.

3. The Interpretation of White Radiation (Laue) Streaks

If the points of the reciprocal lattice for the characteristic radiation, together with the weighted radial rods passing through them, are rotated through the sphere of reflection, reflections will occur wherever a rod cuts the sphere. There will thus be a continuous series of reflections as a rod rotates through the sphere, giving rise to a continuous line of spots, i.e. a streak, on the photograph, with a Kα reflection, if it occurs, standing out on the streak (Fig. 39). Figure 40 shows one such rod, rotating through the sphere as the generating line of a cone, whose intersection with the sphere gives rise to the characteristically shaped curves on the photograph in Fig. 39. In the case of a limited rotation (normally a 5, 10 or 15° oscillation) these streaks will be

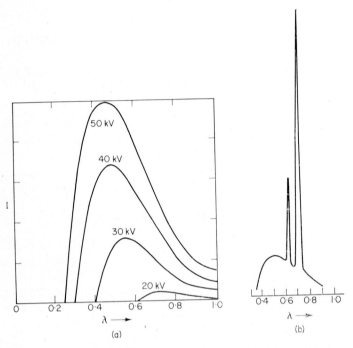

(a)

(b)

Fig. 37. (a) I versus λ for 'white radiation' from a W. target at various voltages. (b) Spectrum of Mo target tube at 35 kV, showing Kα and Kβ lines superimposed on the continuous spectrum. The width of the K lines is due to instrumental broadening—the intrinsic width is less than the thickness of the trace.

hkl   2h, 2k, 2l   3h, 3k, 3l

Fig. 38. The weighted rods (displaced to avoid overlapping) correspond to reflections nh, nk, nl for $n = 1, 2, \ldots, 6$. The relps for CuKα radiation are shown and the diagram is drawn for a voltage of about 30 kV. The rods are not, in fact, uniformly weighted along their lengths, but the thicknesses of the lines are an indication of their relative weights assuming that the intensities of the various orders are equal. The diagram shows the composite nature of the weighted rods in reciprocal space, and therefore the large number of wavelengths involved in a normal Laue reflection.

limited (Fig. 41), and the inner and outer ends can be used to determine the orientation of the crystal. The inner end is usually the more distinct, because the streak tends to fade out towards higher Bragg angles; and it may be difficult to determine whether the outer termination is due to this fade out or to the limit of the oscillation. On the other hand the inner, distinct termination may

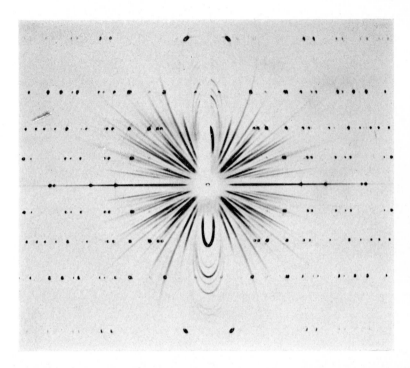

FIG. 39. Laue streaks on the central portion of an over-exposed rotation photograph. The inner ends of the streaks (corresponding to relps for the short wavelength cut-off λ_{min}) can be seen to form 'layer lines', corresponding to those for the characteristic radiation.

also be due either to the limit of oscillation, or to the short wavelength cut-off in the white radiation. For a copper target and a tube run at approximately 35 kV, this cut-off termination will occur at about 1/4 the distance from the origin (where the direct beam strikes the film) to the Kα spot (Figs. 39 and 41). If the inside end of the Laue streak is not due to cut-off, it corresponds to one end of the oscillation. From measurements on this end of the Laue streak (Section 13.2.1.2, p. 150) it is possible to fix the position of the reciprocal lattice vector giving rise to the streak, at a known position

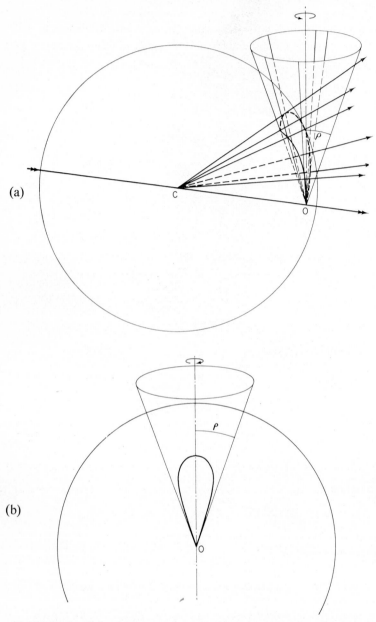

Fig. 40. Weighted radial rod (at an angle ρ, with the axis) rotating through the sphere of reflection, giving rise to a continuous series of reflections on a 'Laue Streak', as in Fig. 39. (a) View nearly perpendicular to the X-ray beam. (b) View looking along the X-ray beam.

of the goniometer, i.e. the end of the oscillation. By contrast, for a Kα reflection, all that is known is that the corresponding RL point was in the reflecting position at some time during the oscillation.

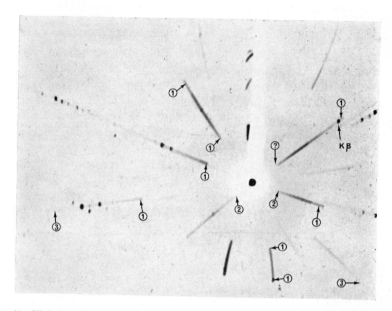

FIG. 41. Slightly enlarged central portion of an oscillation photograph of a randomly oriented crystal, showing Laue streaks with various kinds of termination (1 — end of oscillation, 2 — short wavelength cut-off; 3 — long wavelength fade-out).

4. Stationary Crystal (Laue) Reflections

For a stationary crystal the radial rods will cut the Sphere of Reflection in points, and 'Laue reflections' will occur in the direction of the radii to these points. These reflections constitute a 'Laue photograph' (Fig. 43, pp. 59–62) and correspond to satisfying the three Laue conditions simultaneously by the correct choice of wavelength (Section 2.1.3.1, p. 20). However the wavelengths used for any particular reflection can now be found directly, since they are proportional to the distance from the origin of the point of intersection with the Sphere of Reflection. (**Examples 10** and **11**, pp. 289 and 290).

5. The Interpretation of Laue Photographs

5.1. ZONES OF REFLECTIONS

Since the white radiation rods pass through relps for the characteristic radiation (and are, of course, themselves composed of a continuous series of reciprocal lattices corresponding to all the wavelengths of the white radiation) they have some of the properties of a RL. In particular, they lie in a series of planes. However, since the rods are radial, the planes can only be central lattice planes, i.e. those which pass through the origin. Even so, there are an infinite number of such planes, but only those with an appreciable density of relps will also stand out as planes of white radiation rods. Such a plane will contain at least one important RL rowline, and in general the planes most densely populated with weighted rods will be those containing two RL axes. When such a plane intersects the sphere of reflection (Fig. 42), the points where the rods cut the sphere all lie on the circle of intersection; and the reflected beams lie on the cone with C as apex and this circle as base. The direct beam lies on all such cones. If the reflected beams are caught on a flat photographic film placed perpendicular to the direct beam, the intersection will be an ellipse, a parabola or a hyperbola (depending on whether the cone angle is $<$, $=$, or $> 90°$) passing through the direct beam position. The more densely populated the RL plane the more clearly the curve will be outlined by reflections. The various curves (including the straight lines originating from flat cones) which can be seen in Fig. 43 (particularly (b) and (c)) arise in this way.

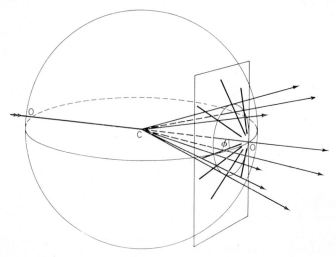

FIG. 42. The Sphere of Reflection and a central lattice plane, showing weighted rods intersecting the sphere on a circle and reflections along generating lines of a cone. The angular coordinate ϕ, common to all the vectors in this plane, is shown.

In Fig. 44 the outer (and inner) ends of the Laue streaks, defining the crescent in the central area of the *cylindrical* film, lie on a distorted ellipse, since if the crystal were held stationary at that end of the oscillation, the Laue reflections would occur there. (**Example 12**, p. 290).

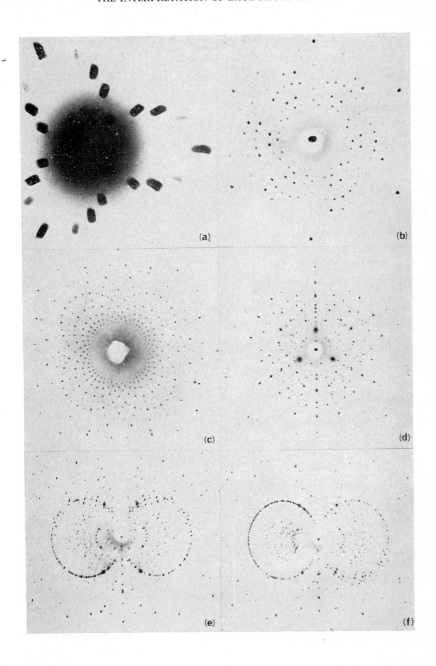

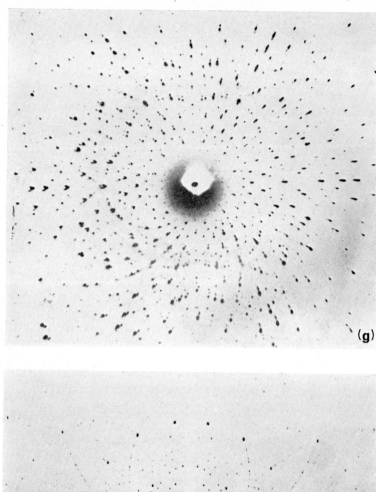

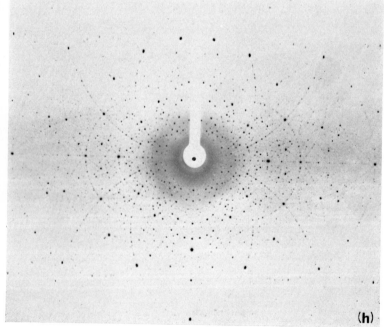

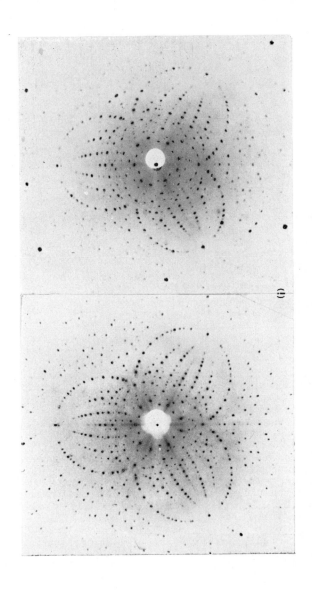

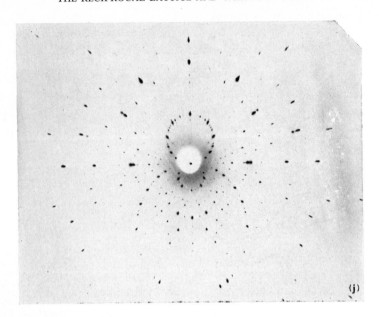

FIG. 43. Forward reflection Laue photographs, showing ellipses of spots and symmetrical arrangements. The symmetry symbol is given in each case.

(a) Early Laue photograph of diamond (beam long [111])—3*m*.
(b) Laue photograph of apatite—6.
(c)   ,,   ,,   ,, Co(CNS)₄Hg—4 (nearly 4*m*).
(d)   ,,   ,,   ,, alite along **c**—pseudo 3*m* (actually *m*).
(e)   ,,   ,,   ,, alite along **a**—*m*.
(f)   ,,   ,,   ,, alite at 120° to (e), showing no symmetry.
(g) Laue photograph of ettringite, showing distorted spots caused by mis-shapen or twisted crystal, which make determination of symmetry difficult—apparently 6*m*.
(h) A beautiful Laue pattern, with the beam along the **a**-axis of a cubic crystal of tricalcium aluminate, showing that it only has 2/*m*3 symmetry.
(i) Two Laue photographs of yttrium anti-pyrine iodide along **c**—3. In one photograph the crystal is slightly mis-set and shows CuKα reflections on one side, apparently destroying the symmetry. Crystal–film distance = 30 mm.
(j) Laue photograph of sucrose, showing a vertical symmetry plane-*m*. This photograph shows two interesting features. (i) The spread of orientation in the mosaic blocks is sufficient to give radially elongated spots, a phenomenon known as 'asterism'. (ii) There are diffuse reflections near some of the sharp Laue reflections. This indicates that a Kα relp is near the point where the weighted rod cuts the sphere of reflection. Thermal vibration gives rise to a volume of weak diffracting power round the relps, which is only detectable for the intense Kα radiation. This volume round the Kα relps touches the sphere of reflection and produces a diffuse reflection, outside the Laue spot if the Kα relp is outside the sphere and vice versa. In one case the Kα relp is so close to the sphere that the diffuse reflection is beginning to swallow up the Laue reflection. Crystal–film distance = 25 mm.

(e), (f) and (h) were taken on cylindrical film, which distorts the ellipses of spots. The rest were taken on flat film perpendicular to the beam. (a)–(f) are reduced in size, (e) and (f) considerably. (i) and (j) are full size. (b), (d), (e), (f), (h) and (i) were taken at 40 kV; (c) and (j) at 50 kV; (a) and (g) unknown.

It has been assumed so far that the film has been placed in the forward position. If it is placed in the back reflection position (with a hole punched for the direct beam to come through) the direct beam, which is a generating line of the cone, runs directly away from the plate and the curves are all hyperbolae (Fig. 45).

All the RL vectors in a central RL plane lie in a **zone**, i.e. they are all perpendicular to a given direction (the normal to the RL plane). The spots on a Laue photograph arising from the weighted rods in such a plane are therefore known as a 'Laue zone' of reflections.

5.2. SYMMETRY

If a crystallographic symmetry direction (an axis or any direction in a plane) lies along the X-ray beam, the Laue zones will show the symmetry of the direction (Fig. 43). One of the main uses of Laue photographs is in demon-

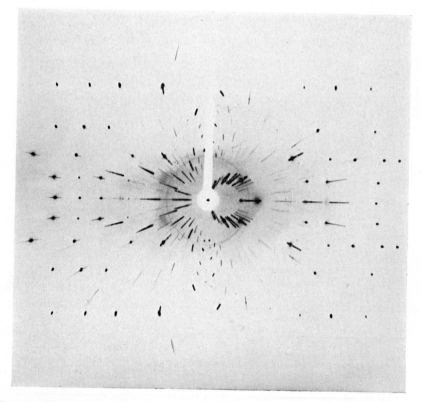

FIG. 44. The central part of an oscillation photograph taken on a cylindrical film, radius 30 mm. A crescent of Laue streaks defines the inner and outer curve of intersection with the cylindrical film of the cone of reflections at either end of the oscillation.

strating symmetry in this way. However, an error of setting of more than a few minutes can cause large differences to appear in symmetry related reflections. This is because corresponding reflections not only appear at different distances from the origin, but also with differing intensities, since different wavelengths, with differing intensities, are being reflected (Fig. 43i). To demonstrate symmetry in this way therefore requires accurate setting methods (Section **13**.2, p. 149). No other method can reveal so much of the symmetry in one photograph;† but where small differences are involved, especially in the case of planes of symmetry, oscillation photographs may

† A cone axis precession photograph reveals the same symmetry elements, but from a much more restricted range of reflections. It has the advantage of being much less sensitive to mis-setting because Kα reflections and Laue streaks are compared. (Section **16**.3.3, p. 222).

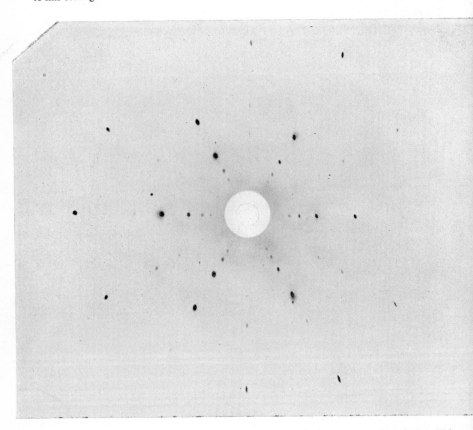

FIG. 45. Back reflection photograph of aluminium showing degenerate hyperbolae. This photograph was taken after correcting the specimen orientation from measurements on Fig. 101 (p. 165) which shows non-degenerate hyperbolae. Incident beam going *downwards* through the film, so that the cut corner is top left.

prove more sensitive. Figure 46 shows an oscillation photograph of a crystal which Laue photographs had apparently shown to possess a horizontal plane of symmetry, i.e. a plane perpendicular to the rotation axis. Small differences between the top and bottom halves of the photograph can be clearly seen, showing that the supposed plane of symmetry is only an approximation. However in this case it is necessary to be sure that the reflections showing differences do not occur at the end of an oscillation, since in that case slight mis-setting may again be responsible for the differences.

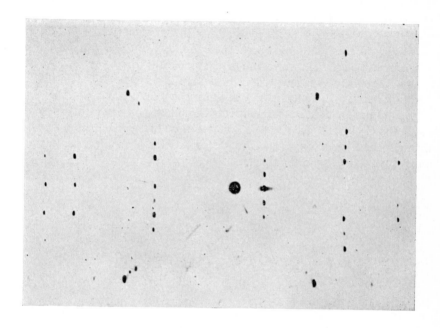

FIG. 46. Part of an oscillation photograph of ettringite about **c**, showing the lack of a horizontal mirror plane and therefore the true symmetry to be 3, instead of 6 as suggested by Fig. 43g (p. 60).

5.3. BLANK AREAS IN THE PHOTOGRAPH

All the Laue photographs of Fig. 43 (pp. 59–62) and Fig. 52 (p. 74) show a blank area round the direct beam. This is particularly prominent in Fig. 43 (b, c and h). Figure 52 also shows the blank areas round the reflections with simple indices, i.e. those at the intersections of a number of zones. The explanation of these blank areas and some quantatitive results are given below.

5.3.1. *Blank areas round the direct beam when a principal crystallographic direction lies along it*

The inner ends of the weighted portions of all the white radiation rods will form the ordinary reciprocal lattice for λ_{min} (Fig. 39, p. 55). This is the reciprocal lattice referred to in subsequent paragraphs of this section.

Consider a crystal with the X-ray beam incident along the **c**-axis. In Fig. 47, *OB* is the trace of the zero plane of the reciprocal lattice perpendicular to **c**. All *hk*0 radial rods lie in this plane which is tangential to the sphere, and therefore no *hk*0 reflections can arise. *DE* is the trace of the *hk*1 plane of the reciprocal lattice. There is no reflecting power lying between the planes *OB* and *DE*. Therefore no reflections can arise from the intersections in the cap defined by the arc P_1O of the sphere. The reflection with the smallest deviation which can arise is CP_1, corresponding to a reciprocal lattice point P_1, actually lying in the sphere. Any reciprocal lattice point in this plane lying outside the circle of radius FP_1 (such as P_3) will not give rise to any reflection, since its radial line is weightless where it cuts the sphere. Any point (such as P_2) inside the circle will give rise to a reflected ray *CA*, which is more deviated than CP_1 and has a longer wavelength (or wavelengths). First, assume that a relp such as P_1 lies in the sphere.

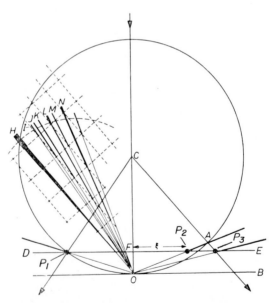

FIG. 47. The sphere of reflection and reciprocal space for white radiation. The dashed lines of the RL for λ_{min} and the circle of reflection identify the part of the diagram which has been rotated 90° about *OH* into the plane of the paper.

Then, since $OF = \lambda_{min}/c$

$$\cos P_1 CF = \cos 2\theta = CF/CP_1 = \left(1 - \frac{\lambda}{c}\right)\bigg/1.$$

Therefore
$$c = \frac{\lambda_{min}}{1 - \cos 2\theta} = \frac{\lambda_{min}}{2\sin^2\theta}. \tag{1}$$

Since $\lambda_{min} \propto 1/V$ (p. 53), $\sin\theta \propto 1/\sqrt{V}$, and the higher the voltage on the X-ray tube, the smaller the blank area.

For $c = 10$ Å and $\lambda_{min} = 0 \cdot 25$ Å, $2\theta = 12°50'$ for a relp such as P_1. If no relp lies in the sphere, the lowest θ reflection on the photograph will arise from a relp such as P_2. 2θ will be larger than for the limiting case considered above, and the value of c derived from (1) will be too small. It is clear, therefore, that the application of (1) can only give a minimum value for c. It is of interest to get some estimate of the error which may arise. Take the case of a cubic crystal with $a = 10$ Å.

The distance, $\xi = FP_2$ is given by

$$\xi = \frac{\lambda_{min}}{a} \cdot \sqrt{(h^2 + k^2)} = 0 \cdot 025 \cdot \sqrt{N}$$

$$FP_1 = \sin P_1 CF = \sin 12°50' = 0 \cdot 222.$$

Since
$$\xi \leqslant FP_1$$

$$\sqrt{N} \leqslant 0 \cdot 222/0 \cdot 025 = 8 \cdot 9$$

therefore
$$N \leqslant 79.$$

The possible values of N in this region are 73, 74 and 80. 80 is too large. For $N = 74$, $\xi = 0 \cdot 215$

$$\tan FP_2 O = \tan\theta = FO/\xi = 0 \cdot 1165$$

$$\theta = 6°39'.$$

Putting this value of θ in (1) gives

$$c' = 9 \cdot 33 \text{ Å}.$$

This is a somewhat unfavourable case, since $N = 80$ is just not taken in, and the gap to $N = 74$ is above the average. However, with a smaller cell size this error of 6·5 per cent will tend to increase since the gaps between the possible values of ξ will increase.

5.3.2. *Blank areas round the direct beam for a randomly oriented crystal*

If the crystal described above is turned through an angle of less than 1° about the appropriate axis, the 100 relp will lie in the sphere. The Bragg angle for the corresponding reflection is given by

$$\sin \theta = \lambda/(2d_{100}) = 0{\cdot}25/20 = 0{\cdot}125;$$

therefore $\theta = 0°43'$ and the deviation $2\theta = 1°26'$.

This is the minimum deviation for all possible orientations, and contrasts with the minimum deviation of $12°50'$ for the same crystal accurately oriented with c along the X-ray beam. For a random orientation of the same crystal the minimum observed deviation would be unlikely to be more than two or three times the absolute minimum, since the positions giving the larger angles of deviation require one of the small number of principle crystallographic directions to be oriented within a cone of semi-vertical angle about 1°, coaxial with the X-ray beam. The probability of this occurring is much less than 1 per cent.

5.3.3. *Summary of quantitative information from the size of the central blank area*

If 2θ is the measured smallest deviation of a reflection on a Laue photograph from a randomly oriented crystal, then $\lambda_{\min}/(2 \sin \theta)$ gives a minimum possible value of the largest interplanar spacing in the crystal, though the actual value may be several times as great.

In the case of an accurately set crystal $\lambda_{\min}/(2 \sin^2 \theta)$ gives the minimum distance between crystal lattice points lying along the X-ray beam. Specimen calculations show that the actual value may in exceptional cases be up to twice as great, but usually not more than about 20 per cent greater.

5.3.4. *Qualitative explanation of the blank areas around Laue reflections with simple indices*

Laue reflections with simple indices derive from an important RL vector, e.g. a cell edge or diagonal. Such a direction is common to a number of important RL planes, each of which gives rise to a zone of reflections on a Laue photograph. Such zones all intersect in the common reflection with simple indices. This reflection is usually intense and is surrounded by a small blank area giving way to reflections along the zones, which are generally weak near the common reflection and gradually increase in intensity as the distance from it increases. Of course, the general effect is modified, sometimes drastically, by the structure factor differences of the various reflections. (Section **1.4.2**, p. 16).

For simplicity and definiteness consider an orthorhombic, near cubic crystal, with a^* in the plane of the paper, c^* along the beam and b^* upwards (Fig. 47, p. 66). OH is the reciprocal lattice line $h0h$. Consider the plane hkh, passing through OH and perpendicular to the paper. This cuts the sphere in a small circle of reflection on OH as diameter. If the plane and small circle are rotated about OH into the plane of the paper we obtain the diagram outlined in dashed lines. Only radial lines passing through relps inside the dashed circle have any weight where they pass through it. The intersections of the weighted radial lines with the circle of reflection are shown at H, I, J, K etc. The zone of reflections on the photograph will be the projection of these points from C (after rotating back to the original position); and distances between reflections on the photograph will be roughly proportional to the distances between the points H, I, J, etc. The intensities of the reflections, neglecting differences in structure factors, will be roughly proportional to the weights of the radial lines at the points of intersection with the sphere, i.e. roughly proportional to the number of overlapping weighted radial lines at these points. This neglects the variation of weight along any line, particularly the fade out at the long wavelength end, and the variation with structure factor; but is sufficient for the following qualitative explanation of the blank areas and weak reflections in the neighbourhood of an axial or other important reflection.

The distance HI is large compared with IJ, JK, etc. because it represents the projection from O of a unit cell side which is nearly perpendicular to the direction of projection, while the others are projections of cell sides which are nearly parallel to this direction. While the sides will rarely be the same length, they will not normally be very different, and the angle of projection will have the greater effect. Thus round the spot with simple indices there is a small blank space which is due to this elementary geometrical effect; but in most cases the area is considerably enlarged by the vanishingly small intensities of

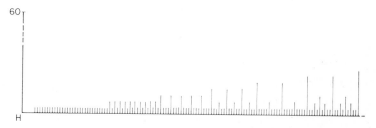

Fig. 48. Diagrammatic representation of the sequences of reflections along a Laue zone from the reflection with simple indices, H. Going away from H, the sequence of reflections corresponds to I, J, K, etc. of Fig. 47 (p. 66) but with a reciprocal lattice net 1/10 the size. The shortest lines correspond to a reflection with only one order, the lengths of the others being proportional to the number of orders in the corresponding reflections and therefore, other things being equal, to the intensities of these reflections.

the near-in reflections. A symmetrical zone has been taken for illustration but the same argument applies to any of the zones containing OH, since the RL net necessarily consists of lines of points parallel to OH.

The reason for the intensity of the reflection with simple indices is obvious from the diagram. Six weighted lines overlap at H. This is equivalent to saying that six orders of reflection coincide on the photograph. On the other hand the next three intersections, I, J, K, have single weight, L just double, M single, N double, nearly triple, and so on. If the RL net were very much smaller relative to the sphere of reflection, as it would be in practice since it corresponds to λ_{min}, the process would go much further. For example, if there were 60 lattice points along OH and the whole net were correspondingly smaller, the sequence of reflections would be represented in Fig. 48. The heights of the lines represent the number of orders contributing to the reflection. The spacing of the lines is only approximately correct. The actual reflections would be irregularly spaced, and, in particular, reflections with many orders would have larger spaces on either side. The whole sequence would only occupy about 10 mm along the zone on a photograph. A study

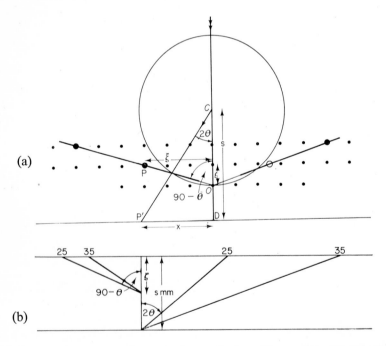

Fig. 49. (a) The relation between the Laue reflections P' on a flat plate perpendicular to the beam and the relps for a particular wavelength. (b) A scale for locating relps from the reflections on a Laue photograph. The figures on the scale are the values of θ for the reflection on one side and the corresponding relp on the other.

of actual photographs shows that most of the single orders do not show up. The tendency for the reflections with many orders to become more widely separated and to stand out more from reflections in between, would continue to increase for some distance along the zone. This accounts for the tendency towards regular spacing of strong reflections round a zone. The most intense of these strong reflections will, of course, have relatively simple indices and also lie at the intersections of a number of zones.

Although this simple picture is drastically modified in practice by the effect of differing structure factors, zone plane nets and reflection circle curvatures, and the variation of weighting along the radial lines, it nevertheless gives a qualitative explanation of many of the general features of Laue photographs.

5.4. INDEXING LAUE REFLECTIONS

It is rare nowadays for Laue reflections to be indexed directly (with the possible exception of photographs of cubic crystals) but the method employed is a good example of the relation between the RL and the weighted rods. Such indexing would in any case only normally be attempted if an important crystallographic direction (usually an axis) had been set accurately along the X-ray beam. The central RL plane normal to the axis would then be tangential to the sphere of reflection, and no reflections would arise from it. We choose an arbitrary wavelength (for the sake of definiteness it can be considered to be the characteristic wavelength) and set out to construct the RL for this wavelength. Suppose c is along the X-ray beam. The zero plane $hk0$ has no effect. The plane $hk1$, however, (Fig. 49a) has weighted rods passing through the points and intersecting the sphere. The corresponding reflections are caught on a flat plate normal to the direct beam and at s mm from the crystal. Let the $hk1$ plane be ζ R.U. from the origin, and a relp P be ξ R.U. from the direct beam. The corresponding reflection P' is x mm from the direct beam. From the figure,

$$\xi = \zeta \tan (90 - \theta) = \zeta \cot \theta, \tag{1}$$

and
$$x = s \tan 2\theta. \tag{2}$$

P' is in the plane defined by the direct beam and OP. By assuming a convenient value for the arbitrarily chosen ζ, it is possible to calculate ξ from the value of x measured on the photograph. P is then plotted on a line from the direct beam D, through P', at ξ R.U. from the origin (for a scale, say, of 100 mm = 1 R.U.). Practically it is convenient to cover half the photograph with a sheet of paper, and take the direction of DP as opposite to that of DP'. This is merely equivalent to rotating the RL through 180° with respect to the photograph. To save time in calculations a scale can be made up as in Fig. 49b. Points labelled with the same θ value are at a distance from D of $s \tan 2\theta$ on the photograph side, and $\zeta \tan (90 - \theta)$ on the RL side. The scale must be made so that s mm is the same on the scale and the X-ray camera, but ζ can be chosen arbitrarily. The scale is then laid across the photograph half covered with paper, with D on the centre spot (direct beam) and the reflections scale over a spot on the photograph. θ is read off and a mark made on the paper at the same θ value on the other end of the scale. This is repeated for all the reflections.

The relps P will be found to lie on a basic net, but there will almost certainly be a considerable number of relps which lie between points of the main net or inside the mesh. Many of these will lie half way between the points of the main net or in the centres of the mesh. They arise from the RL plane $hk2$. The weighted rods for these relps will either coincide with those for $hk1$ (if h and k are both even) or cut the first layer in between relps (if h or k is odd) or in the centre of a mesh (if h and k are both odd). The second and third cases will give rise to Laue reflections as if from relps in the $hk1$ layer, halfway between the main points or in the centre of a mesh (see the point marked with a circle in Fig. 49a.) In the same way the $hk3$ layer will give rise to some points

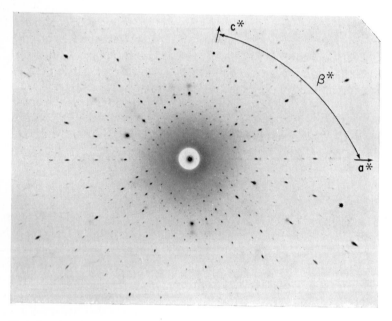

FIG. 50. A flat plate Laue photograph of a monoclinic crystal with **b** along the beam, showing the angle β^* between the **a*b*** and **c*b*** zones. Tube voltage = 50 kV. Crystal to film distance = 25·0 mm. Some reflections at the bottom of the photograph have been obstructed, probably by the arcs or the plasticene used in mounting the crystal.

at $\frac{1}{3}$ or $\frac{2}{3}$ the main net repeat distance and similarly for higher layers. The main net points can be indexed as $n(hk1)$ (since $hk1$; $2h$, $2k$, 2; $3h$, $3k$, 3; ... nh, nk, n; all contribute to the reflection where n may go up to 10 or more orders). The points halfway between can be indexed as $n(hk2)$, etc., h and k being obtained from axes chosen or recognised in the main net. For $hk2$ the distances on the net must be doubled to obtain hk, for $hk3$, trebled, and so on.

Because the second, third and higher layer vectors for the far out relps will have the short wavelength cut-off occurring outside the sphere of reflection, reflections which do occur will correspond to first layer relps. The first layer net can therefore be deduced from the outermost points of the diagram, corresponding to the innermost reflections. (**Example 13**, p. 291).

5.5. LOCATION OF AXES AND MEASUREMENT OF CELL ANGLES BY LAUE PHOTO-GRAPHS

5.5.1. *Flat plate Laue photographs*

A photograph of a monoclinic crystal, with **b** (and therefore **b***) along the X-ray beam, shows not only the two-fold symmetry, but also the two straight lines of spots corresponding to the **a*b*** and **c*b*** zones. If **c** is vertical the **a*b*** zone will be horizontal. If the approximate direction of **c*** has been found from an oscillation photograph (Section **4**.2, p. 34) the **c*b*** zone can be recognised and a more accurate value of β^* obtained. (Fig. 50).

In the triclinic case, if **b*** (but *not* **b**) is along the X-ray beam, the **a*b*** and **c*b*** zones will again be straight lines, and if they can be recognized the angle between them will be equal to the angle between the zone axes **c** and **a**, i.e. β. Similarly if **a*** or **c*** are along the beam, α or γ can be found. The same flat cones give straight lines intersecting the direct beam on back reflection Laue photographs (Fig. 45, p. 64). For a mis-set crystal the resulting offset hyperbolae can be used to measure the angle of mis-setting (Section **13**.5, p. 163).

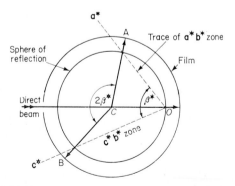

FIG. 51. Diagram showing the relation between important RL directions and the intersection of Laue zones. **b*** is perpendicular to the diagram at O.

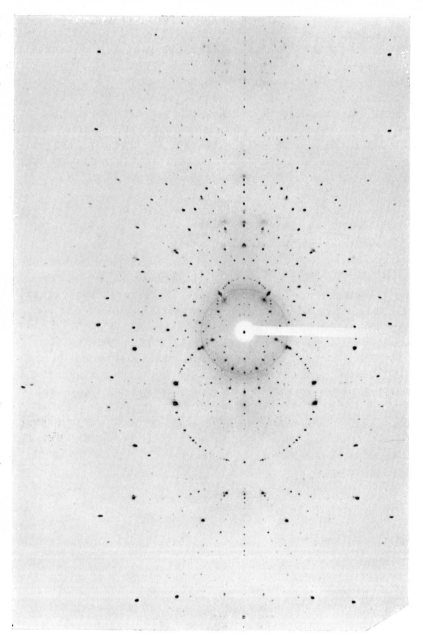

FIG. 52. A cylindrical film (radius 30 mm) Laue photograph of a monoclinic crystal with **b** vertical. The intersections of prominent zones locates possible directions of the **a*** and **c*** axes. The diffuse areas are caused by thermal scattering. The single powder ring is due to the adhesive.

5.5.2. *Cylindrical film Laue photographs*

In the case of monoclinic or triclinic crystals in which one axis has been set along the rotation axis, it may be difficult to recognise the other two RL axes. However if we take a Laue photograph on cylindrical film for the case, say, of a monoclinic crystal with **b** vertical, **a*** and **c*** are in the horizontal plane and the **a*c*** zone gives rise to a horizontal flat cone of reflections (a horizontal straight line of reflections on the unrolled film). Each axis is also on a number of other prominent zones (the most important being **a*b*** and **c*b***). These will give rise to various zones of reflections intersecting the **a*c*** flat cone at A and B (Fig. 51).

The point of intersection of many zones identifies a prominent central RL line, and enables axes to be chosen (or recognised from the angular relationship) from such a photograph (Fig. 52). Measurement of the angle ACB ($= 2\beta^*$) gives a fairly accurate value for β^*, and the ϕ values (Section **13**.2.1.2 (ii) (5), p. 152), for A and B give the positions of the RL axes relative to the X-ray beams.

If **a*** is nearly perpendicular to the beam and β^* is nearly 90°, A may be too near the direct beam to be recognised, and B off the outer edge of the photograph. If no obvious zone axes show up, turn the crystal through 45° and take another photograph.

Chapter 7

Diffraction by Polycrystalline Material

1. The Reciprocal Lattice and a Powder

If the RL for a single crystal and white radiation is considered as a lower form of the point lattice, in which the direction of a RL vector remains, but its length is indefinite, then the form for a randomly oriented powder and mono-chromatic radiation is degenerate in the opposite way—the length of the RL vector remains, but its direction is indefinite. For a completely random distribution of N crystallites a particular relp hkl will occur in N positions, uniformly distributed over the surface of a sphere of radius d^*_{hkl}. If N is large enough this can be considered as a uniform spherical shell of reflecting power in reciprocal space. The reflecting power per unit area will be proportional to that of the relp hkl for a single crystal, and inversely proportional to the square of the radius, d^{*2}_{hkl}. Just as the weighted rod for white radiation is the superposition of rods of many orders, so an actual spherical shell for a powder is the superposition of a number of shells of the same radius. The number of shells is never less than two, since $d^*_{hkl} = d^*_{hk\bar{l}}$, and can be as many as 72 in a cubic crystal, e.g. {432} + {502}†. If all the superposed shells derive from symmetry related relps no difficulty arises, since the experimentally determined reflecting power need only be divided by p'' (the number of superposed shells) which can be derived from the symmetry of the RL (Section **10**.2, p. 128), to give that of the representative shell of the group. p'' is called the 'multiplicity' of the symmetry related group of reflections. Any member of the group can be taken as 'representative', but if possible, one with all indices positive is chosen.

If, however, shells due to non-equivalent points coincide (or overlap, or are so close that they cannot be experimentally resolved) then the reflecting power of the composite shell is the sum of those of two or more equivalent groups. Experimentally only this sum can be measured, and there are no means of separating out the components. It is this aspect of the degeneracy of the RL for a powder which is responsible for the main limitations arising in X-ray

† Curly brackets { } denote not only the set of planes given by the indices in the brackets, but all the other symmetry equivalent sets as well.

work with such specimens, especially with substances of low symmetry or large unit cells, where chance overlap is the rule rather than the exception.

However, the degeneracy has one great advantage, which is responsible for the fact that such a large part of X-ray diffraction work is concerned with

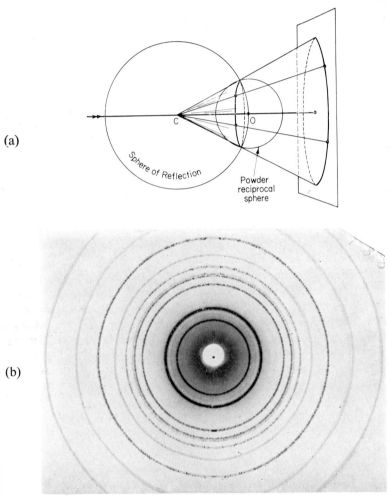

(a)

(b)

FIG. 53. (a) The intersection of one of the reciprocal spheres for a powder with the sphere of reflection, giving rise to a cone of reflected beams which intersect a flat film in a 'powder ring'. (b) Powder photograph on a flat film, showing reflection rings. Specimen–film distance = 25 mm. CuKα radiation. The 'spottiness' of the rings shows that the crystallite size in the specimen was too large, i.e. the number of crystallites in the irradiated volume was too small to give a uniform powder sphere, even although the specimen was rotated. A more divergent beam would probably have increased the number of crystallites reflecting, sufficiently to give smooth lines, but only at the expense of making the lines broader.

powders. The degenerate RL is spherically symmetrical. The diffraction pattern arising from the intersection of the shells with the sphere of reflection (Fig. 53) does not depend on the orientation of the specimen relative to the X-ray beam. This is in contrast to the case of a single crystal, where the diffraction pattern, however it is obtained, varies with the orientation of the specimen. Since powder diffraction for a given substance always gives the same pattern—a pattern which, like a fingerprint, is characteristic of the substance—it can be used as a means of identification [1(266)]. This is, in fact, by far the most extensive use of X-ray diffraction.

It is of course limited to crystalline materials, but within this wide range it is the most definite means of identification, even for a small fraction (1 or 2 per cent in favourable cases) of a mixture. It may be supplemented by information from other sources, particularly infra red spectra. In the case of non-crystalline material X-ray diffraction gives only very limited information, and other methods of identification have to be used.

2. The Geometry of Powder Diffraction

The spherical shells in reciprocal space intersect the Sphere of Reflection in circles perpendicular to and centred on the X-ray beam (Fig. 53a). The reflected rays lie along cones whose apices are at C and which pass through the circles of intersection. The X-ray beam is therefore the common axis of the cones, which intersect in circles, a flat film placed perpendicular to the beam (Fig. 53b). The semi-vertical angle of the cone is 2θ, the angle of deviation, and a cone will be produced for each distinct sphere in reciprocal space, i.e. for all the differing values of d^*_{hkl} in the RL. Each cone of reflected rays necessarily satisfies Bragg's Law, $\lambda = 2d_{hkl} \sin\theta$; and d_{hkl} can therefore be calculated from the semi-angle, 2θ of the corresponding cone. 2θ is obtained from $\tan 2\theta = r/s$ where r mm is the radius of the powder ring and s mm is the distance from the specimen to the film. If the film is in the back reflection position (Section 11.2, p. 135), $\tan 2(90 - \theta) = r/s$. For most purposes, if preferred orientation is known to be absent, a narrow cylindrical film co-axial with the specimen is used for recording. All reflections can then be recorded on the same film, in the equatorial plane, and exposure time reduced by using an X-ray beam with considerable vertical divergence. A number of precautions are necessary to get high accuracy [1(158)].

For most purposes, focussing cameras, employing a convergent beam of radiation from a curved crystal monochromator, have very considerable advantages, although the apparatus and technique is more complicated [1(104)].

Powder diffraction can be produced with a stationary specimen, but rotation helps to even out any irregularities of distribution of relps over the sphere.

3. Preferred Orientation

Some polycrystalline specimens do not have their crystallites oriented completely at random, but RL vectors in at least one direction tend to line up parallel to one another. The relps are therefore not spread uniformly over the sphere, but bunched into areas. The main problem, from the point of view of X-ray diffraction, is to plot the surface density of relps over the sphere in relation to various morphological directions (axis of a wire, rolling direction of a sheet, etc.). Figure 54 shows the principle of the various methods used to achieve this. The intensity of reflections round the powder 'ring' are proportional to the surface density of relps along the circle of intersection with the sphere of reflection (Fig. 55). By rotating the specimen, this circle can be made to intersect different parts of the sphere and the density of relps recorded, either stepwise or continuously. The stereographic projection of

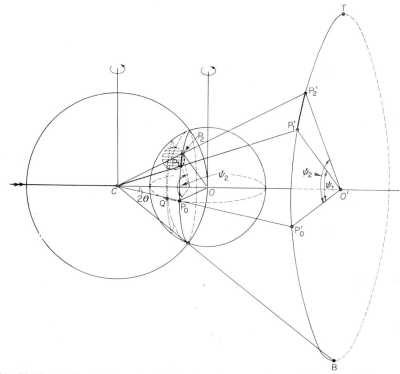

Fig. 54. The relation between the distribution of relps on the powder sphere (when this is not uniform), and distribution of intensity round the powder ring on the photograph. The shaded patch on the *hkl* powder sphere, representing a concentration of relps *hkl*, is cut by the sphere of reflection along $P_1 P_2$. The corresponding heavy arc on the powder ring occurs along $P_1' P_2'$. The dashed and dotted curves and the rotation axes are referred to in Section **17.1** (p. 238).

such a density distribution is known as a 'pole figure' for the reflection concerned. (Section **17.**1 p. 238).

The practical importance of preferred orientation can be seen from the photographs (Fig. 56) of pressings from mild steel, aluminium, brass and copper sheet. In the case of sheet steel the anisotropy in the plane of the sheet is a probably inevitable accompaniment of the development of higher resistance to deformation in the direction of the thickness. This property ('normal plastic anisotropy') is of importance in press forming of car bodies and similar pressed steel components, since it resists thinning of the sheet in the shaping process. The texture with (111) parallel to the surface produces the greatest normal plastic anisotropy, but the (100) texture shown gives the most pronounced earing. In the case of aluminium sheet (Fig. 56a) the left-hand cup with four pronounced 'ears' comes from sheet in which the cube (100) faces tend to line up parallel to the surface, with [100] in the rolling direction, giving considerable mechanical anisotropy to the sheet. However, under other conditions [110] tends to line up along the rolling direction (centre cup). By careful control of rolling and annealing procedures it is possible to get both preferred orientations present in about equal amounts. The right-hand cup is then produced, on which eight small ears can just be seen, but which requires very little finishing compared with the other two (and the finished article is therefore cheaper to produce). It is not possible by any reasonable production process to get an aluminium sheet without preferred orientation. At all stages in the establishment of the conditions for producing

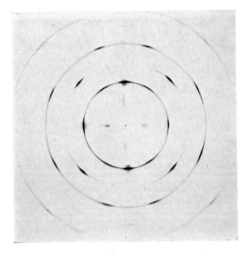

FIG. 55. Centre of a flat film photograph showing preferred orientation in Mo sheet. MoKα radiation. Specimen film distance = 35 mm. Rolling direction (W) horizontal, normal to sheet (N) parallel to X-rays.

the sheet with minimum anisotropy, X-ray diffraction control is essential. In other applications the maximum anisotropy is required, as in magnetic materials, and again, X-ray control is essential.

The properties of drawn wires and extruded fibres also depend on the type and amount of preferred orientation. In this case one direction tends to line up along the wire axis and the directions at right angles to be distributed randomly. The distribution of relps is then in circular patches at the N and S poles, corresponding to the axial direction, and uniform bands on lines of latitude determined by the RL geometry. In the case of a well oriented fibre the distribution is very much the same as would be obtained by rotating a single crystal about the corresponding direction (Fig. 57). In fact, the photograph of a stationary fibre specimen is very similar to that of a rotation photograph of a single crystal (Fig. 58a, p. 84) and can be interpreted in much the same way. (Section **14**.2.2, p. 186). This is of importance in dealing with some mineral specimens which are produced by a secondary, solid state reaction. Usually such specimens are in the form of very fine powders, often poorly crystallised, and the powder photograph cannot be interpreted (except possibly for identification). However, if the original material was in the form of single crystals (50 μm and upwards) it frequently happens that the reaction product is structurally related to the original crystal, and has grown with at least one direction parallel to a direction in the original crystal (Fig. 58b, p.

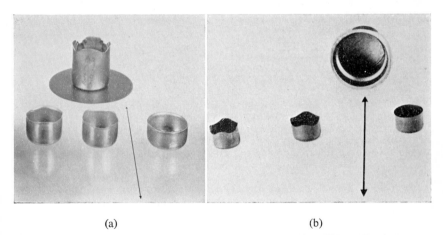

(a)   (b)

FIG. 56. (a) Photograph of cups pressed from (top) low carbon steel sheet, showing strong cubic preferred orientation; (bottom) aluminium sheet showing (left) strong cubic preferred orientation; (centre) strong [110] preferred orientation; (right) equal amounts of cubic and [110] preferred orientation. (b) (Top). Earing in brass pressing for light bulb connector, (bottom) cups pressed from thin copper sheet; (left) 70% cube texture from sheet rolled to 0·012 in; (centre) 40% cube texture in sheet only reduced to 0·018 in; (right) high temperature annealing results in only negligible (10%) cube texture in sheet previously rolled to 0·011 in. The arrows on the two photographs show the rolling direction.

84). In such cases it may well be possible to interpret the fibre or other preferred orientation photograph, although the powder photograph is uninterpretable. Similar considerations apply to organic fibres (Fig. 58c) and even to virus particles, which can be made parallel by flow techniques.

However, in the case of some metal wires, the presence of two preferred orientations (one possibly forming in the interior and the other at the surface) and a considerable amount of misorientation of both preferred axial directions, make it difficult to apply the normal rotation photograph techniques, especially the measurement of layer line spacing. In such cases more sophisticated methods must be employed (Section **17.3**, p. 247).

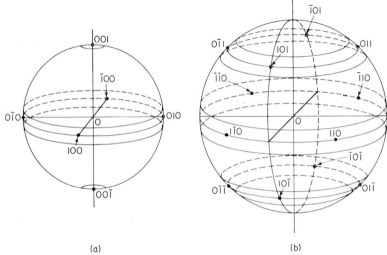

(a)   (b)

FIG. 57. The preferred orientation reciprocal powder spheres for relps (a) 100 and (b) 110 for a cubic crystal with a primitive unit cell. The relps for a single crystal with the c-axis vertical are shown. On rotation about c the relps would trace out the horizontal circles, and where these circles cut the sphere of reflection (not shown—see Fig. 53a, p. 77) a reflection would occur on a rotation photograph. In the case of a fibre the crystallites are arranged with their c-axes nearly parallel, but with the directions perpendicular to c arranged randomly round the fibre axis. The relps therefore lie distributed over a band (densest on the single crystal rotation line and less dense towards the edges) which goes uniformly round the reciprocal sphere. For a stationary fibre these bands will cut the sphere of reflection at the same points as the single crystal rotation ring, and will therefore give reflections at the same positions as on a single crystal rotation photograph, but extended into small arcs of a powder ring on either side. The width of the bands will be proportional to d* and the length of the arcs will therefore increase with θ.

Where powder specimens for X-ray diffraction are made by pressing, preferred orientation may arise if the crystallite shape is very anisotropic (plates or needles). This will alter the relative intensities of the powder rings in the equatorial plane where they are usually measured. Any preferred orientation in the specimen, from whatever cause, will alter the relative intensities. A specimen should always be photographed so as to collect the whole of some rings, in order to check this point before intensity measurements are made in the equatorial plane. A compromise that is useful for some identification purposes, is to photograph a powder specimen using the cylindrical cassette of a single crystal camera (Section 11.2, p. 135). This will give information about some complete (if distorted) rings in the low and high angle regions, as well as recording parts of all the diffraction cones.

4. Line Broadening

4.1. SMALL CRYSTALLITE SIZE

When the number of repeat units becomes less than about 1,000, the width of the diffraction maximum begins to get measurably larger (Ans. 3, p. 453). In reciprocal space the lattice points spread out into small volumes of diffracting power (all points having the same volume) and the powder spheres have a corresponding (equal) thickness. This leads to the broadening of lines on a powder photograph, and this effect can be used to work backwards to the crystallite size.

If $\beta = 2\,\Delta\theta$ (the increase in the width of the diffraction peak) then

$$\beta = \frac{\lambda}{t \cos \theta}$$

where t is the linear dimension of the crystallites (Appendix IV, p. 379) [1(249)] This assumes that the crystallites are all the same size; but in an actual specimen the size range and distribution affect β. $t = \lambda/(\beta \cos \theta)$ is therefore called the 'apparent crystal size'.

4.2. STRAIN AND INHOMOGENEOUS SOLID SOLUTION

Homogeneous strain merely alters the size of the crystal lattice (and therefore the reciprocal lattice). But often the strain in a specimen, especially if due to cold working, is non-homogeneous, and alters the lattice size in different ways in different crystallites and in different directions in the same crystallite. The relps again become volumes, or rather the relps are distributed over volumes round the average position; but the size of these volumes (and the thickness of the corresponding powder spheres) will be proportional to d^*_{hkl}, the distance from the origin.

$$\frac{\lambda}{d} = d^* = 2 \sin \theta$$

Taking logs, differentiating and ignoring the minus sign gives the

strain ε from:
$$\varepsilon = \frac{\Delta d}{d} = \frac{\Delta d^*}{d^*} = \cot \theta \, \Delta\theta$$

84

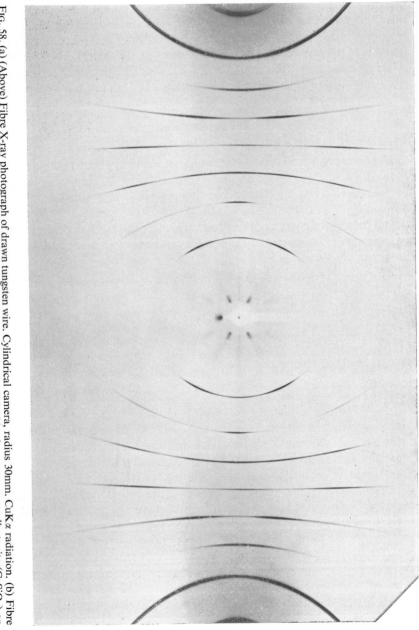

FIG. 58. (a) (Above) Fibre X-ray photograph of drawn tungsten wire. Cylindrical camera, radius 30mm. CuKα radiation. (b) Fibre photograph of okenite (hydrated calcium silicate) before and after heating to 850°C. Okenite on top → wollastonite (CaSiO₃) on bottom, retaining fibre texture. (c) Photograph of a single cotton fibre, taken on a 10 mm. radius camera.

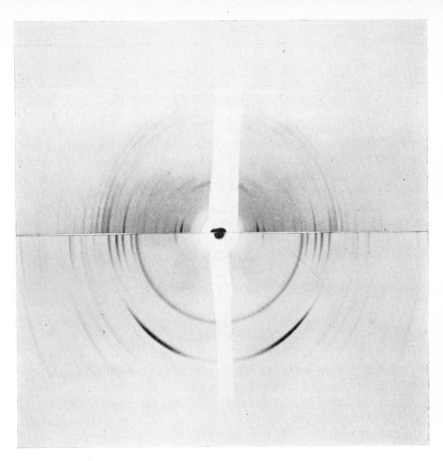

(b)

(c)

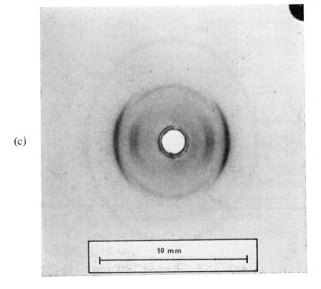

10 mm

$$\therefore \Delta d^* = \varepsilon d^*$$

and
$$\Delta \theta = \varepsilon \tan \theta$$

$$\therefore \beta = 2 \Delta \theta = 2 \varepsilon \tan \theta.$$

Inhomogeneous solid solution alters the crystal lattice and RL size in the same way as strain; but it may be possible to distinguish between them by an annealing treatment, which will not allow appreciable change in solid solution, but will relieve strain. Normally annealing will occur at a temperature well below that at which diffusion in solids would have an appreciable effect on inhomogeneities in solid solutions.

4.3. Strain and Small Crystallite Size Occurring Together

Let
$$\beta = \beta_s + \beta_c = 2 \varepsilon \tan \theta + \frac{\lambda}{t \cos \theta}$$

$$\therefore \beta \cos \theta = 2 \varepsilon \sin \theta + \frac{\lambda}{t}.$$

If $\beta \cos \theta$ is plotted against $\sin \theta$ for a number of reflections, the points should lie on a straight line, with slope 2ε and intercept λ/t.

Simple addition of β_s and β_c assumes that the strain and small size occur in the same crystallite. But they are more likely to occur in different crystallites, and it would probably be more accurate to add the squares of the line broadenings. However, the accuracy with which β can be measured [1(259)] does not usually justify a more complicated expression.

4.4. Disorder, Stacking Faults and Very Anisotropic Morphology

Consider a disordered structure which has started the process of ordering to form a superlattice, without altering the size of the parent lattice. All reflections which are common to both structures will be sharp; but the additional reflections which the new structure alone produces (the superlattice reflections) will arise only from the small nuclei of the ordered structure, and show small crystallite size broadening. The powder photograph thus has some sharp and some broadened lines. The classic case is the copper-gold alloy, $AuCu_3$ [1(262)].

Sometimes a somewhat similar effect shows up in single crystal 'mosaic blocks' (Section 1.2.2.5(b), p. 9) which corresponds to the crystallites of a powder specimen. Wollastonite, $CaSiO_3$, grows with a triclinic lattice. The bc layers have a near approximation to a plane of symmetry, so that when the crystal is growing, the next layer can stack either to the right or the left (the difference being half the lattice spacing). Very little control is thus exercised by the nearest neighbour layer, the next nearest layer being the main controlling factor. This makes it easy for 'mistakes' in stacking to occur. If the 'mistake' persists, a second twinned triclinic form, of opposite hand, grows. Under certain conditions the 'mistakes' become regular, occurring with every layer, and a rare monoclinic form grows. However, the conditions for this growth must be very critical. Usually a mixture of the three forms, right hand and left hand triclinic and monoclinic, occurs, if the monoclinic is present at all. All three forms are related by translations of **b**/2, which correspond to a phase difference of 2π for reflections with k even. Such reflections will therefore have all 3 forms within a single mosaic block co-operating 'in phase' to produce sharp reflections corresponding to the full size of the mosaic block. The odd layers however give weak reflections of all three forms streaked in the common **a*** direction which is perpendicular to the layers, i.e. they correspond to very thin crystals of all three forms (Ans. 14, p. 475). This can be explained in the following way. Within a mosaic block each of the three forms will occur as lamellae, separated by interleaving lamellae of the other forms. The lamellae of any one form can occur in two ways, related by a translation of **b**/2. Two equal lamellae of, say, left-hand triclinic, related by a translation of **b**/2, will give complete destructive interference for reflections with k odd, since the equal resultant amplitudes will have a phase difference of π. In the whole mosaic block, only the difference in volume between the two sets of lamellae of any one form will be effective in producing reflections. This will correspond to a thin plate perpendicular to **a***, giving streaks in this direction.

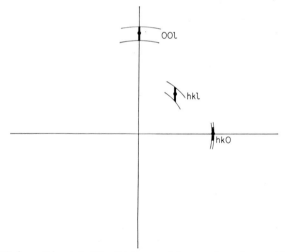

FIG. 59. The RL for a thin plate. The thickness of the powder spheres which would arise from a large number of randomly ordered plates is indicated for a number of reflections. Only the $hk0$ spheres will be thin enough to produce lines above the background on a powder photograph.

Figure 117 (p. 193) shows the Weissenberg photograph of the first layer about **b**, with the relps drawn out into rods in the **a*** direction, to such an extent that the rods bridge the whole gap between relps.

If a powder photograph were taken of such material, the thick powder spheres in reciprocal space would all overlap, and merely give a heavy background to the sharp lines corresponding to the *hkl* relps for $k = 2n$.

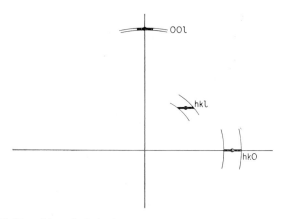

FIG. 60. The RL for a thin rod. Only the 00*l* reflections will be visible on a powder photograph, since the relps have become *discs*, and the effect is cylindrically symmetrical about **c***.

A similar effect is obtained if the powder specimen is composed of small crystallites all having the same anisotropic morphology, i.e. plates or needles all parallel to the same crystallographic directions. Consider plates parallel to **ab**. The relps will be rods parallel to **c***. Figure 59 shows that if the crystallites are thin enough, only the *hk*0 reciprocal powder spheres will be thin enough to give powder lines which will show up above the background. This is the case with calcium silicate hydrate, the main bonding material in hydrated Portland cement, where only a few diffuse *hk*0 powder lines can be recognised. A case combining disorder and thin plates occurs in clay minerals, where the layer structure has such weak bonds between layers that random displacements and rotations occur. However, this does not affect the regular spacing of the layers in the **c**-direction, and the 00*l* reflections are therefore sharp, but all other relps are drawn out into rods, corresponding to the thickness of the parts of the clay particle which have no stacking faults. Since these are very thin, only *hk*0 bands can be seen on a powder photograph, apart from the sharp 00*l* reflections. This is a more extreme example of the wollastonite case.

If the crystallites are very thin needles parallel to **c**, the relps will be discs parallel to **a*****b***. Figure 60 shows that in this case 00*l* reflections will be the only sharp ones. Thin needles occur much more rarely than thin plates.

Chapter 8

Electron and Neutron Diffraction

1. Electron Diffraction

A beam of electrons of uniform energy $eV \equiv \frac{1}{2}mv^2$ has an associated wavelength λ, given by $\lambda = h/mv \simeq \sqrt{(150/V)}$ (V in volts). Since V must be fairly high, at least 50 kV, in order to get appreciable penetration of the electrons, especially for transmission photographs, the wavelength is much shorter than for X-rays, i.e. about 0·05 Å compared with 1·54 Å for CuKα. Even so the crystal plates giving rise to diffraction must be very thin and can no longer be treated as infinitely thick for diffraction purposes, although length and width are usually large enough. The consequence is that the diffraction maxima are broadened in certain directions, and the effect can be represented by extending the reciprocal lattice points into short rods perpendicular to the plate (Ans. 14, p. 475). With the plate lying flat in the untilted specimen holder the rods will also be in the direction of the electron beam.

Since the RL is very small in relation to the radius of the sphere of reflection (because λ is small) the sphere passes through a large number of these rods in the zero-layer, and the reflections project on to a flat plate as an almost undistorted picture of the zero layer (Fig. 61).

There are many other effects, including Kikuchi lines, which are similar to Kossel lines (Section 5.1, p. 40), and arise from electrons scattered at the surface of the crystal to give a beam diverging slightly from the point of impact. The loss of energy on scattering through a small angle is relatively so small that the wavelength change can be ignored, and the small size of the RL allows the formation of numerous lines even with a small angle of divergence. Powder photographs can also be produced. However a picture of the zero layer of the RL is the most usual and useful result; and although it may be difficult to obtain more information by electron diffraction, such a photograph of the zero layer in one direction can be invaluable in helping to sort out the much fuller, but otherwise uninterpretable, information given by an X-ray powder photograph. Some of the modern electron microscope/

89

diffraction cameras have means of tilting and orienting the specimen, so that most of the RL can be investigated directly.

Electron diffraction is limited to very thin specimens which are unaffected by high vacuum and the heating effect of the electron beam. In the latest high voltage (million volt) cameras it may be possible to investigate encapsulated specimens. This will overcome the difficulty of exposure to vacuum.

Because of the much greater interaction of electrons with matter, all the troubles of absorption, extinction, multiple scattering, etc. (Section 9.5, p. 119)

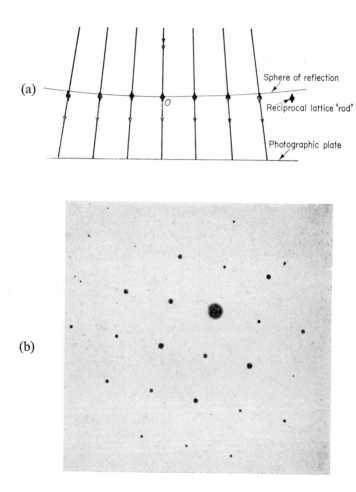

FIG. 61. (a) The interaction of the sphere of reflection with the RL rods of a very thin crystal, to produce an almost undistorted picture of the zero layer normal to the electron beam. (b) Electron diffraction diagram of β-NH_4Br showing the zero layer of the RL.

which make accurate Structure Factor measurements with X-rays difficult, are enhanced in the case of electron diffraction to the point where they may become the major factors in determining intensities. It is common for reflections which should be absent because of the space group symmetry to appear prominently in the diffraction pattern as a result of multiple reflection. In consequence, electron diffraction is, except in special circumstances, an unreliable method for determining space group absences or measuring structure factors.

2. Neutron Diffraction

Neutron diffraction, using thermal neutrons with an associated wavelength of about 1 Å, is in many ways similar to X-ray diffraction. Because of the difficulty of producing a neutron beam of adequate intensity, the method is normally restricted to special cases where it has definite advantages over X-ray diffraction. There are two main special cases of this sort.

2.1. MATERIALS CONTAINING HYDROGEN OR NEIGHBOURING ELEMENTS IN THE PERIODIC TABLE

Because the scattering interaction is with the nucleus it is no longer proportional to the atomic number, as with X-rays, and in fact has no simple relation to it. In consequence many 'light elements' have a scattering power greater than that of 'heavier' atoms, and in particular hydrogen scattering is comparable to that of many other elements. It is therefore possible to locate the position of hydrogen nuclei with much greater accuracy than the centre of its electron cloud can be located by X-rays. Neighbouring elements in the periodic table can also be easily distinguished in most cases because of their widely different scattering powers.

2.2. MATERIALS CONTAINING 'ATOMIC MAGNETS'

Neutrons interact with magnetic fields, whereas X-rays do not. Neutron diffraction has shown up magnetic ordering in many substances. Such ordering is completely inaccessible to direct investigation by X-ray diffraction. There may be small indirect effects which can be detected, such as minor distortions of a unit cell from high symmetry; but where the 'atomic magnets' are arranged in equal numbers in opposite directions, even this is unlikely. Neutrons however will show up the superlattice reflections arising from the magnetic ordering, and enable complicated patterns of 'atomic magnets' to be worked out.

Neutrons have a far lower absorption than X-rays and can therefore be used on large crystals. Since neutron beam fluxes, though continually increasing, tend to be somewhat weak, this low absorption is fortunate, since the intensity of a reflection is proportional to the effective volume of the crystal. However, if the crystal has large perfect domains with a small angular misorientation, which tends to be the case when blocks several mm across can be grown, then extinction (Section 9.5, p. 119) will be severe and must be taken into account.

Chapter 9

The Intensity of X-ray Reflection

1. The Scattering Diagram (Transform)

The amplitude of diffracted beams for given directions of incidence and scattering, relative to the arrangement of atoms in the unit cell, can be calculated or obtained optically and presented in the form of a scattering diagram (mathematically the 'transform' of the electron density distribution in the unit cell).

1.1. CALCULATED SCATTERING DIAGRAMS (TRANSFORMS)

From the results given in Section 1.4.2 (p. 16) the amplitude and phase of scattering from the unit cell contents of a crystal is given by:

$$R = aF = a\sqrt{(A^2 + B^2)}$$

where a = amplitude of scattering by 1 electron.

F = structure factor amplitude.

$$A = \sum_m f_m \cos \phi_m; \; B = \sum_m f_m \sin \phi_m; \; \text{phase angle } \alpha = \tan^{-1} \frac{B}{A}$$

where $\phi_m = 2\pi \, \mathbf{r}_m \cdot \mathbf{n} \dfrac{2 \sin \theta}{\lambda}$.

f_m = atomic scattering factor of the mth atom.

$\mathbf{r}_m$ = position vector of the centre of the mth atom.

$\mathbf{i}$ = unit vector in the direction of the incident beam.

$\mathbf{s}$ = unit vector in the direction of the scattered beam.

$\mathbf{n}$ = unit vector bisecting the angle between $-\mathbf{i}$ and $\mathbf{s}$.

2θ = angle of deviation of the scattered beam.

For a given angle of deviation 2θ, and direction of $\mathbf{n}$, the value of R and α will be the same whatever the directions of $\mathbf{i}$ and $\mathbf{s}$ around $\mathbf{n}$, since f varies only with θ, and ϕ_m is only a function of the direction of $\mathbf{n}$, i.e. the plane containing $\mathbf{i}$ and $\mathbf{s}$ can be rotated about $\mathbf{n}$ and the resultant scattering will be the same in every position. Thus $\mathbf{n}$ and $2 \sin \theta$ fix the resultant amplitude and phase of the scattered beam. We can therefore plot a diagram which will completely define

93

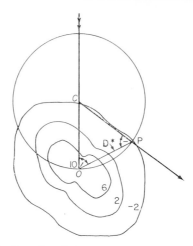

FIG. 62. *R* (or *F*) given by the value of the contour of the scattering diagram at the end of **D***. Draw in the direction of the incident beam to the origin of the diagram *O*. Construct the sphere (circle) of reflection on this as diameter, as for the RL, and draw the scattered direction from the centre of the sphere. Let this line cut the sphere at *P*. *OP* is then the bisector of the angle between $-\mathbf{i}$ and $\mathbf{s}$, and $OP = 2 \sin \theta$ where 2θ is the angle of deviation. *OP* is therefore **D***, and the value of the contour at *P* gives the amplitude of scattering in the direction *CP*.

the scattering power of the unit cell contents, by taking a point on **n** at a distance $2 \sin \theta$ from the origin (at the end of the vector, $\mathbf{D}^* = (2 \sin \theta)\mathbf{n}$) and labelling it with the value of the resultant scattering. If we do this for all directions of **n** and all possible values of $2 \sin \theta$ and join up points of equal scattering to give a three-dimensional contour 'map',† we can immediately say what the scattering for any given directions of the incident and scattered beams will be, by drawing the corresponding **D*** (bisecting the angle between $-\mathbf{i}$ and $\mathbf{s}$ and length $2 \sin \theta$) and noting the value of the contour at its end. The sphere of reflection is a convenient way of doing this (Fig. 62) especially for a given direction of the incident beam, since we can immediately say what the scattering will be in any direction, by using this construction. Because this construction of the scattering diagram from the arrangement of the scattering points (i.e. the electron distribution in the unit cell) is an example of a process which mathematicians call 'Fourier transformation', the scattering diagram is normally referred to as a 'Fourier transform' of the unit cell contents. Figure 63 shows a section of the idealized transform of a benzene ring in its own plane. (**Example 14**, p. 292).

† For a centrosymmetric distribution the phase angle α will be 0 or π (Section **1**.4.1, p. 14). This can be represented on the scattering diagram by labelling the contours $+$ for $\alpha = 0$ and $-$ for $\alpha = \pi$. For a non-centrosymmetric distribution α must be represented by a second, superimposed, contour 'map'.

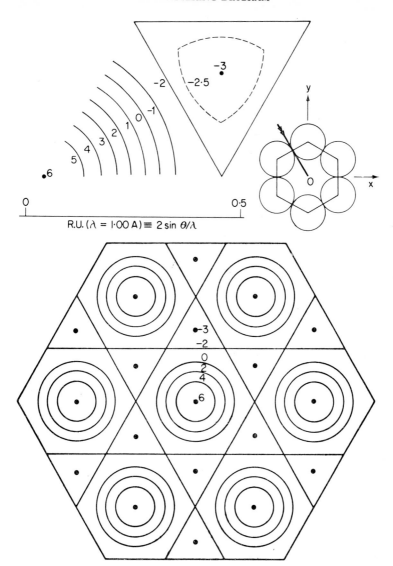

Fig. 63. The idealised scattering diagram (transform) of benzene (section in the plane of the molecule). The atomic scattering factor for the carbon atoms has been taken as 1 (independent of θ), and the slight hexagonal distortion of the circular contours has been suppressed. This only amounts to about 1% variation of the radius, even for the largest circle. The hydrogen atoms have been ignored. The high symmetry of the benzene ring makes the idealised transform periodic; but this periodicity would be lost if actual atomic scattering factors, which fall off with increasing θ, had been used. (After Knott.)

The arrow pointing to the centre of the benzene ring refers to Example 12.

1.2. OPTICAL TRANSFORMS

The production of a Fourier transform of a set of scatterers can be shown optically. (It cannot be demonstrated with X-rays because the energy scattered by a single molecule is undetectable in practice (Section 1.2.2, p. 6)). If we take a set of optical scatterers (small holes in a mask) and direct a beam of monochromatic light perpendicular to the mask, we have a case similar to that of electron diffraction, in which the wavelength is considerably shorter than the distances between the scatterers. The size of the scattering diagram, in the region where detectable scattering occurs, is very small compared with the radius of the sphere of reflection, so that the part of the sphere cutting the scattering diagram is almost flat and all detectable scattering occurs at very small θ angles (Fig. 64a). The scattering is thus a projection of the section of the scattering diagram which is perpendicular to the light beam, (Fig. 64b).

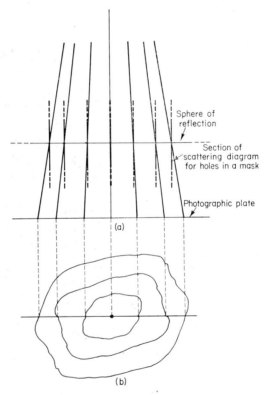

FIG. 64. Scattering of light by holes in a mask. (a) Section containing the light beam, including sections of contour surfaces of the scattering diagram near their intersection with the sphere of reflection. (b) Section normal to light beam, tangential to the sphere of reflection, showing the contours of the continuous distribution (Fig. 65) which is projected on to the photographic plate.

This technique has been developed by Lipson and his collaborators [2]. If the holes in the mask are punched to represent a naphthalene double ring, then the optical scattering will represent the section of the naphthalene scattering diagram in the plane of the ring (Fig. 65).

1.3. THE TRANSFORM OF A CRYSTAL

If we now consider the whole crystal containing M unit cells we know that it scatters only in certain directions, but that in those directions the scattering is M times (in amplitude; M^2 times in intensity) that of the contents of a single unit cell. In other words, not all values of $\mathbf{D}^*$ can give rise to scattering, only those where $\mathbf{s} - \mathbf{i} = \mathbf{D}^*$ is perpendicular to a set of lattice planes (hkl), i.e. in the direction of the reciprocal lattice vector $\mathbf{d}_{hkl}^*$, and where $\lambda = 2d_{hkl}\sin\theta$, i.e. $2\sin\theta = D^* = \lambda/d_{hkl} = d_{hkl}^*$. Therefore only where

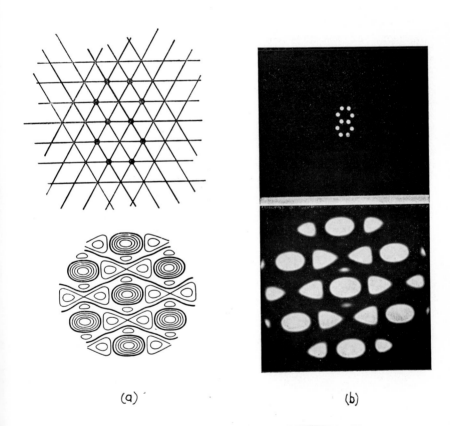

(a)   (b)

FIG. 65. The section of the transform of a naphthalene molecule in its own plane (a) calculated from the net; (b) obtained optically from the mask.

$\mathbf{D}^* = \mathbf{d}^*_{hkl}$ will reflection occur, and its amplitude will be proportional to the contour reading at the end of $\mathbf{d}^*_{hkl}$. This can also be shown optically by repeating the pattern of holes (Fig. 29, p. 38). We thus superimpose the reciprocal lattice on the scattering diagram, and read off the strength of the various reflections from the contours passing through the lattice points. (**Example 15**, p. 293).

If the arrangement of the atoms in a planar molecule, such as a condensed ring system, is known, then the section of its transform in its own plane can be calculated or obtained optically. If there is just one molecule in a flat unit cell, with the short axis perpendicular to the molecule, then the zero layer of the RL perpendicular to this axis will correspond to the section of the transform of the molecule. If we draw out the RL on a transparent sheet to the same scale as the transform, and draw round all the relps discs whose areas are roughly proportional to the intensities of the corresponding reflections, then we can compare the transform section and the weighted RL layer. In this case the only unknown is the angular orientation of the molecule in the unit cell. This can be obtained by placing the origin of the RL layer over that of the transform

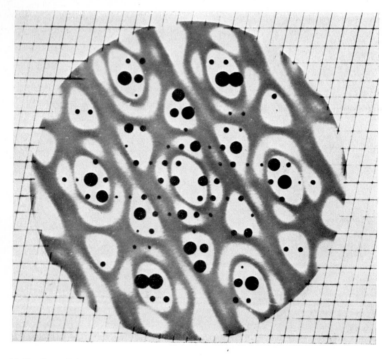

FIG. 66. Section of the scattering diagram of pyrene, produced optically, and superimposed weighted RL layer.

section, and rotating the RL layer until all the heavily weighted relps lie on high areas of the transform (Fig. 66). (**Example 16**, p. 294).

Any departure from the conditions listed above (e.g. more than one molecule in the cell, especially if they are not parallel, or unit cells not stacked directly above one another) leads to difficulties of interpretation. But it has proved possible in a three-dimensional RL, with layers plotted out on transparent sheets, to recognise features of the benzene ring transform for example, and thus obtain a clue to the orientation of part of the molecule, even in complicated structures. It is also possible to obtain optical transforms corresponding to non-zero layers, and extend the use of optical transforms in other ways. [2]

If we attach the values of amplitude and phase obtained from the scattering diagram to each relp, the RL gives us complete information about the scattering of the crystal. From this information, by a reverse transformation process, the arrangement of the scatterers, i.e. the electron density distribution or the atomic arrangement, can be deduced. (This can also be done optically [2]).

Thus we require values of F and α for all relps. F can be obtained experimentally; but since only intensities of reflection can be measured, α is completely lost, and the whole complication of structure analysis is concerned with finding α by indirect means.

1.4. The Structure Factor of a Crystal

Since, for a crystal, scattering only occurs when $\mathbf{D}^* = (2 \sin \theta)\mathbf{n} = \mathbf{d}^*_{hkl}$, we replace $(2 \sin \theta)\mathbf{n}$ in the formula for $\phi_m(16(1))$ by $\mathbf{d}^*_{hkl}$.

$$\therefore \phi_m = 2\pi \frac{\mathbf{r}_m \cdot \mathbf{d}^*_{hkl}}{\lambda} = \frac{2\pi}{\lambda} (x_m \mathbf{a} + y_m \mathbf{b} + z_m \mathbf{c}) \cdot (h\mathbf{a}^* + k\mathbf{b}^* + l\mathbf{c}^*) \quad (1)$$

$$= \frac{2\pi}{\lambda} (hx_m + ky_m + lz_m) \lambda \quad \text{since } \mathbf{a} \cdot \mathbf{a}^* = \mathbf{b} \cdot \mathbf{b}^* = \mathbf{c} \cdot \mathbf{c}^* = \lambda, \text{ and}$$
$$\mathbf{a} \cdot \mathbf{b}^* \text{ etc.} = 0.$$

$$\therefore \phi_m = 2\pi (hx_m + ky_m + lz_m). \quad (2)$$

The Structure Factor $\mathbf{F}_{hkl}$ for a crystal, including amplitude and phase, is given by:

$$F = \sqrt{(A^2 + B^2)}$$

$$\alpha = \tan^{-1} \frac{B}{A}$$

where $A = \sum_m f_m \cos 2\pi (hx_m + ky_m + lz_m)$

$B = \sum_m f_m \sin 2\pi (hx_m + ky_m + lz_m)$.

1.4.1. *The phase angle and lattice planes*

The phase angle ϕ_m, for scattering from the mth atom, can be derived directly from the concept of reflection from sets of planes (hkl) and Bragg's Law. This law implies that the path difference between rays reflected from two adjacent planes is one wavelength, and that the phase difference is therefore 2π. In Fig. 67 ABC is the nearest plane to the origin of the set (hkl). Therefore $OA = a/h$, $OB = b/k$, $OC = c/l$. All rays reflected from the plane ABC at points such as N and M have the same path length and phase. Rays reflected from O, or any other point on the parallel plane through the origin, have a phase lag of 2π compared with those reflected from ABC, i.e. in going from O to any point on ABC a phase increase of 2π occurs. In particular, going from O to A gives a phase increase of 2π. Therefore the phase increase per unit length in the direction of $\mathbf{a}$ is $2\pi/(a/h) = 2\pi h/a$. Similarly the phase increases per unit length in the directions of $\mathbf{b}$ and $\mathbf{c}$ are $2\pi k/b$ and $2\pi l/c$ respectively. Now consider the ray scattered from an atom at Z, whose position vector $\mathbf{r}_m = x\mathbf{a} + y\mathbf{b} + z\mathbf{c}$. To get from O to Z we can go from O to X (a length xa in the direction of $\mathbf{a}$), X to Y (yb in the direction of $\mathbf{b}$) and Y to Z (zc in the direction of $\mathbf{c}$). The total phase increase in going from O to Z is therefore

$$2\pi \frac{h}{a} \cdot xa + 2\pi \frac{k}{b} \cdot yb + 2\pi \frac{l}{c} \cdot zc = 2\pi \, (hx + ky + lz).$$

Looked at in another way, $\phi = \dfrac{2\pi}{\lambda} \mathbf{r}_m \cdot \mathbf{d}^*_{hkl}$ from 99(1)

$$= \frac{2\pi}{\lambda} d^*_{hkl} \mathbf{r}_m \cdot \frac{\mathbf{d}^*_{hkl}}{d^*_{hkl}}$$

$$= \frac{2\pi}{\lambda} \frac{\lambda}{d_{hkl}} \mathbf{r}_m \cdot \mathbf{n}$$

$$= 2\pi \frac{\mathbf{r}_m \cdot \mathbf{n}}{d_{hkl}}.$$

But $\mathbf{r}_m \cdot \mathbf{n}$ is the resolved part of $\mathbf{r}_m$ in the direction of the normal (OS in Fig. 67) and $d_{hkl} = ON$. Therefore the phase angle for the ray scattered from Z is only dependent on the perpendicular distance of Z from the (hkl) plane through the origin. If a parallel plane is drawn through Z, then it is only the relative position of this plane to the set (hkl) which determines the phase, i.e. the ratio of the distance of this plane from the origin to the distance between planes of the set (hkl) $- OS/ON$ in Fig. 67. The phase angle ϕ_m is 2π times this ratio.

This aspect of the phase equation is particularly important in understanding the physical basis of direct methods of phase determination, and in helping to establish a trial structure in the early stages of structure determination by more conventional methods, since it can give clues to the positions of the atoms in the unit cell. If a particular reflection hkl is very strong, then the atoms must lie on or near planes which are parallel to the set (hkl) and are d_{hkl} apart. Notice that the atoms need not lie on the set (hkl), since the positions of this set (as of all others) depends on the choice of origin point from which to construct the lattice. In the case of a centrosymmetric structure, with the origin at a centre of symmetry, the planes of atoms would either lie on the lattice planes or half way between them. The first case corresponds to a resultant phase of 0 and the second to a resultant phase of π. In the non-centrosymmetric case there is no restriction on the phase angle or the position of the planes of atoms relative to the origin.

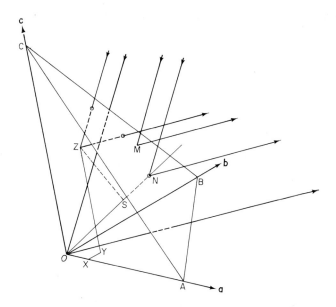

FIG. 67. Diagram of reflection from a set of lattice planes (hkl) showing scattering from a point, Z, between the lattice planes.

The fact that it is only the component of the position vector of an atom in the direction of the normal to a set of planes that affects the intensity of the reflections from these planes, is also important in understanding which reflections are absent as a result of glide planes or screw axes in the space group symmetry of a crystal. (Appendix II, p. 342).

1.4.2. 'Direct' methods of determining the phase angle

All the variations of the 'Direct Method' depend on the relationship $\alpha_{(H_1+H_2)} = \alpha_{H_1} + \alpha_{H_2}$, where $\alpha_{H_n} \equiv \alpha_{h_n k_n l_n}$, the phase angle of the Structure Factor $F_{h_n k_n l_n}$. Thus if one knows the phases of two reflections one can find that of a third, and so on. This relation depends on the fact that the atoms can be taken as the scattering units, and that the phase of scattering for X-rays relative to the centre of the atom is always positive (i.e. the phase angle is zero) so that, for the purpose of calculation, the atoms can be replaced by point scatters at their centres (but with scattering amplitude dependent on the atom type and θ). However the relation is only necessarily true if the structure factors are all very large—nearly the maximum possible. In fact, in the non-centrosymmetric case it is only exactly true if all the atoms in the cell lie exactly on the intersections of the two sets of planes $(h_1 k_1 l_1)$ and $(h_2 k_2 l_2)$. In the centrosymmetric case, because α can only be 0 or π, there will be a very high probability that the relation will be true if all three reflections $(h_1 k_1 l_1; h_2 k_2 l_2$ and $h_1 + h_2, k_1 + k_2, l_1 + l_2)$ are fairly strong. The stronger the reflections are, the greater is the probability that the relationship will hold in a given case; but it is rarely completely certain, and this uncertainty leads to the probability aspect which looms so large in the practical application of the method. However, the physical basis of the method is most easily understood by taking the case when the probability is unity.

If all the atoms in the unit cell lie on the set of planes $(h_1 k_1 l_1)$ and also on the set $(h_2 k_2 l_2)$, they must clearly all lie on the lines of intersection of the two sets. If it can be shown that the set of planes $(h_1 + h_2, k_1 + k_2, l_1 + l_2)$ also pass through these lines of intersection, then the atoms will also lie on this set of planes. In this case the phase angle α will be zero in all three cases. In the centro-symmetric case the only other alternative is that the atoms should lie exactly half way between the sets of planes. α would then be π. If α_{H_1} is 0 and α_{H_2} is π, the atoms will lie half way between the $(H_1 + H_2)$ set, and $\alpha_{(H_1+H_2)} = \pi$. If both α_{H_1} and α_{H_2} are π, the atoms will lie on the $(H_1 + H_2)$ set, and $\alpha_{(H_1+H_2)} = 0$. This can be verified for the case of 302, $10\bar{4}$ and $40\bar{2}$ in **Example 17** (p. 296), but can be demonstrated in the general case and developed for non-centrosymmetric arrangements as follows.

1. Since $\mathbf{d}^*_{(H_1+H_2)} = \mathbf{d}^*_{H_1} + \mathbf{d}^*_{H_2}$ (for proof expand into components) the three sets of planes form a zone and all lines of intersection will be parallel to the zone axis.

2. Let $\mathbf{r}$ be the position vector of a point P, on any line of intersection of two planes from the two sets H_1 and H_2. Then $\mathbf{r} \cdot \mathbf{d}^*_H / d^*_H$ is the resolved part of $\mathbf{r}$ in the direction of the normal of the set of planes H, and is therefore

equal to nd_H, where d_H is the interplanar spacing and n is an integer (P is on the nth plane from the origin).

$$\therefore \frac{\mathbf{r} \cdot \mathbf{d}^*_{H_1}}{\lambda} = n, \quad \frac{\mathbf{r} \cdot \mathbf{d}^*_{H_2}}{\lambda} = m \text{ (since } d^*_H d_H = \lambda)$$

$$\therefore \frac{\mathbf{r} \cdot (\mathbf{d}^*_{H_1} + \mathbf{d}^*_{H_2})}{\lambda} = n + m = \frac{\mathbf{r} \cdot \mathbf{d}^*_{(H_1 + H_2)}}{\lambda}.$$

P therefore lies on the $(n + m)$th plane from the origin of the set $(H_1 + H_2)$. Therefore the set $(H_1 + H_2)$ passes through all the lines of intersection of H_1 and H_2.

3. If $\alpha_{H_1} = \pi$, $\alpha_{H_2} = 0$ and P is an atom half way between the planes of set H_1, but on a plane of set H_2, then

$$\frac{\mathbf{r} \cdot \mathbf{d}^*_{H_1}}{\lambda} = n + \tfrac{1}{2}; \quad \frac{\mathbf{r} \cdot \mathbf{d}^*_{H_2}}{\lambda} = m; \quad \frac{\mathbf{r} \cdot \mathbf{d}^*_{(H_1 + H_2)}}{\lambda} = n + m + \tfrac{1}{2}.$$

P therefore lies halfway between planes of the set $(H_1 + H_2)$, and $\alpha_{(H_1 + H_2)} = \pi$.

In general the phase, as a fraction of 2π, is given by the fractional part of the right-hand side of the equations. Thus if $\alpha_{H_1} = \alpha_{H_2} = \pi$

$$\frac{\mathbf{r} \cdot \mathbf{d}^*_{(H_1 + H_2)}}{\lambda} = n + m + \tfrac{1}{2} + \tfrac{1}{2} \quad \text{and} \quad \alpha_{(H_1 + H_2)} = 0.$$

In the non-centrosymmetric case the atoms lie on planes parallel to the sets H_1 and H_2, but at some general position between them. Consider the plane through the origin and its nearest neighbour of the set. The parallel plane of atoms lies at a distance from the origin of $d_H \cdot \alpha_H/2\pi$ (Section 1.4.1, p. 100).

∴ If P is an atomic position,

$$\frac{\mathbf{r} \cdot \mathbf{d}^*_{H_1}}{\lambda} = n + \frac{\alpha_{H_1}}{2\pi}; \quad \frac{\mathbf{r} \cdot \mathbf{d}^*_{H_2}}{\lambda} = m + \frac{\alpha_{H_2}}{2\pi};$$

$$\frac{\mathbf{r} \cdot \mathbf{d}^*_{(H_1 + H_2)}}{\lambda} = n + m + \frac{\alpha_{H_1} + \alpha_{H_2}}{2\pi}.$$

Therefore $\alpha_{(H_1 + H_2)} = \alpha_{H_1} + \alpha_{H_2}$.

In the centrosymmetric case the atomic positions need only lie near the lines of intersection for the relation given earlier to hold, except in very rare cases. If all three reflections, H_1, H_2 and $(H_1 + H_2)$ are strong this will almost certainly be the case. If however H_1 and H_2 are strong but $(H_1 + H_2)$ weaker, this will indicate that a significantly larger proportion of the atoms lie off the set of planes $(H_1 + H_2)$ than is the case for H_1 and H_2 separately, and the relation may not hold.

Putting $S_H = +1$ if $\alpha_H = 0$ and -1 if $\alpha_H = \pi$ (i.e. the normal sign of a Structure Factor in the centrosymmetric case) the phase relation has the more familiar form

$$S_{(H_1 + H_2)} = S_{H_1} \times S_{H_2}.$$

In the non-centrosymmetric case, as soon as the atoms move at all from the lines of intersection (of the planes of atoms, not of the lattice planes) the value of $\alpha_{(H_1 + H_2)}$ will no longer be $\alpha_{H_1} + \alpha_{H_2}$. One can then only consider the probability that it will not have departed from the formula value by more than a certain amount. This is yet another way of presenting the greater difficulty involved in the structure analysis of a non-centrosymmetric arrangement compared with a centrosymmetric one.

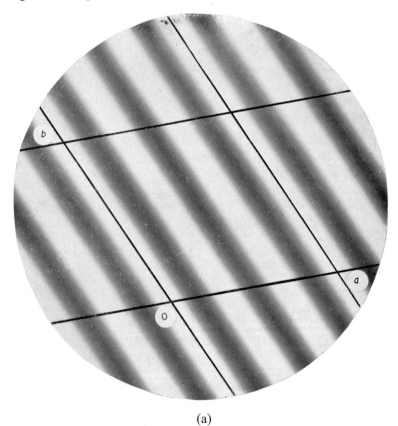

(a)

FIG. 68. Fourier terms for two sets of planes. The blackening should vary sinusoidally. (a) (300), origin at a trough, (b) (210) origin at a peak.

1.4.3. *The structure factor and the electron density distribution*

X-rays are scattered by electrons and therefore the scattering of X-rays, fundamentally, gives information about the distribution of electrons in the crystal. Since this distribution is periodic it must be possible to represent it by a Fourier series, just as a musical sound can be synthesised from the fundamental note and its various harmonics. However, a crystal is periodic in three dimensions and the Fourier synthesis must be three-dimensional also. In every direction in which a crystal is periodic, there must be a Fourier series representing that periodicity. But the directions in which a crystal is periodic are the normals to the various sets of lattice planes, if the electron density is considered to be averaged over the planes. There will be a Fourier series in every direction in which there is a normal to a set of lattice planes (hkl), i.e. in every direction in which there is a RL vector $\mathbf{d}^*_{hkl}$. The first relp

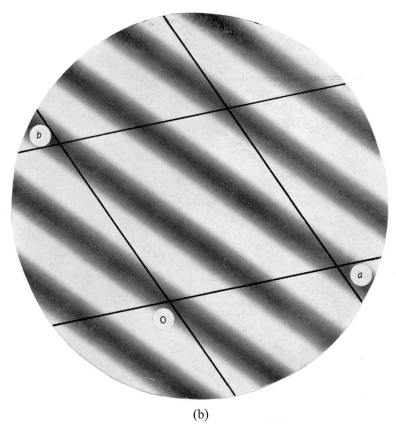

(b)

from the origin (*hkl* with no common factor) will correspond to the first or fundamental term of the series, and the second (2*h*, 2*k*, 2*l*) to the first harmonic, of double the frequency and half the wavelength or spacing, and so on. There is obviously, therefore, a close connection between the terms of the Fourier series and the RL.

It is almost impossible to represent a term of the Fourier series in three dimensions; but a section containing the corresponding RL vector is shown in Fig. 68a for the planes (300), and another for the planes (210) in Fig. 68b. The electron density represented by the blackening on the diagram has to be imagined as extending uniformly to large distances in all directions perpendicular to $\mathbf{d}_{hkl}^*$, including perpendicularly from the plane of the paper. The sum of the electron densities contributed at a point by all these plane

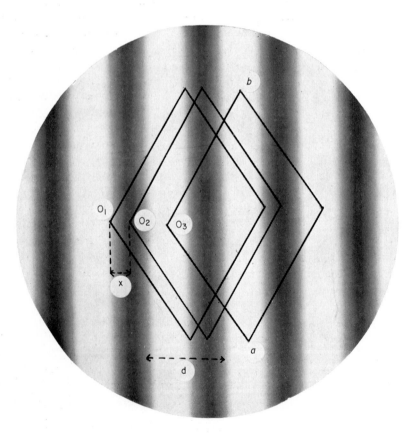

FIG. 69. Various cells outlined with different origins for the (110) term. The problem is to place the 'fringes' in the correct position relative to the cell, but this is most easily illustrated by drawing a number of cells.

waves—a complete Fourier set in the direction of every RL vector—crossing each other in all directions, is the actual electron density at that point in the unit cell. The periodicity of each term and the direction of the wave normal is fixed by the periodicity of the lattice planes and the direction of their normal, i.e. by the RL.

Suppose, as is the case, that it is possible to find the amplitudes of each of these plane waves of electron density, i.e. the maximum value of the electron density for that wave. Figure 69 shows that there is still a fundamental difficulty to be overcome. Where do we place the wave relative to the origin of the unit cell? Figure 68 shows the two possibilities for the centrosymmetric case. We can either place a crest or a trough at the origin. For the non-centro-symmetric case we can place any point from one crest to the next, on the origin. This is one way of stating the famous 'phase problem'. Placing a crest at the origin corresponds to giving zero phase to that particular term of the Fourier series; placing a trough at the origin corresponds to a phase of π. Placing any other point at the origin corresponds to some phase angle α', proportional to the distance of the point from a crest, since the distance between crests corresponds to a phase difference of 2π. If we know the angle α', we can place the term of the series relative to the origin, but until we know it, we cannot build up the Fourier synthesis.

Figure 70 shows two-dimensional syntheses (made after the phases had been determined) which correspond to projecting the electron density on to a plane. It was built up photographically from fringes similar to those of Figs 68 and 69 for the $hk0$ and $0kl$ terms, and shows the concentrations of electron density round atoms.†

The addition of two-dimensional terms can be shown diagrammatically because the third dimension can be used to indicate the magnitude of the electron density. Figure 71 shows one term 240, of the $hk0$ zone, and Fig. 72 a number of separate terms. Figure 73 shows the combination of two terms and the relation to two-dimensional representation by a contour map.

---

† Two-dimensional Fourier summations are produced in this way by the von Eller Photo-sommateur. This consists of a linear light source at one end of a light-tight cabinet, a grating designed to produce sinusoidal variations of intensity falling on photographic paper at the other end of the box, and a rotating holder for the paper. Variation of the distance of the grating from the source produces corresponding variation in the fringe spacing; lateral movement of the grating enables the fringes to be positioned corresponding to any phase angle between 0 and 2π; and rotation of the paper holder enables the unit cell to be position-ed correctly relative to the fixed (horizontal) normal to the fringes. The amplitudes of the terms depend on the exposure time for each fringe set, and this can be automatically set by a timer. The angular position of the paper holder and the distance of the grating from the source are linked with a cursor, so that the correct positions for a given term are obtained when the cursor is over the corresponding relp on a RL net pinned to the front of the paper holder.

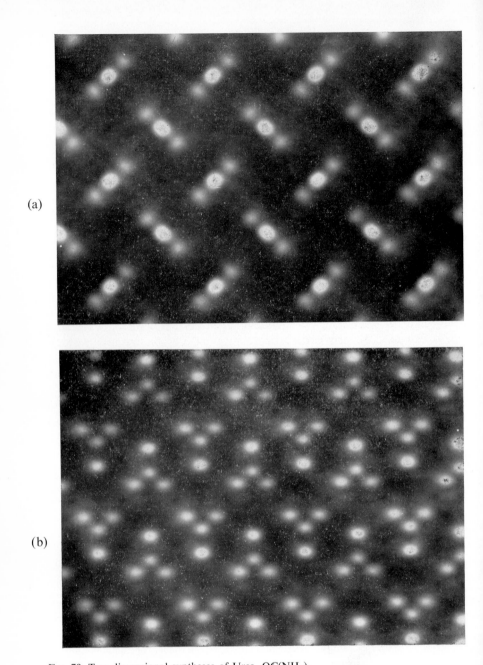

(a)

(b)

Fig. 70. Two-dimensional syntheses of Urea, OC(NH₂)₂.

(a) projected on to (001):— C + O ; (b) on to (100):—

These pictures are negatives. The originals had dense atoms on a less dense background.

Looked at analytically, the Fourier series can be represented by

$$\rho_r = \sum_{hkl} A_{hkl} \cos \left(2\pi \frac{\mathbf{r} \cdot \mathbf{d}^*_{hkl}}{\lambda} - \alpha'_{hkl} \right) \qquad (1)$$

where ρ_r is the electron density at a point given by the position vector

$$\mathbf{r} = x\mathbf{a} + y\mathbf{b} + z\mathbf{c}.$$

$$\frac{\mathbf{r} \cdot \mathbf{d}^*_{hkl}}{\lambda} = \mathbf{r} \cdot \frac{\mathbf{d}^*}{d^*} \times \frac{d^*}{\lambda} = \frac{\mathbf{r} \cdot \mathbf{n}}{d} = \frac{p}{d}$$

is the ratio of p, the projection of the position vector $\mathbf{r}$ on to $\mathbf{d}^*_{hkl}$, to the periodicity d, of the Fourier term $(OS/ON,$ Fig. 67, p. 101). For all points $\mathbf{r}$ for which $\mathbf{r} \cdot \mathbf{n}$ ($=p$) is a constant, i.e. for all points on a plane perpendicular to $\mathbf{n}$, the contribution to the electron density from this term is constant, and varies with p as the cosine of 2π times the ratio p/d. The placing of the plane wave relative to the origin is given by α'. This, therefore, corresponds to the Fourier term as previously described.

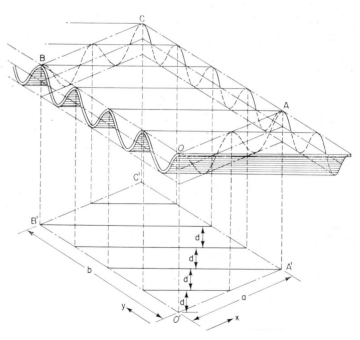

FIG. 71. The surface representing the 240 Fourier term with zero phase angle, i.e. positive. (After McLachlan.)

Since the Fourier terms are so closely related to the RL, which also governs the X-ray reflections, it is reasonable to suppose that the Structure Factor ($\mathbf{F}_{hkl}$) for the reflection hkl has some relation to the Fourier term hkl. In fact, it can be shown (Appendix III, p. 372) that

$$A_{hkl} = F_{hkl}/V \quad \text{and} \quad \alpha'_{hkl} = \alpha_{hkl} = \tan^{-1}\frac{B}{A}.$$

where V is the volume of the unit cell, so that

$$\rho_{\mathbf{r}} = \frac{1}{V}\sum_{hkl} F_{hkl}\cos\left(2\pi\frac{\mathbf{r}\cdot\mathbf{d}^*_{hkl}}{\lambda} - \alpha\right) \qquad (1)$$

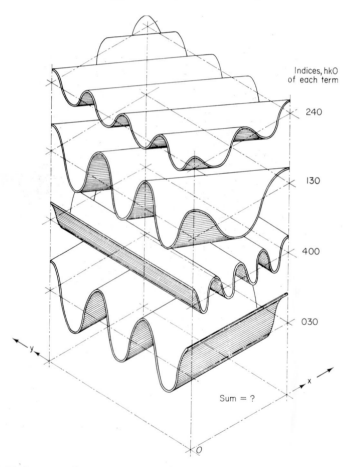

FIG. 72. Various two-dimensional positive Fourier terms in position to be added. (After MacLachlan.)

The amplitude of an X-ray reflection (per unit cell) is thus V times the amplitude of the corresponding Fourier term, and is unaffected by any of the other terms. An X-ray reflection hkl can therefore be thought of as arising from a sinusoidal distribution of electron density, with periodicity d_{hkl} and amplitude F_{hkl}/V, all other terms in the distribution giving destructive interference in this direction.

110(1) is similar in form to 16(1) and the electron density ρ_r in the unit cell is, mathematically, the Fourier transform of the RL weighted with F_{hkl} (c.f. 372(1) and (2)).

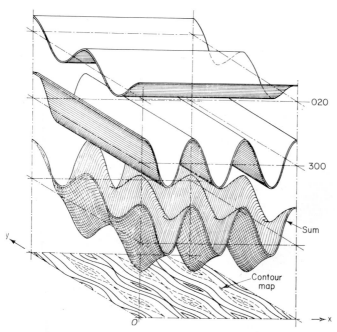

FIG. 73. The addition of two positive terms and the corresponding contour map. (After McLachlan.)

2. The Intensity of Reflection from a Real Crystal

Since, for Bragg reflection, the scattering from all unit cells (or lattice points) is in phase, the total scattered intensity $I = (Ma\,F)^2$ J mm^{-2} s^{-1} where M = number of unit cells in the crystal, and a^2 is the intensity scattered by one electron.

$$a^2 = \frac{I_0(e^2/mc^2)^2}{R^2}$$

where I_0 = incident intensity $(J\, mm^{-2}\, s^{-1})$

R = distance from crystal to measuring device for I (large relative to the dimensions of the crystal)

e, m = charge and mass of electron

c = velocity of light.

Therefore

$$I = I_0\, M^2\, F^2 \left(\frac{e^2}{mc^2}\right)^2 \cdot \frac{1}{R^2} \quad J\, mm^{-2}\, s^{-1}.$$

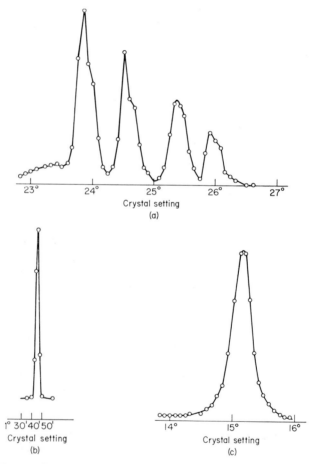

FIG. 74. Reflections from mosaic crystals showing the wide range of angular distribution of the mosaic blocks. The effect of the size of blocks is also contained in the intensity curves and cannot normally be disentangled. However, unless the block size is less than 0·1 μm (which is not usually the case) the size effect is small.

This is the maximum intensity of reflection, which rapidly falls to zero as the crystal is moved away from the reflecting position. However, this measurement cannot be used as it stands, since in almost all practical cases a crystal is a mosaic. This means that it consists of small blocks of perfect texture, separated from each other by fissures or imperfect regions, so that there is misalignment of the blocks; and those which do have exactly the same orientation and therefore reflect together, have only random phase relationships between them. The maximum intensity from such a crystal is thus dependent on the maximum number of blocks reflecting together, and also on the size distribution of the blocks (Section 1.2.2.5(b), p. 9). This may be very different in otherwise similar crystals (Fig. 74). However, if the total energy flux $I' = I \times$ (area of beam) is plotted against the angle of rotation of the crystal ϕ, through the reflecting position, then the area under the curve is constant, although for one crystal the peak may be tall and sharp, while for the other it is low and broad. (Fig. 75). This area is proportional to the total energy reflected during uniform rotation of the crystal through the reflecting position; and it is this quantity which is measured in practice.

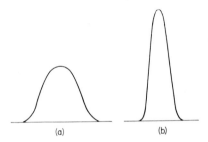

(a)   (b)

FIG. 75. The total energy reflected as the crystals move through the reflecting positions (proportional to the area under the curve) is the same for these two crystals, although the maxima are very different.

2.1. THE INTEGRATED REFLECTION

The derivation of the total energy E, is mathematically somewhat complicated. Appendix IX gives the version for the simplest case, and in this section only a general discussion of the factors involved, together with the formula resulting from the calculation, will be given.

Since the phase relations between mosaic blocks are random, and in any case only a small proportion are in the reflecting position simultaneously, the total energy E' for each block can be calculated, and then the energies for the various blocks added to give the total energy E for the whole crystal, i.e.

$$E = \sum E'.$$

For a single block the maximum intensity, at the Bragg angle, will fall off to zero at the first minimum on either side when the path difference between the waves reflected from the bottom and top planes is equal to λ. (Ans. 3, p. 453). Let the number of planes be N.

Path difference, $l = 2d \sin \theta$ for adjacent planes

$$\therefore \quad \Delta l = 2d \cos \theta \Delta \theta$$

$$\therefore \quad N\Delta l = 2 \, Nd \cos \theta \Delta \theta = \lambda$$

$$\therefore \quad \Delta \theta = \lambda/(2 \, Nd \cos \theta) = \lambda/(D2 \cos \theta)$$

where D is the thickness of the crystal normal to the reflecting planes. There will also be 'spread' of the reflection due to the finite area of the planes. This will also be proportional to λ, and inversely proportional to the dimensions of the crystal in the reflecting planes. In general, therefore, the effect of the 'spread' of the reflection might (with a certain amount of hindsight!) be expected to be proportional to

$$\frac{\lambda}{D_1} \times \frac{\lambda}{D_2} \times \frac{\lambda}{D_3} = \frac{\lambda^3}{\delta v}$$

where D_1, D_2, D_3 are the dimensions of the block in three axial directions, and $\delta v = D_1 D_2 D_3$ is the volume of the crystal.

The total energy will be proportional to the area of the beam at the measuring instrument, which, for a given divergence, is proportional to R^2.

The total energy will also be proportional to the length of time spent in the reflecting position, which is proportional to $1/\omega$, where ω is the angular velocity of the crystal rotating through the reflecting position.

Finally, the total energy might be expected to be proportional to the maximum intensity, I.

$$\therefore \quad E' = KI \frac{\lambda^3}{\delta v} \cdot R^2 \cdot \frac{1}{\omega} = \frac{K}{\omega} I_0 \, M^2 \, F_{hkl}^2 \left(\frac{e^2}{mc^2}\right)^2 \cdot \frac{1}{R^2} \cdot \frac{\lambda_3}{\delta v} \cdot R^2$$

$$= K \frac{I_0}{\omega} \frac{\lambda^3 \, F_{hkl}^2}{V^2} \cdot \left(\frac{e^2}{mc^2}\right)^2 \delta v \quad \text{(since } M^2 \, V^2 = \delta v^2, \text{ where } V \text{ is the unit cell volume).}$$

Reference to Appendix IX shows that there are no other factors involved, and that the constant of proportionality K, is unity. Since the only variable between blocks is δv and $\sum \delta v = v$ (the volume of the whole crystal):—

$$E = \sum E' = \frac{I_0}{\omega} \cdot \frac{\lambda^3}{V^2} \cdot F_{hkl}^2 \left(\frac{e^2}{mc^2}\right)^2 \cdot v$$

$$\therefore \quad \frac{E\omega}{I_0} = J = \frac{\lambda^3}{V^2} \cdot F_{hkl}^2 \cdot \left(\frac{e^2}{mc^2}\right)^2 \cdot v = Q' \cdot v. \tag{1}$$

This quantity J, which is independent of mosaic spread, uniform angular velocity or incident intensity is called the **Integrated Reflection.** Provided only relative values of F are required for structure analysis, E is the only quantity which needs to be measured, and then only on a relative scale, since, if I_0 and ω are kept constant, E and F_{hkl}^2 are the only variables.

In 99 per cent of cases only relative values of E are measured, because of the difficulty of absolute measurements, and the possibility of obtaining an approximate scale factor statistically and refining it during structure determination.

2.2. ABSOLUTE MEASUREMENT OF THE STRUCTURE FACTOR AMPLITUDE

In cases of special difficulty in the structure analysis it may be desirable to have absolute measurements of F. This involves the production of a monochromatic incident beam, either by using crystal reflected radiation (essential for photographic measurement) or indirectly by the use of balanced filters (Section **18**.1.1.2(ii), p. 257) and electronic discrimination, using counter methods. An area of the direct beam, accurately defined by a small hole in a plate placed in the position of the crystal, is required to measure I_0. The alternative is to use a crystal with a flat face larger than the beam, and to measure the whole beam; but this restricts very drastically the number of reflections that can be used. E must be measured on the same scale as I_0. The accurate determination of ω presents little difficulty, but the volume of the crystal v, is difficult to determine accurately unless the crystal is a good approximation to a sphere. The unit cell volume V will have already been determined, and the other factors in the equation are accurately known (fundamental constants and λ). It is thus possible to obtain absolute values of F, but the accuracy will depend on a large number of factors, some of which are likely to contain fairly large errors. However if only a few measurements can be made accurately, these can be used to put a set of relative measurements on an absolute basis.

3. Measurement of the Total Energy E

X-rays produce ionisation in materials on which they impinge. The measurement of X-ray energy is essentially the measurement of this ionisation, either directly as in ionisation chambers or counters, or indirectly by its effect of photographic emulsions.

3.1. IONISATION CHAMBERS, GEIGER AND PROPORTIONAL COUNTERS

The ionisation produced by X-ray photons in a gas was originally measured, by means of an ionisation chamber, as the total charge produced by all the photons absorbed during a given period; but the invention of counters, in which each photon produces an electrical pulse, increased the sensitivity enormously. The total number of photons, registered as a crystal moves through the reflecting position (corrected for absorption and geometrical factors), is a measure of the true integrated reflection. Such counters are potentially more accurate than photographic measurements; and although, in general, only one reflection can be registered at a time, the measurements can be made automatically by suitable electronic equipment. In the case of proportional counters (including scintillation counters) an electronic analyser can exclude pulses above and below the energies set by the 'gate', and produce a considerable degree of monochromatisation (Fig. 76).

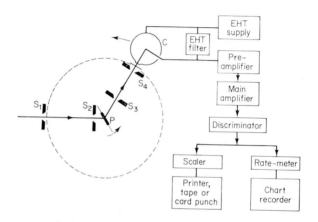

Fig. 76. Diffractometer using proportional counter. S—slits; P—crystal orienter; C—counter requiring Extra High Tension (EHT) supplies; and recording chain.

3.2. Photographic Methods

The optical density of blackening of a photographic plate is proportional to the incident X-ray energy over a limited range. The densities of the X-ray 'spots' are compared by eye with those of a standard scale, or a photometer is used to measure the maximum density. Visual comparison also tends to measure maximum density, although the eye may do a certain amount of automatic integration. Since the distribution of density within a spot, and the size of spot, varies, this is not a very accurate measure of the integrated reflection; but in many cases it is proportional to it within 10 per cent, which is the sort of accuracy usually claimed for such measurements. Integrating ('flying spot') photometers have been developed to overcome this difficulty, but these are complicated and expensive pieces of apparatus giving limited accuracy. The best method for most purposes is the use of the integrating Weissenberg or precession camera and photometry of the uniform centres of the spots (Section 18.1.2, p. 259).

4. Corrections to E

4.1. POLARISATION OF REFLECTED AND INCIDENT BEAMS

The electron scattering formula used above was for a beam of X-rays polarised with the electric vector perpendicular to the plane of reflection (i.e. parallel to the reflecting planes, Fig. 77). For X-rays polarised in the plane of reflection the amplitude of scattering is reduced by a factor of $\cos 2\theta$. For unpolarised X-rays, i.e. X-rays in which the direction of the electric vector is varying randomly with time, the reflected amplitude is a combination of the two extreme cases.

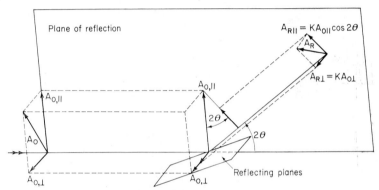

FIG. 77. Diagram showing the effects of the components of the incident amplitude in, and perpendicular to, the plane of reflection.

Consider an incident wave train with electric vector, A_0 (Fig. 77), and its two components, $A_{0\parallel}$ and $A_{0\perp}$. If the reflected component $A_{R\perp} = KA_{0\perp}$, then $A_{R\parallel} = KA_{0\parallel} \cos 2\theta$, since only the resolved part of the electron vibrations caused by the direct beam is effective in producing $A_{R\parallel}$.

But $A_R^2 = A_{R\perp}^2 + A_{R\parallel}^2 = K^2 (A_{0\perp}^2 + A_{0\parallel}^2 \cos^2 2\theta)$.

Since A_0 is varying randomly in direction with time,

$$\overline{A_{0\perp}^2} = \overline{A_{0\parallel}^2} = \tfrac{1}{2}A_0^2 \quad \text{and} \quad \overline{A_R^2} = K^2 \frac{A_0^2}{2} (1 + \cos^2 2\theta).$$

Therefore the intensity observed is reduced by the factor $(1 + \cos^2 2\theta)/2 = p$, from that which would be obtained for X-rays polarised perpendicular to the plane of reflection. p is called the polarisation factor, and the observed E must be multiplied by p^{-1} to give the theoretical value. If the incident beam is partially polarised (e.g. by reflection from a crystal monochromator) the expression is more complicated but is obtained from similar considerations.

(Appendix V, p. 381). In the case of unpolarised incident radiation the correction factor is independent of instrument geometry, and is the same for both powders and single crystals.

4.2. VELOCITY (OR 'LORENTZ') FACTOR

The expression for E was derived for $\theta = 45°$. For any other value of θ the relp passes through the sphere with a smaller velocity along the radius, and the reflection occurs over a longer time. The observed E must be multiplied by a factor which is the velocity of the relp along the radius of the sphere of reflection relative to the velocity for $\theta = 45°$. The inverse factor for zero layer rotation photographs ($1/\sin 2\theta$) was originally given the name 'Lorentz factor' and the symbol L. The symbol has been retained for the general case, and the observed E is multiplied by L^{-1}. Since L and p are both purely geometrical factors, they are usually combined into a single Lp factor, and the observed E is multiplied by $(Lp)^{-1}$. L however is dependent on the instrument geometry (Appendix V, p. 381) whereas p is only dependent on θ unless crystal reflected, i.e. partially polarised, incident radiation is used.

4.3. ABSORPTION

X-rays are absorbed by matter and the energy transformed in various ways. Thus, the parts of a crystal furthest from the X-ray source reflect a beam which is less intense than that incident on the near parts. This reduces the reflected energy below the theoretical value, which was obtained on the assumption that absorption did not occur. The factor A by which it is reduced, is given by $A = 1/V \int e^{-\mu L} \, dv$, where μ is the linear absorption coefficient, L is the path length in the crystal of a beam diffracted from the volume dv, and v is the volume of the crystal. The observed E must be multiplied by $A^* = A^{-1}$ to give the theoretical value. A is difficult to calculate except for simple geometrical forms (Section **18**.6.4, p. 282), although this can now be done by computer if the shape of the crystal can be determined with sufficient accuracy.

4.4. MULTIPLICITY FACTOR

Multiplicity occurs to some extent in rotation photographs, where a number (p') of equal (symmetry related) reflections fall on top of one another. The observed E must be divided by p' (the single crystal multiplicity factor) in each case. If reflections unrelated by symmetry fall on top of one another, it is necessary to take a series of oscillation photographs instead of a complete rotation.

In powder photographs the reciprocal sphere is composed of relps at the corresponding distance from the origin in the RL of a single crystal (Section 7.1, p. 76). If these points are symmetry related, the observed E' (energy

s^{-1}) in the powder ring must be divided by this number (the powder multiplicity factor, p''). If non-symmetry related spheres overlap there is no direct way of separating out the various contributions.

4.5. GEOMETRICAL FACTORS FOR POWDER DIFFRACTION

The total energy E (diffracted during exposure time, t) is measured for a fixed length of powder line, but the size of the ring varies with θ. The observed E must be multiplied by the ratio of the diameter of the ring to that for $\theta = 45°$. The proportion of the total number of crystallites which are in the reflecting position also varies with θ, and a correction factor must be applied to allow for this effect (Appendix V, p. 381).

5. Other Factors which are Difficult to Correct for Experimentally

5.1. EXTINCTION—PRIMARY

This occurs when the mosaic blocks are larger than the maximum size for which simple diffraction theory is a good approximation (Section 1.2.2.5, p. 8). The effect is to reduce the observed value of E for the most intense reflections. (Although the *maximum* intensity of a large block is *increased* compared to that of an equal volume of smaller, parallel blocks, the width of the reflection is reduced to an even greater extent, so that E, the area under the curve, is *reduced*).

5.2. EXTINCTION—SECONDARY

This is due to reflection in the upper mosaic blocks reducing the intensity reaching lower ones, which are also in the reflecting position. The effect is an apparent increase in the normal absorption coefficient, by an amount proportional to the intensity of the reflection, and, for an unpolarised incident beam, to $(1 + \cos^4 2\theta)/(1 + \cos^2 2\theta)^2$. It thus also has most effect on the strong reflections.

It is very difficult either to differentiate between primary and secondary extinction in practice, or to determine the amount present, except by using powder diffraction with very fine powder ($< 1 \ \mu m$) to measure at least some of the low angle reflections, or by using polarised X-rays [3].

The deficiency lines of divergent beam photographs (Section 5.1, p. 40) are due to secondary extinction, and become almost invisible if much primary extinction is present. This enables qualitative judgements to be made of their relative importance.

5.3. THERMAL MOTION OF THE ATOMS

This has the effect of 'smearing out' the atoms and, by causing a certain amount of destructive interference between the equivalent atoms in the various unit cells, reduces the value of the atomic scattering factor f, by the factor $\exp[-B(\sin^2\theta)/\lambda^2]$, thus causing the intensity to fall off more rapidly with θ.

$B = 8\pi^2\bar{u}^2$ where $\bar{u}$ is the r.m.s. displacement of the atom from its rest position. The movement is assumed to be spherically symmetrical, but in most cases the anisotropy of the arrangement of surrounding atoms means that $\bar{u}$ varies with direction. Since for any reflection it is only the displacements perpendicular to the reflecting planes which affect the intensity, the anisotropic temperature factor must be a function of the indices. If the ends of the vectors $\bar{u}_{hkl}$ (where $\bar{u}_{hkl}$ is the r.m.s. displacement in the direction of $\mathbf{d}^*_{hkl}$) are assumed to lie on an ellipsoid centred on the rest position of the atom, the factor has the form $\exp - (B_{11}h^2 + B_{22}k^2 + B_{33}l^2 + B_{12}hk + B_{23}kl + B_{31}lh)$. The meaning of the coefficients B_{ij} and the restrictions on their values are given in Appendix VI (p. 390).

Except in the simplest cases it is not possible to calculate the value of B or B_{ij} directly. But the statistical method used to obtain a scale factor can also give a rough estimate of B; and intensity measurements at a number of different temperatures enable a direct, if partial, analysis to be made. The temperature factor is usually refined by least squares procedures along with the other structure parameters; but the decision to employ anisotropic temperature factors adds five additional parameters per atom to the refinement.

Thermal diffuse scattering (or phonon scattering) also arises from thermal motion of the atoms, and gives a non-uniform background (including large contributions under the Bragg peaks) which is difficult to correct for, and which, if not allowed for, tends to give an apparent reduction of the normal temperature effect on the Bragg reflections. Only use of the Mossbauer effect can, with difficulty, distinguish between Bragg reflection and phonon scattering, utilising the extremely small change of wavelength arising from exchange of momentum between photon and phonon in diffuse scattering. Thermal scattering was responsible for the diffuse areas of blackening in Fig. 52, (p. 74).

5.4. MULTIPLE ('RENNINGER') REFLECTIONS

These occur if two or more relps pass through the sphere of reflection at the same time. In Fig. 78 the two reflections 101 and $30\bar{1}$ occur together. The axes of the RL are shown as full lines centred on the origin 000, at the end of the direct beam diameter. Any reflection occurring in one part of the crystal acts as a weak incident beam in that part of the crystal which it traverses before emerging. Taking the 101 reflection as such a secondary incident beam, we must shift the origin of the RL to a position at the end of this beam, as diameter of the sphere. The new origin (000) and relp coordinates (indices) are shown in parentheses and the axes as dashed lines. Since the RL moves parallel to itself through a distance given by a lattice vector (in this case

$1a^* + 1c^*$) the position of the lattice grid does not change. If no other relp were in the sphere there would be no reflection arising from this secondary incident beam. But if, as in this case, another relp is in the sphere, it will be giving rise to a reflection $30\bar{1}$, from the primary incident beam, and to another reflection in the same direction (from the centre of the sphere C to the relp) from the secondary incident beam. But for this secondary incident beam the relp coordinates, and therefore the indices of the reflection, are referred to the new origin. The reflection is therefore $(20\bar{2})$. In other words, the double reflection of the incident beam, first in the planes 101, and then, in another part of the crystal, in the planes $20\bar{2}$, gives a doubly reflected beam in the same direction as the $30\bar{1}$ primary reflection. (Clearly $30\bar{1}$ can act as a secondary incident beam giving a reflection in the same direction as primary 101). In general the doubly reflected beam will be weak; but if 101 is a strong reflection and $20\bar{2}$ is also strong, but $30\bar{1}$ is very weak, then the doubly reflected contribution may be a very significant part of the measured intensity of $30\bar{1}$. If the primary relp corresponds to a space group absence, the occurence of a double reflection may lead to an incorrect determination of the space group.

If n relps are in the sphere together, each primary reflection will be accompanied by $n - 1$ double reflections, giving rise to multiple reflections.

The relps considered in Fig. 78 were taken in the zero layer for convenience, but the same considerations clearly apply in three dimensions (**Example, 18, p. 296**).

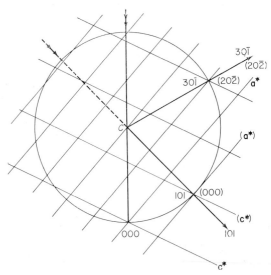

FIG. 78. Diagram showing double reflection occurring as a result of two relps being in the sphere of reflection simultaneously.

On a 4-circle diffractometer it is possible to rotate a crystal about a RL vector such as $d^*_{3\,0\,\bar{1}}$ with the $30\bar{1}$ relp in the sphere. The variation of intensity as other relps come through the sphere can be quite considerable; but the integrated intensity obtained in a position giving a minimum, should have a negligible contribution from multiple reflection.

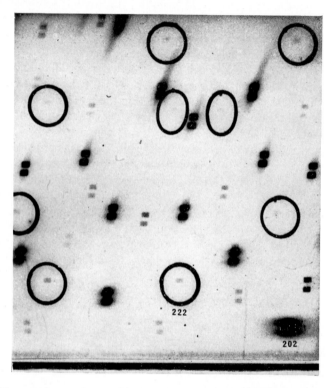

FIG. 79. Two nearly superimposed halves of second layer Weissenberg photographs of paramelaconite; one with the correct equi-inclination angle, μ, and the other (displaced downwards) with μ mis-set by $0.5°$. The rotation axis was a, and the strong 200 reflection in the correctly set photograph has produced the streak at the bottom of the figure (the centre of the complete photograph). All reflections are doubled due to the near super-position of the two photographs, except those ringed, and the single reflection in these cases is due to double reflection, first of the direct beam by the (200) planes and then of the 200 reflection by, e.g. the (022) planes, giving a reflection in the same direction as the 222 reflection from the direct beam. The 222 reflection is of zero intensity and the reflection in the marked ring is entirely due to double reflection. The crystal is tetragonal and 022 has the same intensity as 202, which is the strong reflection marked on the photograph. Two strong reflections can thus give a double reflection comparable to a weak single reflection, and give rise to considerable errors in such reflections. Some of the pairs of weak reflections (not ringed) show differences in intensity which are almost certainly also due to this effect. (The faint central streak from a small residue of the 200 reflection on the 'mis-set' photo-graph has been cut off in reproduction. It was similar in intensity to the air-scattering streak above the heavy central line).

In the use of an equi-inclination Weissenberg camera (Section **15**.2.2, p. 199), if there is an orthogonal crystal axis along the rotation axis, the RL axis will also lie in this direction, and an axial relp will be continuously in the sphere, so that double reflections will occur in non-zero layers whenever another relp comes into the sphere of reflection. In fact there will be triple reflections, because a zero layer relp will also be in the sphere. For this reason, especially if the axial reflection is intense, the equi-inclination angle should be mis-set by $0.5°$. (Fig. 79).

5.5. RESONANCE EFFECTS (ANOMALOUS SCATTERING)

The normal tabulated atomic scattering factors are calculated on the assumption that the binding energies of the electrons are so small that their scattering power can be calculated as though they were free electrons. But bound electrons can have scattering amplitudes that are larger or smaller than for a free electron, and the phase of the scattered radiation may be different. These effects are particularly important when the frequency of the incident radiation is near to a resonance frequency of the atom, i.e. when the wavelength is near to that for an Absorption Edge (usually for the K shell; but for large atomic number it may be the L shell). [1(65)]

If the frequency of the incident radiation is much greater than the resonance frequency of the atom, the amplitude is similar to that of a free electron, and there is a phase lag of π between the oscillations of the electron and those of the electric field of the incident radiation. As the resonance frequency is approached the phase lag decreases to $\pi/2$ at resonance, and below resonance the lag decreases towards zero. Since we are not interested in the phase difference between the incident and scattered radiations, but only in the relative phase differences between the various parts of the scattered radiation, we take the free electron scattering as reference (zero phase angle). At resonance the scattered radiation has a phase *advance* of $\pi/2$, and for an incident frequency well below resonance there is a phase advance of π.

Those electrons in the atom whose resonance frequencies are well below that of the incident radiation will have zero phase angle; those with resonance frequencies below but near the incident frequency will have phase angles between 0 and $\pi/2$; those with resonance frequencies above the incident frequency will have phase angles between $\pi/2$ and π. The amplitude will also be increasing as resonance is approached and falls again on the other side.

For light elements (up to Mg or Al for $CuK\alpha$ radiation) the effect is very small. For slightly heavier elements (up to vanadium for $CuK\alpha$) the scattered amplitude increases, and the phase angle begins to depart appreciably from zero. Figure 80 shows diagrammatically these two cases, and also two examples near to but on either side of resonance, which correspond to elements whose atomic numbers are near to that of the X-ray tube target material.

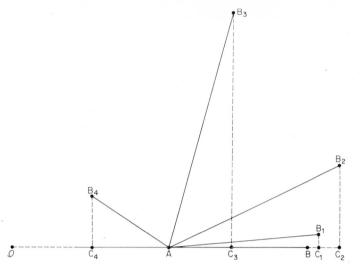

Fig. 80. The complex atomic scattering factor. OA represents on the Argand diagram the resultant contribution of all electrons outside the K-shell of an atom of low to medium atomic number. For incident frequencies a long way above resonance AB is the resultant contribution from the K shell and OB is the normal, tabulated, atomic scattering factor, f_0. For incident frequencies well above resonance there may be a small effect with the K amplitude increasing slightly to AB_1 and a small phase advance B_1AB giving rise to a correspondingly small imaginary part of the structure factor C_1B_1. Not too far above resonance the K contribution is AB_2 with a considerably larger amplitude and phase angle. Near, but still above resonance there is a much larger amplitude AB_3 and a phase angle nearly $\pi/2$. Just below resonance the amplitude decreases considerably to AB_4 and the phase angle approaches π. When the incident frequency is much below resonance for the K shell, the L shell resonance becomes important and the effect is a combination of the two leading to similar effects when the incident frequency comes near to the L resonance frequency. The resultant amplitude OB_i has not been drawn in to avoid confusion.

The tabulated values of the atomic scattering factor f_0, can be modified to take account of resonance effects by adding a real part $\Delta f'$, and an imaginary part $\Delta f''$. The scattering factor then becomes

$$f = f_0 + \Delta f' + i\Delta f'' \quad \text{and is complex.}$$

In Fig. 80,

$$BC_i = \Delta f \quad \text{and} \quad C_iB_i = \Delta f''.$$

Values of $\Delta f'$ and $\Delta f''$ have been tabulated for various wavelengths and $\theta = 0$. Because the electrons concerned are in the inner shells, with only small distances between them compared with the wavelengths of X-rays, variations with θ are negligible. $\Delta f'$ is negative except when the resonance frequency is well below that of the incident radiation, and it is then very small. $\Delta f''$ is always positive, corresponding to a phase advance of between 0 and π.

There are three ways in which these correction terms can be used. The simplest, which is adequate for most purposes, at least in centrosymmetric structures with no atoms very near resonance, is to take account of $\Delta f'$ and ignore $\Delta f''$, i.e. to take the real part of the actual scattering factor. Or, instead of taking the real part as the amplitude, with zero phase angle, the actual amplitude, $+\sqrt{(f + \Delta f')^2 + \Delta f''^2}$ can be taken with zero phase; though it is doubtful whether any real advantage is gained for this extra complication. Or the fact that the phase of the scattering factor is not zero can be taken into account in the calculations of:

$$A = \sum_n f_n \cos \phi_n; \qquad B = \sum f_n \sin \phi_n$$

where ϕ_n is the phase angle of the wave scattered from the nth atom, and A and B are the components of $\mathbf{F}$ (Section 9.1.1, p. 93). These are replaced by (Fig. 81):

$$A = \sum_{n-m} f_n \cos \phi_n + \sum_m (f_m + \Delta f_m') \cos \phi_m + \sum_m \Delta f_m'' \cos (\phi_m + 90)$$

$$B = \sum_{n-m} f_n \sin \phi_n + \sum_m (f_m + \Delta f_m') \sin \phi_m + \sum_m \Delta f_m'' \sin (\phi_m + 90)$$

where $n - m$ atoms are well below resonance and have zero correction factors, m atoms are near or above resonance and require correction, although some $\Delta f''$ may be small.

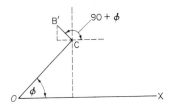

FIG. 81. The effect of a complex atomic scattering factor.
$$OC = f + \Delta f', \qquad CB' = \Delta f''.$$

In the centrosymmetric case

$$\sum \Delta f_m'' \sin \phi_m = \sum f_n \sin \phi_n = \sum (f_m + \Delta f_m') \sin \phi_m = 0$$

so that the only correction affecting A is $\Delta f'$. B would equal zero, apart from resonance, but with resonance $B = \sum \Delta f'' \cos \phi_m$. This will normally be a small term, and unless A is also small for a particular reflection it will have little effect on $F = \sqrt{(A^2 + B^2)}$. Since the effect will be the same for $\phi' = -\phi$, $F_{hkl} = F_{\bar{h}\bar{k}\bar{l}}$ as would be expected, and the only detectable effect will be on the intensity of a few weak reflections, compared with calculated intensities without correction for $\Delta f''$.

In the non-centrosymmetric case, however, significant differences may arise between F_{hkl} and $F_{\bar{h}\bar{k}\bar{l}}$ as shown in Fig. 82. The measurement of the differences resulting from this breakdown of Friedel's Law is the basis for the determination of absolute configurations in asymmetric structures, which crystallise in non-centrosymmetric space groups. The full correction can now be readily programmed for electronic computers, and should be used in all accurate structure work.

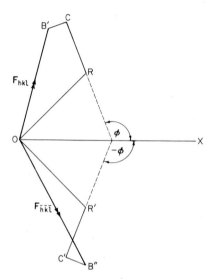

Fig. 82. Structure Factors for hkl and $\bar{h}\bar{k}\bar{l}$ in a non-centrosymmetric structure. OR and OR' are the resultant scattering for those atoms with zero resonance correction, for reflections hkl and $\bar{h}\bar{k}\bar{l}$. $RC = R'C' = f + \Delta f'$; $CB' = C'B'' = \Delta f''$ for the atom with a complex scattering factor. The combined resultant $OB' = F_{hkl}$ is obviously shorter than $OB'' = F_{\bar{h}\bar{k}\bar{l}}$ and the reflections can be distinguished by their differing intensities. The extension to the case with more than one anomalous scatterer is obvious.

6. Use of Corrected Data

When the most accurate possible intensity measurements of indexed reflections have been made and corrected as far as possible, the data, now in the form of relative or absolute F_0^2 values, are ready for use in determining the structure.

Chapter 10

Further Aspects of Lattice Theory

There are two developments of lattice theory which are required in the practical applications of Part II.

1. Modifications to Lattice Theory Due to the Use of Centred Cells

Until now we have tacitly assumed a primitive unit cell, but it is sometimes necessary to use centred $(A, B, C, I, F$ or $R_{Hex})$ cells [4(221); 5(7, 20)] in order that the lattice unit may have the same symmetry as the crystal, or otherwise be more convenient for calculation (Appendix II, p. 342). The penalty is a complication of lattice theory.

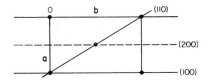

FIG. 83. Outline of a C-face centred cell with the traces of various planes.

Consider a C face centred cell, the base of which is outlined in Fig. 83, together with the trace of the nearest (100) plane to the origin. For the Bragg reflection 100, the path difference between this plane and the one through the origin is λ. But since all lattice points are equivalent, exactly similar planes of points lie through the centring points, half way between planes of the 100 set, giving a path difference between neighbouring planes of $\lambda/2$ and complete destructive interference. In the strict sense the (100) set of lattice planes does not exist, since it does not include all the lattice points. The RL point 100 is also 'non-existent', since no reflection can arise from it. However 200 does exist, since the set of planes contains all the lattice points; 300 does not; 400 does, and so on. Since the inclination of the planes to the c-axis does not affect the issue (if the centring point is included in one unit cell, it must also be in all the others) it follows that $h0l$ reflections are only present for h even. hhl reflections are all present since the $11l$ set of planes contains the centring point. In general relps can be divided into two groups, according as the

127

centring crystal lattice point is included in the corresponding set of planes or not. It is easily verified that the first group has $h + k = 2n$ and the second, which corresponds to 'systematically absent' reflections, has $h + k = 2n + 1$. Such 'systematic absences' may not be noticed until indexing of the reflections has been carried out, but they can be recognised on photographs taken a layer at a time. In Fig. 27b, (p. 36) the layers of relps in the first layer (reflections in the crescent) lie half way between those for the zero layer due to the body centring of the RL. If the 'non-existent' points of the RL are struck out it will be found that the remainder form a centred lattice, corresponding to the centred crystal lattice. For face-centring the RL has the same face centred as has the crystal lattice, but I (crystal) produces F (RL) and F (crystal) produces I (RL). Although such centring of the RL may be recognised on X-ray photographs, it must be emphasised that, from the point of view of indexing, *the RL is always primitive*. The centring of the crystal lattice is expressed in the *systematic absences* of reflections found during indexing on the basis of the primitive RL. For a body-centred (I) crystal lattice, the absent reflections have $h + k + l = 2n + 1$. For the all face-centred (F) lattice reflections are absent if hk and l are a mixture of odd and even numbers, i.e. reflections can only occur for hkl all even or all odd.

For a rhombohedral lattice indexed on a centred hexagonal cell (R_{Hex}), reflections are only present for $-h + k + l = 3n$ (for the standard 'obverse' setting, Fig. 130, p. 207) and $h - k + l = 3n$ for the alternative 'reverse' setting. The justification for these formulae for systematic absences, as well as of those absences caused by translation symmetry elements in the space group symmetry of the crystal, is given in Appendix II, (p. 342).

Another consequence of the use of centred cells is that central lattice lines, containing centring points, will have a point separation which is a half (or a third in the hexagonal case) of the separation calculated for a primitive cell. The corresponding layer lines on a photograph with such a direction as rotation axis will therefore be twice (or three times) as far apart as would be the case for a primitive cell. The relps in the intervening layers all correspond to 'systematic absences'. Nevertheless, taking 'real' relps (i.e. those corresponding to sets of planes which can produce reflections) and including centring points in the crystal lattice, the reciprocal relation between planes of points in one lattice and the perpendicular line of points in the other still holds. In particular, if a crystal lattice axis (c say) is along the rotation axis, then $c = \lambda/\zeta$ where ζ is the interplanar spacing obtained from the rotation photograph, *whether the cell is centred or not*.

2. The Symmetry of the Reciprocal Lattice

Three cases can be considered: (1) the RL consisting of points of equal weight; (2) the RL of points weighted with the values of F_{hkl} or F_{hkl}^2; and

(3) the RL of points weighted with the values of F_{hkl} (i.e. amplitude and phase angle).

2.1. POINTS OF EQUAL WEIGHT

The lattice is periodic, and has the same symmetry as the crystal lattice from which it derives, i.e. through each lattice point pass the symmetry elements of the highest class of the system to which the crystal belongs. The symmetry elements through the origin are important in considering multiplicity in powder diffraction (Section 7.1, p. 76) in conjunction with the symmetry in Section 2.2.

2.2. POINTS WEIGHTED WITH F OR F^2.

This case is important in connection with the problem of deciding how much of the RL needs to be investigated, and of finding the multiplicity of reflections in single crystal rotation and in powder photographs. The lattice is no longer periodic and all symmetry elements pass through the origin. The symmetry elements are those of the class to which the crystal belongs, with the addition of a centre of symmetry if it is not already present. This symmetry is known as the Laue symmetry of the RL (or of the crystal) because it is the symmetry which shows up in X-ray photographs.

The Laue symmetry of the RL as a whole can be at least partially reconstructed from the two-dimensional symmetry of the nets perpendicular to a given axis, which is all that Weissenberg or retigraph photographs can show (Section 15.3, p. 202).

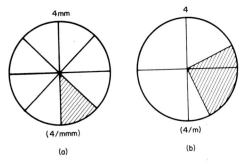

FIG. 84. Laue symmetry and the Reciprocal Lattice. Point group symmetry above, Laue symmetry in parentheses underneath.

As an illustration of the use of the Laue symmetry of the RL consider two crystals belonging to the classes $4mm$ and 4. The Laue symmetry will be $4/mmm$ and $4/m$, since the addition of a centre to an evenfold axis produces a plane perpendicular to the axis. Figure 84 shows the RL within the limiting sphere, together with the symmetry elements present. Since there is a plane of symmetry in the plane of the paper, only the top half of the sphere need be considered. In case (a) only a 45° segment need be investigated, *but it must be the segment between two vertical symmetry planes*. In case (b) twice as many reciprocal lattice points must be investigated, but any 90° segment will suffice.

For determining multiplicity, which varies with the co-ordinates of the relp, all the Laue symmetry related relps are found; and the number of such relps (in the individual layer for a rotation photograph and in the lattice as a whole for a powder) is the multiplicity. In practice the relps are divided into sets, e.g. those not on a symmetry plane or axis; those on a particular symmetry element; those at the intersection of two symmetry elements, etc. Each set will have a specific multiplicity.

It is then necessary to consider the geometrical symmetry (Section 2.1), to see whether there are other relps which are related by the geometry of the lattice, and therefore have the same d^*_{hkl}, but are not related by the Laue symmetry. These two sets of reflections will fall on top of one another and cannot be separated out by any direct method. (**Example 19**, p. 296).

2.3. POINTS WEIGHTED WITH F

In this case the RL is a complete description of the structure, and can be transformed mathematically into the crystal lattice and electron density distribution in the real cell (Section 9.1.4.3, p. 111). Its symmetry is of importance when this transformation is being carried out by two- or three-dimensional Fourier synthesis. The symmetry can be divided into two parts. That of the absolute amplitude is, of course, identical with that of Section 2.2, but that of the phase angle depends on the space group symmetry. It there are no translational elements of symmetry (glide planes or screw axes) and all space group symmetry elements pass through lattice points (except, of course, the derived elements half way between lattice points), then the symmetry of phase angles is the same as that of the crystal class (i.e. without the addition of a centre of symmetry which occurs in the case of the amplitudes). If translational symmetry elements occur, or any symmetry element does not pass through the origin, then the symmetry of the phase angles will depend on the coordinates of the relps (i.e. the indices). For relps related by inversion through the origin (i.e. hkl and $\bar{h}\bar{k}\bar{l}$) the phase angle relation is always $\alpha_{hkl} = -\alpha_{\bar{h}\bar{k}\bar{l}}$.† The reason for this is easily seen in terms of the Argand diagram. If d^*_{hkl} in 99(1) becomes $d^*_{\bar{h}\bar{k}\bar{l}} = -d^*_{hkl}$, then all the phase angles for scattering from the atoms change

† Anomalous dispersion (Section 9.5.5, p. 123), which causes this relation ($F_{hkl}=F_{\bar{h}\bar{k}\bar{l}}$; $\alpha_{hkl}=-\alpha_{\bar{h}\bar{k}\bar{l}}$—known as Friedel's law) to break down, is excluded from this discussion.

sign. The Argand polygon turns into its image, reflected in the line OX. The amplitude of the resultant is the same, but its phase angle is also reflected in OX, i.e. changes sign.

If a centre of symmetry exists at the origin of the space group, then $\alpha_{hkl} = \alpha_{\overline{hkl}}$ and therefore α_{hkl} is either 0 or π. Buerger [8] has considered the problem of the phase angle symmetry and derived the relationships which define it. As an example consider the simple case of Pm and Pc. The Laue symmetry $2/m$ and point group symmetry m, are the same for both. In Fig. 85 the symmetry plane relates a point hkl to $h\overline{k}l$, and a centre of symmetry relates this pair to $\overline{hk}\overline{l}$ and $\overline{h}k\overline{l}$. It follows that in both cases,

$$F_{hkl} = F_{h\overline{k}l} = F_{\overline{hk}\overline{l}} = F_{\overline{h}k\overline{l}} \quad \text{and for } Pm \quad \alpha_{hkl} = \alpha_{h\overline{k}l} = -\alpha_{\overline{hk}\overline{l}} = -\alpha_{\overline{h}k\overline{l}}$$

since the space group symmetry plane is non-translational and passes through the origin.

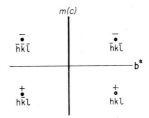

FIG. 85. One set of symmetry related relps in the space groups Pm and Pc.

For Pc the symmetry plane is translational and the following relationships arise:

$$\alpha_{hkl} = \alpha_{h\overline{k}l} = -\alpha_{\overline{hk}\overline{l}} = -\alpha_{\overline{h}k\overline{l}} \quad \text{for } l \text{ even, and}$$

$$\alpha_{hkl} = \pi + \alpha_{h\overline{k}l} = -\alpha_{\overline{hk}\overline{l}} = \pi - \alpha_{\overline{h}k\overline{l}} \quad \text{for } l \text{ odd.}$$

These relationships, derived analytically, are listed, in part, in Vol. I of the International Tables of Crystallography [5(374)].

Part II

Instrument Geometry and Interpretation

Chapter 11

Single Crystal Cameras

1. Hazards and Safety Precautions

The X-ray beam from the open window of an X-ray tube is extremely dangerous. An overdose of radiation can be received in a matter of seconds. Even if a camera is in front of the window scattered radiation will be intense, although the direct beam may be stopped a few inches from the window. In such circumstances the fingers are especially vulnerable, since they are likely to be manipulating near the front of the camera, and will intercept the beam at its most intense point, just outside the window. Even a fraction of a second in the beam here will produce a bad skin burn which will take months to heal, even if it does not produce any long term effects.

It is therefore essential that interlocks be provided to prevent an X-ray tube window being open (with the H.T. set on) except when a properly shielded piece of apparatus is against it. A mechanical interlock is most reliable, and details of one form of this, together with a scheme for the optical adjustment of cameras to give the most intense beam of X-rays through the collimator, and also allow immediate interchange of cameras, is detailed in Appendix VIII (p. 408).

In Britain the use of X-rays in industry is governed by *The Ionising Radiations (Sealed Sources) Regulations* 1969, and the Department of Employment and Productivity (DEP) publishes a useful booklet *Ionising Radiations; Precautions for Industrial Users* (HMSO), which contains a section on X-ray diffraction. Teaching is not covered by the legal requirements of the Regulations, but the DEP has established a *Code of Practice for the Protection of Persons Exposed to Ionising Radiations in Research and Teaching*, and operates a voluntary inspection system. Those engaged in research or teaching in X-ray crystallography should make sure that their conditions of work conform to the Code of Practice or to similar directives in other countries.

134

2. Rotation (and Oscillation) Camera

The requirements for the simplest form of X-ray camera are as follows:—

(i) A spindle (normally vertical) driven at uniform angular velocity in rotation or oscillation, with a scale and vernier to measure the rotation angle.

(ii) A collimator to define the beam of X-rays (0·25–0·5 mm in diameter), adjusted so that its axis intersects the spindle axis at right angles.

(iii) A pair of arcs attached to the spindle which are capable of angular adjustment over a limited range (±20–30°), with vernier measurement to 0·1°. The bottom arc has its axis perpendicular to the rotation axis. The top arc is carried on the slide of the bottom arc, and its axis is only perpendicular to the rotation axis if the bottom arc is at zero (Fig. 86). The specimen is attached to a slider on the top arc.

(iv) Means of adjusting the position of the specimen (which should normally be small enough to be bathed in the beam) so that its centre lies, (a) on the spindle axis, and (b) on the collimator axis.

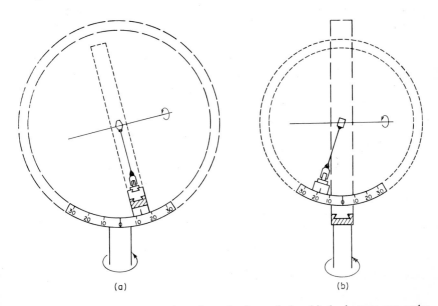

(a)   (b)

FIG. 86. Diagrammatic representation of a pair of arcs facing (a) the bottom arc scale; (b) the top arc scale. The 'arcs' are parts of complete circles (shown by the dashed lines). In deciding which way to move the slides for a given angular correction it is helpful to visualise the complete circle. The section of the arc which is perpendicular to the plane of the paper is shaded and the rotation axis for that arc is drawn through the crystal. In (b) the circle is not in the plane of the paper unless the bottom arc is reading zero. For most purposes the inclination of the top 'arc' can be ignored. The error so introduced will only be 2% for a bottom arc reading (i.e. a top arc inclination) of 10°, while for 25° it is still only 10%.

(a)

Fig. 87. Photographs of an oscillation (and rotation) camera:— (a) with a cylindrical film holder (1) in position, the microscope (2) in the normal position opposite the collimator, and the camera withdrawn from the tube to enable the coupling arm to be lowered for removal of the collimator.

The spindle height adjuster (3) can only be used after the set screw under the scale (19) (Fig. 87c) has been released with the removable key (20). For oscillation, the jockey pulley arm (4) is locked back as shown, to slacken the belt (5), and the oscillation arm (6) replaced on the rod projecting from the collar above the scale, if it has been removed. The cam follower (7) is locked at the desired height, opposite one of the 5, 10 and 15° cams (8) which are driven by a synchronous motor under the base. The cord (9) with its attached weight, is hooked to the oscillation arm, and runs over the pulley (10) to hold the cam follower against the lightly greased cam. A paper cylinder for the weight is a useful precaution. An oscillation should be started with the cam follower on the point of the cam as shown, and the collar above the scale locked to the spindle at the desired reading of the scale.

(b) with a flat plate film holder (11) in position on the rail (12) replacing the microscope which has been moved to the L.H.S. The lens (13) is used to read the scale and vernier, illuminated by the lamp (14). The camera is in position against the tube with the shutter open, locking the camera to the tube. The collimator (15) and the arcs (16) can be clearly seen.

For this illustration the oscillation arm (6) has been withdrawn, and the jockey pulley (4) released on to the belt which drives the spindle in continuous rotation. With a flat plate film holder the arm would normally be used to hold the spindle in a specified position.

(c) with the back reflection film holder (17) and collimator (18) screwed into the plate, replacing the normal collimator bracket. The scale (19) and removable key (20) for locking the spindle height or oscillation collar can be seen.

An optical collimator (21) can be attached to either side of the camera, and the front lens of the microscope raised to turn it into a telescope. The instrument can then be used as a single circle optical goniometer.

(b)

(c)

(a) can be achieved by two linear slides, giving translations at right angles to each other and (at least approximately) to the spindle axis. The slides can either be between the spindle and the arcs or between the top arc and the crystal. The former is easier to construct, has a larger range of adjustment and the arcs need not be restricted to having their centres coincident. However, every arc adjustment will move the crystal off the spindle axis, and requires to be followed by 're-centring', using the linear slides. If the linear slides are placed above the top arc, and the arc centres are coincident and on the spindle axis, the arcs are said to be 'eucentric'. Adjustment of the crystal to the right height above the arcs (by the use of a gauge, and an auxiliary height adjuster provided above the slides) and on to the spindle axis, automatically brings it to the coincident centres of the two arcs. Subsequent arc adjustment does not cause linear movement of the crystal, so that 're-centring' is not required.

Eucentric arcs are more difficult to use in the early stages, and are unnecessary for a simple rotation camera. But for special purposes (for use on Weissenberg and especially precession cameras and some single crystal diffractometers) where fouling of the arcs on the rest of the apparatus is much more of a problem, they are almost essential.

Adjustment (b) of the crystal on to the collimator axis is achieved in both cases by a vertical movement of the spindle.

(v) A microscope (to aid in adjusting the crystal) which can be attached either parallel to, or on either side, perpendicular to the collimator axis.

(vi) Film holders.

1. Cylindrical, coaxial with the spindle. Normal radius 30 mm. The gap between the ends of the flat film rolled into a cylinder allows the collimator to project into the cylinder.

2. Flat plate, perpendicular to the direct beam, which registers the low angle, 'forward' reflections.

3. Flat plate, perpendicular to the direct beam but arranged to register the high angle, 'back' reflections. The direct beam must pass through the plate before hitting the crystal. The collimator is normally fastened to the film holder, and a hole must be punched in the film to accommodate it.

(vii) Where the direct beam would otherwise hit the film, a beam trap must be fitted or a hole made in the film through which the direct beam can escape, without giving rise to scattered radiation. It is desirable that the beam trap should be movable, so that the direct beam can be registered for a short time on the film. If the time can be made short enough (with reduction of the tube current and voltage for low absorption crystals) a radiograph of the crystal will be registered on the film. The centre of this radiograph provides the most exact possible origin for measurements on the film. (Fig. 39, p. 55).

Figure 87 shows a rotation camera with the different film holders in position. The details of the adjustment of such a camera are given in Appendix IX (p. 422). The production of oscillation and rotation photographs will be described in terms of this instrument, but the description can be easily adapted to other designs. Unless otherwise specified, it will be assumed that the cylindrical film holder will be used.

3. Weissenberg Camera

This is a development of the rotation camera, designed to photograph the RL one layer at a time. The essential additional features are:—

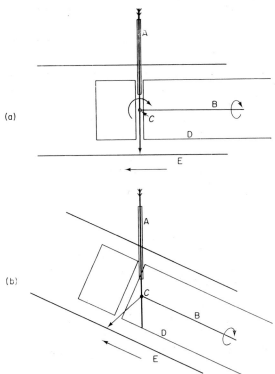

FIG. 88. The essentials of a Weissenberg camera; A—the collimator, attached to the main base; B—the rotation axis carrying the crystal C; D—the cylindrical layer line screen, arranged (a) to let through only the flat cone of reflections from the zero layer and (b) to let through the cone of reflections from a non-zero layer; E—the cylindrical film holder which moves synchronously with the rotation of the crystal. B, C, D and E are attached to the secondary base which can be rotated about an axis perpendicular to the diagram through C. The direct beam stop has been omitted for clarity.

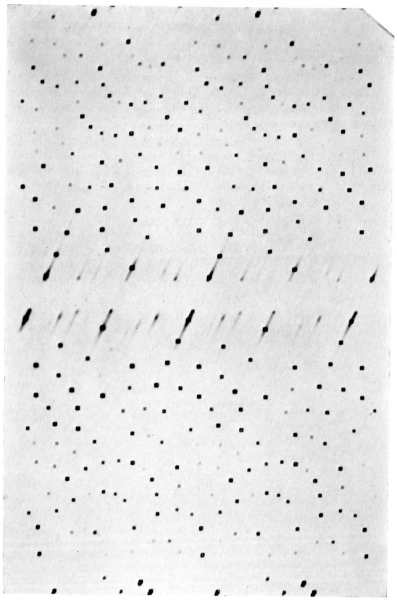

FIG. 89. An integrated Weissenberg photograph of the zero layer about **c** of a tetragonal crystal. The integration produces large rectangular spots which are more easily visible than the non-integrated reflections. Non-integrated reflections would be similar to those on a rotation photograph (Fig. 4, p. 5) but slightly drawn out in the direction of motion (Fig 167, p. 271). Figure 120 (p. 196) shows the top half of the photograph with the reciprocal lattice axes, etc. marked.

(i) a 'layer line screen' arranged to allow through only the reflections from the RL layer to be photographed (Fig. 88); (ii) means of translating the cylindrical film holder parallel to its axis, coupled to the rotation of the crystal, so that the reflections arising from the interaction of the RL net with the sphere of reflection are spread out in a two-dimensional pattern on the film. Mechanically this is usually achieved by having the rotation axis horizontal and the film holder running on horizontal rails, driven by a lead screw and nut. Gears on the end of the lead screw drive the spindle, and a synchronous motor turns the lead screw. The result is a rather badly distorted picture of the RL layer (Figs. 89 and 120. p. 196), but one which, as is shown in Section **15**.2, (p. 190), can be easily interpreted. For non-zero layers of the RL this interpretation is made very much easier (and there are important additional advantages) if the direct beam makes the same angle with the rotation axis as the reflected beams. This is known as the 'equi-inclination' position. To achieve this the rotation axis, film holder rails, lead screw, etc. must be capable of rotation relative to the collimator. The collimator is therefore attached to the main base, and the rest of the instrument to a secondary base, which can rotate about a vertical axis passing through the point of intersection of the spindle and collimator axes. The crystal, which is

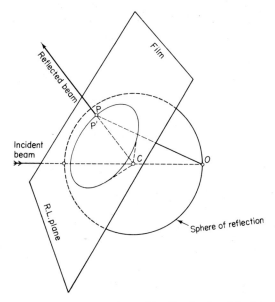

FIG. 90. The principle of a retigraph camera. The film is placed in the same position as the RL layer to be photographed and moved mechanically so that it follows the movement of the RL layer, remaining in contact with it. When a relp *P* is in the sphere of reflection, the reflection *P'* is registered at the same point on the film. A layer screen has to be provided to prevent reflections from other layers reaching the film.

at this intersection, does not suffer translation when the angle between the rotation axis and collimator is altered by such a rotation, and remains in the X-ray beam (Figs. 88(b) and 225, pp. 139 and 411).

Finally, if the instrument is to be used for accurate intensity measurements, it is desirable to have mechanical integration of the reflections. To achieve this the carriage for the film holder must have provision for rotating and translating the film during the exposure. A commercially available instrument with integrating facilities, designed by Wiebenga and Smits, is described in Appendix VII (p. 398), and the details of setting up procedures for accurate intensity measurements in Appendix IX (p. 429).

4. Retigraph Cameras

These cameras produce an undistorted picture of a RL layer. Essentially it is a 'contact print' of the layer. The film is placed in the plane of the RL layer to be photographed and moves with it, so that as a relp passes through the sphere of reflection, the reflected beam flashes out through the relp and marks a coincident spot on the film (Fig. 90). The mechanical means of doing this are described in Chapter 16 (p. 213).

Chapter 12

Initial Stages in Photographing the Reciprocal Lattice

1. Log Book and Film Marking

It is essential to record adequate information about every photograph taken. The photographs must be numbered serially and marked for orientation. Suggested headings for a log book are given in Fig. 91. If the film is held (unrolled if cylindrical) so that edges which were horizontal or vertical on the camera remain so, *and so that the direction of the incident beam through the film is towards the observer*, then a standard corner (e.g. R.H. top) cut off will mark the orientation. This is best done when the film is removed from the holder, because this is usually an easier operation than loading and therefore there is less likelihood of mistakes arising; and also because it is easier to load an uncut film so that no light fogging can occur. If at the same time initials and a serial number are scratched below the cut corner or marked with a hard pencil, they will develop up during processing. (In the case of a square film the identification mark will prevent any ambiguity of orientation.)

2. Mounting and Centring the Crystal

2.1. MOUNTING

The crystal is first mounted on a fibre, which should be a fine tube (c. 50 μm diameter with 10 μm walls) of borosilicate glass to minimise scattering and absorption. This procedure is carried out under the microscope, after all possible optical and morphological information has been used to determine the desired direction of mounting. The fibre is fastened by plasticene to a rod which is conveniently the spindle of a simple stage goniometer (Fig. 92). The end of the fibre is dipped into the adhesive and transferred quickly on to the crystal. For quick drying adhesives such as shellac which tend to form a skin, it is better to have a drop on the slide near the crystal so that the dipped fibre can be moved directly on to the crystal. For suitable adhesives refer to Vol. III of the International Tables [7(22)].

143

LOG BOOK HEADINGS

No.	Specimen	Type of Photo and Flat Plate Distance	Top Arc	Bottom Arc	Osc. limits Laue setting or Precession (and Dial) angles	Time On	Off	Exp.	Camera	Tube	Rad. KV ma.	Filter	Date	Remarks
1	KH₂PO₄ No. 5	Setting osc. c vert.	3·7R	0·3L	155+15°	20·50	21·10	20m.	3125	B	Cu 35,18	None	10/1/71	Bottom arc ⊥ beam. Correction 1·5° A.C.
2	KH₂PO₄ No. 5	Setting osc. c vert. Double osc.	3·7R	1·2R	200+15° 20+15°	21·30 21·45	21·45 21·50	15m. 5m.	3215	B	7Cu 35,18	None	10/1/71	Nearly set T0.5°C, B0.1°A.C.
3	KH₂PO₄ No. 5	Setting osc. c vert. Double osc.	3·2R	1·3R	200+15° 20+15°	9·30 9·45	9·45 9·50	15m. 5m.	3125	B	Cu 35,18	None	11/1/71	Check—O.K
4	KH₂PO₄ No. 5	Rotation	3·2R	1·3R	—	10·30	12·30	2h.	3125	B	Cu 35,18	Ni	11/1/71	Back stop requires adjusting
5	KH₂PO₄ No. 5	Laue 25·0 mm	3·2R	1·3R	127°	14·30	15·00	30m.	2183	C	Cu 50·18	None	11/1/71	a nearly along beam Correction +0·55°.
6	KH₂PO₄ No. 5	Precession	3·7R	1·3R	10(218·3)°	17·10	19·10	2hr	70	D	Mo 45,16	Nb	12/1/71	a*c* net nearly set
7	KH₂PO₄ No. 5	Int. Weissenberg L.L. Screen check	3·2R	1·3R	39–64	9·10	10·5	55m.	66053	E	Cu 35,18	None	13/1/71	Screen RH 1·3; LH 2·0 Altered to 1·4; 1·9
8	KH₂PO₄ No. 5	Zero layer	3·2R	1·3R	80–280	10·55	?	2 cycles 279041 s	66053	E	Cu 35,18	Ni	13/1/71	Automatic shut down by cycle recorder and timer
9	KH₂PO₄ No. 5	L.L. Screen check	3·2R	1·3R	39–64	18·20	19·10	50m.	66053	E	Cu 35,18	None	10/1/71	O.K.

Fig. 91. Examples of log book entries for various types of photographs. If the arc have provision for rotating the specimen the ...

The mounted crystal may first be used for further optical investigation before the fibre is transferred to the camera arcs. On top of the general purpose arcs there should be a small manually operated slider carrying a split pin. A 'pip' is held firmly on this pin, but can be rotated on it by means of a pair of strong tweezers (a small rotating stage may replace this arrangement with little advantage). The glass fibre carrying the crystal is fastened to the 'pip' by a small blob of plasticene. (Fig. 86, p. 135).

(a)   (b)

FIG. 92. Photograph of a stage goniometer ('twiddler') (a) on the microscope for optical investigation of the crystal, and (b) with the spindle separated from the base. The spindle itself is used in the initial mounting of a crystal under the microscope. A glass fibre is is attached to the tip with plasticene, and a drop of adhesive on the end of the fibre is brought into contact with the crystal. An impossibly large crystal has been used for the illustration. Normally the use of an eye-glass is necessary to see the crystal (or to see that it has dropped off!)

The function of the slider is twofold. (i) If during the preliminary setting of the crystal the fibre has to be bent over in the plasticene at a large angle, it may be impossible to centre it because one of the adjustable slides comes to the end of its travel. In that case, after centralising the main slide, the pip is turned so that the plane of the pin and fibre is parallel to the slide, and the slider moved to bring the crystal as near as possible to the central position.

Fine adjustment will then be possible on the main slide. (ii) If it is desired to preserve the orientation of the crystal after it is set, but to free the arcs for another crystal, then the slider can be removed and stored with the information necessary to replace it in position.† A number of sliders will, of course be necessary for this purpose; and if a mechanical safety device to prevent the normal removal of the slider is incorporated, this must be taken off.

2.2. REMOUNTING

If the crystal has to be picked up more or less at random in order to transfer it subsequently to a second fibre in a specified orientation, then the first adhesive used to attach a fibre to the side of the crystal must be soluble in a liquid which will not attack the crystal. A second fibre must subsequently be attached in the desired position, and the adhesive of the first fibre dissolved off.

The crystal is oriented on a rotation camera (Chapter 13, p. 149) with the desired rotation axis direction (normally a crystal lattice axis for a Weissenberg camera or a RL axis for a precession camera) in the equatorial plane, and perpendicular to the collimator axis.

A second microscope, mounted in the position perpendicular to the collimator axis, has a fibre attached by plasticene to the front lens. This microscope is racked in towards the crystal, and the arc slides and height adjuster moved so that the axis of the fibre intersects the centre of the crystal when they touch. The fibre is retracted, wet at the end with the second adhesive (which must be insoluble in the solvent for the first), and rapidly brought into contact with the crystal. If insufficient adhesive has remained on the end of the fibre, it may be necessary to retract it again and repeat the procedure. When the second fibre is securely attached the adhesive is allowed to dry thoroughly, and then a third microscope is attached on the other side, with a pointed spill of paper attached to the lens. This spill is loaded with a few drops of solvent, and brought up to the crystal until a drop of liquid surrounds it and is held in position by surface tension. These operations are observed through the normal microscope, or by looking down on the crystal through a watchmaker's eyeglass. The second fibre is retracted so that the first is just visibly bent when observed in the microscope. When the first adhesive dissolves sufficiently the first fibre will pull away from the crystal, and the remaining adhesive can be dissolved by submerging the crystal in a

---

† The cork of a 20 mm specimen tube has a dovetail cut in it to take the slider, which must have one end marked with a dot. If L(eft) or R(ight) refer to directions facing the relevant arc scale (the top scale for the slider), then the information can be written on a disc attached to the top of the glass cover as follows: R(or L), $T\psi_1$ R(or L). $B\psi_2$ R(or L) where T and B refer to the top and bottom arcs and ψ_1 and ψ_2 are the scale readings. If arcs with both possible arrangements of scale are used, it will also be necessary to put R(or L) to indicate which side the top scale is on when facing the bottom scale.

bath of solvent. The crystal, end mounted on the second fibre, is then ready for final setting.

2.3. CENTRING

The crystal must be moved by the linear arc slides so that its centre does not move when the spindle is rotated. This is accomplished while observing the crystal in the microscope, and can take a surprisingly long time unless done systematically and with a full understanding of the process.

First, there will be a certain amount of play in the spindle bearings. The movement of the crystal in the field of view of the microscope which is possible from this cause should first be investigated, since this will determine the accuracy with which it is possible to centre the crystal. Secondly, although the vertical crosswire of the microscope should be near the image of the rotation axis, it is unlikely to be exactly coincident with it. So the image of the crystal can be brought on to the crosswire to start with, as an approximate adjustment, but the final adjustment must be made using the crosswire only as a *reference* position. As the spindle is rotated, the image of the centre of the crystal must remain stationary relative to the crosswire, within the limits set by play in the bearings. If this stationary position is more than one tenth of the width of the field from the crosswire, then the microscope requires adjustment, but the axis should not be expected to be exactly on the crosswire. Thirdly, when adjusting a slide, there are only two positions of the spindle at which it is useful to observe the crystal—the two positions at 180° at which the slide is perpendicular to the axis of the microscope. The arcs can only be set at these positions by taking one's eye away from the eyepiece. The common sight of a student, with his eye glued to the microscope, madly twiddling the spindle and blindly groping for the adjusting screws of the slides to bring the crystal back on to the crosswire, only to see it go off again at the next twiddle, is due to ignoring this fact, as well as to treating the crosswire as marking the axis instead of just providing a reference mark. By numbers

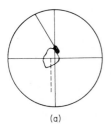 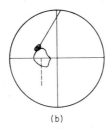 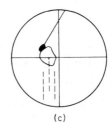

| (a) | (b) | (c) |

FIG. 93. The view in the eyepiece of the microscope when a crystal is being centred. (a) One slide perpendicular to the microscope axis. (b) after a rotation of 180°. (c) After adjustment to move the crystal so that its centre is halfway between the positions in (a) and (b). The size of the crystal and the distance of the crosswire from the rotation axis are both exaggerated

the process is: (i) Bring slide (1) perpendicular to the microscope axis. Focus, and adjust the slide to bring the crystal approximately on to the crosswire. (ii) Bring slide (2) perpendicular to the microscope axis, refocus if necessary and adjust the slide to bring the crystal approximately on to the crosswire. (iii) Observe the position of the crystal carefully (Fig. 93a). (iv) Turn the spindle through 180° and observe again (Fig. 93b). (v) If necessary return to position (iii) to refresh recollection of the position of the crystal, and move the slide adjustment to bring the crystal to a position midway between the first two positions (Fig. 93c). (vi) Repeat the process for slide (1). (vii) Repeat the process for both slides to see if a final adjustment is required. (viii) twiddle as madly as you like to make sure that the crystal does not wander by more than is due to play in the bearings.

This process should take five minutes at the outside. If your crystal is not centred at that time, re-read this section.

Finally, shine a light beam from an extended source down the collimator, and adjust the height of the spindle to bring the crystal into the centre of the beam. If this is not possible the collimator is out of alignment or the vertical adjustment is insufficient or faulty (Appendix IX, p. 423).

Chapter 13

Setting the Crystal and Measuring Cell Parameters

1. The Nature of the Setting Problem

In most cases there are two parts to the problem of setting a crystal:—
(i) to bring a particular RL plane perpendicular to a given direction, defined for the purposes of this discussion as a *vertical* rotation axis. This is equivalent to bringing the crystal lattice vector, which is perpendicular to the RL plane, parallel to the rotation axis (Equatorial Setting);
(ii) to bring a given RL vector, in the equatorial plane, into a position making a given angle with the plane containing the rotation axis and the X-ray beam (ϕ Setting—Section 4, p. 160).

2. Equatorial Setting

2.1. PRELIMINARY SETTING

2.1.1. *Setting with help from crystal morphology and optics*

Crystals with well defined morphology or optics can usually be set so that the selected crystal lattice vector makes an angle of less than 10° with the rotation axis.

An obvious example is a prismatic crystal with a prism edge sufficiently well defined to be set along the rotation axis, with the help of the microscope. If it is certain that such a crystal is less than 5° mis-set, then all that is required is a final setting by X-rays (Section 2.2, p. 153). If however this is not certain, then a 15° unfiltered [1(68)] oscillation photograph (Section 14.1, p. 175) should be taken, with one of the arcs perpendicular† to the beam at the mid-point of the oscillation. The zero layer curve may not be easy to trace over most of the photograph; but near the centre it should be well defined by the pronounced Laue streak (Fig. 179, p. 301). The angle which the tangent to the curve at the origin (direct beam) makes with the horizontal, can be

---

† The plane of the arc is referred to, since this is the obvious feature on the apparatus. The axis in this case would be parallel to the X-ray beam (or in the plane of the beam and rotation axis in the case of the top arc when the bottom arc is not at zero).

measured with a protractor to within a degree or so. The arc perpendicular to the beam is then moved through this angle in the direction required to make the tangent, which can be thought of as attached to the crystal, horizontal. The process is repeated for the other arc. The crystal will then be ready for final setting.

2.1.2. Setting crystals whose morphology and optics are of little or no assistance

(i) Unknown cell dimensions

From an unfiltered oscillation photograph, taken with one arc perpendicular to the beam, a possible zero layer curve at the origin is selected, and the crystal adjusted (if necessary by pushing the fibre over on the plasticene (Section 2.1.2 (ii) (8) p. 153) to bring the tangent horizontal, as above. This is repeated as often as necessary for both arcs, until the crystal is ready for final setting. If layer line curves cannot be recognised at all, individual reflections will have to be chosen and brought approximately into the equatorial plane, by the method given in Section 2.1.2 (ii).

The crystal setting device described by Kulpe [9], enables the shadow of an adjustable rim to trace out the position of the zero layer curve of a mis-set crystal, so that the mis-setting is recorded on the adjusting arcs of the rim. With this device it is possible to trace a zero layer curve which would otherwise be lost in the overlapping reflections from non-zero layers.

(ii) Cell dimensions known

If the first oscillation photograph looks promising it may be easier to proceed as in Section 2.1.2 (i), especially if the Kulpe device is available. But if not, then the positions of at least three indexed RL vectors must be found, using an unfiltered oscillation photograph which has had sufficient exposure to bring out the low angle Laue streaks. The procedure is as follows:—

1. Calculate all the possible values of d_{hkl}^* for the Kα wavelength out to $d^* = 0.8$ R.U., and tabulate representative indices hkl, Bragg angle (θ), and an indication of the intensity of the hkl reflection, if known, for each value of d_{hkl}^*.

2. Use the θ chart (Fig. 94) to measure the Bragg angles of all Kα reflections lying on a distinct Laue streak out to $\theta = 25°$. Number the reflections on the photograph or a traced copy.

3. Index the reflections by comparison with the table. If there are insufficient Kα reflections it may be necessary to repeat (1) and (2) for the Kβ wavelength. (Where only one reflection lies near the end of a Laue streak, this end of the streak must lie between the Kα and Kβ reflections. If this is the inner end the reflection is Kα; if the outer end it is Kβ.)

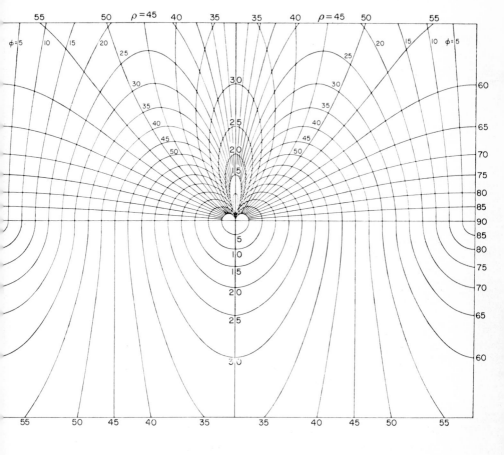

FIG. 94. θ and ρ, ϕ charts for a cylindrical camera. The constant θ curves on the bottom half of the chart correspond to powder specimen 'rings' on a cylindrical film for the values of θ marked against the curves. By superimposing the film and chart, the θ value of any reflection can be read to $\pm 0.25°$. On the top half of the chart the constant ρ curves correspond to Laue streaks (Fig. 39, p. 55), and the constant ϕ curves correspond to vertical zones on a Laue photograph (Fig. 52, p. 74) or the ends of Laue streaks in Figure 44 (p. 63), which are essentially the same thing. These two sets of curves can be used to measure ρ and ϕ for a particular RL vector in the same way as the bottom set is used to measure θ.

Earlier charts similar to this Figure had the numbering of the constant ϕ curves reversed, i.e. they were $(90 - \phi)$ charts. If such an earlier chart is used, the values of $\phi' = 90 - \phi$ should be recorded, and the supplement calculated subsequently.

The chart is printed for a film of radius 30 mm and has been reduced in reproduction here.

4. Choose 3 or 4 reflections whose indices are not in doubt and which are as simple as possible (2 zeros, 1 zero and 2 equal indices, etc). In each case mark the position of the Laue reflection derived from the associated weighted vector, when the crystal is at one end of its oscillation (specified by the corresponding goniometer scale reading ω). This Laue reflection will be one end of the associated Laue streak, *not* the $K\alpha$ (or $K\beta$) reflection. For the purpose of deciding which end of the Laue streak should be measured, the crystal planes can be regarded as optical mirrors, in order to relate the end of the streak to the specified end of the oscillation. If this is the inner end on the right-hand side of the photograph, it will be the outer end on the left-hand side and *vice versa*. It is necessary to reject any cases of short wavelength cut-off on the inner end of the streak, and of fade-out on the outer end (Section **6**.3, p. 53). If the desired end of the streak cannot be measured for either of these reasons, the other end may be measured and an adjustment made by adding or subtracting the oscillation angle.

5. The positions of the Laue reflections for the goniometer reading ω fix the directions of the corresponding RL vectors for this position of the crystal. Since we are only concerned with directions, the most suitable method of defining these vectors is by the two angles of spherical coordinates. ρ is defined as the angle between the rotation axis and the RL vector (Fig. 40, p. 56). As this angle does not alter with rotation, the whole of the Laue streak will lie on a constant ρ curve (Fig. 39, p. 55). The chart (Fig. 94) has such constant ρ curves plotted every 5°. ϕ is defined as the angle between the plane containing the rotation axis and the RL vector, and the plane containing the rotation axis and the X-ray beam (Fig. 42, p. 58). ϕ of course varies with the rotation of the crystal; but at any specified goniometer reading it is determined by the position of the reflection. In particular at ω it is determined by the position of the end of the Laue streak. Figure 42 shows a constant ϕ circle outlined by the intersections of a vertical plane of weighted RL vectors with the sphere of reflection. In Fig. 44 (p. 63) the inner and outer curves defined by the Laue streaks arising from the oscillation of such a plane, are constant ϕ curves for the two values of ϕ, corresponding to the two ends of the oscillation. Such curves are also plotted at 5° intervals on the chart (Fig. 94) and both ρ and ϕ can be measured by interpolation (over most of the area) to the nearest 0·5° by superimposing the photograph and chart. This is done for the selected ends of the Laue streaks (Fig. 178, p. 298), the relevant descriptions, U(p) or D(own) and L(eft) or R(ight), being attached to the readings. For reflections on the bottom half of the photograph ρ is measured from the downward direction of the rotation axis. (**Example 20**, p. 299).

6. The positions of the RL vectors are marked on a stereogram, with the rotation axis at the centre and the X-ray beam coming towards the observer

(Fig. 272, p. 490). A RL vector with given ρ and ϕ then lies on the intersection of a small circle of radius ρ about the centre of the stereogram, and a vertical great circle (i.e. a straight line through the centre) making an angle ϕ (L or R) with the X-ray beam. Since all RL vectors cutting the sphere of reflection make an acute angle with the negative direction of the X-ray beam, they all project in the unshaded part of the stereogram. Reflections on the bottom half of the film give RL vectors on the bottom half of the stereogram. The opposite direction of such a vector, which will have the same representative set of indices, can then be plotted on the top of the stereogram, making subsequent working easier.

7. If the axial directions are required, the angles which they make with the plotted RL vectors must be calculated (Appendix I, p. 330; in most cases more than one angle will be possible), and all the small circles with these angles as radii drawn round the relevant RL vectors. The two, three or four which intersect in one point fix the axial direction (Fig. 274, p. 492). The reason for choosing simple indices as in (4), is to reduce the number of possible angles.

Alternatively an important pole can be brought to the centre of the stereogram, and the other poles moved correspondingly, using the Wulff net or construction. This is particularly useful with cubic crystals, and an example of the location of a 111 pole is shown in Fig. 275, p. 494).

8. Draw in the direction of the axes of the camera arcs, and work out on the stereogram the adjustments required to bring the RL plane into the equatorial position (Fig. 276, p. 495) (**Example 21**, p. 299). If these adjustments are out of range of the arcs the fibre must be pushed over on the plasticene. It should be possible to do this within 5° with the following procedure. Use both arcs to bring the fibre to the upright position first; offset it by the required angle on the arc being corrected; and then bring the fibre upright again (relative to both arcs) by pressure on the plasticene. Finally, return both arcs to their original setting. If this procedure is not possible the error is likely to be considerably larger, unless an eyepiece incorporating an angular scale is used. (**Example 22**, p. 300).

2.2. FINAL SETTING

2.2.1. *General method*

When the setting error is known to be less than 5° on both arcs, the final setting is best achieved in almost all cases by using the method of Weiss and Cole as modified by Davis. The modification is important, since it enables the arc corrections to be treated independently. A 15° oscillation photograph is taken with the arcs at 45° to the beam at the middle of the oscillation (Fig. 95). An exposure is given which will produce fairly dense reflections with unfiltered radiation. The crystal is then turned through 180°, and a second 15° oscillation registered on the same film, but with an exposure time less than half the previous one. On the developed film there will be the strong zero layer curve due to the first exposure, and a weaker reflection of it in the horizontal line through the origin, due to the second exposure (Fig. 180, p. 302). The form of this double layer line can be understood by reference to Fig. 96.

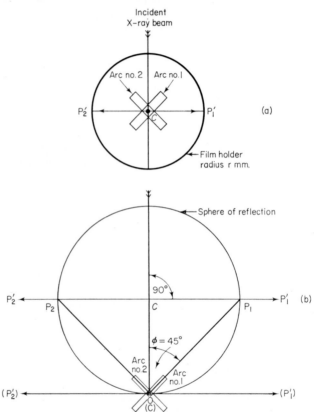

FIG. 95. (a) The relation between the direct beam, the arcs and the reflections P_1' and P_2' (above or below the equatorial plane but making a projected angle of deviation of 90°). (b) The RL vectors OP_1 and OP_2, with relps P_1 and P_2 (above or below the plane) giving rise to reflections P_1' and P_2'. It is required to bring OP_1 and OP_2 (and consequently CP_1' and CP_2') into the equatorial plane. For this purpose the origin of the RL is taken at the crystal and (C) (P_1') and (C) (P_2'), parallel to CP_1' and CP_2', are the actual reflected beams, corresponding to CP_1' and CP_2' in (a).

At or near $2\theta = 90°$ the separation of the two layer lines, x_1 and x_2 mm, is measured. The arc corrections, α_1 and α_2, can then be calculated from the expression derived below.

From Fig. 95a, the angle of elevation from the horizontal of the reflection P_1' at $2\theta = 90$, is $x_1/2r$, where r mm is the radius of the camera. The height of the relp giving rise to the reflection above the equatorial plane is therefore (Fig. 95b)

$$\zeta = CP_1 . x_1/2r = x_1/2r.$$

Therefore the angle of elevation of the RL vector OP_1 is

$$\alpha_1 = \zeta/OP_1 = x_1/(2\sqrt{2}r). \text{ Similarly } \alpha_2 = x_2/(2\sqrt{2}r) \text{ rad.} \tag{1}$$

$$\text{For } r = 30 \text{ mm}, \quad \alpha = 0.675x \approx 0.7x \text{ degrees.} \tag{2}$$

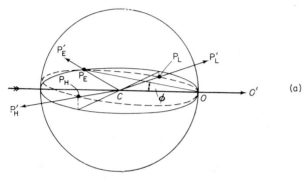

(a)

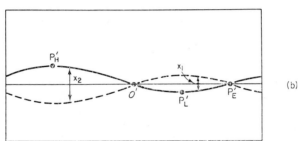

(b)

FIG. 96. (a) The dashed curve represents the intersection of a tilted RL zero layer with the sphere of reflection. The line of tilt must always pass through the origin O and intersect the sphere in a second point P_E, which is on the equatorial circle. The dashed curve will be projected on to the cylindrical film by reflections from C, those with the greatest deviation from the horizontal lying in the vertical plane perpendicular to the 'hinge' OP_E.

(b) The form of the projected curve, outlined by $K\alpha$ and $K\beta$ reflections and Laue streaks, as it appears on the unrolled cylindrical film. There are always two points of the curve, O' and P_E', on the equatorial line, but the position of P_E' depends on the direction of the tilt line, OP_E. There are two special cases:— (i) the tilt line is perpendicular to the X-ray beam, O and P_E coincide, and there is no second 'cross-over' point (or both are coincident at O'). The curve is horizontal at O' and symmetrical on either side of it, curving away to maximum deviation at the outside of the film ($2\theta = 180°$); (ii) the tilt line is along the X-ray beam. P_E' is at the edge of the film and the curve is anti-symmetrical about O', having its maximum deviations at $2\theta = 90°$.

If the crystal is rotated through 180°, the line of tilt is in the same position, but the tilt is now reversed in direction. (a) will become its mirror image in the equatorial plane, and the curve on the film (shown dashed) will be imaged in the horizontal line. The measurements of x_1 and x_2 mm are made, as shown, where the projected angle of deviation is 90°. x_1 and x_2 will normally be much smaller than on this diagram.

To apply the corrections the goniometer is set to a reading near the centre of the oscillation which had the longer exposure; and the direction of application of the correction is obtained from the position of the strong zero layer curve at $2\theta = 90°$, when the photograph is held upright with the beam coming towards the observer. The correction from the right-hand side of the photograph is applied to arc No. 1, whose plane is NE–SW (N away from the observer along the X-ray beam). The direction of movement of the arc must be such that the RL vector, imagined as a wire sticking out of the crystal in the NE direction (the RL origin is transferred to the crystal for this operation), will be brought into the horizontal plane. The correction from the left-hand side is applied similarly to arc No. 2 whose plane is NW–SE.

It is essential to distinguish clearly between the crystal as the origin for the reflected beams at right angles to the incident beam, and the crystal as the origin for the RL vectors which are at 45° to the incident beam. Confusion may arise, because this is one of the few applications where the origin of the RL is taken at the crystal. (**Example 23**, p. 300)

2.2.2. *Setting for the precession camera*

If the crystal is being set for transfer to a precession camera it is the RL axis which must lie along the rotation axis. In orthogonal cases the two are the same; but otherwise the crystal must be set in the normal way, and the arcs then corrected to bring the RL axis parallel to the rotation axis. A stereogram of the positions of the axes and arcs is the simplest method of finding the corrections to within 0·5°. Final setting can be done on the precession camera (Section **16**.3.6, p. 228). (**Example 24**, p. 300).

2.2.3. *Setting rhombohedral crystals about* **c**

In setting rhombohedral crystals about the trigonal axis it often happens that there are too few zero layer reflections for the employment of the normal 180°, double oscillation method. However there are many more reflections on first or second layers which are near enough to the zero layer for the same approximations to be used in deriving the corrections; but these repeat at 120° instead of at 180°. The separation of these equivalent reflections at, or near, $\theta = 45°$ in a double oscillation, where the second oscillation is at 120° to the first, can be used to calculate the arc corrections in the following way.

The 'set' position is obtained when the three-fold axis lies along the rotation axis, and the zero layer RL plane is in the equatorial plane perpendicular to the rotation axis. Any displacement from the 'set' position can be represented as a small rotation ρ, about an axis (or 'hinge') lying in the equatorial plane. The trigonal axis will move through ρ from the 'set' position, in the plane perpendicular to this 'hinge' (Fig. 97). It is required to find the arc corrections to bring the RL zero layer into the equatorial plane.

The first oscillation is registered with the central positions of the arcs making angles of 75° and 15° with the X-ray beam (i.e. 30° clockwise from the normal 45° position) as shown in the diagram. The second oscillation is registered after a further rotation of 120° clockwise from the position shown. P_1 and P_2 are the reciprocal lattice points giving reflections $P_1{}'$ and $P_2{}'$ at $2\theta \approx 90°$ and near the equatorial plane; and Q_1 and Q_2 are the points at 120° to P_1 and P_2 which will give the corresponding reflections in the second oscillation, at 120° to the first. In the set position these points and reflections would coincide. The measured separation of the reflections provides the basis for calculating the arc corrections.

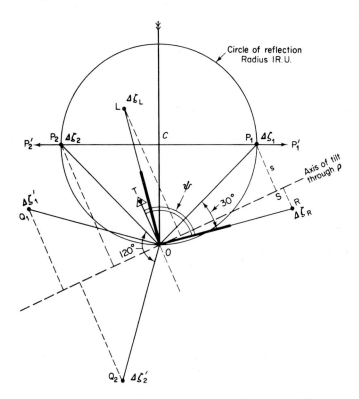

FIG. 97. Diagram showing the positions of the arcs and RL vectors in the setting method for rhombohedral crystals. The arcs OR and OL are set 30° clockwise from their normal 45° positions. OS is the axis of tilt (or 'hinge') for the mis-setting of the $\mathbf{a^*b^*}$ equatorial plane (cf. OP_E, Fig. 96 (a), p.155). OP_1 and OP_2 are the RL vectors giving reflections with a projected angle of deviation of 90°. OQ_1 and OQ_2 replace OP_1 and OP_2 after a clockwise rotation of 120°. P_1, P_2, Q_1 and Q_2 are all out of the plane of the paper and shifted from their 'set' positions by varying amounts $\Delta\zeta$. The plane containing the rotation axis (perpendicular to the paper at O) and the RH arc (OR) makes an angle ψ with the plane containing the rotation axis and the c-axis of the crystal (OT). OT is necessarily at right angles to OS.

We assume that the ζ displacements (+ve upwards) of the near equatorial relps from the 'set' position are, to a first approximation, equal to the lengths of the arcs through which they move on tilting the RL about the 'hinge' through the small angle ρ; i.e. $\Delta\zeta = s'\rho$, where s' is the distance of the relp from the axis of tilt. Since the relps are near equatorial, s' can be taken as equal to the projected distance s, i.e. $P_1 S$ in Fig. 97 for relp P_1. Therefore the movement of P_1 is given by

$$\Delta\zeta_1 = -\rho s = -\rho\, OP_1 \cos OP_1 S$$

$$= -\rho\sqrt{2}\cos TOP_1 \tag{1}$$

since the projected value of OP_1 is $\sqrt{2}$ R.U., and OT (the projection of the trigonal axis) is perpendicular to OS.

R is the point in the zero layer of the reciprocal lattice at a projected distance of $\sqrt{2}$ R.U. from O in the plane of the right-hand arc. In the normal method of 180° double oscillation, R would be a relp giving rise to an equatorial reflection from which the arc correction could be determined. But since R is not a relp, the equatorial reflection does not occur, and the corrections required to bring it and the corresponding point L for the left-hand arc into the equatorial plane, have to be found indirectly.

Let $TOR = \psi$. Then

$$\Delta\zeta_1 = -\rho\sqrt{2}\cos(\psi - 30) \quad \text{from (1).}$$

The reflection will be registered at a distance

$$\frac{\Delta\zeta_1}{CP_1} \times r = \Delta\zeta_1 r$$

from its 'set' position, where r is the radius of the camera.

Similarly $\Delta\zeta_2 = -\rho\sqrt{2}\cos(ROL + LOP_2 - \psi)$

$$= -\rho\sqrt{2}\cos(90 + 30 - \psi) = -\rho\sqrt{2}\cos(90 - (\psi - 30))$$

$$\Delta\zeta_1' = -\rho\sqrt{2}\cos(ROP_1 + P_1OQ_1 - \psi)$$

$$= -\rho\sqrt{2}\cos(30 + 120 - \psi)$$

$$= -\rho\sqrt{2}\cos(180 - (\psi + 30)).$$

$$\Delta\zeta_2' = \rho\sqrt{2}\cos(\psi + 108 - ROP_2 - P_2OQ_2)$$

$$= \rho\sqrt{2}\cos(\psi + 180 - 240)$$

$$= \rho\sqrt{2}\cos((\psi + 30) - 90).$$

Let $\Delta\zeta_R' = \Delta\zeta_1 - \Delta\zeta_1'$ and $\Delta\zeta_L' = \Delta\zeta_2 - \Delta\zeta_2'$

$$\therefore \quad \Delta\zeta_R' = -\rho\sqrt{2}\left(\cos(\psi - 30) + \cos(\psi + 30)\right)$$
$$= -\rho 2\sqrt{2}\cos\psi\cos 30 \tag{2}$$
$$\Delta\zeta_L' = -\rho\sqrt{2}\left(\sin(\psi - 30) + \sin(\psi + 30)\right)$$
$$= -\rho 2\sqrt{2}\sin\psi\cos 30. \tag{3}$$

The displacement of the 'relp' R is given by

$$\Delta\zeta_R = \rho\sqrt{2}\cos(180 - \psi) = -\rho\sqrt{2}\cos\psi$$

and the normal double oscillation would give a separation of the two relps at 180° of

$$2\Delta\zeta_R = -\rho 2\sqrt{2}\cos\psi = \Delta\zeta_R'\sec 30 \quad \text{from (2)}$$

Similarly

$$2\Delta\zeta_L = \Delta\zeta_L'\sec 30 \quad \text{from (3).}$$

If the separation of P_1' and Q_1', the two reflections corresponding to P_1 and Q_1, is equal to x_1 mm, the angle between them is x_1/r and the separation of the relps is

$$\Delta\zeta_R' = \frac{x_1}{r} \times CP_1 = \frac{x_1}{r}\,\text{R.U.}$$

$$\therefore \quad 2\Delta\zeta_R = \Delta\zeta_R'\sec 30 = \frac{x_1}{r}\sec 30$$

and the correction on the arc OR is:

$$\alpha_1 = \frac{\Delta\zeta_R}{OR} = \frac{x_1 \sec 30}{2r \cdot OR} = \frac{x_1}{2\sqrt{2}r}\sec 30.$$

$$\text{Similarly } \alpha_2 = \frac{x_2}{2\sqrt{2}r}\sec 30.$$

Thus the correction is made in exactly the same way as for the normal double oscillation (155(1)), except that the measured separation of the equivalent reflections is multiplied by $\sec 30 = 1\cdot155$.

However, the direction in which the correction must be applied is not so directly obvious as in the case of zero layer 180° double oscillations. For example, in Fig. 97, (p. 157) the right hand reflection for the first oscillation would be too low, whereas the RL vector in the plane of the arc is too high. But the reflection for the second oscillation is lower still, so that the relative positions of the two reflections is the same as it would be for zero layer reflections in a normal 180° double oscillation. Inspection of the relations between $\Delta\zeta_R$ and $\Delta\zeta_R'$ (and $\Delta\zeta_L$ and $\Delta\zeta_L'$) shows that this holds for all values of ψ; so that if the first oscillation reflection is above that for the second oscillation, the correction on the corresponding arc is downwards and *vice versa.*

If the arcs for 180° double oscillation setting are normally placed perpendicular and parallel to the beam, the same procedure (first oscillation 30° clockwise, second 30° anti-clockwise from the normal positions) will give spot separations which, if multiplied by 1·155, can be used to calculate the arc corrections exactly as if the normal zero layer 180° double oscillation had produced the results. However, because each correction is now a function of the measurements on both sides of the film, the possibility of making an error in the sign of the correction is much greater.

3. Orientation Required of Known Crystal Lattice Relative to the Morphology of the Crystal

The procedure is almost exactly the same as for Section 2.1.2 (ii), (p. 150), but directions from the morphology of the crystal (e.g. the normal and edge of a flat surface) are plotted on the stereogram instead of the arc axes. The microscope is used to obtain the necessary angles by a procedure similar to that used in Section 2.1.2 (ii) (8) (p. 153); or if specular reflection can be obtained from the face, the optical collimator attachment (Fig. 87c (21), p. 137) can be employed to get accurate values for the angular coordinates of the face normal, without removing the arcs from the camera. If the specimen is too large to be bathed in the X-ray beam it will be necessary to take glancing angle oscillation photographs from a surface adjusted to contain the rotation axis. The minimum angle of glancing incidence should be about 10°, and the smallest collimator available should be used in order to reduce the size of the spots on the film (a fine focus tube is desirable in such cases). The position of the direct beam can be registered either by displacing the backstop for the whole of the exposure, if the specimen is not too thick, but thick enough to prevent halation occurring, or by turning the specimen parallel to the beam at the end of the exposure and displacing the backstop in the usual way (Section 11.2 (vii) (p. 138).

4. φ Setting

It is first necessary to identify a RL vector in the equatorial plane. If this has been done as in Section 2.1.2 (ii) (p. 150) or 6.5.5.2 (p. 75), the ϕ angle of such a RL vector will be known within a degree or two; otherwise a modified version of Section 2.1.2 (ii), or interpretation of an oscillation photograph using the RL (Section 14.1.4, p. 183), must be used to obtain it. If the ϕ angle is required more accurately, it is necessary to take a Laue photograph at a distance of at least 40 mm on a plate camera. Since the approximate orientation is known, the crystal can be set to give an equatorial Kα reflection at about the middle of a 5° oscillation. If the crystal is held stationary at one end of the oscillation, a Laue reflection of known indices (checked from the θ value of the Kα reflection) will be produced, whose angle from the direct beam can be calculated accurately from the measured distances:— crystal to plate ($CO' = y$ mm) and central spot to Laue spot, $O'P' = x$ mm. (Fig. 98).

Then $\tan 2\theta = x/y$ and $\phi = 90° - \theta$.

We can estimate the accuracy of this setting by differentiating:—

$$\Delta 2\theta = \frac{y\,\Delta x - x\,\Delta y}{x^2 + y^2}.$$

If $x = 40$ mm, $y = 40$ mm, $\Delta x = 0\cdot05$ mm, $\Delta y = 0\cdot1$ mm,

then

$$\max \Delta\phi = \max \Delta\theta = \frac{1}{2}\,\frac{40 \times 0\cdot05 + 40 \times 0\cdot1}{40^2 + 40^2} = 0\cdot001 \text{ rad} = 0\cdot06°.$$

If y is increased to 80 mm the same final accuracy will be produced with $\Delta x = 0\cdot1$ mm and $\Delta y = 0\cdot2$ mm.

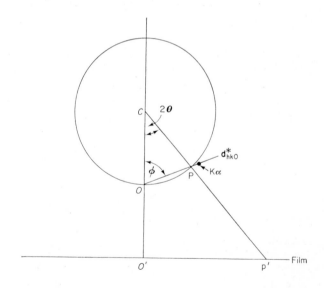

FIG. 98. The crystal is held stationary at one end of a 5° oscillation, designed to include the Kα reflection to check the identification; and the Laue reflection P' is used to obtain an accurate measurement of ϕ, the angle the RL vector makes with the X-ray beam, for this setting of the goniometer.

If the Laue symmetry of the crystal includes a vertical reflection plane, or a horizontal two-fold axis, and if this plane or axis can be set approximately parallel to the X-ray beam, then measurements can be made on a Laue photograph, as above, for corresponding equatorial spots on each side of the centre. If δ is the angle between the symmetry plane or axis and the X-ray beam (Fig. 99), we have:—

$$\tan 2(\theta + \delta) = x_1/y, \qquad \tan 2(\theta - \delta) = x_2/y;$$

$$\therefore \quad \delta = \frac{1}{4}\left(\tan^{-1}\frac{x_1}{y} - \tan^{-1}\frac{x_2}{y}\right), \qquad (1)$$

where θ is the Bragg angle when the symmetry plane is parallel to the beam.

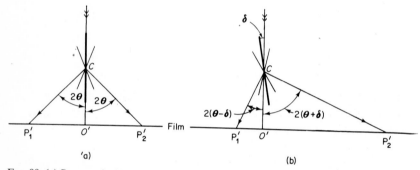

FIG. 99. (a) Symmetrical lattice planes about a symmetry line set accurately along the X-ray beam, produce reflections P_1', P_2', equally spaced on either side of the direct beam O'. If the crystal is rotated through an angle δ, the reflected beams move through 2δ to give the situation shown in (b). If corresponding reflections P_1', P_2' can be recognised (very unlikely in the case shown, because δ is too large) they can be used to calculate an unknown δ.

For similar values of x and y, this method has a smaller probable error, but the same maximum error. However, if the plate is near enough (normally at 25–30mm or less), a number of major zones make identification of corresponding spots easier than the identification of a single spot in the previous method, and this usually more than compensates for some increase in the maximum error.

If δ is small ($< 0.5°$) we can use differentials to simplify the expression.

Differentiating $\tan 2\theta = \dfrac{x}{y}$, we have:—

$$\sec^2 2\theta \cdot 2\Delta\theta = \frac{\Delta x}{y} \qquad \therefore \delta = \Delta\theta = \frac{\Delta x}{2y}\cos^2 2\theta.$$

Since Δx is positive on one side and negative on the other,

$$x_2 - x_1 = 2\Delta x$$

$$\therefore \delta = \frac{x_2 - x_1}{4y} \cos^2 2\theta \text{ rad} \equiv 14\cdot3 \cos^2 2\theta \frac{x_2 - x_1}{y} \text{ degrees.} \qquad (1)$$

$\cos^2 2\theta$ need only be evaluated roughly (5%) since $(x_2 - x_1)/4y$ is small and δ cannot be used to correct the arcs to better than $0\cdot05°$. By contrast the two values of $\tan^{-1}(x/y)$ in 162(1) have to be evaluated accurately, because the difference is required.

If suitable equatorial reflections are not available, spots a few degrees above or below the equatorial plane can be used without introducing appreciable error. (**Example 25**, p. 303).

This method can be used for small arc corrections where a plane of symmetry is to be brought horizontal, by taking corresponding reflections above and below the direct beam to calculate the correction for the arc parallel to the beam.

5. The Use of Back Reflection Laue Photographs

5.1. GENERAL

In most cases a back reflection (i.e. high angle) Laue photograph will have a large number of reflections and will be difficult to interpret. The reason for this can be seen from Fig. 100, in which the usual method of showing variation in wavelength, by a sphere of reflection of radius 1 R.U. to a fixed scale and a RL with d^*_{hkl} proportional to the wavelength, is reversed. The RL is drawn to scale for $\lambda = 1$ Å, and the sphere of reflection is drawn to the same scale with a radius of $1/\lambda$. If the spheres are drawn for $\lambda = \lambda_S$ (the short wavelength cut-off [1(52)]), and for $\lambda = \lambda_L$ (the effective long wavelength limit or 'fade-out') then reflections will occur corresponding to all the relps between the two spheres. For a maximum deviation from the direct beam, forward or backward, of 45°, the volume containing relps for back reflections is far larger than that for forward reflections.

However, in the case of high symmetry crystals with small unit cells, especially cubic and hexagonal metals, it becomes possible to recognise zones of reflections (Section **6.5.1**, p. 57) and identify them by measurement with a simple chart. This is particularly important in metallurgy because metal specimens are often large, and back reflection, stationary (i.e. Laue) photographs are the only practical possibility.

5.2. PRODUCTION

The back reflection camera is described in Section **11.2** (p. 135) and its adjustment in Appendix IX (p. 428). The specimen is fastened to the holder, which can conveniently be a stepped platform on top of the spindle with the

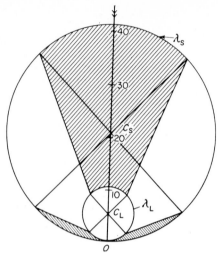

Fig. 100. The spheres of reflection for λ_S (short wavelength 'cut-off') and λ_L (long wavelength 'fade out').

The number of relps in the coarse hatched volume (obtained by rotation about the X-ray beam), all of which would give back reflections within 45° of the X-ray beam, is clearly much greater than those in the fine hatched volume, which would give forward reflections. The number of relps along the X-ray beam is given for a spacing in this direction of 5 A. The circle for λ_S is correct assuming that the tube is run at 50 kV, but in order not to obscure the diagram the circle for λ_L is shown about three times the correct size. This will clearly make little difference to the relative volumes corresponding to forward and backward reflections.

step containing the axis, so that a flat surface can be accurately located. If a single linear slide parallel to the step is incorporated, this, together with the height adjustment, enables different areas to be selected. The area which is to be investigated is marked and brought on to the collimator axis. This can be located by shining a powerful beam of light down the collimator. For very small areas a lens or auxiliary microscope may be required. Small specimens can be fastened to ordinary arcs, and the required surface area located on the rotation axis perpendicular to the X-ray beam, using the normal microscope and adjustments. A jig is required for punching a hole in the film and black paper wrapping. The film is placed in the holder with the hole over the collimator, and is held flat and light tight by a screwed collar.

5.3. INTERPRETATION

The most usual problem is to find the orientation of a metal specimen relative to its morphology.

Back reflection Laue zones (Section 6.5.1, p. 57) arise from central RL planes which make a small angle with the X-ray beam. The intersections of the weighted rods in the central RL plane with the sphere of reflection give

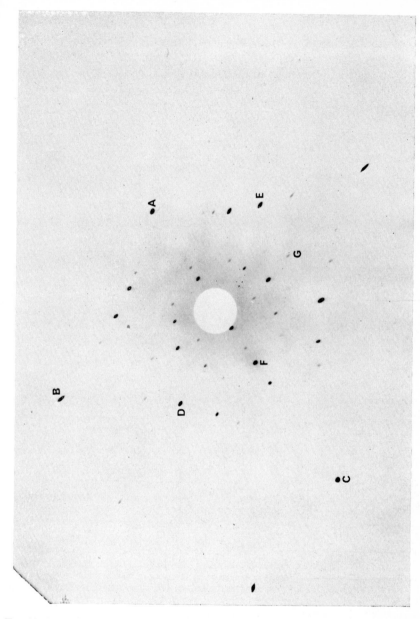

Fig. 101. Back reflection Laue photograph from a randomly oriented grain of aluminium. Some of the reflections on the intersections of a number of zones, which correspond to important RL vectors, have been marked. Specimen-film distance, 30 mm. CuKα radiation. Goniometer scale reading, $\omega = 259\cdot5°$. Incident beam going downwards through the film, so that cut corner is top left.

rise to Laue reflections on a nearly flat cone. These reflections strike the flat film along a hyperbola. If the crystal is set so that an important RL vector is along the beam, many of the cones will be completely flat, and the degenerate hyperbolae will appear as straight lines through the origin, i.e. the point where the X-ray beam passes through the film, as in Fig. 45 (p. 64). Where in the more normal case the orientation is random, we require to interpret a photograph such as Fig. 101.

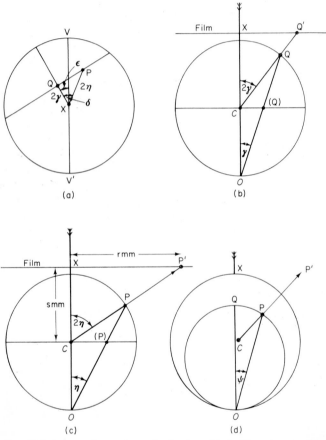

FIG. 102. (a) A stereogram of the sphere of reflection with the X-ray beam going downward in the centre. The crystal C is at the centre of the projection sphere and the S projection pole is the origin of the RL, O. OQP is the central RL plane whose weighted rods are giving rise to reflections where they intersect the sphere. (b) shows the section through the X-ray beam and Q. This section is normal to the RL plane. The section of the film is shown here and in (c), (P) and (Q) are the projections of CP and CQ, i.e. they are the points P and Q on the stereogram (a). (c) is the section through the beam and the RL vector OP. (d) is a semi-elevation looking in the direction QX. C is at the centre of the sphere and CP is not in the plane of the small circle OPQ.

The method of measuring such a photograph can best be seen from a stereogram of the sphere of reflection, with the beam going downwards in the centre, i.e. viewed from the X-ray tube as a temporary exception to the normal procedure. We take the sphere of reflection as the projection sphere, with the crystal C at the centre, and the origin of the RL, O, as the S pole. The projection plane is the equatorial plane through C, perpendicular to the direct beam.

In Fig. 102a the straight line PQ is the projection of the small circle through O, in which the central RL plane cuts the sphere of reflection. (Any small circle through the projection pole projects as a straight line. This is the intersection of the plane of the small circle and the projection plane). X represents the backward direction of the X-ray beam and XQ is perpendicular to QP. XQ, the angle from the centre C, equals 2γ, where γ is the angle which the RL plane makes with the X-ray beam. ε is the angle which the plane containing the X-ray beam and the normal to the RL plane makes with the vertical VV'. $\widehat{POQ} = \psi$, the angle between the central RL vector OQ and some other vector OP in the RL plane. The angle between P and Q on the small circle is 2ψ, but the angle between the directions CP and CQ is not. Figs. 102, (b) (c) and (d), containing the X-ray beam, help the interpretation of the stereogram. CP, CQ (represented on the stereogram by the points P, Q) are the directions of the reflections arising from the intersection of the RL vectors OP, OQ with the sphere of reflection. All such directions lie on a cone with apex O, which intersects the film in a hyperbola.

The point P', in which the reflection CP strikes the film, can be found in terms of γ, ε and ψ. The directions OX, OQ and OP have known angles γ, between OX and OQ, and ψ between OQ and OP. On a normal projection, *with O at the centre*, XPQ is a right angled spherical triangle. This can be solved by Napier's rule [4(183)] for η, the angle between OX and OP, to give:—

$$\cos \eta = \cos \psi \cos \gamma;$$

and for δ, the angle between the planes containing OX, OP and OX, OQ, to give:—

$$\cos \delta = \cot \eta \tan \gamma.$$

δ is also the angle between XQ and XP, in the stereogram with O at the S pole (Fig. 102a). The angle which the plane XCP makes with the vertical VV', is $\delta - \varepsilon$, and P' on the film will also lie in this plane. The distance r mm, of P' from the origin (the direct beam) is given by

$$r = s \tan 2\eta$$

where s mm is the distance between crystal and film. (Fig. 102c).

From a knowledge of γ, ε and ψ it is thus possible to plot the position of the corresponding reflection on the film. If γ and ε are kept constant and ψ varied, the points will lie on a hyperbola whose nearest distance to the origin will be the reflection Q', with $r = s \tan 2\gamma$.

Varying ε will merely swing the hyperbola around the origin. Keeping ψ and ε constant and varying γ will give points on a curve, corresponding to reflections on a series of hyperbolae, all of whose RL vectors have the same angle POQ, i.e. the constant value of ψ.

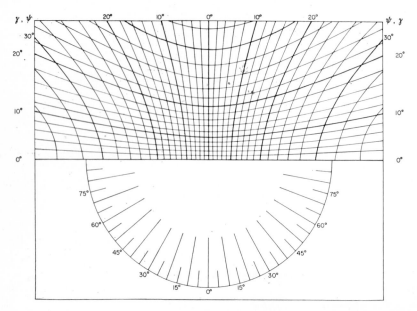

Fig. 103. Greninger chart, showing hyperbolae and curves of constant ψ at 2° intervals. The horizontal line at the centre is a degenerate hyperbola corresponding to $\gamma = 0$, i.e. a flat cone of angle 180°.

By taking fixed values of γ at 2° intervals, and plotting the corresponding hyperbolae by varying ψ, we obtain one set of curves of the Greninger chart (Fig. 103); and by taking fixed values of ψ and varying γ we get the other set. The protractor on the other half of the chart measures ε, i.e. it measures how much the chart has to be rotated from the vertical in order to fit it to a hyperbola of spots on the film. The second set of curves (constant ψ) enables measurements to be made of the angle between the RL vectors corresponding to two reflections on a hyperbola to which the chart has been fitted.

In using the Greninger chart we put the centre on the direct beam, fit the chart to a hyperbola on which the reflection lies, measure ε with the protractor, γ from interpolation of the hyperbolae curves, and ψ from the constant ψ curves. The RL vector is then plotted on a normal stereogram, with O at the centre, as in Fig. 104. Since the central RL plane containing OQ and OP now passes through the centre of the projection sphere, all the poles similar to OP (i.e. for constant γ) lie on a great circle, sloping at $\gamma°$ from the X-ray beam, and cutting the primitive in points making an angle $\varepsilon°$ with the horizontal. P is $\psi°$ from Q along the great circle. P can therefore be easily plotted using a standard Wulff stereographic net.

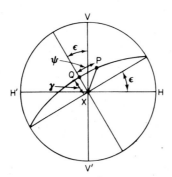

Fig. 104. Stereogram showing the directions of the RL vectors, i.e. with the X-ray beam downwards but with O at the centre of the projection sphere, instead of C as in Fig. 102(a).

Since the X-ray beam is *downward* at the centre of the stereogram it is necessary to view the film with the beam going down through it, away from the observer. This means that the application of corrections derived from the stereogram must be made with care, taking this unusual circumstance into account.

It is only useful to plot important RL directions so that the relationships between them can be recognised, and indices assigned. With a little experience this can often be done for the cubic case from the photograph.

Reflections corresponding to important RL vectors can be recognised by virtue of being at the intersection of a number of zones. Some of these are labelled in Fig. 101 (p. 165). Although the symmetry elements in the direction of the RL vector are distorted unless it lies along the X-ray beam, indications of possible symmetry can help in indexing. A major aid in identification is comparison of the angles between RL vectors (the algebraic sum of the ψ angles on the parabola joining the reflections) with a table of angles between normals. [6(120)].

When sufficient vectors have been indexed the orientation of the specimen is determined, and re-orientation, if required, can be obtained from the stereogram, as in Answer 21 (p. 489). (**Example 26**, p. 303).

6. The Determination of the Lattice Parameters and Unit Cell Contents

6.1. LATTICE PARAMETERS

If definite or probable axial directions can be found from the morphology and optics, rotation photographs about these directions will give layer lines from which the axial lengths can be found (Section **14**.2.2, p. 186). Alternatively, when one axial direction has been found (**c**, say) the zero layer of the rotation photograph can be interpreted (Section **14**.2.2, p. 186), and a^*, b^* and γ^* determined. If a moving film camera is available, the easiest and most certain method is to use it to photograph the zero layer and measure the two RL axes in this layer (a^*, b^*) and the angle (γ^*) between them, (Section **15**.2, p. 190). If the symmetry shows that the lattice is orthogonal or pseudo-orthogonal, i.e. hexagonal, refinement of the parameters can then be carried out. In the general case assuming **c** vertical, the crystal should be set with a^* along the beam and a flat plate Laue photograph taken. The horizontal zone is a^*b^*, and a prominent straight line zone as nearly vertical as possible is chosen as the a^*c^* zone. The angle between the lines of reflections is equal to the angle between these two zones, and therefore to the angle α between their zone axes, **c** and **b**. Similarly, by taking a Laue photograph along b^*, β can be found. In the monoclinic case, if **b** is set along the X-ray beam and an oscillation photograph taken with this as starting position, an almost undistorted picture of the a^*c^* net will be obtained (Section **4**.2, p. 34) from which an approximate value of β^* can be measured. More accurate values can then be obtained using Laue photographs as above, and in this and the general case, the angle in the zero layer perpendicular to the rotation axis can be refined using a cylindrical Laue photograph (Section **6**.5.5.2, p. 75). With the approximate values of the RL parameters (α^* and β^* can be found from α, β and γ^*, see Appendix II, p.362) a reciprocal net can be drawn, and more accurate values found by interpretation of oscillation photographs or measurements on Weissenberg or retigraph photographs. Finally, accurate values for the parameters can be obtained using high angle reflections.

6.2. ACCURATE LATTICE PARAMETERS

6.2.1. *Theoretical*

In general we have to find accurate values of d^*_{hkl} for various special sets of hkl (d^*_{h00} to find a^*, etc.) and so we measure the θ angles of the corresponding reflections, for use in the equation:—

$$d^* = 2 \sin \theta. \tag{1}$$

Take logs of both sides of (1) and differentiate:—

$$\frac{\Delta d^*}{d^*} = \frac{\Delta(\sin \theta)}{\sin \theta} = \frac{\cos \theta \, \Delta\theta}{\sin \theta} = \cot \theta \, \Delta\theta. \tag{2}$$

$\Delta\theta$ is the error in the measurement of θ. This tends to decrease slightly in the back reflection region, because of the separation of the α_1, α_2 doublet [1(56)] (Fig. 181, p. 304) and of a certain amount of focussing of the reflections in that region. The product of $\Delta\theta$ and cot θ gives the relative error in d^*. Since cot $\theta \rightarrow$ 0 as $\theta \rightarrow 90°$, the relative error in $d^* \rightarrow 0$ as $\theta \rightarrow 90°$. A determination of a^* from a reflection at $\theta = 85°$ will be at least 10 times as accurate as from a reflection at 50°. Ideally a number of determinations from $h00$ reflections should be plotted against a function of θ which gives a straight line plot, and extrapolated to $\theta = 90$. The function which has been derived semi-empirically for powder photographs is $\frac{1}{2}[\cos^2\theta/\sin\theta + \cos^2\theta/\theta]$, and the tabulated values [1(316), 6(228)] can be used in the single crystal case. However in most cases a single high angle (80°) measurement is sufficient to obtain a cell parameter with an accuracy of 1 in 5,000. This accuracy is consideably higher than that of the fractional coordinates of the atoms obtained in structure analysis, and this is the main requirement. If a comparison is required with a well characterised powder, e.g. to check for variation in solid solution, then extrapolation should be employed.

A variation of the method using high angle reflections takes the difference between $d^*(\alpha_1)$ and $d^*(\alpha_2)$, i.e. the separation of the α_1 α_2 doublet.
    Let

$$d^* = 2 \sin \theta \text{ for } \lambda(\alpha_1)$$

and $d^* + \Delta d^* = 2 \sin (\theta + \Delta\theta) \text{ for } \lambda(\alpha_2)$

$$\therefore \quad \Delta d^* = 2 \left(\sin (\theta + \Delta\theta) - \sin \theta \right)$$

$$= 4 \cos\left(\theta + \frac{\Delta\theta}{2} \right) \sin \frac{\Delta\theta}{2}$$

$$= 4\left(\cos \theta \cos \frac{\Delta\theta}{2} - \sin \theta \sin \frac{\Delta\theta}{2} \right) \sin \frac{\Delta\theta}{2} .$$

putting

$$\cos \frac{\Delta\theta}{2} = 1 \text{ and } \sin \frac{\Delta\theta}{2} = \frac{\Delta\theta}{2} :\text{-}$$

$$\Delta d^* = 2\Delta\theta\left(\cos \theta - \sin \theta \frac{\Delta\theta}{2} \right) . \tag{1}$$

$$d^* = \frac{\lambda}{d} .$$

Taking logs and differentiating with respect to λ :—

$$\frac{\Delta d^*}{d^*} = \frac{\Delta \lambda}{\lambda} = \frac{\Delta d^*}{2 \sin \theta} = \Delta \theta \left(\cot \theta - \frac{\Delta \theta}{2} \right) \quad \text{from 171} \quad (1)$$

$$\therefore \quad \cot \theta = \frac{\Delta \lambda}{\lambda} \cdot \frac{1}{\Delta \theta} + \frac{\Delta \theta}{2}.$$

If the radial separation (perpendicular to the rotation axis direction) of the α_1, α_2 reflections is t mm, then $\Delta(2\theta) = 2\Delta\theta = t/r$, where r mm is the radius of the film holder.

$$\therefore \quad \cot \theta = \frac{\Delta \lambda}{\lambda} \cdot \frac{2r}{t} + \frac{t}{4r}.$$

θ, $\sin \theta$ and $d^* = 2 \sin \theta$ can be evaluated from this expression, to give an accuracy of 1 or 2 in 1,000. This is sufficient for most purposes.

6.2.2. *Experimental*

The RL net, drawn from the approximate values of the parameters, is used to design 5° oscillations which will give the desired reflection, first on one side of the collimator and then on the other (Section **14**.1, p. 175). A 6 mm hole is punched through the centre of the film and black paper wrapping, level with the collimator hole, and a thin collet and collar fitted in the hole to exclude light. The film is loaded the reverse way round to normal, with the collimator projecting through the hole in the film. Since only back reflections are of interest, the collimator guard slit can be removed, giving more clearance in the hole. Spring clips are placed about 10 mm above and below the collimator, to press the film against the film holder. The camera must be in good adjustment (Appendix IX, p. 422) especially as regards the concentricity of the film holder and camera axis.

The reasons for this experimental arrangement are similar to those for the corresponding powder technique [1(87, 158)]; i.e. they minimise the effect of errors in the measured radius of the camera, allowing for half the thickness of film and paper envelope, and, in this case, the effect of shrinkage. For the highest accuracy, shrinkage should be measured by printing a scale on the edge of the film before processing, or calculated from the known average shrinkage for the film and processing used.

The arc whose plane is nearest to the RL vector being measured is mis-set by 0·25°, and the two designed oscillation photographs registered on the same film, together with the two at 180° to the first pair. The arc mis-setting gives a slight vertical separation to the two pairs of reflections so that they can be measured separately, but has negligible effect on their horizontal separation.

The horizontal separations of the α_1 and α_2 reflections for the two pairs are measured by travelling microscope, averaged and converted to θ values, as though they were high angle powder reflections [1(85)], and the values of $d^* = 2 \sin \theta$ for $\lambda(\alpha_1)$ and $\lambda(\alpha_2)$ are calculated.

This is repeated until accurate values of a^*, b^* and c^* have been determined. α^*, β^* and γ^* are obtained by triangulation. The highest angle $h\bar{k}0$ reflection for which ha^* and kb^* are not too unequal is chosen, and $d^*_{h\bar{k}0}$ measured. This, together with ha^* and kb^*, form a triangle with γ^* between the two axes. The cosine rule gives an accurate value of γ^*. Similarly α^* and β^* can be determined. Crystals mounted about the three axes are required in the general case. For orthogonal crystals only two axes are required, and the same applies to the monoclinic case, as long as one of the axes is **b**.

Figure 181, (p. 304) shows a photograph taken to determine $d^*_{10, 0\bar{8}}$ for sucrose.

(**Example 27**, p. 304; but not on first reading—the results of Section **14**.1 (p. 175) are required.)

An alternative procedure is to measure all the values of d^* above $\theta = 75$ or 80° in the 3 zones, and refine the approximate values of the parameters by least squares, using a computer. This can lead to inaccurate values for one or more parameters. If for example, all the d^*_{hk0} measured had low h and high k, b^* would be obtained more accurately than a^*, and γ^* would also be less accurate.

In the method using the separation of the α_1, α_2 doublet there are no special experimental conditions. The ordinary zero layer non-integrated Weissenberg photograph, normally taken to estimate very weak intensities, even when integration is employed, is sufficient. This is because at the lower level of accuracy aimed at, the error due to eccentricity or an inaccurate value of the film radius is negligible, compared with the error due to the inaccuracy in the measurement of the separation t.

The measurement of t is of crucial importance, and must be done on a low power travelling microscope to get an accuracy approaching 0·025 mm, if the parameter accuracy of 1 in 1,000 is to be obtained. It is, however, possible to evaluate a considerable number of different d^*_{hkl} values from all the reflections with significant α_1, α_2 separations on a zero layer film. The approximate parameters which will have been determined earlier (Section **13**.6.1, p. 170) can then be refined by a least squares procedure [10]. The method has been extended to non-zero equi-inclination layers, so that all the parameters can be obtained from one crystal setting [11].

6.3. THE CONTENTS OF THE UNIT CELL, THE CALCULATED DENSITY AND THE 'MOLECULAR' WEIGHT

From the accurate cell parameters the volume $V \text{ Å}^3$ of the unit cell is calculated (Appendix II, p. 361). The observed density D_0 is normally measured by a flotation method, preferably a density column [12]. Then:—

$$\text{mass of unit cell} = V \times 10^{-24} D_0 \text{ gm.}$$

Let the 'molecular' weight of the substance be M on the O/16 scale. (In ionic crystals there may not be any recognisable molecule, and the 'asymmetric unit' of structure in such cases is included in 'molecule.')

The mass of the 'molecule' $= M \times 1\cdot660 \times 10^{-24} \text{ gm.}$

6.3.1. *Number of 'molecules' per unit cell*

If there are Z' 'molecules' per unit cell

$$Z' \times M \times 1\cdot660 \times 10^{-24} = V \times 10^{-24} D_0$$

$$Z' = \frac{V D_0}{1\cdot660 \, M}.$$

For a reasonably perfect crystal the number of 'molecules' per unit cell is an integer to a high degree of accuracy, and is also subject to space group restrictions. The approximate value found above, if it is within $3\sigma(c\cdot0\cdot1)$ of an integral value allowed by the space group, is put equal to the nearest integer Z. If the divergence is greater than 3σ the reason must be found and allowed for.

6.3.2. *The calculated density*

$$D_c = \frac{ZM \times 1\cdot660}{V} \qquad (Z \text{ integral}).$$

This value is normally quoted for comparison with D_0.

6.3.3. *The 'molecular' weight*

If the 'molecular' weight is unknown, or only known approximately, then

$$nM = \frac{V D_0}{Z'' \times 1\cdot660}$$

where Z'' is the smallest integral value allowed by the space group and n is an integer. If an approximate value of M is known, then n can be found and an accurate value of M calculated.

Chapter 14

The Production and Interpretation of Oscillation and Rotation Photographs

1. Oscillation Photographs

Refer to Section **11**.2 (p. 135) for a description of the camera.

1.1. PRODUCTION

Except in very special circumstances oscillation and rotation photographs are taken using the cylindrical film holder. Charts are available to enable the various angular and linear coordinates of the RL vector, corresponding to a given reflection, to be read off directly from the photograph. The range of the oscillating mechanism is chosen (usually 15°, but 10 or 5° for special purposes) and the mechanism set at one end of the range, preferably the anti-clockwise end. For the cam operated oscillation arm (Fig. 87, p. 136) the height of the cam follower is adjusted to the required cam, and the motor stopped when the follower is on the point of the cam. The spindle is then locked to the oscillation mechanism, either at random or at some specified reading of the goniometer scale (in Fig. 87, by tightening the set screw in the collar above the scale). The oscillation motor is started, the tube shutter opened, and with a normal sealed Cu-target X-ray tube and a crystal about 0·25 mm in linear dimensions, an exposure of the order of 30 mins. is made. Tank development of the exposed film and the use of a hypo-eliminator bath are both desirable.

1.2. THEORY OF INTERPRETATION

It is now necessary to relate in more detail the positions of the reflections on the photograph to those of the corresponding relps. If the unit cell of the crystal is unknown this stage may be required before ϕ setting (Section **13**.4, p. 160) is possible.

The coordinates of the relps relative to $\mathbf{a}^*$, $\mathbf{b}^*$ and $\mathbf{c}^*$ are the indices hkl of the reflections, and it is these indices we wish to find. To do this it is necessary to measure the RL coordinates on an absolute basis, and then relate them to the coordinates in terms of the RL axes. Since we are concerned with the coordinates of points which are rotating round an axis, they can most conveniently be given as cylindrical coordinates. In Fig. 105 the cylindrical coordinates of the relp P are (a) the radius of the cylinder ON (N is the projection of P on to the equatorial plane); (b) the height above the equatorial plane PN; (c) the angle NOC, between the two planes containing the rotation axis and C and P respectively. ON is given the symbol ξ, and PN the symbol ζ. NOC is the angle ϕ, used previously in defining the direction of OP.

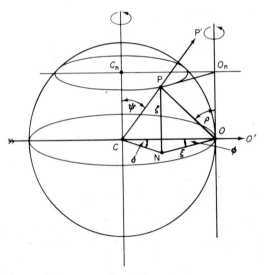

Fig. 105. The relation between the cylindrical coordinates ξ, ζ, and ϕ of a relp P in the sphere of reflection, and the angles ψ and δ defining the direction of the reflected beam, CP'. The angle ρ, one of the spherical coordinates of P, is also shown. The other two spherical coordinates are d^*_{hkl} (i.e. OP) and ϕ.

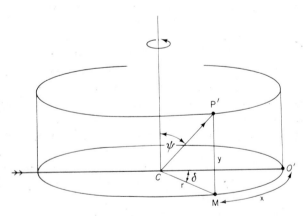

Fig. 106. The relation between ψ, δ and the camera radius r mm, and the coordinates x, y of the reflection P', on the film.

In Fig. 106 the coordinates x, y, of the corresponding reflection P', on a cylindrical film of radius r, are defined. From this figure:

$$\cos \psi = \frac{y}{\sqrt{r^2 + y^2}}; \qquad \sin \psi = \frac{r}{\sqrt{r^2 + y^2}}; \qquad \delta = \frac{x}{r}.$$

In ∇ONC, Fig. 105:—

$$\xi^2 = ON^2 = OC^2 + CN^2 - 2OC \cdot CN \cos \delta$$

$$= 1 + \sin^2 \psi - 2 \sin \psi \cos \delta.$$

$$\zeta = \cos \psi$$

$$\therefore \quad \zeta = y/\sqrt{r^2 + y^2} \tag{1}$$

$$\xi = \sqrt{1 + r^2/(r^2 + y^2) - 2r \cos (x/r)/(r^2 + y^2)^{\frac{1}{2}}}. \tag{2}$$

From these expressions, ξ and ζ can be calculated from x, y and r. Bernal, in 1926, constructed a chart giving lines of constant ξ and ζ at intervals of 0·05 R.U. By laying such a 'Bernal chart' (Fig. 107) on the photograph, the ζ and ξ coordinates of the relp, corresponding to a given reflection, can be read off. (**Example 28** p. 305).

The straight lines of constant ζ across the chart correspond to layer lines (Fig. 24, p. 33) and present no problem. The curves of constant ξ are more difficult to understand, but they correspond to the intersection of a cylinder of radius ξ with the sphere of reflection. The projection of this intersection curve by reflections from C on to the film gives a constant ξ curve as shown in Fig. 108.

It should be noted that the third coordinate ϕ can be read off using the ρ, ϕ chart (Section **13.2.1.2** (ii) (5), p. 152); but the value is only known $\pm 7·5°$ (for a 15° oscillation) since, in the absence of an associated Laue streak, there is no means of knowing at what point in the oscillation the reflection occurred.

These cylindrical coordinates in R.U. must now be related to coordinates in terms of the RL axes, **a* b* c***. It is assumed that the lattice has been determined (Section **13.6**, p. 170), and the crystal set with $+$ **b** upwards along the rotation axis. The set of RL planes parallel to **a* c*** will therefore give rise to the layer lines on the photograph (Fig. 109, p. 180). The k index of any reflection will then be determined by the layer line (0 for the zero layer line, $+1$ for the line above zero, $\bar{2}$ for the second below zero, etc., Section **3.2**, p. 28). The h and l indices have to be found for each reflection. In the case of a complete rotation photograph there is likely to be considerable overlapping and

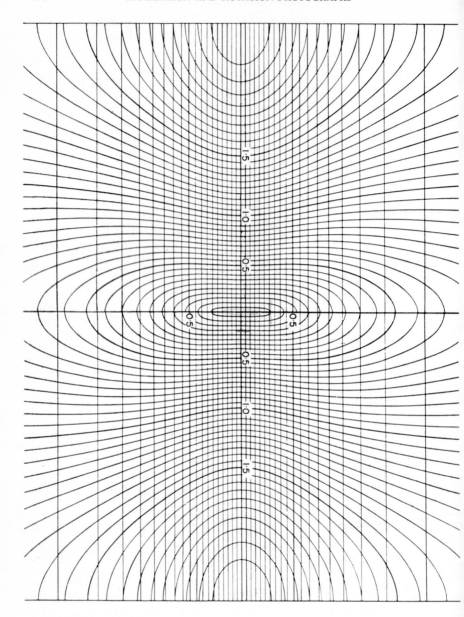

FIG. 107. The ξ, ζ (Bernal) chart for a cylindrical film. The horizontal constant ζ lines correspond to layer lines on a rotation photograph. The ζ reading for a layer line is the height of the corresponding RL layer above the equatorial plane. The constant ξ curves are the projection on to the film of the line of intersection of the sphere of reflection and a cylinder of radius ξ (Fig. 108).

ambiguity, so that an oscillation of 5, 10 or 15° is usually employed instead of a complete rotation of the crystal.

The angular position of the X-ray beam, relative to the axial directions **a*** and **c***, must be known at the two ends of the oscillation. A section through the sphere of reflection containing a particular RL layer can then be drawn, and the problem reduced to two dimensions. The 'Circles of Reflection' cut by such sections have different radii, but all have centres C_n, vertically above one another (and are therefore 1 R.U. from the rotation axis) and the diameters which cut the rotation axes are all parallel (Fig. 105, p. 176). The **a*****c*** net is first drawn (Appendix II, p. 352) to a convenient scale and the 'Origin

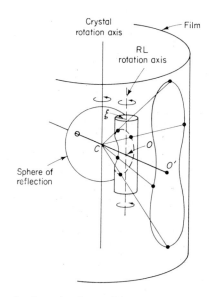

FIG. 108. Diagram showing how the shape of the constant ζ curves arises from the intersection of the sphere of reflection and a cylinder of radius ζ.

of the Net' L, marked. The point where the rotation axis cuts the net for a particular layer (the 'Origin of Construction' O_n) is found and marked. The directions of the X-ray beam at the two ends of the oscillation are drawn from O_n. In this connection it is easier to deal with a stationary crystal and an X-ray beam rotating relative to it. But it must be remembered that the sense of rotation is reversed, i.e. a crystal rotating clockwise from a given goniometer reading corresponds to the X-ray beam rotating anti-clockwise. In the case of the zero layer, L and O_n are coincident, but this is not necessarily so for non-zero layers. For the zero layer L is also the same as O, the origin

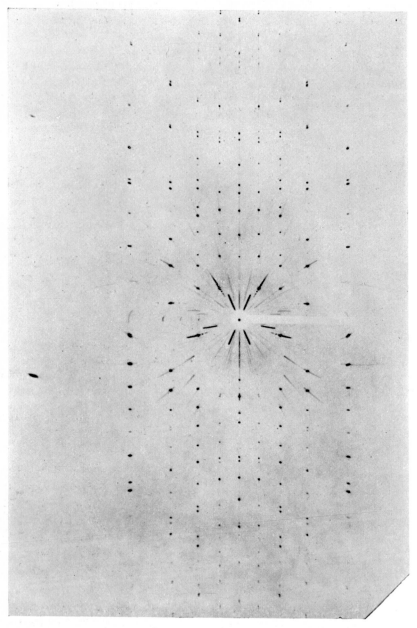

FIG. 109. A 15° oscillation photograph of sucrose, taken with CuKα radiation and a cylindrical camera of radius 30 mm. The row lines of reflections following constant ξ lines on the Bernal chart show that the vertical **b** axis is orthogonal to **a** and **c**, i.e. **b*** is parallel to **b**.

of the whole RL. The circle of reflection for the zero layer passes through O, but this is not the case for O_n and the circle for any other layer.

1.3. Scheme for Indexing an Oscillation Photograph

1. Construct the RL net (100 mm = 1 R.U.) which is perpendicular to the rotation axis (Fig. 110b). It is essential to draw an accurate net. Errors should not be greater than 0·1 mm, and the axial distances ha^* etc. must be calculated to 3 decimal places for the various values of h. On no account must the axial points be 'stepped off' with compasses or dividers set to a^*, since this accumulates errors and can give rise to discrepancies of 1 or 2 mm for the points furthest from the origin. Draw a small, freehand diagram to determine the best position for the RL origin and axes on the paper (Fig. 110a).

2. Mark the origin of construction, O_n. This is the point where the rotation axis cuts the RL layer. For all zero layers it is the same as the origin of the RL net. This is also the case for all layers about an orthogonal axis (Fig. 110b). But in the non-orthogonal case the origin of construction for the non-zero layers is no longer at the net origin, and its position has to be found (Appendix II, p. 367).

3. From the origin of construction O_n, draw the limiting directions of the beam on the RL plane (beam rotation in the opposite direction to that of the crystal) as in Fig. 110. Orientate the paper so that the beam is directed more or less towards the observer, as it normally is in practice and when a film is being viewed in the standard orientation (p. 143).†

4. Draw the circle of centres, radius one R.U. from the *origin of construction* O_n, cutting the two lines representing the extreme positions of the beam.† (Fig. 110b).

5. With the points of intersection as centres construct two circles of reflection, radius $\sqrt{1 - \zeta^2}$, where ζ is the height of the layer being investigated above the zero layer. (Measure ζ from the photograph (Fig. 109) with the Bernal chart). The points which lie in the two lunes will have passed through the sphere of reflection during the oscillation; and the corresponding reflections, if of detectable intensity, will appear on the photograph, those corresponding to the left-hand lune on the left-hand side of the photograph and similarly for the right-hand lune.

6. Using the Bernal chart measure and list the ξ values of the reflections (Left and Right with the beam coming up through the photograph—

---

† Points (3) and (4) must be repeated for each layer in the non-orthogonal case.

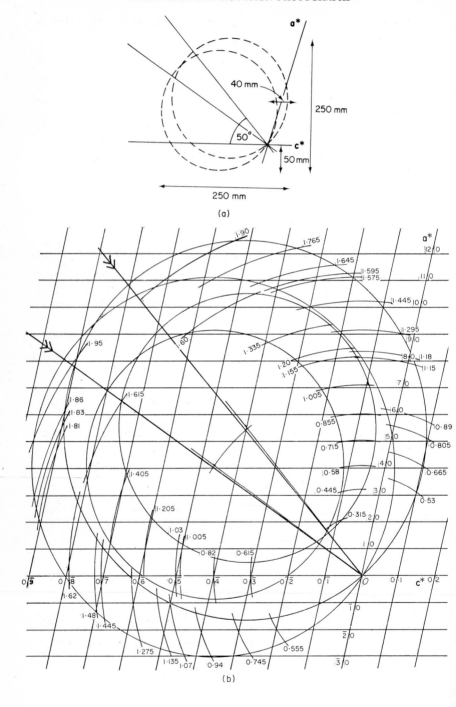

(a)

(b)

clipped corner top right), and add a rough estimate of intensity to help in identification. Leave a column for indices (Fig. 111).

7. Set a pair of compasses to the ξ values in turn (100 mm = 1 R.U.) and, with the origin of construction as centre, mark an arc through the appropriate lune in each case (Fig. 110b). The relp in the lune which is nearest to the arc (within 1 or 2 mm) will correspond to the reflection having the given ξ value. Its indices are given by the coordinates of the point (Fig. 110b). Put a ring round each point which corresponds to a reflection, read off its indices and put them into the table. (**Example 29, p. 305**).

1.4. PROCEDURE FOR UNKNOWN ϕ

If the limiting directions of the X-ray beam relative to the RL axes have not already been determined, they can usually be found from the photograph itself. For this purpose the circles of reflection are drawn on a transparent sheet which has the measured ξ values marked as arcs through the lunes. A hole at the origin is placed over a drawing pin stuck upwards through the origin of construction on the net. The circles of reflection can then be swung round until a position is found in which the arcs all pass near relps. Attention should first be concentrated on the small ξ values, and possible solutions checked and final adjustments made using the larger values. In favourable cases, especially if the Laue streak method (Section 13.2.1.2 (ii) (5) p. 152) can be used to increase the accuracy, it should be possible to obtain the directions of the X-ray beam to $\pm 0.5°$. The procedure for indexing is then the same as in the previous section.

If one possible position can be found on the zero layer, there is always another related to the first by the two-fold rotor at the origin. This is a consequence of the centre of symmetry necessarily at the origin of a lattice of uniformly weighted points, and produced by the operation of Friedel's law for a RL weighted with intensities. However if the zero layer axes are non-orthogonal it may happen that no reasonable match can be found at all. This would be so if the net (**a* c*** say) had been drawn with + **b*** upwards, whereas in the crystal it was, in fact, downwards. If the net has been drawn on tracing

---

FIG. 110. The interpretation of the zero and fourth layer lines of Figure 109. The reproduction is reduced to nearly $\frac{1}{2}$ scale and trimmed to the limit. (a) Small freehand sketch to decide the size of net and position of origin. (b) The accurate drawing for the interpretation. The directions of the beam were slightly in error for the zero layer, and were corrected for the fourth layer interpretation. The points corresponding to reflections have not been ringed, because of the small scale of the diagram, but the ξ values for the arcs are given. (The measurements were made from the original photograph. Those from Fig. 109 are slightly smaller due to shrinkage in reproduction.)

paper it can be inverted and a correct solution found. A less desirable procedure (because more confusing, especially in the later stages) is to invert the photograph instead.

When the RL axis is not along the rotation axis (**a** or **c** along the rotation axis in the monoclinic case or any axis in the triclinic case) there are two possible positions for the origin of construction in the non-zero layers. These two positions can be tested by trial and error, and the correct one determined.

The correct way up for the zero layer and the correct position of a non-zero layer origin of construction, determines the positive directions of the three axes in the triclinic case. In the higher symmetry systems a choice of positive direction must be made for at least one axis (Appendix II, p. 352).

	L.H.S.			R.H.S.	
ξ	I	hkl	ξ	I	hkl
0·555	m	$\overline{2}0\overline{2}$	0·53	s	301
0·745	s	$\overline{2}0\overline{3}$	0·665	vs	401
0·94	s	$\overline{2}0\overline{4}$	0·805	s	501
1·07	s	$\overline{1}0\overline{5}$	0·89	s	600
1·135	s	$\overline{2}0\overline{5}$	1·15	m	$80\overline{1}$
1·275	s	$\overline{1}0\overline{6}$	1·18	m	800
1·445	m	$00\overline{7}$	1·295	s	$90\overline{1}$
1·48	s	$\overline{1}0\overline{7}$	1·445	s	$10,0\overline{1}$ (10,0$\overline{2}$?)
1·62	vw	$10\overline{8}$ ($20\overline{8}$?)	1·575	vw	$11,0\overline{2}$
1·81	vw	$40\overline{9}$ ($30\overline{9}$?)	1·595	s	$11,0\overline{3}$
1·83	vw	$50\overline{9}$	1·645	vw	$11,0\overline{4}$
1·86	m	$60\overline{9}$	1·765	w	$12,0\overline{4}$
1·95	s	$80\overline{9}$	1·90	w	$12,0\overline{6}$ (11,0$\overline{7}$?)

FIG. 111. Table of ξ values measured from the photograph (Fig. 109, p. 180) by means of the Bernal chart (Fig. 107, p. 178). An indication of intensity (weak, medium, strong) is added to help in identification when referring back to the photograph. The indices have been filled in subsequently as a result of the interpretation. (The ξ values were measured from the original photograph. Those measured from Fig. 109 are slightly smaller, due to shrinkage in reproduction.)

A circle of reflection drawn on a transparent sheet is also useful in designing oscillations to bring in given reflections. Figure 112 shows a more elaborate version, which can save a great deal of time in such cases. (**Example 30,** p. 305).

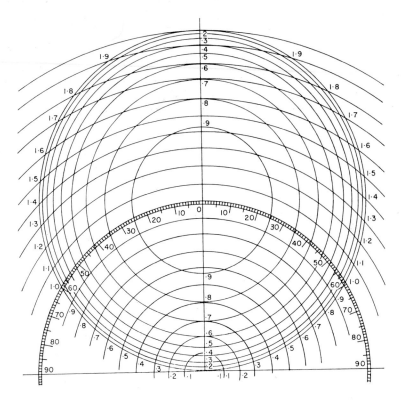

Fig. 112. A chart engraved on perspex designed for interpretation of oscillation photographs. The circles of reflection for ζ values from 0 to 0·8, in steps of 0·1 are combined with constant ξ arcs about the origin, also in steps of 0·1, and an angular scale in degrees. This is especially useful for checking interpretations in orientation problems. The original was drawn to the normal scale of 1 R.U. = 100 mm.

2. Rotation Photographs

2.1. PRODUCTION

The procedure is essentially the same as for an oscillation photograph, except that the oscillation mechanism is uncoupled and the uniform rotation drive substituted (in Fig. 87, p. 136, the oscillation arm is taken off and the sprung jockey pulley released on to the belt drive).

2.2. INTREPRETATION

(a) The layer separation ζ (obtained from $n\zeta$, measured with a Bernal chart from the pair of highest order layers) gives the separation of crystal lattice points, g along the rotation axis, from the relation $g = \lambda/\zeta$ (Section 3.3, p. 30). Usually $\mathbf{g}$ will be one of the crystal lattice axes, and this relation gives the axial length.

(b) If an important RL direction is along the rotation axis, there will be vertical rows of relps parallel to it. All the relps of one row will have the same ζ value, and the reflections will all lie on a constant ξ curve on the rotation photograph. These more or less vertical lines of spots are known as 'row lines'. If, as will normally be the case, a crystal lattice direction is also along the rotation axis, layer lines will also be present and the ξ value of the whole row line will be the same as that of the zero layer reflection which lies on it. It is thus possible to use upper layer reflections, which have greater dispersion (the constant ξ curves on the Bernal chart become wider apart at higher levels) and are less obscured by Laue streaks, to help in measuring the ξ value of the zero layer reflections. It is possible, in this way, to detect an accidently absent reflection on the zero layer and obtain the ξ value of the relp.

Row lines are formed whenever an orthogonal crystal is rotated about an axis, or a monoclinic crystal about the $\mathbf{b}$ axis.

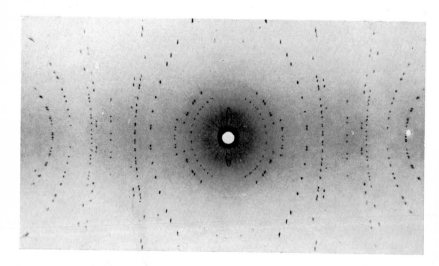

FIG. 113. A multi-rotation photograph of cuprite (Cu_2O), a cubic mineral, which also occurs in corrosion deposits. The identification is easily made from one photograph without any problems of orientation. Enlarged ($2\times$) from cylindrical film in a 10 mm radius camera. 60 μm crystal. Ehrenberg–Spear fine-focus X-ray tube.

(c) The zero layer of a rotation photograph can be treated as a simplified powder photograph and interpreted similarly. [1(120)]. In particular, if the zero layer axes are known or suspected to be orthogonal, a special Bunn chart is available for indexing and finding the ratio of the axial lengths. In the general case an application of Ito's method should enable the two RL axes and the angle between them to be determined.

But moving film cameras are generally available with which the zero and other layers can be photographed and indexed, without the ambiguities and overlapping which occur on almost all rotation photographs. It is therefore rare for rotation photographs to be used other than as a means of measuring the RL interplanar spacing ζ, and for checking that no weak intermediate layers exist which might be missed on oscillation photographs. (**Example 31, p. 308**).

(d) There is however one application of rotation photographs which is sometimes useful for identification purposes. If the crystal is randomly oriented and a rotation photograph taken, followed by another on the same film after one of the arcs has been given the maximum angular displacement, followed by a third after displacement of the other arc, then an approximation to a powder photograph is produced, with very spotty lines (Fig. 113). This can be used for identification, by comparison with an actual powder photograph or the data in the A.S.T.M. or other index [1(266)]. This technique is particularly useful for cubic crystals where the powder lines are well separated. In the cubic case it is also possible to differentiate between P and I lattices, which give very similar powder diagrams, by counting the number of spots on each of the first three lines. The number of spots on a line is roughly proportional to the multiplicity (Section **10**.2.2, p. 129), and the multiplicities are very different for the two cases. (Answer 19, p. 483).

(**Example 32,** p. 308).

Chapter 15

The Production and Interpretation of Weissenberg Photographs

1. Production

Refer to Section **11**.3 (p. 139) for a Description of the Camera

1.1. CRYSTAL AND SCREEN SETTING

The crystal will normally have been set on a rotation camera and the arcs transferred to the Weissenberg camera. Care must be taken to see that the arcs are not more than a few degrees from the zero position, and that the translation slides are also nearly central, otherwise the arcs may foul the layer line screen. There should be only a few mm of fibre between the plasticene and crystal, otherwise significant bending may occur, because the Weissenberg rotation axis is normally horizontal. Final setting should be done on the Weissenberg camera itself, using the same methods, and the highest possible accuracy achieved. This is important because the more accurately the crystal is set, the narrower the screen slit can be made, and this keeps scattered radiation and therefore fog on the film to a minimum. This is especially the case for the higher layers which are particularly sensitive to setting errors. Final setting can be accomplished using a strip of film in a light tight envelope, held round the screen by a rubber band. The screen has a wide slit which, after the setting is complete, is narrowed until the edges are nearly touching the reflections. This is checked by an oscillation photograph, with film round the screen as for setting. The background fog defines the edges of the slit. For non-zero layers the slit is moved a calculated distance s mm, and a similar check made. For the higher layers it may be necessary to widen the slit slightly.

1.2. FILM CHOICE AND LOADING

For accurate intensity measurement a multiple film pack is required, since the greatest range that can be measured accurately by photometer is 100 or 120 to 1. For eye estimation against a standard scale the range is nearer 40 to 1.

188

Since the range of intensities to be measured is often 1,000 to 1 or more, it is necessary to have a sensitive film nearest the crystal to measure the weak reflections, and a less sensitive film underneath to measure the strong ones. A factor of 8 or 10 to 1 between the two films enables the range to be covered by photometer, and still leaves sufficient reflections measurable on both films to enable the correlation factor to be obtained. The factor varies with the obliquity of the reflection on the film, and with the film batch. It must be deter-

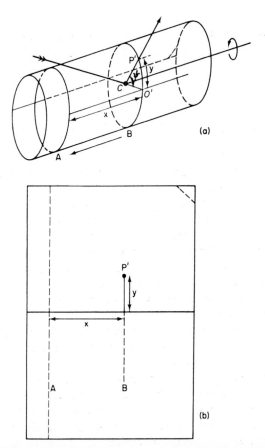

FIG. 114. Schematic diagram of a Weissenberg camera, showing the crystal at C giving rise to a zero layer reflection at P'. The layer line screen is omitted for clarity. The line at O' parallel to the direction of traverse, is the trace of the direct beam, and the opposite line represents the join of the two ends of the film. In practice a gap must be left to take the collimator. The angle of deviation ψ (for the zero layer this is 2θ, but for non-zero layers it is the projected angle and no longer 2θ), and the coordinates, x and y mm, of the reflection are shown. (b) shows the reflection and coordinates on the unrolled, flat film. The cut corner of the film locates its position in the film holder. (The corner should be cut off immediately after unloading the film).

mined for each layer. For eye estimation a smaller factor and 4 or 5 films in a pack are required.

For a multiple film pack, the inner films must be progressively shortened to take account of the smaller cylinder on which they will lie, and a carefully tailored envelope of *uniform*, light tight material must be made to contain them.

1.3. DESIGNING THE WEISSENBERG OSCILLATION

Weissenberg photographs, especially if integration is employed, usually require long exposures, and it is essential to make sure that the required relps are photographed and that unnecessary ones are not. The determination of the oscillation required from the symmetry of the RL (Section **10**.2, p. 128) and from the geometry of the sphere of reflection is essentially the same as for similar problems on the rotation camera, taking due account of the differences arising from equi-inclination settings, as detailed below.

2. Interpretation of Weissenberg Photographs

2.1. NORMAL BEAM METHOD

2.1.1. *Zero or equatorial layer*

Figure 114 shows the coordinates x and y mm, of a reflection on the film. x is measured parallel to the cylinder axis from the section A (defined by the layer line screen opening—Fig. 88, p. 139) which is opposite the crystal at the beginning of an anti-clockwise oscillation looking from the left.† (This corresponds to looking down on the crystal in the case of the rotation camera). y is measured along the arc of the circle from the centre line of the film, where, but for the back stop, the direct beam would hit it. When the film is unrolled y becomes the perpendicular distance from the centre line.

The problem is to relate the coordinates x, y mm of the reflection on the film to the coordinates X, Y, R.U. of the relp corresponding to it. Figure 115 shows the geometry of the RL and the sphere of reflection. OX is along the beam at the beginning of the oscillation and OY perpendicular to it. After a rotation ω, the relp P is in the sphere of reflection, and gives rise to the reflection P'. CE is parallel to OX.

† If the relation of translation to oscillation is the opposite to that shown in Fig. 114, A will be at the R.H. end of the film and x measured from right to left. If the unrolled film from such a 'reversed' Weissenberg camera is turned over about the y axis, the geometry will be identical to that described, except for the elongation and shortening of spots on upper layers (Section **18**.2.2.2, p. 272) which will be interchanged, top and bottom.

From Fig. 114:—

$x = C\omega$ where C is a camera constant which is normally $57 \cdot 3/2$, i.e. $1 \text{ mm} \equiv 2°$.

$y = r\psi$ where r mm is the radius of the film, normally $5 \cdot 73/2$ mm.

From Fig. 115, with PDE and CF parallel to OX and CE parallel to OY:

$$X = PE - DE = PE - CF.$$

$$\begin{aligned} X &= CP \cos (\omega - \psi) - CO \cos \omega \\ &= \cos (\omega - \psi) - \cos \omega, \text{ since } CP = CO = 1 \text{ R.U.} \end{aligned}$$

$$Y = OF - DF = OF - CE.$$

$$\begin{aligned} Y &= CO \sin \omega - CP \sin (\omega - \psi) \\ &= \sin \omega - \sin (\omega - \psi). \end{aligned}$$

Substituting for ω and ψ, we have

$$X = \cos (x/C - y/r) - \cos x/C.$$
$$Y = \sin x/C - \sin(x/C - y/r).$$

From these equations it is possible to construct the RL from the coordinates of the reflections on the film. Fortunately this is not usually necessary and for most purposes a coordinate chart can be used. The construction of such charts can best be understood by starting with central lattice lines, which will usually be the RL axes.

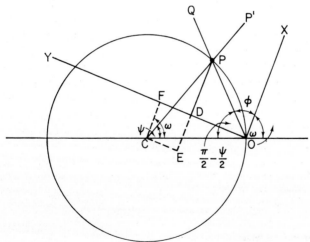

FIG. 115. The equatorial section of the RL, showing the axes OX, OY, (OX along the positive direction of the beam at the start of the rotation) to which the position of the relp is referred. PD is perpendicular to OY and therefore $X = PD$, $Y = OD$.

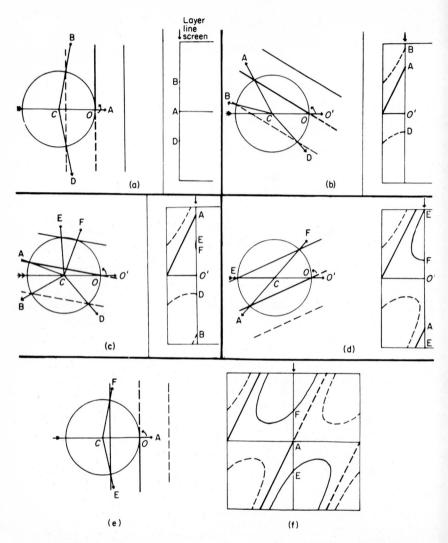

Fig. 116. The relation between lines of points in the RL and the corresponding curves (or slanting straight lines) of reflections on the film. (The relps can be considered to be so close together that the lines, both of relps and reflections, can be treated as continuous). The position of the RL in relation to the sphere of reflection is shown for a number of angles of rotation ((a) 0°; (b) 60°; (c) 80°; (d) 112·5°; (e) and (f) 180°), and the corresponding curves on the film up to that point. The positions of the layer line screen are marked with arrows and the reflections are lettered to correspond with the RL. The final diagram of the photograph (f), includes the next 180°, which is obtained from the first 180° by interchanging the dashed and full lines. In practice mechanical considerations normally limit the rotation to about 200°.

In Fig. 115 consider the central lattice line OQ making an angle ϕ, with OX.

$$\omega = \pi - \left(\frac{\pi}{2} - \frac{\psi}{2}\right) - \phi = \frac{\pi}{2} - \phi + \frac{\psi}{2}$$

$$\therefore \quad x = C\left(\frac{\pi}{2} - \phi + \frac{\psi}{2}\right).$$

Substituting $\psi = y/r$

$$x = C\left(\frac{\pi}{2} - \phi + y/2r\right)$$

$$= \frac{C}{2r}y + C\left(\frac{\pi}{2} - \phi\right)$$

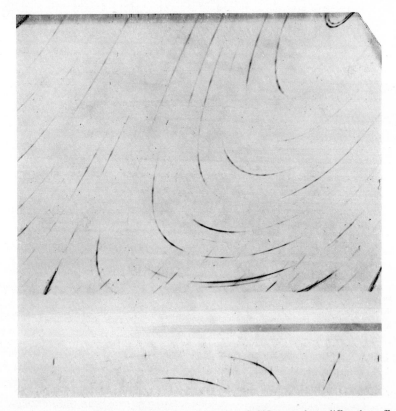

FIG. 117. Stacking faults in a crystal of wollastonite, $CaSiO_3$, produce diffraction effects equivalent to the extension of relps into rods parallel to $\mathbf{a}^*$ (Section 7.4.4, p. 86). These rods then trace out the continuous curves shown in the top half of a 1st layer Weissenberg photograph about $\mathbf{b}$. Such rods are easily distinguished from white radiation rods which are always radial.

or

$$y = \frac{2r}{C} x - \left(\frac{\pi}{2} - \phi \right) 2r.$$

Since, for any given central lattice line the final term is constant, the spots (x, y) on the film which arise from relps on the central lattice line, will lie

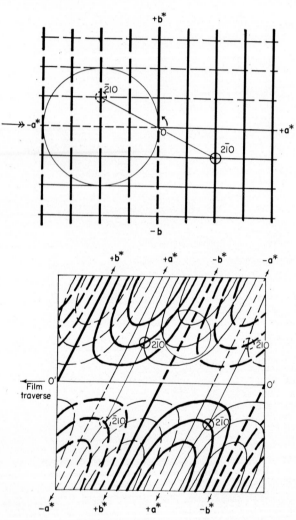

FIG. 118. A RL layer with orthogonal axes and the corresponding photograph. An area of minimum distortion is marked with a circle.

Relps related by a 2-fold point at the origin of construction always come out on the characteristic sloping line at equal distances on either side of the centre line. Two such points and reflections are marked.

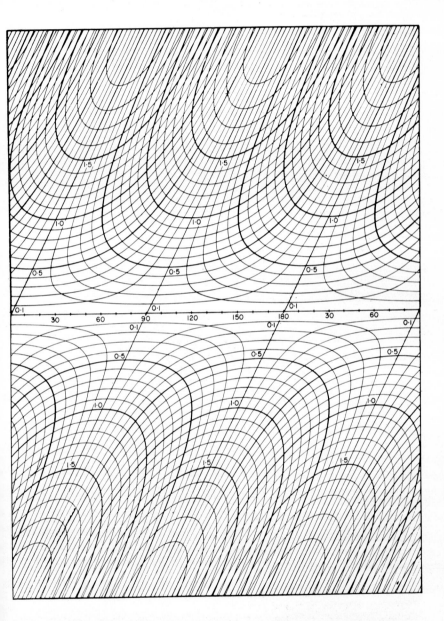

FIG. 119. A coordinate chart for a Weissenberg photograph corresponding to continuous lines in a RL layer with orthogonal axes of length 0·1 R.U. Slightly reduced in reproduction.

on a straight line making an angle of $\tan^{-1}(2r/C)$ with the x axis, i.e. the centre line of the film. For the standard Weissenberg constants, $r = C$ and $\tan^{-1}2 = 63\cdot5°$. Important central RL lines, i.e. axes or diagonals, will therefore show up on the film as well populated straight lines of reflections, making this characteristic angle with the centre line of the film (Fig. 89, p. 140).

In Fig. 116 (p. 192) the curves joining reflections which correspond to lines of relps (one axis and one parallel row-line on either side of it) are traced on the film as the crystal rotates through 360°. (Very occasionally a crystal will itself trace continuous curves on the film. Fig. 117, p. 193).

In Fig. 118 (p. 194) a second axis and parallel row-lines perpendicular to the first are added, and the corresponding curves marked on the film. This is done by noticing that, after a rotation of 90°, the second axis is in the same position as the first was at the begining of the oscillation. All that is required,

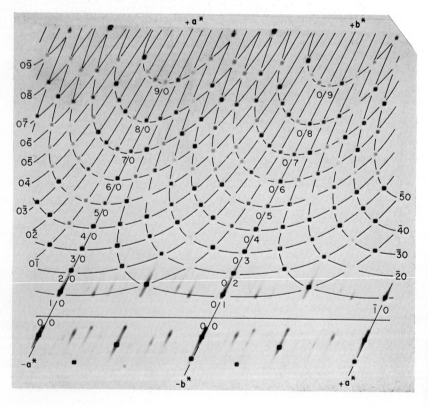

Fig. 120. The top half of Figure 89 (p. 140) with the RL axes and parallel lines of relps marked in. The area on either side of 040 and 050 (or 400 and 500) shows the RL net with minimum distortion. CuKα radiation. Cylindrical cassette, diameter 57·3 mm.

therefore, is to move the axes and curves of Fig. 116 (p. 192) to the right through a distance equivalent to 90° rotation. The last 90° section of the first set becomes the first 90° of the second. If this procedure is carried out by calculation, and the curves plotted for lattice lines which are 0·1 R.U. apart, we obtain a coordinate chart for orthogonal axes. (Fig. 119, p. 195).

When this has been done the distortion introduced by this method of photographing the RL net can be appreciated. It is best to take the top and bottom halves separately. Looking first at the top half of the photograph (Fig. 118, p. 194) and the $+ \mathbf{a}^*$ half of the RL net, one can imagine taking $+ \mathbf{b}^*$ in the left hand, $- \mathbf{b}^*$ in the right, breaking the two apart at the centre and bringing them round parallel to $+ \mathbf{a}^*$, distorting the net row-lines into U-shaped curves. The three axes are then slanted over at 63·5° and that part of the top half of the photograph between $+ \mathbf{b}^*$ and $- \mathbf{b}^*$ is produced. If the hands are reversed and the process repeated for the $- \mathbf{a}^*$ half of the net, we reproduce the next 180° of the photograph, to the right of $- \mathbf{b}^*$ (or to the left of $+ \mathbf{b}^*$). If, however, this distortion is tried on the bottom half of the photograph, it does not work until the RL net is turned over. The top half of the film photographs the front of the RL net, the bottom half the back of it!

When this distortion is understood a zero-layer Weissenberg photograph can be indexed almost by inspection. A coordinate chart may be required to decide which are the axial directions, and which reflections lie on a U-curve

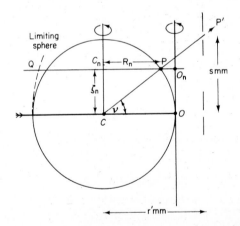

FIG. 121. The reflections for the nth layer, making an angle v with the normal to the rotation axis, are isolated by the layer line screen, which must be moved through a distance, s mm from its zero layer position opposite the crystal.

The diagram is drawn to correspond to similar rotation camera diagrams, but it has to be turned over about the X-ray beam to correspond to Figure 114 (p. 189) and most actual instruments.

The relp P and reflection P' have been drawn in the plane of the diagram, but could be anywhere round the cone of semi-angle $90 - v$ on CC_n as axis.

corresponding to a given RL row-line. But when the axes and curves have been marked, as in Fig. 120, the indices of any reflection can be read off from the photograph. The coordinate chart can be moved in the x-direction until one of its central lattice lines coincides with an axial line of reflections on the film, since the origin for x is essentially arbitrary. The parallel row lines will then lie on the corresponding curves of the chart; and if the net is orthogonal, the other axis and parallel row-lines will lie on the other set of axes and curves on the chart. If the net is non-orthogonal the chart will have to be moved to coincide with the second axis and the set of curves for lines parallel to it. (**Example 33**, p. 310).

2.1.2. Non-zero layers

The layer line screen is shifted by $s = r' \tan \mu$, where $\sin \mu = \zeta_n$ and r' mm is the radius of the layer line screen (Fig. 121). (The film cassette must be moved by $S = r \tan \mu$ for the pattern to start at the same distance from the edge of the film).

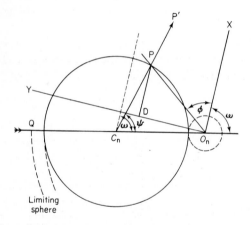

FIG. 122. The nth layer diagram corresponding to Figure 115 (p. 191). The angle ψ is now the projected angle and no longer equal to 2θ. O_n is not on the circle of reflection and the dashed circles mark the 'blind area' round the axis and beyond the reflection circle.

In Fig. 122, with the same symbols as in Fig. 115 (p. 191), but with $C_n P$ and $O_n P$ the projections of the reflected beam and RL vector onto the nth layer of the RL we have:

$$X = PD = C_n P \cos(\omega - \psi) - C_n O \cos \omega$$
$$= R_n \cos(\omega - \psi) - \cos \omega$$
$$Y = O_n D = \sin \omega - R_n \sin(\omega - \psi)$$

where $R_n = (1 - \zeta_n^2)^{\frac{1}{2}}$ (Fig. 121).

The coordinates x and y of the reflection P' on the film have the same relation to ω and ψ as for the zero layer; but the reciprocal lattice coordinates are a function of ζ as well as of x and y. The curves corresponding to RL row-lines are different for every layer. In particular, since $\widehat{PO_n C_n}$ is no longer equal to $\pi/2 - \psi/2$, but is a function of ζ_n, the central lattice lines are no longer straight lines on the photograph, but are curves whose exact shape varies from layer to layer.

Finally there are some relps which cannot be recorded by this method. They lie in the 'blind area' inside the inner dashed circle and outer annulus of Fig. 122.

Fortunately all these difficulties can be overcome and a number of other advantages achieved, by using the equi-inclination method. Since all modern Weissenberg cameras incorporate this facility, as described in Section **11.3** (p. 139), the further development of the interpretation of 'normal beam' upper layer Weissenberg photographs will not be pursued.

2.2. EQUI-INCLINATION METHOD

2.2.1. Zero layers

The X-ray beam is perpendicular to the rotation axis and there is no difference from the normal beam method.

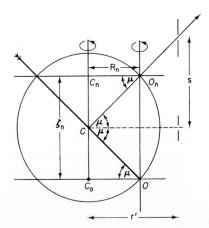

FIG. 123. Equi-inclination geometry for photographing the nth layer. For diagrammatic purposes it is easier to consider that the X-ray beam has been rotated clockwise through μ, although in practice it is the direction of the rotation axis which is altered. The rotation takes place about C (through which the real rotation axis passes) and O moves round the circle from its zero layer position on the dashed line. The zero and nth layer positions of the layer line screen are shown. As for Figure 121, the diagram should be turned over about the beam to correspond to the arrangements in most instruments. This does not affect the geometry.

2.2.2. Non-zero layers

The essential characteristic of the equi-inclination method is that the camera axis should be set at such an angle to the direct beam, that the point where the rotation axis cuts the RL layer being photographed should lie on the sphere. Fig. 123 shows that in this case the diffracted beams make the same angle with the rotation axis as the incident beam—hence the name 'equi-inclination.'

The movement of the layer-line screen from its position opposite the crystal for the zero layer is different from that for the normal beam method, and is given by: $s = r' \tan \mu$ where $\sin \mu = \zeta_n/2$. The cassette movement $S = r \tan \mu$. The direction of movement of the layer line screen must be such that the collimator, where it enters the screen, intersects the layer line opening. The cassette must be moved in the same direction, but the amount can be judged by eye, so that the point where the collimator enters the cassette slot is about the same distance from the end of the cassette as it was for the zero layer. The radius of the circle of reflection R_n is also different and equals $\left(1 - (\zeta_n/2)^2\right)^{\frac{1}{2}}$. (**Example 34,** p. 313).

The analysis of the equi-inclination non-zero layer photograph is very similar to that for the zero layer. The same figure (Fig. 115, p. 191) can be used if we change C and O to C_n and O_n. Since $C_n O_n = C_n P = R_n$ we have:

$$X = C_n P \cos (\omega - \psi) - C_n O_n \cos \omega$$

$$= R_n \left(\cos (\omega - \psi) - \cos \omega \right)$$

$$Y = R_n \left(\sin \omega - \sin (\omega - \psi) \right).$$

If we define $X' = X/R_n$ and $Y' = Y/R_n$, the equations for X' and Y' are exactly the same as for X and Y on the zero layer. The zero layer RL coordinate chart can therefore be used; but the coordinates measured will be X' and Y', which must be multiplied by R_n to give the true coordinates. Alternatively, if we expand the RL net isotropically by the factor $1/R_n$, the non-zero layer can be interpreted as though it were the zero layer, i.e. the reflections will lie along the same type of U-shaped curves as for the zero layer, but further out on the chart. Otherwise the interpretation proceeds exactly as for the zero layer, except that the central lattice lines, which come out as straight lines on the photograph, are lines through the *origin of construction* and *not* the origin of the net, where these are separate. In non-orthogonal systems, or for centred cells, there may not be any reflections on some central lattice lines for the non-zero layers. (Fig. 187, p. 314.)

The additional advantages gained are considerable. Whereas the practical limit for ζ_n in the normal beam method is 0·8 R.U., for an equi-inclination angle of 45° (which is about the physical limit) $\zeta_n = 1·4$ R.U. It is thus possible to photograph nearly twice as many layers about one axis. But each layer has a larger area photographed. In fact the maximum area theoretically possible is photographed, since relps which are furthest out are 2 R.U. from the origin, and there is of course no 'blind area.' It is not surprising therefore, that the equi-inclination method is now in universal use. (**Example 35** p. 313).

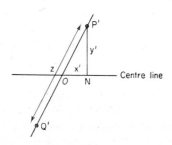

FIG. 124. The diagram of the **c*** axis on the zero layer of a Weissenberg photograph, with two $00l$ reflections, P' and Q', at equal distances on either side of the centre line. An accurate measurement of z in mm allows a fairly accurate determination of c^* to be made.

2.3. PARAMETER MEASUREMENT FROM A ZERO LAYER WEISSENBERG PHOTOGRAPH

In Fig. 124 P' and Q' are the reflections $00l$ and $00\bar{l}$ respectively, and $P'Q'$ is the sloping straight line corresponding to the **c*** axis in the RL. The equation of the line is $y = 2x$ (Section 2.1.1, p. 194).

Let ON be x' mm, $P'N$, y' mm and OP', $z/2$ mm. Then $y' = 2x'$ and

$$z/2 = \sqrt{5}\,x' = \sqrt{5}\,(y'/2).$$
$$\therefore \quad y' = z/\sqrt{5}$$
$$d^*_{00l} = lc^* = 2\sin\theta = 2\sin(y'/2r).$$

If $2r = 57\cdot3$mm, as in a standard instrument,

$$d^*_{00l} = 2\sin(z/\sqrt{5}) \text{ where } z/\sqrt{5}\text{ is in degrees.} \tag{1}$$

The procedure is therefore to measure in mm along the central lattice line on the photograph, from the reflection on the top of the film to the corresponding one on the bottom ($00l$ to $00\bar{l}$ in the example above), and convert to d^* by equation (1). As for other methods of parameter determination (Section **13**.6.2, p. 170) the highest angle reflections give the greatest accuracy, so the furthest out reflections should be used, preferably with resolved α_1, α_2 doublets, both of which should be measured. This method gives greater accuracy than the use of a coordinate chart (Section 2.1.1, p. 197), because the interpolation on a mm scale is more accurate. (**Example 36** p. 313).

3. Symmetry in Weissenberg Photographs and its Relation to Three-Dimensional Laue Symmetry†

Although the RL net suffers the large distortion already described, all the two-dimensional Laue point group symmetry elements that may be present in the original net can be easily recognised in the photograph.

The minimum distortion occurs about one third the length of an axial line of reflections from the centre line of the photograph (Figs. 118, p. 194, and 120, p. 196). It is here that inspection shows the type of two-dimensional unit cell present in the net.

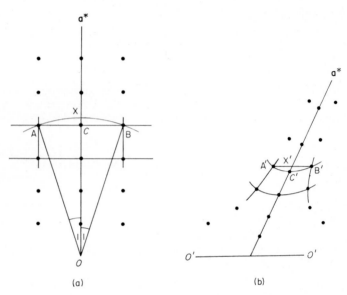

Fig. 125. The relation between angles and distances (a) in the RL, and (b) on the film.

But first it is necessary to relate measurements on the photograph to those on the RL net. Equal distances in the direction of traverse of the film, parallel to the direct beam centre line, correspond to equal angular separations of RL vectors in the net. Reflections which are equidistant from the centre line have equal projected angles of deviation, and correspond to relps which are equidistant from the origin of construction. In Fig. 125a, the orthogonal net on either side of the RL axis a* is distorted as shown in Fig. 125b on the photograph. A, B and X are all at the same distance from the RL origin O. The corresponding reflections A', B' and X' (X' can be considered to be a white radiation reflection lying on the Laue streak along a*) are therefore all at

† A knowledge of point group symmetry elements and their symbols is assumed in this and the following section.

the same distance from the centre line, i.e. A' X' B' is parallel to the centre line. Since $\widehat{AOX} = \widehat{XOB}$ the distances A' X' and X' B' are equal. Notice that it is not the distances to the axial reflection C' which are equal, but the distances to the point X' on the axial line where the straight line A' B' cuts it. If all pairs of reflections such as A' and B', at equal distances from the centre line and with A' B' bisected by **a***, are equal in intensity, then a mirror line of symmetry lies along **a***. Any RL mirror line, which must be a central lattice line, can be recognised in the same way. Such lines can clearly only occur in an orthogonal net. Figure 167 (p. 271) shows the effect of *mm* symmetry on a Weissenberg photograph.

Figure 126 shows some of the other possible arrangements of the RL net and the corresponding positions of the reflections.

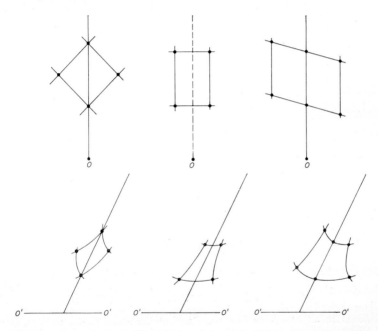

FIG. 126. The appearance, in the RL and on a photograph, of (a) a diamond net; (b) an *mm* net with no axial reflections (only possible on a non-zero layer); (c) a non-orthogonal net.

The other Laue symmetry elements that can occur in a two-dimensional net are 2, 3, 4 and 6-fold rotation points (rotors) at the origin. All except 2-fold rotors demand an orthogonal net (**a*** and **b*** axes for 3 and 6-fold nets can be chosen to produce a centred orthogonal cell with $a^* : b^* = \sqrt{3} : 1$) and can therefore be accompanied by mirror lines. The combination of a 2-fold rotation point and a mirror line produces a second mirror line at right angles to the first, and is therefore equivalent to *mm* symmetry.

Rotational symmetry can be easily detected on a Weissenberg photograph. The pattern repeats itself along the film in the direction of the centre line, after distances corresponding to $360/n°$ ($= 360/2n$ mm for a standard camera) for an *n*-fold rotation point. Fig. 127 illustrates this for trigonal symmetry.

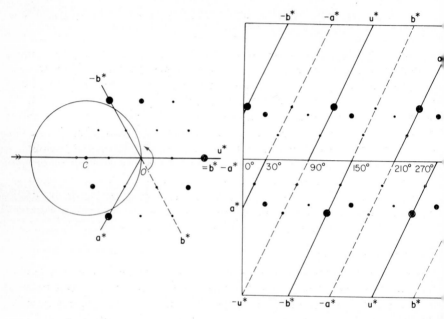

FIG. 127. Trigonal symmetry in a RL net and its appearance on a Weissenberg photograph. The size of the spot corresponds to the intensity of the reflection. Axes have been chosen in the RL to bring out the trigonal symmetry, but **u*** is *not* used for indexing (see Fig. 131, p. 209).

It is important to distinguish between the symmetry of zero and non-zero layer nets. In the Laue symmetry of the RL all symmetry elements pass through the origin, and a centre of symmetry is always present at the origin. The centre at the origin produces no two-dimensional symmetry except in a zero-layer net, and here it is equivalent to a 2-fold rotation point. Therefore every zero-layer net contains a 2-fold rotation point. However, only if a 2-fold axis is perpendicular to the net will the rotation point also appear in the non-zero layers. If a 3-fold axis is perpendicular to the layers, the additional 2-fold rotation point in the zero layer will produce 6-fold symmetry, but the non-zero layers will only contain a 3-fold rotation point.

Also a mirror line in the zero-layer may be due to a mirror plane perpendicular to the layers, or to a 2-fold axis lying in the zero layer (the centre will ensure that a mirror plane is perpendicular to this). The zero layer will have *mm* symmetry, but this will only be reproduced in the non-zero layers if there are two mirror planes perpendicular to the layers. If one of the symmetry lines in the zero-layer is due to a 2-fold axis, only the mirror plane perpendicular to it will produce a symmetry line in the non-zero layers. (**Example 37** p. 316).

4. The Deduction of the Three-Dimensional Lattice Type from the Stacking of Layers

Finally, the way in which the layers stack above one another and the type of three-dimensional lattice can be deduced from three photographs (the zero, first and second layer equi-inclination photographs). There are so many possibilities of combining the lattice types and symmetry elements that the description will be limited to the main relationships of the three layer photographs. It will be assumed (unless specified otherwise) that a properly chosen **c** axis is the rotation axis, and that if the net is hexagonal (trigonal), tetragonal or non-orthogonal, a primitive (*p*) unit cell has been chosen for the 0-level net. It will be recalled (Section **10**.1, p. 127) that in the three-dimensional lattice (ignoring the relps of zero weight) a single face centring of the RL corresponds to the same face centring in the real lattice, but that *I* (reciprocal) ≡ *F* (real) and *vice versa*. There are four main possibilities.

4.1. THE SAME PATTERN

The three photographs show essentially the same geometrical pattern apart from the scale effect. **c** and **c*** are therefore parallel, and the nets stack vertically above one another. The unit cell is either *P* or *C* depending on whether the net is (*p*) or (*c*). It is orthogonal if the net is orthogonal, hexagonal (or trigonal) if the net is hexagonal, otherwise monoclinic (first setting).

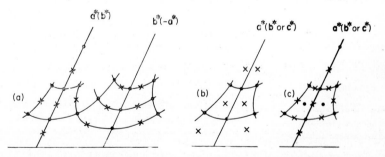

Fig. 128. The stacking of RL nets. (In all diagrams dots denote zero layer points or reflections; crosses denote the same for the 1st layer and circles for the 2nd layer. The scale effect which is very small for the first few layers ($< 5\%$ up to $\zeta = 0\cdot6$) has been ignored unless specifically referred to). (a) shows the result of *B* (or *A*) face centring; (b) is *I* (reciprocal) ≡ *F*(real); (c) *F*(reciprocal) ≡ *I*(real).

4.2. Alternating Patterns

The 0- and 2-layers are essentially the same pattern, but the 1-layer reflections come halfway between in at least one axial direction, or in the middle of the 0-layer net. c and c^* are parallel, and if the 0-layer net is (p), the 1-level

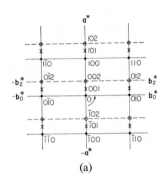

(a)

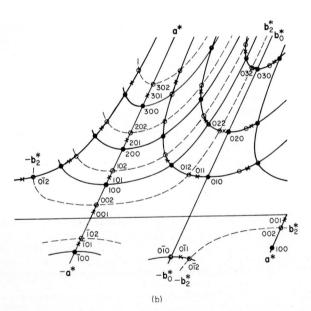

(b)

Fig. 129. The stacking of nets in the monoclinic case and the effect on the photograph. The first layer relps and reflections have not been linked in the b^* direction to avoid crowding the diagram. The reflections move in the direction of $\mathbf{a}^*$ by $c^* \sin \beta^*$ (App. II, p. 370) as the level goes from n to $n + 1$. The origin of the net (002 for the second layer) moves off the centre line and the $\mathbf{b}_2^*$ axis moves to a U-shaped curve passing through 002. If the shift is nearly $a^*/2$ it may be difficult to decide from the photographs which is the positive direction of $\mathbf{a}^*$. ($\bullet$ — 0 layer; X — 1 layer; $\bigcirc$ — 2 layer).

relps (and reflections) can correspond to A (or B) face centring or body cent-ring of the RL cell, depending on the position of the reflections in the 1-layer (Fig. 128). If the 0-level is (c), the 1-level reflections must correspond to A and B centring, giving an all face-centred RL cell.

4.3. THE NON-ORTHOGONAL CASE

One or both axial lines of the 0-layer net move to become curves, corres-ponding to non-central lattice lines parallel to the 0-level axes in both 1 and 2 level photographs. This implies that c^* is not parallel to c, and that the origin of the net has moved off the rotation axis. In the monoclinic case the move-ment of the origin of the net will be in the direction of a^*, which will therefore remain a central lattice line with m symmetry. The corresponding line on the non-zero layer photograph will coincide with the a^* direction on the zero layer, but the origin of the net will no longer be on the centre line. b^*, in an upper layer, will no longer be a central lattice line and will appear as a U-shaped curve on the photograph (Fig. 129).

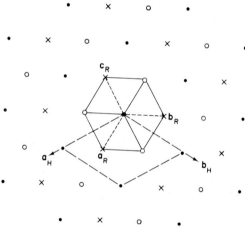

FIG. 130(a). A rhombohedral lattice, showing the relation between the primitive rhombo-hedral cell, defined by a_R, b_R, c_R, (all coming up out of the paper) and the doubly centred hexagonal cell, defined by a_H, b_H, $c_H = (a_R + b_R + c_R)$. The orientation shown, which produces systematic absences for reflections for which $-h + k + l \neq 3n$, is the standard 'obverse' setting. Another hexagonal cell can be chosen by rotating the one shown through 60°. This produces absences for $h - k + l \neq 3n$ and is known as the 'reverse' setting. A third axis, u_H at 120° to a_H and b_H, can be chosen to conform with the symmetry, and the intercepts on this axis are given by a fourth index $i = -(h + k)$. Permutation of the first three indices hki then produces symmetry related faces. For most diffraction purposes this additional index is an undesirable complication, and it is usually replaced by a dot ($hk.l$).

A crystal structure based on a rhombohedral lattice can be thought of as made up of identical layers perpendicular to c_H, each of height $\frac{1}{3}c_H$. Adjacent layers are shifted relative to one another by $\frac{1}{3}$ of the long diagonal of the base of the hexagonal unit cell. The direction of shift for the reverse setting is the opposite of that for the obverse setting. ($\bullet$ — 0 layer; $\times$ — 1 layer; $\circ$ — 2 layer).

If both axial lines become curves we either have a triclinic lattice, or a rhombohedral lattice rotated about a rhombohedral axis (or of course an incorrect camera setting!).

4.4. THE RHOMBOHEDRAL CASE

A rhombohedral lattice is usually referred to hexagonal axes with a doubly centred cell (Fig. 130a). The RL corresponding to this cell has two-thirds of its points with zero weight (the systematic absences). The position of the points corresponding to reflections can be most easily worked out by remembering that a rhombohedron is a distorted cube (squashed or extended along the body diagonal), and that the RL cell will be a parallel cube (extended or squashed respectively) in the same direction. This reciprocal rhombohedron can, in turn, be referred to a doubly centred hexagonal cell, parallel to the real one. However, it must be remembered that for indexing purposes the RL cell is *always* primitive. It must therefore be small enough for all the relps referred to in this discussion as 'centring points' to become lattice points. In the present case (as in all the others) this is the RL cell derived from the centred real cell. The relation of the various cells is shown in Fig. 130b.

For rotation about hexagonal c_H the 0-layer is hexagonal, with the true hexagonal a_H^* axis as the second most populated central lattice line. The

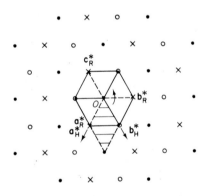

FIG. 130(b). The RL corresponding to Figure 130a. The rhombohedral RL is defined by a_R^*, b_R^*, c_R^* (all coming up out of the paper). The large, doubly centred RL cell, (which has the same relation to the rhombohedral RL cell as the hexagonal real cell has to the rhombohedral real cell) has not been outlined, but can be readily picked out by reference to Figure 130a. The RL cell corresponding to the hexagonal real cell is defined by a_H^*, b_H^*, and c_H^* (which is the height of *one* layer). Two-thirds of the relps defined by this cell have zero weight and correspond to the systematic absences for a rhombohedral lattice referred to hexagonal axes. Since a_H^* and b_H^* are at 60° it is not possible to choose a third axis to bring out the symmetry, and the index i therefore has no direct relation to the RL ($\bullet$ — 0 layer; X — 1 layer; $\circ$ — 2 layer).

1-layer is trigonal, and corresponds to shifting the 0-level net 1/3 of its repeat distance (which equals $3a_H^*$) in the a_H^* direction (or, of course, either of its trigonal equivalents $-b_H^*$ or $(b_H^* - a_H^*)$) (Fig. 130b). The 2-level corresponds to a further shift of 1/3. The 3-level would be shifted a whole repeat

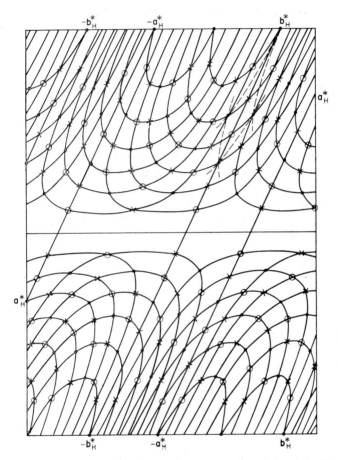

F<small>IG</small>. 131. Superimposed zero, first and second layer Weissenberg photographs of a rhombohedral crystal rotated about c_H. It is derived from the RL of Figure 130b, assuming that the X-ray beam is coming from the left at the beginning of the oscillation. All layers have the same RL net, as shown, but the systematic absences cause only the marked relps to give rise to reflections. For the zero layer a diamond cell is outlined based on b^*. If all the zero layer points are moved through $-b^*$ ($\equiv +a^*$) they come to crosses corresponding to weighted relps of the first layer. The diamond cell for the first layer is easily picked out. Another movement of $-b^*$ brings the points to the circles of the 2nd layer. A third movement brings the zero layer into coincidence with the third layer. This relation between photographs of the various layers is typical of rhombohedral crystals. ($\bullet$ — 0 layer; X — 1 layer; $\bigcirc$ — 2 layer).

distance, and would therefore correspond geometrically to the 0-level, but with weights which reduce the symmetry to trigonal.

For rotation about $\mathbf{a}_H$ the net has a doubly centred orthogonal cell which, for the 1-level, is shifted $1/2$ the repeat distance in the $\mathbf{b}_H^*$ direction (Fig. 132). The 2-level is therefore the same as the 0-level. This is a special case of Section 4.2, (p. 206). In the 0-level the diagonal through the origin containing the centring points is the rhombohedral $\mathbf{a}_R^*$ axis.

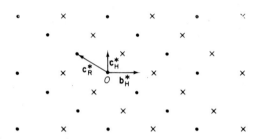

Fig. 132. The RL perpencidular to $\mathbf{a}_H$. Only the relps giving rise to reflections are shown, i.e. two thirds of the relps are missing, including those at the ends of $\mathbf{b}_H^*$ and $\mathbf{c}_H^*$. The direction which will show up most clearly on a photograph is clearly the $\mathbf{c}_R^*$ direction, with the hexagonal axes not even being the next most populated central lattice lines. $\mathbf{a}_H^*$ makes an angle of $60°$ with $\mathbf{b}_H^*$. ($\bullet$ — 0 layer; X — 1 layer; $\bigcirc$ — 2 layer; the zero and second layers coincide in this case if all the relps are considered, but for relps giving rise to reflections, coincidence does not occur until the sixth layer).

4.5. General Comments

The discussion above shows that the type of lattice, P, A (B or C), I, F, R, can be found from the 0-, 1-, 2-layer photographs about one of the axes. However, it is not always possible to distinguish the crystal system in this way; and if a mistake is made in putting a non-axial lattice direction along the rotation axis, a primitive triclinic cell may be deduced from the three photographs. If the RL nets are drawn out accurately on tracing paper, and superimposed with the axes parallel and the origins of construction above one another, a few trials should show whether a more symmetrical unit cell can be chosen. If so, this will either mean taking another set of photographs about the correct direction, or will involve considerable difficulties in indexing and subsequent processing. It is therefore of great importance to sort out the crystal system and the correct axes before starting to take Weissenberg photographs, which often require long exposures. Optical evidence will in most cases allow cubic crystals to be recognised (no birefringence), and the unique optic axial direction to be found for uniaxial (hexagonal (trigonal) and tetragonal) crystals. The orthorhombic axes are parallel to the directions of the acute and obtuse bisectrices and the optic normal; and the monoclinic

b axis is along one of these three directions. Morphological evidence and Laue photographs along the symmetry axes will check the optical deductions. In the cubic case a powder or multi-rotation photograph (Section **14**.2.2, p. 186) will not only check the system, but show the lattice type (P, I or F) and give the lattice constant a.

In the rhombohedral case the most characteristic photograph comes from a rotation about the trigonal axis. Where relps are present on the zero layer, others occur above them on the 3rd, 6th, 9th etc. layers. Those present on the first layer have others above them on the 4th, 7th etc. layers, but there are relps with the same ξ value on the 5th, 8th, etc. layers (Fig. 130b, p. 208) whose reflections come on the same row line (constant ξ line) on the rotation photograph. There are therefore two sets of row lines on such a photograph, one set with reflections on 0, 3, 6, 9 etc. layers, and the other set with the remaining reflections, i.e. 1, 2, 4, 5, 7, 8, etc. This characteristic pattern is well shown in Fig. 133.

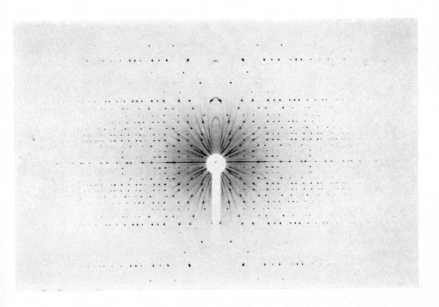

FIG. 133. A rotation photograph of a rhombohedral crystal of yttrium anti-pyrine iodide with c_H (the trigonal axis) along the rotation axis. The characteristic alternation of row lines having reflections at 0, 3, 6, 9 etc. layers, with others having the reflections at 1, 2, 4, 5, 7, 8 etc., shows up clearly. Cylindrical film holder, radius 30 mm. CuKα radiation. The central portion of the film is shown, full size. Large thermal vibrations of the loosely bound anti-pyrine groups and I$^-$ ions cause the rapid fall-off in intensity with increasing θ.

The number of different possibilities for all systems and types of lattice, even if the axes have been correctly identified beforehand, is considerable. But if the symmetry of the crystal and the rotation axis are known in a particular case, the interpretation of the Weissenberg photographs should be straightforward. If a mistake has been made in identifying the rotation axis, the fact of the mistake and its nature will be easily determined in most cases.

It must be emphasised again that centred RL cells can only be discussed by ignoring the relps of zero weight. These centred cells are always larger than the true RL cell derived from the centred real cell, and for the purposes of indexing, the true RL cell must be used. The RL will then include all the points of zero weight which correspond to the 'systematic absences' in the analytical method of determining centring. *The true RL cell is always primitive.*

Chapter 16

The Production and Interpretation of Retigraphs

The principle of a retigraph camera is very simple (Section **11**.4, p. 142). But the mechanical problem of keeping a flat film in contact with the RL plane to be photographed, as it moves through the sphere of reflection, can be more complicated.

1. The de Jong–Bouman Camera

This first retigraph was an extension of the rotation camera. A second physical rotation axis, coincident with the conceptual RL rotation axis, carries a circular flat film at the same height as the RL layer to be photographed (Fig. 134). The crystal rotation axis (and therefore the RL rotation axis) and the film rotation axis rotate synchronously; and a layer screen allows only the reflections from the layer being photographed to fall on the film, where they make a contact print of the RL layer. The position of the layer screen, s mm from the crystal, is simply calculated from its radius r, and the relation $s/r = \zeta_n/R_n$ (Fig. 134).

This camera suffers from the same disadvantages as the normal rotation camera or normal beam Weissenberg camera, with the additional grave disadvantage that the zero layer cannot be photographed, because the film would have to pass through the crystal and the reflections would be parallel to the film. However, by utilising equi-inclination geometry many of these disadvantages can be overcome, at the cost of considerable mechanical complication.

2. The Rimsky Retigraph Camera

This instrument not only uses equi-inclination and anti-equi-inclination geometry to overcome the disadvantages of the de Jong–Bouman camera, but is built round the idea of the sphere of reflection, and incorporates a number of automatic adjustments.

2.1. PRINCIPLE OF THE DESIGN

In Fig. 135 SO is a collimated beam of X-rays; C is the crystal specimen; CO $(= X$ mm) is the radius of the sphere of reflection (the equatorial section through which is indicated by the dashed circle); and O is the origin of the RL.

An annular diaphragm DE, normal to and concentric with the axis CC', allows through only those reflections from the RL layer being photographed. The plane of the photographic film, which is represented in the diagram by its projection $O_n P'$, is coincident with the RL layer.

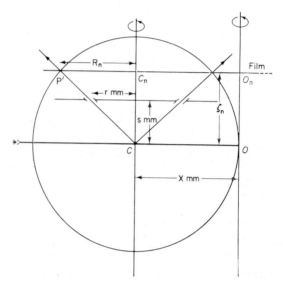

FIG. 134. Diagram of the de Jong–Bouman camera. The radius of the sphere of reflection, CO equals 1 R.U. $= X$ mm, i.e. the distance between the two rotation axes. The circular film holder, centred on O_n, is raised up the spindle and locked at a height above the equatorial plane of ζ_n R.U. $= \zeta_n X$ mm.

The crystal specimen rotates about CC' and the RL rotates simultaneously about the parallel axis OO'. The circular cassette holding the photographic film rotates with the RL layer, which is therefore recorded as a contact print.

2.2. CONSTRUCTION

In Fig. 135 the cross-hatched areas represent objects above the plane of the paper, and outlined areas objects in or very near the plane of the paper. Dashed straight lines represent the paths of rays.

In the lower plane, OO' and CC' are two parallel bars. They pivot on an optical bench about the points O and C which define the direction of the incident X-ray beam. They are connected by a linkage $O'C'$ equal in length to the radius X mm of the sphere of reflection, thus forming an articulated parallelogram $O'C'CO$. The angle of inclination μ can be read from a scale at F. CB is a link, also equal to X, pivoting about C and sliding at its other extremity along a guide below and in the plane of the film O_nP'. CB is therefore the projection of the horizontal generator of the cone of reflections hitting the

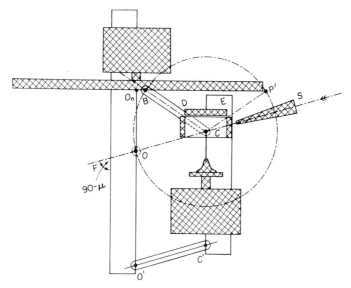

FIG. 135. A plan of the construction of the Rimsky rotigraph camera.

film when O_nP' is set coincident with a RL layer. A pin on the mount of the annular diaphragm immediately below the gap D engages a slot along the length of CB. It keeps the diaphragm in its correct position relative to the crystal and the film, no matter what inclination angles may be chosen. Synchronous motors are used to keep the film and RL layer rotating together.

2.3. PHOTOGRAPHING THE RECIPROCAL LATTICE

2.3.1. The zero layer

Anti-equi-inclination geometry is used (Fig. 136) with the film passing through the origin O. The smallest glancing angle of incidence on the film $(\widehat{P'OC})$ which still gives reasonable definition is about 25°. The largest radius OP' of the circular area of the zero layer that can be photographed is thus $2\cos 25° = 1 \cdot 8$, so that the high angle reflections $(\theta > 90 - 25 = 65°)$ on the zero layer cannot be photographed. However, in the case of the de Jong–Bouman camera the zero layer cannot be photographed at all.

2.3.2. Non-zero layers

Equi-inclination geometry (Fig. 137) is almost exactly the same as for the Weissenberg camera. OO_n is set to the ζ value for the layer to be photographed (to a scale of 1 R.U. $= X$ mm) and the equi-inclination angle, $\mu = 90 - O_n\,OC$, is calculated in the same way (Section **15**.2.2.2, p. 200). This will photograph the maximum possible area of the RL plane, providing the cassette radius is

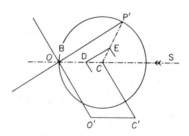

FIG. 136. The arrangement of the Rimsky retigraph camera for photographing the zero layer using anti-equi-inclination geometry.

equal to $O_n\,P'$ (less a small amount for mechanical clearance of the collimator). However, only a limited number of cassettes of graded sizes are available, and it may be desirable to reduce the inclination angle slightly to accommodate a larger cassette, at the cost of creating small 'blind areas' near O_n and outside P'. Also for small values of $\zeta = OO_n$ the film, with an equi-inclination setting, may be too near the crystal, and a smaller or even an anti-equi-inclination angle may have to be set, at the cost of creating similar 'blind areas.'

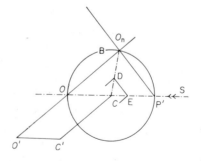

FIG. 137. Photographing an upper layer with the Rimsky retigraph camera, using equi-inclination geometry.

It is the great merit of the design of this instrument that once OO_n has been set equal to ζ, the inclination angle can be given any convenient value, and a part of the layer will automatically be photographed as a contact print. The layer screen (DE, Fig. 135, p. 215) is automatically aligned for any setting. However the Lp corrections will only have the normal values for equi-inclination settings. For any other inclination they will have to be specially calculated. (Appendix V, p. 381).

Fig. 138. A photograph of the Rimsky retigraph with the aligning microscope attached. The synchronous motor rotating the crystal on its arcs (and the RL) is on the further bar of the parallelogram linkage, and that rotating the cassette on the nearer bar, with its axis coincident with the conceptual rotation axis of the RL.

The Rimsky retigraph is best suited to photographing high order non-zero layers. With MoKα radiation the limitation due to the small range of cassette sizes does not apply, since there will be few or no reflections at the outer edges of the film. Equi-inclination geometry can be applied, and the only adjustments are setting $OO_n = \zeta$ and $O_n OC = 90 - \mu$. But for zero and lower order non-zero layers ($\zeta < 0.8$) the obliquity of incidence on the film becomes too great, and equi-inclination geometry cannot be applied. However for $\zeta = 0.4$ the 'blind area' arising from the larger inclination angle required to give $CBP' = 25°$ would only be about 0·1 R.U. radius. This is the largest blind area that need arise. A greater difficulty would be the necessity for calculating special Lp corrections; but it would be relatively simple to arrange to calculate these during programmed data reduction.

Figure 138 shows a photograph of the instrument.

The rotating cassette can be replaced by (a) a stationary flat film holder, which can be set perpendicular to the direct beam for taking Laue photographs; or (b) a stationary cylindrical film holder for taking normal or inclined oscillation and rotation photographs.

Preliminary investigation and setting of the crystal must be done using these accessories or on another instrument.

3. The Precession Camera

3.1. GENERAL

This is the only camera which is not a modification of the rotation camera. It was designed by Buerger to give a photograph with cylindrical symmetry about the X-ray beam, to bring out the symmetry of the RL.

3.2. CONSTRUCTION

The crystal and photographic film holder are mounted on gimbals, described for convenience in terms of a fixed vertical axis and a horizontal axis carried in a frame which rotates on the vertical axis (Fig. 139). This ensures

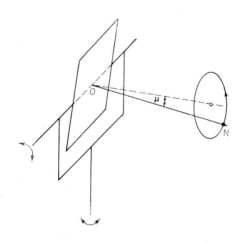

FIG. 139. The gimbal mounting employed in a procession camera.

that any horizontal line in the crystal (or in the RL which moves with it) remains horizontal during the precession movement. The two gimbal axes intersect at O (and C, Fig. 140), both points of intersection being on the precessing axes CN_0 and ON. The collimator axis is adjusted to coincide with CO. When the precession angle μ, is set at zero, the film holder ON_0 is perpendicular to CO.

With the precession angle μ equal to zero,† a crystal lattice axis (or, in general, a central lattice line) is adjusted along CO (Section 3.6, p. 228). The zero RL plane perpendicular to this axis will then be normal to CO, and coincident with the flat film at O. The crystal lattice axis CN_0 and the film normal ON are maintained parallel by a parallelogram linkage. A U-arm maintains connection with the crystal gimbals, while keeping the centre of the instru-

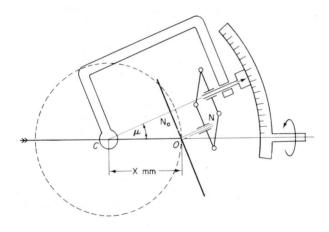

Fig. 140. Schematic drawing of a precession camera, showing the relation of the sphere of reflection (dashed) to the construction of the camera. O and C are the points of intersection of the axes of the two gimbals.

ment clear for the movement of the film on its gimbals. The precession angle μ is set up on the circular arc which is driven round the rotation axis with uniform angular velocity. (On later instruments an improved parallelogram linkage of the gimbals themselves obviates the need for this arm, and the normal ON, instead of CN_0, is set on an arc and driven round, Fig. 141).

The crystal is mounted on arcs which are attached to the 'dial' axis, whose rotation relative to the gimbal axis is measured on a dial. The dial axis can be locked to the horizontal gimbal axis, with which it is co-axial.

† Buerger uses $\bar{\mu}(= 90 - \mu)$ to maintain a formal relationship with the equi-inclination angle of a Weissenberg camera. Since the significant relation appears to be the movement of the apparatus in setting the equi-inclination or precession angle, μ is used here for both angles.

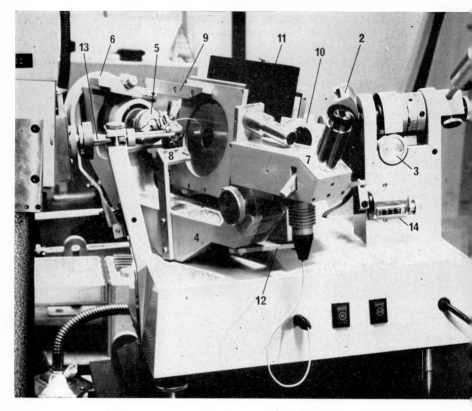

FIG. 141. A photograph of a precession camera with the improved linkage between the gimbal mountings. (1) is the counterweight to balance the moving parts when they are offset on the circular arc whose short end can be seen at (2). This arc is driven round on the rotation axis by the motor driven worm wheel (3). The heavy gimbal frame (4) carries the arcs (5) on its horizontal axis at one end. The arcs can be rotated independently on a coincident inner spindle (the 'dial axis') attached to the dial (6), which can be locked to the gimbal axis. Longitudinal movement of the dial axis is provided for centering the crystal in the beam. The other end of the frame (4) carries a microscope with retractable objective (7) for viewing the crystal in the mirror (8). The horizontal gimbal axis carries the layer screen holder (9) which can be set at varying distances from the crystal and locked by the knurled screw.

One end of the rear gimbal frame (10) can just be seen above the microscope. This carries the film holder (11) on its horizontal axis. One of the parallelogram links between the two frames can be seen at (12). The horizontal axes also have a parallelogram linkage at the further end, with ball joints in the link to allow for the varying orientation of the parallelogram. The collimator (13) carries a removable beam trap. A ratchet and counter (14) records the number of revolutions. The layer screen in front of the film holder (11) would normally have an annular slot. The screen shown with a circular aperture is used for setting purposes (Section 3.6, p. 228).

The length, X mm, of the radius of the sphere of reflection, gives the scale on which the RL will be registered as a contact print on the film, i.e. X mm $\equiv$ 1 R.U. The length, X mm, is the distance between the two vertical axes of the gimbals. This is continuously variable on the older instruments, with a scale and vernier giving the distance. On some of the newer instruments there are a small number of fixed settings, on others just one fixed value of X.

To register non-zero layers all that is necessary is to have a movable film holder clamped to the frame which is mounted in the gimbals. It must be fixed in the position shown in Fig. 140 for the zero layer, but advanced into the sphere so that the film coincides with the non-zero RL layer which is to be photographed. In Fig. 142, ON_0 is the film position for the zero layer and $O_n N_n$ for the nth layer, where OO_n equals ζ_n, the distance between zero and nth layers, to the scale 1R.U. $= X$ mm.

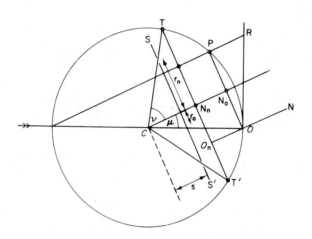

Fig. 142(a). The section through the sphere of reflection which contains the X-ray beam and the normal to the RL plane. The section makes an angle $90 - \psi$ with the vertical. The position of the layer screen is shown. This has an annular slot of radius r_n, to let through the nth layer cone. The radius required for the zero layer would be r_0. A circular area on the nth layer of radius $O_n T'$ remains inside the sphere during the whole precession movement, and therefore the relps in this area are not registered; i.e. non-zero layers have a 'blind area' at the centre.

Finally, in order to exclude reflections other than from the layer being photographed, a frame to hold an annular screen is fixed to the horizontal axis of the crystal gimbals so that only the cone of reflections defined by the circle of intersection of the RL plane and the sphere can reach the film (SS'—Fig. 142a). The radius of the annular gap in the screen r mm, and the distance of the screen from the crystal s mm, are related by $s = r \cot v$, where v is the

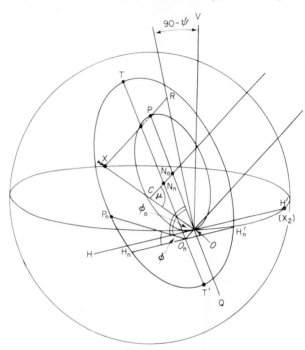

FIG. 142(b). A perspective drawing of the sphere of reflection, the zero and nth RL layers and the horizontal gimbal axis HH'. If the crystal has been correctly adjusted a RL axis will lie along HH', and therefore HH' is always contained in the plane of the zero layer. Similarly, H_nH_n', the parallel line through O_n, is contained in the plane of the nth layer. A relp P_n, in the circle of reflection of the nth layer, is defined by the angle ϕ which O_nP_n makes with O_nH_n. At the same time the position of the centre of the circle of reflection is defined by the angle $N_nO_nH_n = N_0OH = \phi_0$. These angles are used in the calculation of the velocity factor (App. V, p. 00). The section XRO is shown in Figure 142(a).

cone angle of the reflected rays. s is made variable so that a small number of screens (i.e. a discrete set of values of r) can cover the whole range of possible values of v. A cassette for taking cone axis photographs can also be placed in the layer screen frame.

3.3. CONE AXIS PHOTOGRAPHS

The layer screen moves with the crystal so that its centre remains on CN_0, and the cone of reflections passes through the annular ring as the normal 'precesses' round the axis CO. If a film in a light-tight envelope is attached to the side of the screen facing the crystal, this cone of reflections will be registered as a ring of spots on the film. The cones from all the other layers will also be registered, since the screen does not cut them off in this position of the film. The result is a 'cone axis photograph' which, because of the almost exact cylindrical symmetry of the motion of the circle of reflection round the layer normal $O_n N$, shows up any axial symmetry of the RL along the normal, or any planes of symmetry containing the normal. The white radiation streaks, which are also registered on the film, considerably enhance the effect of such symmetry elements, and make them more visible. For demonstrating symmetry such a photograph has many of the advantages of a Laue photograph, although the effect is not as striking as the symmetry related Laue zones. However it avoids the difficulties arising from an apparent loss of symmetry in Laue photographs, due to the effect of $K\alpha$ radiation. (Section **6.5.2**, p. 63).

A cone axis photograph can also be used to obtain the spacing ζ of RL planes perpendicular to ON.

In Fig. 142a:
the radius of the nth layer cone $r_n = s \tan v_n$;
the zero layer cone radius $r_0 = s \tan \mu$;

$$\zeta_n = OO_n = \cos \mu - \cos v_n.$$

Therefore $s = r_0 \cot \mu$ and s can be obtained from the known value of μ and the measured zero cone radius.

$\tan v_n = r_n/s$ and v_n can be obtained from the measured nth layer cone radius.

$$\therefore \ \zeta = \zeta_n/n = \frac{\cos \mu - \cos v_n}{n}$$

The accuracy is 1–2% at best.

The zero layer cone can be identified by the Laue streaks which all lie in the circle on the film.

Figure 143 shows a cone axis photograph of sucrose.

3.4. THE PRECESSION MOTION

At first sight the precession motion seems relatively simple. The circle of reflection cut by the RL plane appears to rotate uniformly with respect to the plane, about the normal to the plane. For a considerable time after the introduction of the precession camera this was assumed to be the case, and velocity factors were calculated which depended only on the radial distance of the relp from the normal. However when careful intensity measurements were made, systematic discrepancies were revealed which showed that the relative motion of relps and sphere of reflection was not cylindrically symmetrical. A more detailed investigation of the geometry revealed that the constraints exercised by the gimbal mounting caused rotation of the RL plane about its normal (faithfully followed by the film under the same constraints). The motion of the relps due to this rotation—actually a double oscillation in one complete precession cycle—has to be added to the relative motion of the RL plane and circle of reflection in calculating the velocity factor, and this gives rise to the loss of cylindrical symmetry.

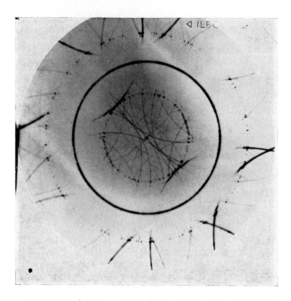

FIG. 143. A cone axis photograph showing 2-fold symmetry. Some of the outer reflections were obscured by the arcs, so that there is an apparent lack of 2-fold symmetry in parts of the outer ring. The zero layer Laue streaks are confined to the zero ring of $K\alpha$ reflections, and almost completely obscure them.

The origin of the oscillation can be most easily demonstrated by means of a stereographic projection. In Fig. 144 VV' is the vertical gimbal axis which remains fixed. HH' is the position of the horizontal axis when the precession angle is zero. O is the direction (from the source) of the X-ray beam and the rotation axis. H (and H') are constrained to move along the great circle HOH'. With the precession angle μ equal to zero, the zero layer RL plane is represented by the primitive; and its normal N is at O. On setting up a precession angle ($\mu = ON$) the RL plane tilts about the axis KK' ($\perp$ to ON) to KQK', where $OQ = 90 - \mu$. At the same time if there is no rotation about N, H' will move on the small circle about K to X_1. But H' is constrained to move on HOH' and therefore actually moves to X_2, causing a rotation of $X_1 X_2 = \delta$. If δ were constant for a given μ, there would be no problem, since we are not concerned with rotation as μ is changed. What concerns us is whether, for a fixed μ, δ changes with the angle of rotation ψ. δ is clearly not constant with change of ψ since it is zero for $\psi = 0$ and $90°$, but, as Fig. 144 shows, not for intermediate values. To calculate the velocity factor we

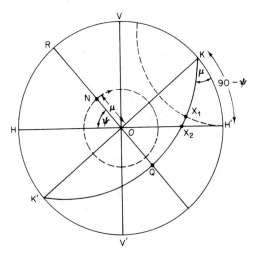

FIG. 144. Stereogram showing the angular relations of the gimbal axes (VV' — vertical; HH' — horizontal); the X-ray beam and rotation axis O; the normal to the RL plane (the precessing axis) N; the precession angle μ; and the angle of rotation of the precessing axis from the horizontal ψ.

require to find the value of the instantaneous angular velocity $d\delta/dt$, in terms of μ and $\psi = \Omega t$ where Ω is the constant angular velocity of rotation. This is done in Appendix V (p. 384).

3.5. THE LAYER SCREEN

3.5.1. *Adjustment of the layer screen*

From Fig. 142a (p. 221) if r is the mean radius of the annular slot in the layer screen, and s its distance from the crystal

$$r/s = \tan v \quad \text{or} \quad r = s \tan v. \tag{1}$$

Also

$$CN_n = \cos v = CN_0 - N_0 N_n = CN_0 - OO_n$$

$$\therefore \quad \cos v = \cos \mu - \zeta_n. \tag{2}$$

The screen holder incorporates a scale from which s can be read directly. For most purposes the precession angle μ is made as large as possible, in order to photograph the maximum area of the RL layer. The mechanical limit is about $\mu = 30°$. Having decided on the value of μ, v is calculated from (2) and the known value of ζ_n in R.U. The maximum value of s can be calculated by subtracting about 5 mm from $CN_n = \cos v$ (on the scale of 1 R.U. $= X$ mm. Fig, 140, p. 219). Alternatively, after setting the precession angle to μ and the cassette in the position O_nT (i.e. forward from the zero layer position by $OO_n = \zeta_n$, to the same scale) a screen can be placed in position, and the maximum convenient value obtained by trial. The value of r' is calculated from (1) for this value of s', and the screen with the nearest radius *smaller* than this value is chosen. For this value of r (which may be considerably less than r' since only a comparatively small number of screens are supplied with the instrument) the value of s is calculated from (1), and the screen placed in this position by means of the scale. In the case of the zero layer there is no need for a central disc in the layer screen, if the $\bar{1}$ layer does not enter the sphere. This will be so if (Fig. 140):

$$1 - CN_0 < \zeta$$

$$1 - \cos \mu < \zeta$$

or

$$\cos \mu > 1 - \zeta.$$

These expressions can be used to determine whether a central disc is needed for a given μ, or to find the value of μ which will make a central disc unnecessary.

3.5.2. The Width of the slot in relation to μ and ζ

In the general case, for both zero and non-zero layers, it is necessary to make sure that, for the width of annular slot being used, no reflections from the layers on either side can be recorded. Since the limit occurs for small values of ζ, we can use the differentials Δr, Δv and $\Delta \zeta$ to obtain an approximate solution of the problem.

Differentiating 225(1) and 225(2),

$$\Delta r = s \sec^2 v \, \Delta v; \qquad -\sin v \, \Delta v = -\Delta \zeta$$

$$\therefore \quad \Delta r = \frac{s \, \Delta \zeta}{\sin v \cos^2 v}. \tag{1}$$

The denominator has a maximum value of $0 \cdot 386$ at $v = 35°$. This is the worst case, giving the smallest half width of the annular slot for a given distance between RL layers, $\Delta \zeta$. Working the other way round, for a slot width of 2 mm and allowing $0 \cdot 5$ mm margin

$$\Delta \zeta = \frac{1 \cdot 5 \times 0 \cdot 386}{s} = \frac{0 \cdot 579}{s}.$$

For MoKα radiation, $\lambda = 0.71$ Å, and the lattice point spacing along the precession axis

$$= \frac{\lambda}{\Delta \zeta} = \frac{0.71 s}{0.579} = 1.23 \, s.$$

Thus the axial length that can be accompanied by a 2 mm layer screen is equal to rather more than the value of s in mm. For $\mu = 30°$, s, for this worst case, can be about 48 mm (for $X = 60$ mm, Fig. 140, p. 219) and the maximum spacing about 60 Å.

For spacings larger than this it is necessary to reduce the value of μ. As an example, for a protein crystal with a spacing of 200 Å,

$$\Delta \zeta = \frac{0.71}{200} = 0.00355.$$

For $\Delta r = 1.5$ mm, from 226(1)

$$1.5 = \frac{48 \times 0.00355}{\sin v \cos^2 v}$$

$$\therefore \quad \sin v \cos^2 v = 0.1135. \tag{1}$$

For small values, such as this, $\cos^2 v \approx 1$ and

$$\sin v \approx 0.1135$$
$$v \approx 6°3'.$$

Using this value for $\cos^2 v$ in (1) gives $\sin v = 01.150$. $\therefore v = 6°36'$.
    For the first layer:

$$\cos 6°36' = \cos \mu - 0.00355 \text{ (from 225(2))}$$
$$\therefore \quad \cos \mu = 0.9934 + 0.00355$$
$$= 0.99695$$
$$\mu = 4°28'$$
$$v_1 = 6°36'$$
$$\therefore \quad r_1 = 48 \tan 6°36' = 48 \times 0.1157$$
$$= 5.55 \text{ mm}.$$

We can check that this gives the correct solution by calculating r_0 and r_2, which should differ by 1.5 mm from r_1. For the zero layer

$$v_0 = \mu \text{ and } r_0 = 48 \tan 4°28' = 48 \times 0.0781 = 3.75 \text{ mm}.$$

For layer 2,

$$\cos v_2 = \cos \mu - 0.0071 = 0.99695 - 0.0071$$
$$= 0.98985$$
$$v_2 = 8°12'$$
$$\therefore \quad r_2 = 48 \tan 8°12' = 48 \times 0.1441$$
$$= 6.92 \text{ mm.}$$

Thus even at these small angles the approximate calculation is within 0·1 mm of the exact calculation. A 2 mm slot would exclude the zero and second layers, with a margin of 0·4 mm, when the first layer is being photographed. It is probable that the radius available would be 5·0 mm and then

$$s = 5.0/\tan 6°36' = 5.0/0.1157$$
$$= 43.4 \text{ mm.}$$

This would slightly reduce the margin to 0·35 mm which should still be adequate. But it might be advisable to have a reduced slot width of 1·0 or 1·5 mm.

3.5.3. *Checking the functioning of the layer screen*

To check that the layer screen is functioning properly it is advisable, after adjusting the camera, to take a test exposure. A piece of film in a black paper envelope is clipped to the side of the screen away from the crystal, and a photograph taken on this film with sufficient exposure for the fogging to define the annular slot. The reflections from the layer being photographed should lie in the centre of the slot and no others should be visible at the edges. This is essentially a cone axis photograph (Section 3.3, p. 222) with the film on the far side of the screen. (**Example 38,** p. 317).

3.6. FINAL CRYSTAL SETTING ON A PRECESSION CAMERA

If the crystal has been set with a RL axis (or other important central RL line) exactly or approximately along the dial axis, and the orientation of a second RL axis found relative to the planes of the arcs (Section **13**.4, p. 160), then final setting can be easily and accurately done on the precession camera itself.

The method utilises the recording on the photograph of the weighted radial rods due to the white radiation (Section **6**.1, p. 53). The outer ends of these 'Laue streaks' define the circular area of the RL zero layer, of radius OP (Fig. 142a, p. 221) which is recorded in a precession photograph (Fig. 145a). If the crystal is not properly set the RL plane will not coincide with the film, and a distorted photograph of the RL zero layer will be obtained. In particular the Laue streaks outline an area which is no longer exactly circular, and, more importantly, is no longer centered on the direct beam (Fig. 145b). The eccentricity is a measure of the mis-setting and can be used to calculate the required arc and dial axis corrections.

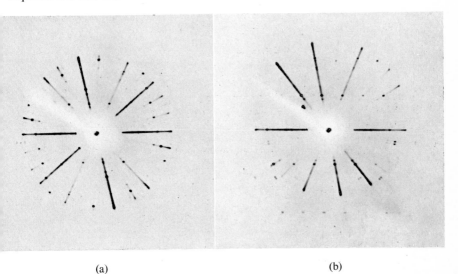

(a)   (b)

FIG. 145(a). A setting precession photograph (taken with $\mu = 10°$, $X = 60$ mm and MoK radiation) showing the circular area outlined by the ends of the Laue streaks. This is centred on the direct beam as it should be when the crystal is properly set. (b) A setting photograph showing considerable vertical mis-setting (on the dial axis) and slight horizontal mis-setting (on the vertical arc axis).

The blank outer area of both photographs has been cut off. Each film is approximately 5″ square, but with $\mu = 10°$ only the central position is used (But see Figure 148, p.233).

Figure 146 shows a central section through the sphere of reflection, containing the precessing axis CN_0 at the two positions 180° apart. OP' and OP'' are the traces of the film for the two positions. The RL zero layer would coincide with them if the crystal were set. The mis-setting in this plane is about an axis through O, perpendicular to the diagram, by the angle α. Mis-setting about an axis in the plane of the diagram will only have second order effects on the dimensions in the plane. OP_1 (and OP_2) are the traces of the mis-set RL plane. P'_1 (and P'_2) are the points where the corresponding reflections hit the film and define the ends of the Laue streaks. $P_1 N_1$ (and $P_2 N_2$) are drawn perpendicular to OP' (and OP''). If α is small enough for α^2 terms to be neglected, we have:

$$OP_1 = 2 \sin (\mu + \alpha); \qquad OP_2 = 2 \sin (\mu - \alpha)$$

$$\frac{N_1 P'_1}{N_1 P_1} = \tan (\mu + 2\alpha) \approx \tan \mu$$

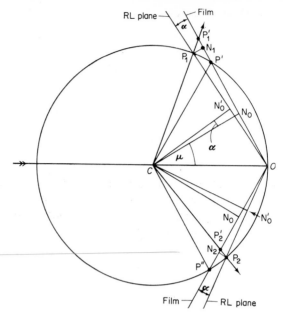

FIG. 146. The section through the sphere of reflection containing the precessing axis CN_0 at the two positions 180° apart. The traces of the film OP', OP'' and of the mis-set RL plane OP_1, OP_2 are shown, together with the normals to OP_1 and OP_2.

(Terms containing α in the expansion of $\tan(\mu + 2\alpha)$ can be neglected because they would be multiplied by α in subsequent development).

$$N_1 P_1 = OP_1 \alpha = 2\alpha \sin(\mu + \alpha)$$

$$\therefore \quad N_1 P_1' = N_1 P_1 \tan \mu = 2\alpha \sin(\mu + \alpha) \tan \mu$$

$$OP_1' = ON_1 + N_1 P_1' = OP_1 + N_1 P_1'$$

$$= 2 \sin(\mu + \alpha) + 2\alpha \tan \mu \sin(\mu + \alpha)$$

$$= 2 \sin(\mu + \alpha)(1 + \alpha \tan \mu)$$

$$= 2(\sin \mu + \alpha \cos \mu)(1 + \alpha \tan \mu)$$

$$= 2(\sin \mu + \alpha \sin \mu \tan \mu + \alpha \cos \mu)$$

$$= 2 \sin \mu + 2\alpha \left(\frac{\sin^2 \mu + \cos^2 \mu}{\cos \mu} \right)$$

$$= 2 \sin \mu + 2\alpha \sec \mu.$$

Similarly $OP'_2 = 2 \sin \mu - 2\alpha \sec \mu$

$$\therefore \quad OP'_1 - OP'_2 = \Delta = 4\alpha \sec \mu.$$

$$\therefore \quad \alpha = \frac{\Delta \cos \mu}{4}.$$

Δ is in R.U. and α in radians. In millimetres and degrees

$$\alpha = \frac{\Delta}{X} \cdot \frac{\cos \mu}{4} \times 57\cdot3°. \tag{1}$$

The angular correction about an axis perpendicular to any plane containing the direct beam can, therefore, be obtained by measuring the two distances OP'_1 and OP'_2 along the corresponding line of the photograph (Fig. 147), and multiplying the difference Δ mm, by

$$\frac{57\cdot3 \cos \mu}{4X} = \frac{14\cdot3}{X} \cos \mu.$$

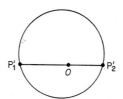

FIG. 147. Diagram of a setting precession photograph. The circle (which will be slightly distorted) is outlined by the ends of the Laue streaks. O is the point where the direct beam intersects the film. The mis-setting about a vertical axis causes the distances OP_1' and OP_2' to be different. Similarly, the mis-setting about a horizontal axis causes the vertical off-centering.

The factor $\cos \mu$ in the expression for α is unimportant since it only varies by about 15% over the range from $\mu = 0-30°$. The precession angle can there-fore be chosen from other considerations. Since the Laue streaks are strong-est near the centre, and their ends define the zero layer area, μ should not be too large. A small value of μ, say 10°, will also reduce the exposure time. In addition if the photograph is taken without a screen, in order to avoid any possibility of the screen cutting off the Laue streaks on one side, a small value of μ will reduce the overlapping of reflections and Laue streaks from other layers. If a screen is used it must be larger than normal and without the central disc (Fig. 141, p. 220). To accommodate the mis-setting the radius of the hole must be $r' = s' \tan \mu + \Delta/2$.

From 231(1), for $X = 60$ mm and $\mu = 10°$, $\alpha = 0.234\,\Delta°$.   (1)

For

$$\alpha = 2°, \qquad \Delta = 8.6 \text{ mm.}$$

$$\therefore \quad r' = s' \times 0.176 + 4.3 = 5.3 + 4.3 = 9.6 \text{ mm for } s' = 30 \text{ mm.}$$

A convenient setting screen is therefore of radius 10 mm, placed at 30 mm from the crystal. If the mis-setting is greater than 2° the screen can be used with a smaller value of s', but is likely to allow non-zero layer reflections to be registered. The appearance of a flattened edge to the 'bulge' of the central area is an indication that the screen is cutting off part of the streaks.

The axes on which corrections can be made directly in this way must be approximately perpendicular to a plane containing the X-ray beam. The possible axes are the horizontal dial axis, and the axis of an arc which has been set vertical by means of the dial axis.

In practice the first requirement is to check and, if necessary, adjust the arcs. With $\mu = 0$, adjust the dial axis so that the plane of one of the arcs is horizontal and its axis therefore vertical. Set $\mu = 10°$ and take an unfiltered photograph. Since a RL axis (**a*** say) is nearly along the dial axis, the Laue streak from this weighted vector will be nearly horizontal, and the measurements on the Laue streak can be used to obtain Δ and α even if no zero layer can be recognised in any other direction (Fig. 148). The correction is made on the arc, in the direction which will either bring the RL vector out of the sphere on the side with the longer streak, or alternatively bring a Laue reflection on the longer streak in towards the centre. These are of course equivalent, and since the commonest setting error is to make the correction in the wrong direction, they can usefully be used as a double check.

If the **a*** Laue streak is more than 1 or 2° off the horizontal in the above photograph (Fig. 148), measure the angle with a protractor or measuring circle and put the correction on the other arc whose axis is horizontal. Re-centre the crystal if necessary, turn the dial axis through 90° and repeat the procedure for the second arc.

If the bottom arc is off zero, the top arc axis will not be exactly vertical, but the difference in the correction will be negligible for small angles. Because of the approximate character of the equation, 231(1), if α is more than 1 or 2° the setting must be rechecked, and if necessary a final adjustment made on the arcs.

The dial axis is then adjusted to bring the second RL axis, e.g. **b***, nearly into the vertical plane (when $\mu = 0$) and another setting photograph is taken. This time a nearly circular area defined by the **a*b*** Laue streaks should be easily recognised, and the *vertical* measurements used to calculate α for the dial axis correction. When this correction has been applied the crystal is set, unless the correction is large, when a second photograph and adjustment may be necessary.

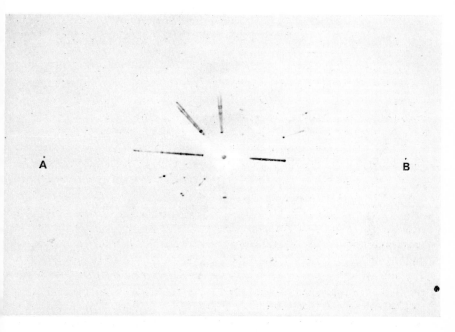

FIG. 148. A setting precession photograph taken with MoK radiation and with the top arc axis vertical; $\mu = 10°$, $X = 60$ mm. The only recognisable feature is the nearly horizontal Laue streak, corresponding to the RL axis which is to be set along the dial axis. The correction for the top arc can be obtained by measurement of OP_1' and OP_2' (Fig. 147). That for the bottom arc can be obtained somewhat less accurately by means of a measuring circle with vernier reading to 0·1°. The horizontal line is defined by the two fiducial marks A and B on the film. Light is allowed to fall on two small holes drilled in the back of the cassette for just sufficient time to give black dots on the film. Part of the blank area top and bottom has been removed during reproduction.

It is possible, when the process of carrying out the corrections singly has become familiar, to do all three corrections from one photograph, with considerable saving of time. The **a*b*** plane is adjusted as nearly as possible perpendicular to the X-ray beam (with $\mu = 0$) and a setting photograph is taken. The value of Δ for the vertical plane gives the dial axis correction. The value of Δ for the horizontal plane gives the correction α_V for rotation about a vertical axis; and the angle with the horizontal of the **a*** Laue streak (obtained accurately on a measuring circle) gives the correction α_H, for rotation about a horizontal axis along the X-ray beam.

If the arc axes are vertical and horizontal all three corrections can be applied directly. But in general the axes will have been moved through an angle ϕ from the position with the bottom arc axis vertical. ϕ is positive for clockwise rotation, looking along the dial axis at the free end of the crystal. If clockwise rotations are counted as positive, looking in the directions of the single arrows in Fig. 149, the arc corrections are given by:

$$\alpha_B = \alpha_V \cos \phi + \alpha_H \sin \phi$$
$$\alpha_T = \alpha_V \sin \phi - \alpha_H \cos \phi$$

providing the corrections are small enough for the rotations to be represented by vectors, and manipulated accordingly.

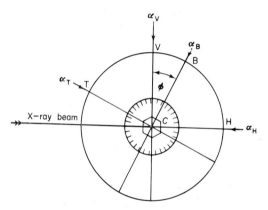

FIG. 149. The directions of the vertical, horizontal and arc rotation axes, looking along the dial axis at the free end of the crystal.

Again, if the corrections are more than 1–2° the check setting photograph may show that final small corrections are needed, followed by a further check. (**Example 39** p. 317).

3.7. THE EFFECT OF MIS-SETTING ON A PRECESSION PHOTOGRAPH

If a precession photograph is taken with a mis-set crystal, the relps will no longer be in contact with the film (Fig. 146, p. 230). The reflected X-ray beam will go across the gap from relp to film in different directions, depending on whether the relp is entering or leaving the sphere. The reflections will therefore strike the film at different points and form a doublet, except for the outermost reflections whose relps enter and leave the sphere at the same places. The splitting of reflections in this way on a zero layer is a sign of mis-setting. The axis of mis-setting can be found from the direction on the film in which splitting does not occur. On upper layers splitting can be caused in all directions by an incorrect setting of the film cassette; and in all cases film bulging will cause splitting over the area of the bulge.

3.8. A COMPARISON OF PRECESSION AND ROTATION CAMERAS

It is possible to start with a completely unoriented crystal of unknown cell dimensions, and investigate it on a precession camera. Whether the initial orientation of the crystal can be accomplished more easily on a precession camera or a rotation camera, is probably largely a question of the experience of the crystallographer in the two methods. But once a major RL plane has been picked up on the precession camera, the cell dimensions can be rather more easily determined than in the case of a rotation camera with the crystal 'set' about a major crystal lattice vector. In the latter case the crystal would have to be transferred to a moving film rotation (i.e. Weissenberg) camera or several photographs taken on the rotation camera (Answer 30 p. 512). For an unoriented crystal of known cell dimensions however, the orientation can be done more systematically on a rotation camera because of the greater simplicity of crystal movement, and the consequent availability of charts to aid in the process.

The precession camera has the advantage over the Weissenberg camera, in that layers about two axes can be photographed with the same setting of the crystal, merely by rotating the crystal about the dial axis. It is true that the non-zero layers have a 'hole' of radius $O_n T' = \sin v - \sin \mu$ (Fig. 142a, p. 221) round the origin, in which no relps are recorded; but the missing relps can be obtained by a series of photographs of a 'sheaf' of zero layers, each containing the RL axis which is along the dial axis and a RL vector corresponding to one of the 'missing' reflections. These all have to be correlated, and velocity factors calculated for each orientation, but this may be preferable to attempting to remount a precious crystal about another axis, as would be required in the case of a Weissenberg camera. On the other hand, if several crystals are available it is always desirable to obtain data from more than one, since imperfections of one kind or another may be affecting the intensities, and this would not show up on one crystal. In this case the collection of data on the Weissenberg camera may be simpler.

A Weissenberg camera can collect data almost out to the limiting sphere, i.e. $d^* \approx 2$, whereas a zero layer precession photograph can only collect data to $d^* \approx 1$. However this advantage is more apparent than real, because with MoKα radiation the precession camera can collect as much data as the Weissenberg camera with CuKα radiation. The use of MoKα radiation with a Weissenberg camera rarely produces much additional data, because thermal vibrations render undetectable the intensities of reflections for relps outside the CuKα sphere. The exceptions are inorganic materials whose atoms are strongly bound together, and specimens investigated at low temperatures. For both of these atomic vibrations will be small. In these cases the additional data obtained with MoKα radiation are likely to be particularly significant.

If the unit cell of the crystal is large, the Weissenberg reflections, even with CuKα radiation, will be very close together in the high angle region, difficult to index and impossible to integrate. With MoKα radiation, the average distance between reflections on a precession photograph will be about the same as that on the Weissenberg photograph with CuKα radiation, because the crystal to film distance for the precession camera is normally about twice that of the Weissenberg camera. However, there is no crowding together at higher angles, indexing is simple and integration only becomes impossible with much larger unit cells. The use of MoKα radiation and a larger crystal to film distance, leads to longer exposure times (because the intensity is proportional to λ^3 and $1/R^2$). This is partially offset by a smaller absorption correction factor, and may be accompanied by a smaller error in intensity measurement due to absorption. Nevertheless, the long exposure times for precession photographs may be the determining factor leading to the use of a Weissenberg camera. Multiple films can only be used in a precession camera if a special, semi-circular, 'Zoltai' layer screen is used which cuts out the reflections from the second penetration of the sphere, so that doubling of the spots does not occur on all those films which are not at exactly the right distance from the crystal. This doubles the exposure time and alters the velocity factor. Otherwise multiple exposures must be used instead of multiple films, further increasing the time required.

The undistorted photograph of the RL given by a precession camera makes it easier to find the relation between two lattices in difficult twinning problems. However, the same information is contained in a Weissenberg photograph; and with a certain amount of trouble, it can be presented in essentially the same form by plotting out the relps, using a coordinate chart or measuring table.

In general, one uses the camera with which one is most familiar or which happens to be available, but there are some cases when one or the other has definite advantages.

4. The Interpretation of Retigraphs

The discussion of Section **15**.3 (p. 202) on symmetry in Weissenberg photographs and its relation to three-dimensional Laue symmetry, can be applied to layer photographs from retigraph cameras without the complication introduced by the Weissenberg distortion of the RL net. In particular, the various possible two-dimensional symmetry combinations given in Fig. 188 (p. 315) can be recognised directly.

The indexing of retigraph photographs is simple, the only problems being the recognition or choice of axes. Since the scale of the RL layer contact print is known, the determination of axial lengths is a matter of simple measurement with a ruler. The considerations applying to choice of axes (Appendix II, p. 352) are not peculiar to any particular recording method.

The origin of construction will be the point where the normal from the RL origin to the net being photographed cuts the net. This, in the case of the precession camera and the Rimsky rotation retigraph, will be the centre of the photograph, defined on the zero layer by the relp at the position of the direct beam. If the films are properly located in the cassette, or if fiducial marks are put on them (Fig. 148, p. 233), they can be superimposed to determine any shift of the origin of the net from one layer to the next, or the type of centering if any is present.

Chapter 17

Measurement and Interpretation of Preferred Orientation

1. The Production of Pole Figures

Refer to Section 7.3 (p. 78) and Figs 54 and 55 (pp. 79 and 80).

1.1. PHOTOGRAPHIC METHODS

Photographs are taken with a stationary specimen and flat film perpendicular to the beam, either in the forward direction, as in Fig. 54, or on a back reflection camera (Section 11.2, p. 135). The principles of interpretation for both are very similar and only the low angle case will be dealt with.

The powder rings on which the preferred orientation arcs lie are first measured and indexed. Reflections with simple indices ($h00$; $hh0$ and hhh for preference) are chosen for further investigation. Information about the distribution of each RL vector should be plotted on a separate stereogram. In Fig. 54 the RL vector OP_0, corresponding to the equatorial reflection P_0', makes an angle $\widehat{P_0OC}$ with the negative direction of the X-ray beam. Since $\widehat{OCP_0} = 2\theta$, $\widehat{P_0OC} = (90 - \theta)$, and all RL vectors corresponding to reflections on the same powder ring, such as OP_1 and OP_2, make the same angle with the X-ray beam. On a stereographic projection therefore, the directions of such RL vectors all lie on a small circle of radius $(90 - \theta)$ about the X-ray beam.

OP_2 and OP_2' both lie in the plane which contains the X-ray beam and makes an angle ψ_2, with the horizontal. OP_2 therefore lies in the great circle on the stereogram with the beam as diameter and making the angle ψ_2 with the horizontal. The intersection of this great circle with the small circle centred on the beam gives the pole of OP_2 on the projection. There are four possibilities, corresponding to OP_2' being R(ight) or L(eft), U(p) or D(own), but the correct one is easily distinguished.

Notice that only the directions of the RL vectors are plotted on the stereogram, and that the length of the vector is not involved except in so

238

far as it determines θ. Thus information derived from the reflection $2h$, $2k$, $2l$, or in general nh, nk, nl, can be plotted on the same stereogram as that for hkl, since the same directions are involved. This assumes that the intensities of the various orders are approximately the same. This is generally true for metals, but not for more complicated structures. For accurate measurements only one order should be used.

ψ is measured on the photograph, and the RL vectors corresponding to the extreme positions of heavy arcs on the powder ring are plotted on the stereogram, and joined by a heavy line along the small circle. This heavy line is a section through a patch of relps on the reciprocal powder sphere. To plot the extent of this patch we must obtain other sections through it. The simplest way to do this is to take powder photographs with the beam along three mutually perpendicular directions in the specimen. (**Example 40** p. 319). However, this does not really give sufficient coverage, so that to plot the patches adequately requires a step by step method.

1.1.1. *Step by step recording*

The specimen (and therefore the reciprocal sphere) is rotated through a small angle, e.g. $5°$ anti-clockwise, about the vertical rotation axis. This will bring the dashed curve through Q (Fig. 54, p. 79) into the intersection with the sphere of reflection, so that a powder photograph in this position will give information on the relps lying along this curve. The information is temporarily plotted as before; but the heavy lines on the stereogram are then rotated $5°$ clockwise for final plotting, so that the information from the second photograph is given with the specimen in the same orientation as for the first photograph. This is repeated for a further rotation until sufficient of the powder sphere has been covered.

The boundaries of the patches on the sphere are drawn on the stereogram and the result is a 'pole figure' for the reflection hkl concerned, i.e. a map of the patches on the powder sphere where the $\mathbf{d}_{hkl}^{*}$ vectors are concentrated.

1.1.2. *Continuous recording*

The recording process can be made continuous by rotating the specimen synchronously with a translation of the film. This is moved horizontally behind a screen which has a semi-circular slot, allowing half the powder ring ($T P_0' B$ in Fig. 54) through. A continuous series of semi-circles is recorded on the film, every $5°$ of which corresponds to half the powder ring produced by the previous method. The concentrated patches of relps come out as areas of blackening on the film. The result can be treated as a distorted stereogram, and a chart can be used to transfer the concentrated patches of relps to a normal stereogram for proper evaluation.

However, the method has two limitations. It is not possible to investigate the whole of the reciprocal sphere with one setting of the specimen (there are blind patches top and bottom), and it is difficult to make quantitative evaluations. The need for quantitative measurements, particularly in the metal fabrication industries, has led to the use of counter methods, and ultimately to the automatic plotting of pole figures.

1.2. COUNTER RECORDING WITH 'TEXTURE GONIOMETERS'

1.2.1. *Transmission goniometer*

In Fig. 54 (p. 79), if a counter is placed at P_0', information is obtained about the reflecting power, i.e. the density of relps, at P_0. If the specimen (and therefore the reciprocal sphere) is rotated about a horizontal axis along the X-ray beam, then all the reflections around the powder ring, and therefore all the relp densities along the circle of intersection $P_0 P_1 P_2$, are investigated in turn. The intensity of the reflection is plotted along the corresponding small circle on the stereogram.

(i) Step by step recording

To follow the next step let us be more specific about the specimen, and suppose that it is a plate whose normal (and therefore the horizontal rotation axis which rotates it in its own plane) is along the X-ray beam, and that at the beginning of the specimen rotation the rolling direction (W) is horizontal. Consider the point Q, as the point initially at P_0 rotates round the reflecting circle. Q will rotate round the dotted circle centred on the horizontal rotation axis, and all the points round this circle will take the place of Q in turn. When W is again in the horizontal position, the specimen and its rotation axis are turned about the vertical axis until Q reaches P_0. The specimen rotation axis is no longer along the X-ray beam but makes an angle $\phi = P_0 OQ$ with it. During rotation about this axis the relp density along the dotted circle through Q is investigated by the fixed counter at P_0'. It is thus possible to investigate a series of small circles centred on the specimen rotation axis (the normal to the plate) and making an angle of $(90 - \theta - \phi)$ with it. (Compare this with the series of small circles of constant radius $(90 - \theta)$ rotated about the vertical axis in Section 1.1, p. 238).

By making ϕ negative and equal to θ we investigate the great circle in the plane of the plate, and by increasing ϕ through zero to positive values, we investigate circles on the powder sphere of smaller and smaller diameters. The theoretical limit occurs when the reflected beam (CP_0' in Fig. 54 $\equiv OC$ in Fig. 150(b)) is in the plane of the plate, i.e. $\phi = (90 - 2\theta)$, but the practical limit is some 20° earlier. The smallest circle which can be investigated therefore makes an angle of $(90 - \theta - (70 - 2\theta)) = (20 + \theta)°$ with the normal to the plate. Even with the use of MoKα radiation θ cannot be much less than 10°, so that the powder sphere cannot be investigated in directions of less than about 30° with the normal, i.e. the stereogram has a fairly large 'blind area' at the centre. By decreasing ϕ beyond $-\theta$ we investigate the other half of the powder sphere, but come up against the same limit when the direct beam makes an angle of 70° with the plate normal.

The apparatus which enables this recording to be accomplished is called a transmission texture goniometer. Fig. 150 shows diagrammatically the construction of the instrument. The angle of specimen rotation α, and the angle ϕ which the normal ON makes with the X-ray beam, enable the counter recording to be plotted on a stereogram.

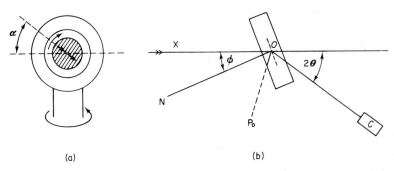

(a)   (b)

FIG. 150. Diagrammatic representation of a texture goniometer used for investigating preferred orientation in specimens thin enough to transmit X-rays. (a) the specimen holder, which enables the specimen to be rotated in its own plane about a horizontal axis (ON in (b)). The rotation is defined by the angle α, which a designated direction in the specimen (e.g. the rolling direction) makes with the horizontal. The whole housing for the specimen holder can be rotated about a vertical axis to vary ϕ, the angle ON makes with the X-ray beam. (b) Plan of the arrangement. The counter C is rotated on its arm to make an angle $2\theta_{hkl}$ with the direct beam. OP_0, bisecting the angle XOC, is the direction in which the relp density is being measured at C. If $\phi = -\theta$, the plane of the specimen contains OP_0, and as the specimen rotates all the directions in its plane are investigated. These correspond to the points round the primitive of the stereogram. The maximum practical value of ϕ occurs when OC makes a glancing angle of about 20° with the plane of the specimen. The two rotations can be geared together so that the specimen makes, for example, one revolution while ϕ increases by 5°.

(ii) Continuous recording

To investigate one circle at a time still involves changing ϕ for each circle on the powder sphere. The next step is to gear the two rotations together, so that points on a spiral about the specimen normal are brought in turn to the P_0 position and investigated by the counter.

(iii) Automatic recording

Finally, the output from the counter can be made to operate a multiple pen head, the base pen of which is tracing out a spiral which corresponds in stereographic projection to the spiral on the powder sphere. As the X-ray

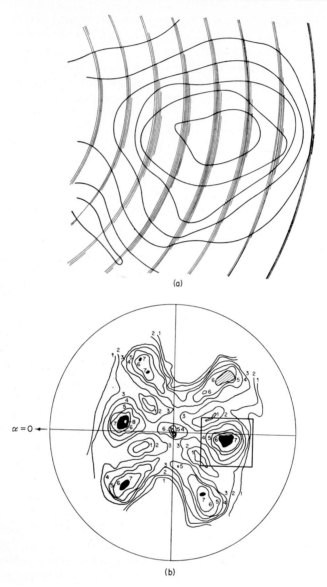

FIG. 151. The automatic production of pole figures. (a) An example of the multiple pen recording of intensities (relative relp densities). The ends of the nth line, projected on to the basic spiral, are points on the nth level contour. (b) The whole 200 pole figure for rolled aluminium sheet, contoured from automatic recording in this way. The outlined portion is shown enlarged in (a).

This recording was done with a *reflection* texture goniometer, whose limitations are opposite to those of the transmission instrument.

intensity reaches certain preset amounts, so first one and then, with increasing intensity, up to 10 pens are brought into contact with the paper (Fig. 151a). Each pen, coming into or out of operation, corresponds to the intensity reaching a certain point and then dropping below it. [13]. The contour lines of the pole figure are then drawn by joining up the ends of corresponding lines (Fig. 151b).

There are a number of difficulties. When the normal to the plate makes an angle of more than 60–70° with the X-ray beam, the corrections due to obliquity (larger irradiated volume, greater path length and absorption) become so large that even corrected values are very inaccurate. This is the main reason for the practical limit of 30° from the centre of the stereogram below which the figure cannot be plotted. Even up to 70° obliquity, corrections must be applied. These can either be calculated, or preferably a run can be made with an otherwise exactly similar specimen, but of known unoriented material, to give a base line from which to measure the variation due to preferred orientation. For the fully automatic case these corrections must be applied as the recording is being done, either by altering the speed of recording to compensate for the corrections, or by applying automatic corrections to the output.

1.2.2. *Reflection goniometer*

So far we have assumed a thin plate used in transmission. But many specimens are massive and cannot be prepared in this way. In that case a reflection goniometer is used, in which the flat face of the specimen rotates in its own plane with the rotation axis initially parallel to OP_0 (Fig. 54, p. 79). For this zero value of ϕ, specimen rotation has no effect, but with ϕ increasing, the vectors passing through the position P_0 will trace a spiral from the centre of the stereogram outwards. (Fig. 152a). If one spiral does not give sufficient resolution, two runs with interlocking spirals can be made by rotating the specimen in its holder, through 180° about the normal. (Fig. 152b).

However, with this arrangement, the rotation about the vertical axis from $\phi = 0$, is limited to $\theta°$, at which point either the incident or the reflected beam would lie in the surface of the specimen. In practice the rotation is limited to about $\theta - 20°$ so that for low angle reflections very little of the sphere could be investigated. It is therefore necessary to provide a vertical rotation circle, so that the normal to the plate can be moved synchronously with the rotation of the specimen holder, from the position OP_0, through an angle in the vertical plane (Figs 153a and b). P_0 then traces out the spiral without coming up against the limitation which rotation in a horizontal plane produces. The angles which the incident and reflected beams make with the plate remain equal, but get smaller as ϕ increases, becoming zero for $\phi = 90°$. Because of this increasing obliquity on the surface of the specimen, it is

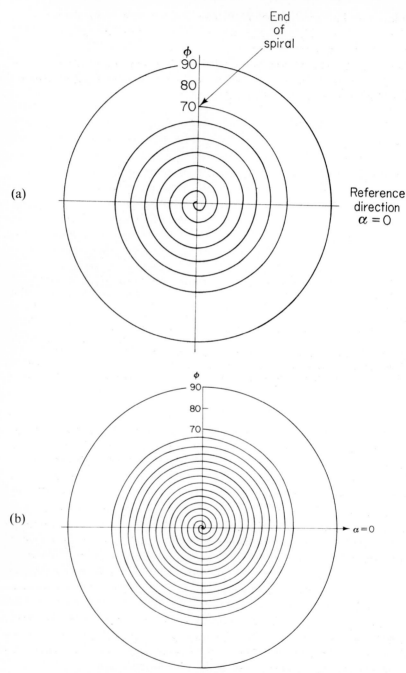

FIG. 152. (a) The spiral on the stereogram corresponding to the directions on the powder sphere, which are investigated as the specimen rotates and the angle from the centre ϕ increases synchronously. (b) Increased resolution of two interlocking spirals, obtained by remounting the specimen at 180° rotation from the initial mounting.

impractical to increase ϕ beyond 70°. However, up to $\phi = 40°$ the obliquity on the specimen is changing very slowly and corrections are not necessary. Above 40° they must be made.

The apparatus shown diagrammatically in Fig. 153 (a and b) is called a reflection texture goniometer, and the same automatic recording can be arranged as in Section 1.2 (iii), (p. 240).

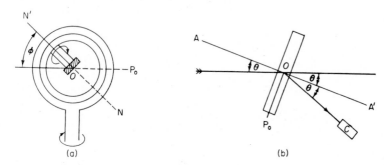

FIG. 153. Diagrammatic representation of a reflection texture goniometer. (a) The specimen holder, which rotates about the axis NN' normal to the face of the specimen. The rotation spindle is mounted in the inner ring of the vertical circle, which rotates about a horizontal axis (AA' in (b)). (b) The plan of the arrangement. The counter arm OC is set to the angle 2θ for the reflection whose pole figure is to be determined. The horizontal rotation axis of the vertical circle AA' is set to θ, so that OP_0, the direction in which the relp density is being investigated, is in the plane of the vertical circle. The rotation of the specimen about ON is geared to the rotation of the vertical circle so that, for example, $\alpha = 36\phi$ or one revolution of the specimen occurs for every 10° increase of ϕ. This produces the spiral of Figure 152(a).

Usually 70° from the centre of the stereogram is sufficient in the reflection case; and if it is not, reorientation of the specimen, cutting a fresh surface if necessary, enables the 'blind area' near the primitive to be filled in.

Because of their greater versatility most automatic 'texture goniometers' are of the reflection type. Fig. 153c is a photograph of one such instrument.

If the crystallite (grain) size is small compared with the cross-section of the beam, the reflection intensities obtained will represent a reasonable average for the material. But if not, the specimen must be oscillated linearly in its own plane as well as rotated, so that an average output is obtained from many tens and preferably hundreds of crystallites. Even so the result obtained in the reflection case only applies to the depth of penetration of the X-ray beam. For this reason MoKα radiation should be used.

It must be emphasised that a preliminary X-ray photograph of a specimen should always be taken to check the grain size, and pick up any peculiarities such as partial ordering, etc. (Section 7.4.4, p. 86) which would be impossible or difficult to detect with the goniometer.

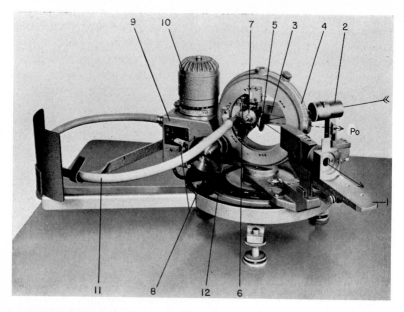

Fig. 153(c). A Siemens texture goniometer set up for reflection. The detector is carried on the arm (1) which is set at an angle of deviation of 2θ. The incident beam is defined by slits in the housing (2) and a horizontal slit diaphram (3). The vertical ring (4) is set with its axis at θ to the direct beam. The specimen in its holder (5) can be moved in its own plane by a slide (6), in order to scan the surface of specimens with large grain size. At the end of each movement of the variable eccentric operating the slide, a ratchet (7) turns the specimen holder through a small angle α, about its normal. At the same time another ratchet mechanism (8) rotates the inner vertical circle through an angle ϕ. With ϕ zero as shown the normal to the specimen is along OP_0 and the relp density at the centre of the stereogram is being investigated. The gearbox (9) allows the selection of a number of combinations of speeds of the two angular motions, which are driven by the motor (10), directly for ϕ and via the flexible drive (11) for α (and the slide).

The goniometer can be used in transmission by replacing the specimen holder (5) with a frame parallel to the vertical circle. This is then rotated by a separate fast drive (12) for varying α (Fig. 150(a), p. 241), but the angle ϕ (Fig. 150 (b)) has to be set by hand, so that a spiral scan is not possible.

2. The Interpretation of Pole Figures

2.1. ROLLED SHEETS

The morphology of the specimen (e.g. normal to and rolling direction in a sheet) is plotted on the pole figure, and the mean positions of the patches of relps are plotted as a single cube orientation.

The normal to the sheet is usually plotted in the centre of the stereogram, especially in the case of automatic recording. The commonest relations of morphology and texture in rolled metals are: (a) (111) [$\bar{1}$10] i.e. a (111) face in the plane of the sheet and [$\bar{1}$10] in the rolling direction (the so-called 'cube on corner' orientation). This produces a peak in the centre of the 111 stereogram, and on the 110 stereogram the peaks are in the rolling direction and at 60° around the primitive. (b) (100) [001] or (100) [011]—'cube on face' orientations, with either the cube axis or [011] in the rolling direction. (c) (110) [001] or (110) [1$\bar{1}$0] or (110) [1$\bar{1}$1]—'cube on edge' orientations, with three possibilities for the rolling direction. The positions of the poles on the stereograms for these orientations are easily worked out for comparison with the observed pole figure, but in practice slight variations and mixtures of orientations tend to occur.

2.2. ROLLED RODS

The specimen is usually cut normal to the rod axis, and in the ideal case of a single fibre orientation, the pole figure consists of circular bands centred on the normal, or a circular patch at the normal. If a specimen is cut longitudinally, the bands will lie on small circles about the axial direction which lies in the primitive. Fig. 57 (p. 82) shows the type of orientation diagrammatically. Such orientation is easily plotted on a stereogram for comparison with an actual pole figure.

However the rolling of rod, unlike the drawing of wire, is rarely an axially symmetric process, and often produces a mixture of orientations which makes the pole figure very difficult to interpret. In this and other cases of hot or cold working it may be more useful to investigate the preferred orientation in a single direction in the specimen, as in Section 4 (p. 251).

3. Fibre Diagrams

Refer to Section 7.3 (p. 81) and Figs. 57 and 58 (pp. 82 and 84). A fibre diagram of a metal wire, as in Fig. 154, often cannot have the layer line separation measured because of the extent of the misorientation, and possibly because of the simultaneous presence of two preferred orientations. In such a case it is a relatively simple matter, for a cubic metal, to draw to scale the likely patterns, and by comparison with the photograph determine the orientations present. From examples which can be unequivocally interpreted as a single preferred orientation, it is known that [100], [110] and [111] are the three likely axial directions. Diagrams on tracing paper are therefore drawn for these three possibilities, as follows:

1. Treat the photograph as a cubic powder diagram and decide the type of unit cell, i.e. P, I or F.

In Fig. 58(a) (p. 84) the d^{*2} (= equatorial ξ^2) values for the first two lines are in the ratio 2:1; the lines form a regular sequence and the seventh line is present. The cell is therefore I and $N = 14$ for the seventh line. (Appendix II, p. 364) [1(120)].

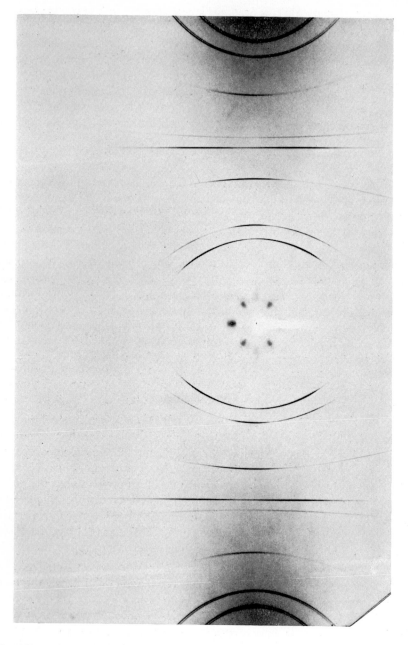

FIG. 154. A photograph of Cu wire, taken on a cylindrical film of radius 30 mm, using CuKα radiation.

2. Use the Bernal chart to measure the equatorial ξ values of the line with highest N ($d^{*2} = Na^{*2}$) and find a^*.

For $N = 14$, $\xi = 1 \cdot 82$ and $a^* = 1 \cdot 82/\sqrt{14} = 0 \cdot 485$.

3. Calculate the RL interplanar spacings, ζ_{pqr}, perpendicular to the lattice vectors g_{pqr}, for $pqr = 100$; 110; and 111.

For tungsten, $\zeta_{100} = a^* = 0 \cdot 485$

$$\zeta_{110} = \lambda/g_{110} = \lambda/(\sqrt{2}a) = a^*/\sqrt{2} = 0 \cdot 343$$

$$\zeta_{111} = a^*/\sqrt{3} = 0 \cdot 280$$

From the layer line spacing measured with a Bernal chart ($\zeta = 0 \cdot 34$) the fibre axis is [110]; but even in this simple case comparison with the calculated pattern will be a useful check.

4. Draw the powder rings for the top half of the photograph on tracing paper, copying the photograph and using the constant θ chart where necessary. Draw and label the horizontal lines of constant ζ for the three axial directions, [100]; [110] and [111]. Fig. 155 shows all three on one diagram.

5. Draw up a schedule of the reflections which are possible on the layer lines and their multiplicities for the three axial directions. For a crystal lattice axial direction g_{pqr}, the relps hkl on the nth layer are given by $hp + kq + lr = n$ (Appendix II, p. 367). It is important to remember that the indices of a powder ring are representative indices, and that they include all the others of the symmetry related set. For cubic metals, this means all possible permutations of hkl (klh, lkh, etc.) and all possible combinations of positive and negative indices. Thus for g_{110} (i.e. the [110] direction) along the axis, the 200 reflection will not appear on the zero layer ($n = 0$) since $hp + kq + lr = 2$ in this case. But the 002 and 00$\bar{2}$ reflections will appear on the zero layer, and the 200 powder ring will therefore have an arc of multiplicity two on the zero layer. Similarly, 211 will not occur on the first layer, but 2$\bar{1}$1 (and 2$\bar{1}\bar{1}$, $\bar{1}$21, $\bar{1}$2$\bar{1}$) will be present. Figure 156 gives the schedule for tungsten.

6. Draw small arcs where reflections will be present in each of the three cases, preferably on separate sheets, or with different colours on one diagram, and compare with the photograph. In the case of tungsten, a traced copy of the photograph (Fig. 58(a), p. 84), laid over Fig. 155, clearly verifies the axial direction as [110]; but since the photograph was taken with CuKα radiation, for which the reflected beam is reduced to 1% by a penetration of only about 10 μm, the information only tells us about the outer layers of the wire, and not about the inner core. (**Example 41**, p. 321).

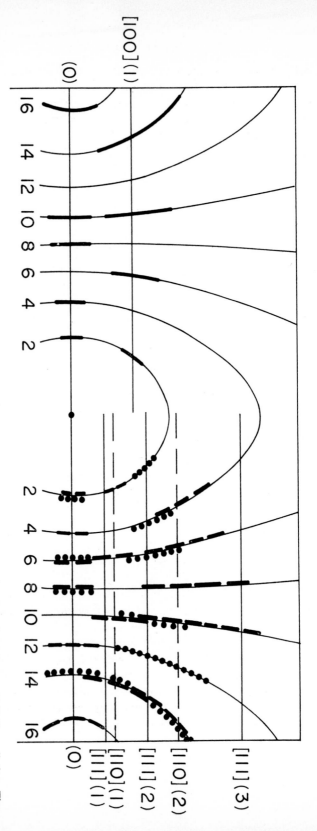

Fig. 155. The powder rings for W on a cylindrical film of radius 30 mm, which would be produced with a randomly oriented specimen and CuKα radiation. The layer line levels for rotation photographs of a single crystal of W about [100], [110] and [111] are marked. The arcs which would appear on a fibre photograph of W are shown: (a) full lines on the LHS for [100], (b) dashes on the RHS for [110], and (c) dots on the RHS for [111]. Overlapping has meant that on lines 10 and 14 the dotted [111](2) arcs are broken up. The second layer for [100] and the third for [110] are off the top of the photograph. The arcs shown correspond to a 15° range of misorientation, i.e. the axes of the crystallites, which, in the case of a single crystal, would be along the wire axis, actually make angles of up to 7½° with it.

This procedure is particularly useful if a number of specimens of the same metal have to be investigated, and allows a rough estimate of the amounts of the different orientations to be given. However for more accurate estimates, for the general case as well as for fibres, the counter methods described in the next section are required.

layer	ζ	pqr	$N \rightarrow$ 2 $\sqrt{N}\,a^* \rightarrow$ 0·69 $hkl \rightarrow$ 110	4 0·98 200	6 1·20 211	8 1·38 220	10 1·55 310	12 1·69 222	14 1·83 321	16 1·96 400
0	0	[100]	4	4	0	4	8	0	0	4
0	0	[110]	2	2	4	2	0	4	0	2
0	0	[111]	6	0	6	6	0	0	12	0
1	0·49	[100]	4	0	8	0	4	0	8	0
1	0·35	[110]	4	0	4	0	4	0	8	0
1	0·28	[111]	0	0	0	0	0	0	0	0
2	0·69	[110]	1†	2	2	4	2	0	4	0
2	0·56	[111]	3	3	6	0	6	3	6	0
3	0·84	[111]	0	0	0	0	0	0	0	0

† On the axis and too far away for the cap to cut the sphere of reflection.

FIG. 156. The schedule of possible reflections (up to $N = 16$) and their multiplicities on the layer lines of a rotation photograph from a single crystal of W about [100], [110] and [111]. Layer 2 for [100] and layer 3 for [110] are too high to be recorded. Layers 1 and 3 of [111] could have been omitted, since the body centering (I) of W means that the difference between crystal lattice points in the [111] direction is halved. Therefore the layer lines will be twice as far apart as for a primitive (P) lattice (Alternatively, all the reflections on odd layers are 'systematic absences', Sec. 10.1, p. 127).

4. Inverse Pole Figures

In many applications it is important to have a quantitative estimate of the amount of preferred orientation in one or a small number of directions in the specimen. Electrical, magnetic, thermal and mechanical properties in a rod or sheet, may depend mainly on the preferential orientation along the axis of the rod or the normal to the sheet, without being much affected by the kind or amount of preferred orientation perpendicular to these directions. In any case it is possible to investigate a number of directions by the technique to be described, providing a specimen with a flat surface perpendicular to the specified direction can be prepared. (Sheet can be cut into square strips, to give either the normal direction or a direction in the plane of the sheet). The method can be used for cubic materials, but is particularly useful for lower symmetry substances such as orthorhombic α-uranium, because there are more independent RL vectors of differing lengths and therefore more powder reflections.

The method is the 'inverse' of the production of a normal pole figure, because instead of investigating the relp density in all directions over one reciprocal sphere, it investigates the relp density in one direction for as many spheres as possible. The one direction is the normal to the specimen surface, so that we investigate the relp density at the centre of a reflection pole figure (Fig. 151b, p. 242) for as many reflections as possible.

The relp density at N, the centre of the stereogram, is measured at the counter when the normal ON to the specimen surface lies along OP_0 (Fig. 54, p. 79) for whichever reciprocal sphere is being investigated. This arrangement is simply achieved by the usual powder diffractometer geometry. The normal to the specimen surface is set to bisect the angle between the incident and diffracted directions, and the counter arm rotates at twice the rate of the specimen holder. As each powder reflection comes up on the chart recorder, the intensity is dependent on the relp density of the powder sphere in the direction of the specimen normal ON. It is thus possible to take the usual powder trace with such a specimen and extract the required information.

Quantitatively we have to find the ratio p_{hkl}, of the density of the hkl relps at N, to the relp density in a specimen with completely random orientation (or the ratio of the hkl relp density at N to the average density over the sphere). The procedure for finding p_{hkl} from the diffractometer trace is dealt with below. First the method of presenting the information has to be worked out.

4.1. PRESENTATION OF THE RESULTS

The results for the various RL vectors are obviously not completely independent. If a concentration of $\mathbf{d}^*_{001}$ vectors from a certain fraction of the crystallites in the specimen occurs in the direction of ON, the same crystallites could not give a concentration of $\mathbf{d}^*_{010}$ vectors in that direction. There would normally be a deficiency of such vectors there for the specimen as a whole[†]. Similarly $\mathbf{d}^*_{014}$ would make a small angle with $\mathbf{d}^*_{001}$; and if a high density cluster of $\mathbf{d}^*_{001}$ vectors occurred around N, one would expect the corresponding patch of $\mathbf{d}^*_{014}$ vectors to spread enough to give an above average relp density at N as well. Since the relationship is connected with the directions which the various RL vectors make with one another in a single crystallite, the values of p_{hkl} are plotted on a stereogram of the normals to the various sets of lattice planes. The stereogram is usually drawn in the standard orientation. Contour lines can be drawn for constant values of p_{hkl}, indicating the RL directions which tend most to line up with ON, and therefore

[†] 'Normally' because two modes of preferred orientation may, and often do, occur together, and the second mode might conceivably line up another set of crystallites to give a concentration of $\mathbf{d}^*_{010}$ vectors at N.

the crystal lattice planes which tend most to line up parallel with the surface. Figure 157 shows two such 'inverse pole figures' for sections of uranium rods. Since α-uranium is orthorhombic, the stereogram has three planes of symmetry, perpendicular to the three axes, and therefore only one octant need be plotted. In Fig. 157 the indices of the normals are given in one octant, and the p_{hkl} values at the symmetry related points in two others, one for each section.

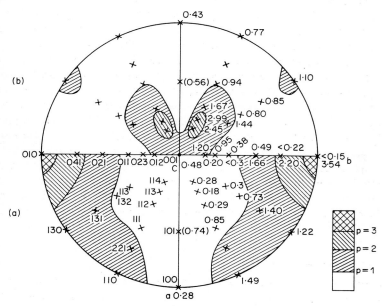

FIG. 157. Inverse pole figures for a lightly hot-rolled α-uranium rod, showing the concentrations of relps in a direction (a) normal to a transverse section, i.e. along the axis of the rod, and (b) normal to a section containing the rod axis (After Harris).

In a randomly oriented specimen all the p_{hkl} values would be unity, and the average for all directions in a preferentially oriented specimen would also be unity. If the number of reflections n that have been measured is sufficient to give an adequate spread of points on the stereogram we have:

$$\frac{1}{n}\sum p = 1 \qquad (1)$$

to a good approximation.

4.2. DETERMINATION OF p_{hkl}

The list of measured intensities (the areas under the peaks on the chart recording or the total counts, less background, in a step by step record) for as many resolved reflections as possible from the preferentially oriented specimen, must be supplemented by a similar record from a randomly oriented specimen. For a flat plate, reflecting the whole spread of the beam incident on it at the Bragg angle θ, the measured reflected intensity is given by:

$$I_{hkl} = K I_0 \frac{B}{2\mu} p(hkl) F_{hkl}^2$$

$$= K C p_{hkl} F_{hkl}^2$$

where I_0 is the constant incident intensity; K is an instrumental and geometrical constant, which is a function of θ, but the same for different specimens; B is a function of the perfection of the crystallites and the spread of the reflections which, for crystallites that are not both very small and platy or needle shaped, should be the same for all reflections; μ is the linear absorption coefficient, which may vary from specimen to specimen, depending on compaction; F_{hkl} is the normal structure factor. C is therefore a constant for a particular specimen.

For the randomly oriented specimen

$$I'_{hkl} = K C' F_{hkl}^2, \text{ since } p_{hkl} = 1$$

$$\therefore \qquad \frac{I_{hkl}}{I'_{hkl}} = \frac{C}{C'} p_{hkl}$$

$$\therefore \quad \text{from 253(1),} \quad \frac{1}{n} \sum \frac{I_{hkl}}{I'_{hkl}} = \frac{C}{C'} \cdot \frac{1}{n} \sum p_{hkl} = \frac{C}{C'} .$$

Therefore for any particular reflection:

$$p_{hkl} = \frac{I_{hkl}}{I'_{hkl}} \bigg/ \frac{1}{n} \sum \frac{I_{hkl}}{I'_{hkl}} .$$

Thus, to get p_{hkl} for one reflection it is necessary to measure at least ten and preferably twenty other reflections to obtain the factor C/C'. It is usually worth calculating p_{hkl} for all the reflections and drawing the inverse pole figure, since the size and shape of the contour peaks give information about the mode of preferred orientation—single or multiple, a small proportion sharply lined up with a large amount of random orientation or a large proportion with slight preferred orientation, etc. This can supplement the information from a normal pole figure for one reflection. However, to follow the effect of different amounts of reduction in rod diameter or sheet thickness, it may be useful to plot the value of p_{hkl} for a particular set of planes known to be orienting perpendicular to the rod axis or parallel to the sheet surface. Figure 158 shows such a plot for reduction of mild steel sheet, in which the value of p_{111} (which must necessarily be the same as p_{222}) reaches a value of 5 times random.

4.3. ACCURACY

Since all preferred orientation measurements are essentially statistical, it is necessary that the volume sampled by the beam should be representative of the whole specimen. In a homogeneous, fine grained material this will normally be the case; but many specimens are coarse grained and have preferred orientation which varies from place to place, e.g. from the surface inwards. The necessity of scanning a large surface in material that preliminary photographic investigation shows to be coarse grained, limits the possibility of investigating the variation of preferred orientation. In an extreme case this may lead to the averaging of two modes, if the type of preferred orientation is changing with position in the specimen.

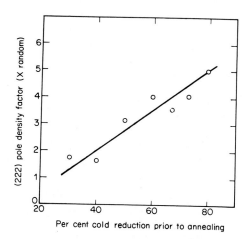

FIG. 158. The effect of cold reduction, prior to annealing, on the concentration of 222 relps in the direction of the normal to stabilised steel sheet. (After Atkinson.)

The accuracy of the method is further limited by the comparatively small number of reflections which are measurable, giving errors in the value of C/C' derived from 254(1). In the best circumstances, p_{hkl} is unlikely to have much better than 10% accuracy. However, as values ranging for 0–6 are found, significant results can normally be obtained even in less favourable circumstances.

Chapter 18

The Measurement and Weighting of the Intensities of X-ray Reflections and their Reduction to F_0^2 Values

The various factors affecting the intensity of X-ray reflections have been discussed in Chapter 9. The present chapter is concerned with the practical aspects of the matter, primarily with measurement of single crystal reflections, although a good deal will apply, with some modification [1(186)] to powder specimens. Counter methods will only be dealt with sufficiently to enable a comparison to be made with photographic methods, since the use of counter diffractometers is a highly specialised technique which has been adequately dealt with elsewhere [14].

1. Methods of Measurement and Recording Errors

1.1. COUNTERS

1.1.1. *Proportional and scintillation counters and discrimination*

Only proportional and scintillation counters need be considered in detail, since Geiger counters cannot provide discrimination, and are limited to counting rates some hundreds of times less than those of the other counters. However, in cases where they can be used, the simplicity of their electronic requirements is a strong reason for doing so.

Electronic discrimination is based on the fact that the pulses of proportional and scintillation counters are not just 'triggered' (as in a Geiger counter, where they are all the same height) by the absorbed photon, but that the pulse height is proportional to the energy of the photon. But the pulses are several orders of magnitude smaller than for a Geiger counter, and require elaborate amplifiers and high stability H.T. packs. In addition electronic apparatus is needed to cut off pulses which are above or below the set 'gate'. The gate is centred on the Kα pulse height; but the spread inherent in the ionisation process in the proportional gas counter (or the production of light pulses and their amplification in a scintillation counter) means that the gate cannot be very narrow, or too many $K\alpha$ pulses will be lost. $K\beta$ pulses can just about be cut out but this is the most that can be achieved. However, this discrimination is sufficient to cut out most of the white radiation and reduce the background to small proportions, thus greatly improving the peak to background ratio and therefore the accuracy of the measurements. This is the primary reason for discrimination, but it has two other, more specialised uses.

1.1.2. *Discrimination and absolute measurements*

(i) Crystal monochromators

If measurements are being made using crystal reflected radiation, the incident beam will contain not only the desired $K\alpha$ component, but also components of $\lambda/2$, $\lambda/3$, etc. from reflections of higher orders. These occur at the same Bragg angle as the first order $K\alpha$ reflection from the monochromator. (This is easily demonstrated using a Guinier powder camera, with a specimen which gives strong, low angle lines. Each strong line will be accompanied by a weak line, nearer the direct beam. This weak line is due to the $\lambda/2$ radiation, and can cause considerable confusion if not identified as such). In the RL, the relps due to $\lambda/2$ will occur at exactly half the spacing of the $K\alpha$ relps ($\lambda/3$ at a third, etc.) and the even (third etc.) orders will coincide with a $K\alpha$ relp. Absolute measurements made photographically (or by Geiger counter), using crystal reflected radiation, will be subject to errors (usually small) due to the measurement of $\lambda/2$ ($\lambda/3$ etc.) photons. These will occur both in the direct beam and in the $K\alpha$ (hkl) reflections, due to the $2h$, $2k$, $2l$ ($3h$, $3k$, $3l$ etc.) reflection of the $\lambda/2$ ($\lambda/3$ etc.) radiation. Electronic discrimination however can easily prevent such photons being measured and thus eliminate this source of error.

(ii) Balanced filters

The other use of discrimination occurs when balanced filters (first introduced by Ross in 1926 and sometimes called 'Ross filters') are being used instead of crystal reflection for monochromatisation. Balanced filters are a development of ordinary β filters [1(68)]. Two measurements are made of each reflection (or of the direct beam). One measurement is made with a filter in position whose absorption edge is just on the short wavelength side of the $K\alpha$ radiation. This filter lets most of the $K\alpha$ radiation through. The second measurement is made with a filter whose absorption edge is just on the long wavelength side. This stops most of the $K\alpha$ radiation. The difference between the two measurements is due to the difference in the $K\alpha$ radiation passed by the two filters. The effect is shown in Fig. 159. The thicknesses are arranged by rolling and etching so that the absorptions of both filters are 'balanced', i.e. as near as possible the same on the two sides of the $K\alpha$ peak, and the difference in the Kα intensities is a maximum. There will also be a difference in the background counts for the small range within the pass band, but this will be negligible compared with the Kα counts, especially for copper radiation.

The main advantage of balanced filters is that the (difference) intensity is only reduced to about 70% of the incident Kα intensity, compared with 10% or less with crystal reflected radiation.

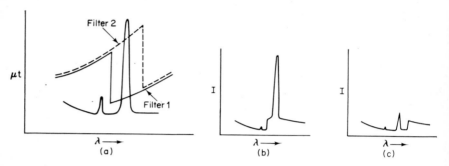

FIG. 159(a). Schematic diagram of the X-ray spectrum near the Kα peak, with the absorption curves for a pair of 'balanced filters' superimposed. (b) The spectrum with filter (1) in position. (c) The spectrum with filter (1) replaced by filter (2). The Kα peak is relatively much higher than that shown on the diagrams. Filter (1) will be made of the same material as the corresponding β filter [1(68)] and filter (2) will normally be the next element down in the periodic table (Ni and Co for CuKα).

By differentiating the expression for the difference in Kα counts it can easily be shown that there is a maximum for a thickness

$$t = \frac{\ln \mu_1 - \ln \mu_2}{\mu_1 - \mu_2},$$

where μ_1 is the linear absorption coefficient for the material at the Kα wavelength within the pass band, and μ_2 is the absorption coefficient which the material would have at this wavelength if extrapolated from the other side of the absorption edge. Since $\mu_1 t$ for one filter $= \mu_2' t'$ for the other, from the requirement for 'balance', this formula gives the correct thickness when applied to either material, both for balance and maximum intensity.

The 'pass' intensity decreases rapidly for thinner filters, but only slowly with increase in thickness. However, the thickness for maximum intensity is also near the optimum for constant transmission within the pass band. This is not a critical requirement for X-ray crystallographic applications. But in the study of scattering from non-crystalline material, where it is desired to discriminate against Compton scattered radiation [1(66)], with its change in wavelength, this constant transmission is important. (**Example 42**, p. 321).

The difference in the counts obtained with the two filters in position will be due to the difference in the $K\alpha$ contributions. This will be true for the direct beam as well as the reflections, and enables absolute measurements to be made. Unfortunately if the filters are 'balanced' near the $K\alpha$ peak, they will not be balanced at the white radiation 'hump', and this limits the accuracy obtainable with balanced filters if a Geiger counter (or ionisation chamber) is used. If, however, discrimination can be employed, none of the white radiation photons in this region will be counted, so that the lack of balance does not matter.

Since, in measuring the direct beam, we are measuring a difference between two much larger counts (the rejected white radiation photons must be recorded before they can be rejected) it is necessary to make sure that the direct beam with the low absorption filter does not overload the counting chain. It is an advantage of monochromatisation by crystal reflection that it does not tend to overload the counting chain in the same way; but this becomes a disadvantage when the beam is too weak to get near the maximum count rate. Also, it is much more difficult to get a uniform beam of crystal reflected radiation, so that balanced filters are preferable for most purposes when counter recording is used. For photographic recording balanced filters are hardly feasible.

1.1.3. *Accuracy obtainable by counter methods*

Accuracy in this field is usually thought of largely in terms of counting statistics, presumably because the method was developed from radio-active counting, where counting statistics are the main limitation on the accuracy of the results. However, the accuracy we are interested in is that of the F_0^2 values obtained from the measurement of E, the X-ray energy. If other errors in deriving F_0^2 are much greater than counting errors, or less reducible, then they are more important. As is shown in Section 2.1 (p. 261), absorption errors are likely to be less reducible than counting errors, and the latter should always be considered in relation to absorption errors. Only for very weak reflections or very short counting times are counting errors likely to be the main source of inaccuracy in F_0^2 values. Where, however, other errors can be reduced below 1% the possibility of making counting errors as small as desired becomes important.

1.2. Photographic Recording

1.2.1. *General*

The methods normally used for intensity measurement by photographic means may be classified as:
1. Eye estimation against a standard scale, preferably made from a reflec-

tion of the crystal concerned, and followed by spot area correction. The best accuracy that can be expected is not much better than 10%.

2. Flying spot photometry of a normal (non-integrated) reflection. This may be necessary if minimum exposure times are required for any reason, but the apparatus is complicated and expensive and the accuracy only about 3–4%. The contouring photometer may serve the same purpose for a few reflections, but is mainly useful for following changes taking place in a specimen.

3. The photometry of integrated reflections produced by an integrating Weissenberg or precession camera. This involves increasing the exposure time by a factor of 2 or 3; but as against this, spots with a uniform central area whose optical density is proportional to the integrated reflection, can be measured with a simple photometer of moderate cost (Appendix X, p. 446). Such an instrument gives an accuracy of better than 1% in measurement of optical density over the required range. It is this third method that will be discussed, primarily for the case of an integrating Weissenberg camera using CuKα radiation.

1.2.2. *Errors due to photographic recording*

In an area of film uniformly blackened by exposure to X-rays, the optical density $D = \log (i_0/i)$, where i_0 is the intensity of the incident photometer light beam and i the intensity of the transmitted beam. D is proportional to $E = It$, the X-ray exposure producing the film blackening, where $I = $ X-ray intensity and $t = $ time. If t is constant $D = KI$. This linear relationship holds (within experimental error, or with slight but measurable deviations, depending on the type of film) up to about $D = 1.0$. If measurable deviations exist it may be necessary to use a calibration curve for the highest accuracy, but in most cases the deviation can be ignored.

The experimental requirements for accurate photographic recording are described in Appendix IX (p. 429). Assuming that these have been satisfied, and include specifically a photometer with a beam diameter of 0·5 mm approximately and a 500 mm scale, then on the top film of a two film pack (typically Ilford Industrial G) there is a constant error of 0·006 in optical density (D), and an independent error of 0·3% of the measured D. On the second film (typically Ilford Industrial B) absorption in the top film and lower sensitivity in the second give a ratio of about 10–1 in D values, and a constant error of 0·008 plus 1·5% of D. The errors proportional to D are due to irregularities in the emulsion of the film, the second film being affected by non-uniform absorption in the top film, as well as the inhomogeneity of its own emulsion.

One can roughly summarize the results on photographic measurement of intensity as follows. The errors can be reduced to the order of 1% by arrangng to obtain an optical density of around 1·0, but for normal purposes with a wo-film pack the errors in a range of intensities from 1 to 1,000 are as given n Fig. 160. Very weak reflections, 1–5 on this scale, will normally be mea- ured to, at worst, 10% by eye estimation of a non-integrated Weissenberg hotograph. This, taken before the integrated one, also allows a check on spot ize for integration.

ntensity	1–5	5	60	120	120	1000
Error	10%	10%	1%	0·5%	5%	2%
					(2%)	(1%)
	Top film	Top film			Bottom film	
Method	Eye estimation	Photometry			(Second exposure)	
					Photometry	

FIG. 160. Summary of film errors in measuring X-ray intensities photographically, using photometry of integrated reflections where possible. A second exposure of 1/10 duration gives smaller errors than a 2 film pack.

In most cases these errors are less than those arising from crystal shape and texture, but there is one exception which can provide an important limitation. The white radiation streak through low angle reflections derives, to a con- siderable extent, from higher order reflections than the one being measured. Therefore the background density must be measured in the streak on either side of the reflection. If the background is varying linearly, little additional error is caused. But this is by no means always the case; and where the slope of the background changes rapidly in the neighbourhood of the reflection, errors of up to 30% can be caused by taking it to be linear. For Mo radiation the position is probably worse, because the Laue streaks are far more intense and affect more reflections. Moreover, the absorption edges for Ag and Br on either side of MoKα introduce further irregularities, and the β filter has more effect on the Laue streak than is the case for Cu radiation. It is one of the real advantages of proportional counters and discrimination that such problems can be effectively dealt with. However, only a few reflections are usually affected with Cu radiation, although the problem is more serious with Mo. Such reflections must be given a lower weight in least-squares procedures.

2. Errors Common to all Recording Methods

2.1. ABSORPTION ERRORS

2.1.1. *General*

Absorption errors are generally underestimated, partly for historical rea- sons (as long as eye estimation of intensities with an error of 10% was the main method, comparable absorption errors could be ignored and have largely continued to be); and partly from an error in reasoning from the A* (absorption correction factor) tables for cylindrical crystals. The process is as follows. A rough measurement is made of the dimensions of the crystal from which the diameter of the equivalent cylinder is estimated. From the calculated value of μ (the linear absorption coefficient) μr is found, and A* looked up in the tables. If A* is only slightly greater than 1·0 and hardly varies across the table with θ, it is assumed that absorption errors will be negligible. This is almost a complete fallacy. If the crystal were a perfect cylinder, then (except for absolute measurements) absorption could be ignored in such a case. But no crystal is a perfect cylinder, despite all efforts to make it so. Consequently for each reflection there is an equivalent perfect cylinder, the radius of which varies from reflection to reflection, depending on the aspect of the crystal presented to the X-rays.

What matters therefore is not how A* varies with θ, but how it varies with r. Thus it is the variation in the vertical columns of the table, rather than in the horizontal rows, which is significant, and this is quite large everywhere. At $\mu r = 0·1$ the slope is still half as great as it is at 1·0. This explains why a crystal may have no significant variation of the absorption correction with θ, but nevertheless have quite large variations from one symmetry related reflection to another of the same θ.

Finally it cannot be too strongly emphasised that only the measurement of symmetry-related reflections can give a full indication of absorption errors. Measurement of the same reflection about two different axes only presents partially different aspects of the crystal to the X-rays (related by rotation about the plane normal), and in general these aspects are unlikely to be as different as those corresponding to different reflections. Therefore the measure of the error in intensity should be taken as the 'coefficient of variation' (percentage standard deviation from the mean) of the intensity measurements on a set of symmetry-related reflections.

Because of the tendency to underestimate absorption errors, it is all the more essential to have some means of estimating what they are likely to be in various cases. In favourable cases, combined with recording errors, such estimates can provide a basis for an objective weighting scheme in subsequent least-squares processes.

A square prism is taken as one example, to give some indication of the effect of considerable irregularity; and an imperfect sphere with small random irregularities as another, to show the size of error in the case where the crystal has been shaped to minimise absorption errors. The errors for the most perfect sphere which it is possible to grind will be of the same order as the errors for a crystal bounded by more or less plane faces, whose dimensions have been measured as accurately as possible, so that individual absorption corrections can be calculated by computer for each reflection.

μa	3.2		1.6		0.8		0.4		0.2		0.1	
	A*	μr	A*	μr	A*	μr	A*	μr	A*	μr	A*	μr
Case (i)	7·7	1·66	3·51	0·866	2·00	0·438	1·44	0·221	1·204	0·1114	1·098	0·0545
Case (ii)	11·2	2·19	4·03	0·984	2·11	0·476	1·47	0·236	1·217	0·1178	1·104	0·0580
Diff. %	37	27·5	13·8	12·8	5·4	8·4	2·4	6·6	1·1	5·6	0·5	6·2
Δr for	0·166		0·078		0·048		0·038		0·032		0·035	
ρ { $a = 1.0$ mm }	0·60		0·58		0·57		0·57		0·57		0·56	

The equivalent values of μr for a cylinder are taken from the A* tables [6(295)]

FIG. 161. Values of the absorption correction factor A*, for the two cases of reflection shown in figure 162 for various values of μa (a mm is the side of the square prism).

2.1.2. *Errors in the case of a square prism*

In all but the most transparent crystals absorption modifies the reflected intensity, by an amount which depends on the crystal shape and orientation and the Bragg angle. The magnitude of this effect can be calculated for simple shapes (Section 6.4, p. 282). The results of calculations of the absorption correction factor A* for a square prism are given in Fig. 161. ΔA*, the difference between the factors for the two cases considered (Fig. 162), gives an indication of the error involved in ignoring such corrections.

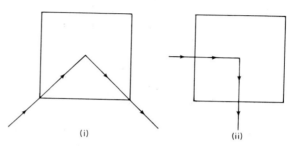

Fig. 162. The directions of incident and reflected beams in relation to the square prism, for which the absorption correction factors of Figure 161 were calculated.

The crystal is assumed to be bathed in a uniform incident beam, and the absorption correction factors A* were calculated by the method of Albrecht (Example 44, p. 321). The value of the equivalent μr is taken from the A* table for cylindrical crystals, [6(295)]. The differences are given as percentages of the average A* or μr.

If we assume a constant value of $a = 1\cdot0$ mm, then μ for the various columns is easily found. Dividing $\Delta\mu r$ and μr in each case by the corresponding μ gives Δr and $\hat{r}$. These values are given in Fig. 161, and for the lower values of μ remain remarkably constant. This means that at least up to $\mu r = 0\cdot5$ the equivalent perfect cylinders for the two cases remain constant in size, and the increase in $\Delta A*$ (or the error, if corrections are not applied) is due entirely to the increase in μ. From $\mu r = 1$ upwards the differences between the radii of the two equivalent cylinders increase rapidly with μ, but the average rises only very slowly.

The percentage difference in A* can, in the case of a square prism, be taken as a measure of the error incurred by ignoring absorption corrections (or using a correction factor for a cylinder). But the figures quoted make no allowance for errors arising from the terminations of the prism, and apply only to the two extreme cases for one Bragg angle, $\theta = 45°$. It is difficult to assess the errors for all values of θ and all aspects of the crystal. However, the experimental results for spheres suggest that for small μa one underestimates, and for $\mu a > 1$ overestimates the errors, by using the percentage difference in A* in place of the coefficient of variation of intensity.

2.1.3. *Errors in the case of spheres and cylinders*

The absorption correction factors for spheres and cylinders are tabulated, but actual crystals, even when specially prepared, depart from these exact geometrical shapes. Thus the corrections made from the tables are never exact. At each reflecting position the true absorption factor (A* + ΔA*) for an approximately spherical crystal will differ slightly from the absorption factor (A*) for a perfectly spherical specimen with the same volume. For this true absorption factor there corresponds an ideal sphere radius R which has the same absorption factor A* + ΔA*. R will vary from reflection to reflection. For a set of crystallographically equivalent reflections, the standard deviation of the corresponding radii $\sigma(R)$ is an indirect but convenient measure of the imperfection of the crystal shape. This model may be used to estimate the absorption errors arising from measured departures from a truly spherical crystal specimen. The necessary relations are developed in the following paragraphs.

The observed integrated intensity I is inversely proportional to the absorption correction factor A*, and so,

$$\sigma(I)/I = \sigma(A^*)/A^*.$$

Since A* is a function of R

$$\sigma(A^*) = \left(\frac{\partial A^*}{\partial R}\right) \cdot \sigma(R).$$

These expressions lead directly to

$$\frac{\sigma(I)}{I} = \frac{R}{A^*}\left(\frac{\partial A^*}{\partial R}\right) \cdot \frac{\sigma(R)}{R}$$

or

$$v(I) = \frac{R}{A^*}\left(\frac{\partial A^*}{\partial R}\right) \cdot v(R) \tag{1}$$

where $v(I)$ and $v(R)$ are the coefficients of variation of I and R respectively. The factor which converts $v(R)$ to $v(I)$ can be evaluated from the standard tables for A*, and it remains to relate $v(R)$ to the shape of the crystal specimen.

Measurements of crystal shape can be made by taking photomicrographs, and from these the average crystal radius $\hat{r}$, and an estimate of $\sigma(r)$, the standard deviation of the crystal radii, can be obtained. For an approximately spherical crystal $\hat{R} \simeq \hat{r}$. The value of $v(I)$ can be obtained from a diffraction photograph, by measuring the densities of spots in symmetry-related sets. After adjustment for film errors, the root mean square deviation of the densities in a set is an estimate of $\sigma(D)$; and $v(I) = 100 \, \sigma(D)/D$, where D is the average density of the related set. From equation 265(1) the value of $v(R)$ and thus $\sigma(R)$ can now be calculated for comparison with $\sigma(r)$.

Experience shows that the model of a randomly imperfect sphere is a satisfactory one, and that $v(r)$, the measured variation, is approximately linearly related to $v(R)$, the variation derived from the model, i.e.

$$\sigma(R)/\sigma(r) = v(R)/v(r) = \alpha, \qquad (1)$$

where α is about 3.

Thus the contribution of absorption to the coefficient of variation of the integrated intensities of reflections from an imperfectly spherical crystal is, from 265(1) and (1):—

$$v(I) = \frac{\alpha R}{A^*}\left(\frac{\partial A^*}{\partial R}\right) \cdot v(r) = \beta \cdot v(r) \qquad (2)$$

As it stands, the factor β is a function of μ and R, but, by a simple

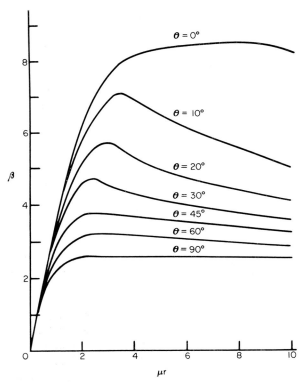

Fig. 163. Graph of $\beta = v(I)/v(r)$ against μr for various values of θ.

transformation,

$$\beta = \frac{\alpha R}{A^*} \cdot \mu \cdot \left(\frac{\partial A^*}{\partial \mu R}\right)$$

(where μ is the linear absorption coefficient). β is thus a function of $\mu R(\simeq \mu r)$ and can be plotted against it (Fig. 163).

The plot shows that for many organic crystals, where μr is less than 0·4 and β is a linear function of μr, absorption errors can be estimated from the much simpler relation:—

$$v(\mathrm{I}) = 4\mu \cdot \sigma(r) \times 10^2. \tag{1}$$

This rule breaks down for larger μr; and $v(\mathrm{I})$ then becomes dependent on Bragg angle and r.†

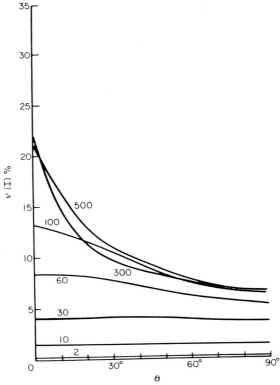

FIG. 164. Graphs showing values of $v(\mathrm{I})$ against θ for various values of μ (marked on the curves) and spheres of radius 0·15 mm. The value of $v(r)$ is taken as 2·5% corresponding to an average sphere.

† μ is given in cm^{-1} in most tables, and must be converted to mm^{-1} if crystal measurements are in mm.

The best sphere ($r \approx 100 \, \mu$m) which can be ground by present methods has $\sigma(r) \approx 1 \cdot 0 \, \mu$m; and the average has about $3 \cdot 0 \, \mu$m, so that even for $\mu r = 0 \cdot 1$, $v(I) = 1 \cdot 2\%$ for the average case (which will be difficult to achieve with soft organic crystals). Only in the most favourable case ($v(I) = 0 \cdot 4\%$) does the absorption error fall below the photographic recording error ($0 \cdot 006 + 0 \cdot 003D$) at all intensities. However, for the average sphere and low absorption ($\mu = 10 \, \text{cm}^{-1}$) photographic recording errors are greater than absorption errors for $D < 0 \cdot 5$. Thus if there are a large number of weak reflections to be measured in such a favourable case, counter recording would be desirable; and this would be particularly true if MoKα radiation were being used.

The effect of increasing μ can best be shown by calculations for a particular case. Figure 164 shows examples of $v(I)$ calculated for several typical cases for absorption errors only. The radius of the sphere is taken as $0 \cdot 15$ mm in all cases, and $v(r)$ has the value of $2 \cdot 5\%$, corresponding to an average sphere. Figure 165 gives values of μ for various typical cases.

μ		Substance
MoKα	CuKα	
1·5	10	Molecular organic compounds, no atom heavier than O.
4	30	Na salts of organic acids, *etc.*; inorganic crystals composed of atoms of low atomic number and containing large amounts of water of crystallization.
12 80*	100	Organic bromides, *etc.*; inorganic crystals composed of light elements or heavier elements with water of crystallization.
35 200*	300	Organic crystals containing fairly heavy atoms; inorganic crystals composed of elements in the intermediate range.
200	500	Crystals containing very heavy elements.

* See Sec. 18.5 (p. 279).

FIG. 165. Values of μ for MoKα and CuKα together with an indication of the kinds of material having such absorption coefficients.

For spherical crystals of a given perfection of shape the absorption errors increase initially, as substances of higher linear absorption coefficient are examined (linearly up to $\mu r = 0 \cdot 4$). When, however, a value of μr of about 2 is reached there is little further increase in the errors. It is always an improvement to reduce $v(r)$ if r is constant; but Fig. 163 shows that it is not advantageous to reduce r, even if $\sigma(r)$ can be reduced in the same ratio $(v(r)$ constant), unless μr can be made less than about 2. If μr is less than 0·4, $v(\mathrm{I})$ becomes independent of r, and for a given substance depends only on $\sigma(r)$.

For spheres which are not randomly imperfect (e.g. an ellipsoid of revolution) the estimate of errors from 266(2) will probably be too large by a factor of up to 2, but will still be better than guesswork. For cylinders the experimental comparison to find α from 266(1) has not been made, but if a value for $v(r)$ can be obtained, the magnitude of $v(\mathrm{I})$ derived from the expressions for a sphere should be of the right order. For irregular crystals the values of ΔA^* in Fig. 161 (p. 263) will give the right order of magnitude of errors likely to arise from absorption.

2.1.4. *Summary of the discussion on absorption errors*

The discussion on the magnitude of absorption errors can be summarised as follows:—

(i) In most cases where accurate measurements are made, either photographically or by counter methods, the errors due to absorption in the crystal are greater than recording errors.

(ii) With copper radiation (i) is almost always true. Only for organic crystals with no atom heavier than oxygen (i.e. $\mu \sim 10\ \mathrm{cm}^{-1}$) and with linear

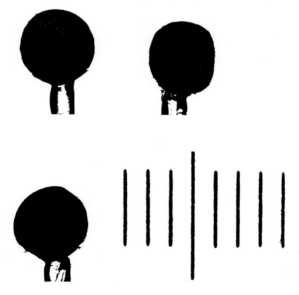

FIG. 166. Photomicrographs of a benzil crystal which had been ground as nearly spherical as possible. The scale has 0·1 mm divisions.

dimensions less than 0·1 mm, will absorption errors be less than 0·5%, which is less than photographic recording errors in all cases. However, the introduction of a sodium atom giving $\mu \sim 30\,\text{cm}^{-1}$, or the increase in size of the specimen, will bring absorption errors above recording errors for a considerable proportion of reflections.

For example, the benzil $(C_6H_5CO)_2$ crystal of Fig. 166 had a diameter of 0·3 mm, $\mu = 12·9\,\text{cm}^{-1}$, and gave a variation in intensity for symmetry-related reflections of nearly 3%. This is for an approximation to a sphere; not a very good approximation, but the best that could be ground with such soft crystals. Unground shapes would certainly give greater errors; and if the linear absorption coefficient had been $30\,\text{cm}^{-1}$, the errors would have increased to 8–10%.

(iii) In the case of molybdenum radiation, absorption errors can be ignored for a larger class of crystals. For those materials with no atom heavier than oxygen, crystals up to 0·5 mm can be used without introducing errors greater than 0·5%. In the case of organic materials containing sodium ($\mu \sim 5\,\text{cm}^{-1}$) the crystals should not be larger than 0·1–0·2 mm for errors to be within this limit. When the linear absorption coefficient gets beyond $10\,\text{cm}^{-1}$, then absorption errors, as for copper radiation, begin to become greater than photographic recording errors.

There are two groups of elements for which the linear absorption coefficients for MoKα are only slightly less than for CuKα radiation (Section 5, p. 279). Organic bromides, as a particular example, will have absorption errors nearly as large for MoKα as for CuKα for this reason.

2.2. OTHER ERRORS

2.2.1. *Errors occurring with all recording methods.*

Of the other sources of error discussed in Section 9.5 (p. 119):—(i) Extinction. (ii) Thermal diffuse scattering. (iii) Multiple reflections. (iv) Resonance effects—there is little more to be said from an experimental point of view. Mostly they are ignored or dealt with during structure analysis.

(i) Extinction can be minimised by thermal shock treatment to the crystal, e.g. dipping into liquid air, if the risk of loss can be tolerated. A parameter defining the extent of secondary extinction, if small, can be refined in the LS structure analysis.

(ii) Thermal diffuse scattering is still in general ignored, because of the impossibility of correcting for it, although the error in a simple cubic case has been shown to be up to 30%. Fortunately, in the cubic case, the effect is as stated in Section 9.53 (p. 119), to produce a structure corresponding to a lower temperature than the one at which the measurements were made; and this is probably true to a good approximation in less symmetric structures. The analysis therefore gives a smaller temperature factor than the true one,

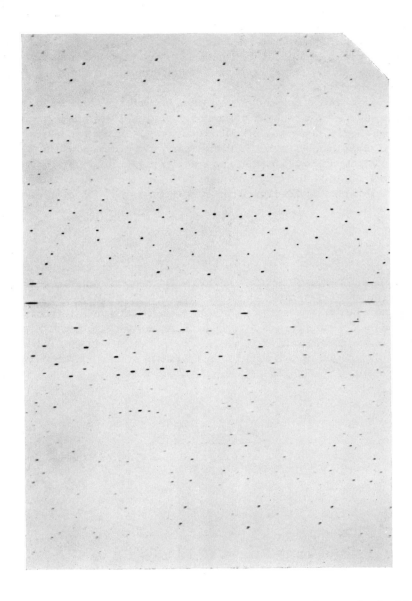

FIG. 167. Upper layer ($\mu = 19°$) equi-inclination Weissenberg photograph of calcium chloro-carbonate hydrate ($CaCl_2 \cdot 2CaCO_3 \cdot 6H_2O$), showing the extension of spots below the centre line and contraction above. Taken with $CuK\alpha$ radiation and a cassette of diameter $57 \cdot 3$ mm.

but is otherwise unaffected. The extent of the error involved in this assumption for non-cubic structures has still to be thoroughly investigated.

(iii) Multiple reflections can be minimised for orthogonal crystals by slight mis-setting of the equi-inclination angle, to prevent it occurring systematically for all reflections. It is less likely with large RL unit cells, and therefore with longer wavelengths.

(iv) Resonance effects should if possible be accurately measured, and not treated as errors. For non-centrosymmetric crystals this will involve separate consideration of the sets of reflections related by the Laue symmetry, but not by the point group symmetry. A LS refinement program which can deal with complex atomic scattering factors, is required in all cases. Where the structure can exist in two enantiomorphic forms, the absolute configuration can be obtained by LS refinement of both forms, to find which gives the lower residual.

2.2.2. *Errors specific to non-integrated equi-inclination Weissenberg photographs*

On upper layer photographs taken with an equi-inclination Weissenberg camera, the non-integrated reflections are extended on one half of the film and contracted on the other (Fig. 167 and Fig. 187, p. 314). For eye-estimation of intensities this creates errors, since the peak optical density (or the average, whichever eye-estimation is thought to measure) will be higher on the contracted than on the extended spots. Correction factors, proportional to the areas of the spots, are required, and formulae for calculating these have been worked out. [15]. The effect is due to the divergence of the incident beam. For

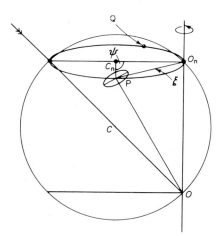

FIG. 168. Diagram showing the extension of the relp *P* into a circular cap perpendicular to *OP*, as a consequence of the divergence of the incident beam.

purposes of calculation a point relp is taken, with spheres of reflection corresponding to the limits of divergence of the X-ray beam. However, the geometry of the effect can be more easily understood by taking a single sphere of reflection, and giving the RL vectors all the directions corresponding to the divergence of the X-ray beam. If we take the RL corresponding to the central axis of the divergent X-ray beam as reference, this will produce small spherical caps perpendicular to each RL vector, of radius $d^*\alpha$, where α is the maximum semi-angle of divergence of the X-ray beam (Fig. 168). Figure 169a is the

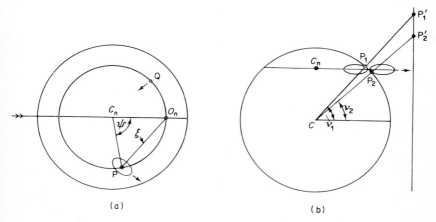

(a)   (b)

FIG. 169. (a) Projection looking down the rotation axis. (b) A vertical section through $C_n P$ and the rotation axis, showing the two positions of the cap on entering and leaving the sphere of reflection. The movement is not entirely in the plane of the diagram (see (a)) but has a normal component, so that P_1 and P_1' are above the plane of the diagram, and P_2 and P_2' below it. For a normal Weissenberg camera as described in Section **15.2.1.1** (p. 190) the movement of the film corresponding to this relp movement would be 'upward', so that P_1' would be carried upward from the position shown by the time P_2' occurred. This spot would thus be extended and would occur on the bottom half of the film.

projection down the rotation axis, and Fig. 169b is the projection perpendicular to the plane $CC_n P$, with the divergent beam 'relp' in the two positions corresponding to the beginning and end of the reflection. It is clear that the angle of inclination of the reflected beam v, is decreasing as the 'relp' goes through the sphere in the outward direction, and that the reflection starts at P_1', and moves down to P_2' in the direction parallel to the rotation axis and film movement. If the sense of the film movement is the same as that of the reflection, P_1' will have been carried down towards P_2' by the time the latter reflection actually occurs, and the spot will be shortened, and *vice versa*. If P is moving outwards, a relp such as Q, in Fig. 169a, giving a reflection on the other half of the film, will be moving inwards, with the opposite effect. Therefore the spots will be contracted on one side and extended on the other.

Since the direction of movement of both the relps and the film reverse at the end of the oscillation, the effect will be the same for both directions of the oscillation.

In Fig. 169b P_1 is below the plane of the paper and P_2 above it (compare Fig. 168). Therefore ξ is less for P_1 than for P_2 i.e. the angle of deviation ψ increases from P_1' to P_2', and as a result the reflection will not be parallel to the centre line of the photograph.

Even when corrections are applied, this variation in shape is an additional cause of error in eye-estimated intensities, and a further reason for using integration methods.

3. Computation of Total Error and Weights for Individual Reflections from Near-Spherical Crystals

3.1. Estimation of Absorption and Double Film Error

To obtain an estimate of the errors in intensity due to absorption, it is first necessary to measure the coefficient of variation of crystal radius, and thus obtain $v(I)$ as detailed below. For each density, $\sigma(D) = D \cdot v(I)/100$, is then easily obtained for combination with the other errors.

To obtain the coefficient of variation of the radius, the crystal should be photographed from at least three, and preferably five, directions, at a magnification of a few hundred times. The directions should be typical of the whole crystal, but need bear no special relation to the axes. (Normally the crystal will be rotated about an axis through 30° between exposures, and the axis should make an angle of about 45° with the microscope axis if weighting of the measurements is to be avoided (Fig. 170)). On each photograph a point, corresponding to the crystal centre, is chosen by finding the centre of the circle which is the best fit to the profile of the crystal. The crystal centre so obtained will not necessarily be the same in each photograph; but if the crystal is essentially spherical the discrepancy is not important. Measurements of crystal radii can now be made and $v(r)$ calculated. To obtain a reliable result about 50 radii should be measured.

The method assumes that deviations from spherical shape are random. In particular, if there is any tendency for the shape to approximate to an ellipsoid of revolution about the rotation axis, $v(r)$ will be over-estimated. If sufficient sets of symmetry-related reflections are available in the data, they should be used to check, and if necessary correct, the value of $v(r)$ obtained from the photomicrographs. The value of μr is calculated and used with Fig. 163 (p. 266) or equation 267 (1) to determine $v(I)$. If μr is greater than 0·4, it will be necessary to work with several values of β corresponding to ranges of Bragg angle. Three values, 8° to 15°, 15° to 30° and over 30° are generally convenient for desk calculation.

When two films are used it is necessary to determine the ratio k of the densities D_1 and D_2 of the same reflections on the top and second films, and then to convert the densities on the second film to their top-film equivalents by the relation $d_1 = kD_2$. This process by itself introduces little error if there is a fair number N of measurable reflections, common to both films, from which

FIG. 170. Jig for holding the arcs to enable various aspects of a crystal to be photographed through the microscope.

k can be determined (Fig. 171). The relative error in k is of the order of $\sqrt{(2/N)}$ of the error in D_2, so that the additional error in $d_1 = kD_2$ is $(1/N)\sigma(D_2)$. Since $\sigma(D_2)$ is normally less than 5% of D_2, this source of error can generally be ignored.

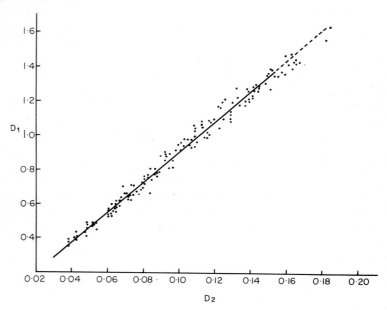

FIG. 171. Plot for the $hk0$ reflections of $Co(CNS)_4$ Hg, of D_1, the density of a reflection on the top film against D_2 for the second film. The linear relationship begins to break down for $D_1 \sim 1.4$. Top film: Ilford Industrial G; bottom film Ilford Industrial B.

If the integration limits have for some reason not been set so that α_1, α_2 doublets can be photometered either separately or sufficiently overlapped (Appendix VII, p. 402), there will be additional errors arising from partial overlapping. The additional error here can only be roughly estimated from experience in the photometry of such spots. Fortunately, for both this special case and the case of low-angle reflections on strong Laue streaks, the geometrical correction factors are small, and the absolute errors are thereby reduced.

3.2. COMPUTATION OF EXPECTED TOTAL ERROR

The calculation of expected errors, including film errors, is easily adapted to electronic computing. The program forms errors by computing the expression:—

$$\sigma(D) = \left(a^2 + (bD)^2 + (cD)^2\right)^{\frac{1}{2}}$$

where a is the constant film error (normally taken as 0·006) and b is a film factor. This is 0·003 for most reflections; but for the difficult cases mentioned above it includes an estimate of the additional error, and is put in with the corresponding intensity data. Values of b for these cases have been taken in the range 0·03–0·07, including 0·03 for values of $D > 1·0$, which allows for the additional error in the D-values from the second film. (Section 1.2.2, p. 260). The high value of 0·07 applies to reflections on the second film which also lie on a strong Laue streak, or have partial α_1, α_2 separation. The values have been deliberately estimated on the high side, since the errors are ultimately used to form weights for a least-squares program. (For counter recording the counting errors would be computed to replace these first two terms.) c is the main factor and is derived from a curve, plotted from the data of Fig. 163, of $\sigma(D)/D$ versus d^*. The ordinates of this curve are fed in as part of the data; the program adds the three terms to form $\sigma(D)$, and multiplies both this and the value of D by $(Lp)^{-1}$ and A^* correction factors to give D' and $\sigma D'$. The weight W of D' is then obtained from the relation $\sqrt{W} = \text{const}/\sigma(D)$ for use in least-squares correlation of layers. In subsequent processing to produce a correlated and averaged set of F values, $W' = k(\Sigma W \times D')$ is formed where W' is the weight for the corresponding F value.

4. Comparison of Counter and Photographic Recording

Photographic recording has the great advantage that the primary record is permanent, and the relationship between intensities is exhibited on the record, together with other indications, such as shadows in the background, the effects of white radiation, etc. These allow checks on the functioning of the apparatus to be made by inspection of the primary record. These visual checks are lost with counter recording, and very elaborate instrumental checks are required as a substitute.

These checks do not always function properly, as those will know who have read *The Report of the International Union of Crystallography Commission on Crystallographic Apparatus Single Crystal Intensity Project.*

With the crystals sent out to the participants in this project, was a statement of the cell parameters. In order that parameter differences should not affect the collection of intensity data, the parameters were determined to 1 part in 10^5 on a Bond diffractometer. One parameter was given to ± 1 in the fourth decimal place, and when the participants came to use it, some (but not all!) found that it was out in the first decimal place! There was nothing wrong with the instrument, which was capable of producing the claimed accuracy; but the crystal was composed of two or more closely aligned individuals (a state of affairs very familiar to X-ray photographers!) and the instrument had measured peaks of two different individuals.

This is an illustration of the difference between using a single photon counter as detector, and a film with millions of reliable photon detectors arranged all round the crystal, recording everything that is going on. Of course the emulsion grains will not discriminate, and cannot give back immediate information which can be used to modify the programme of an on-line 4-circle diffractometer. But as the example above shows, before a crystal is put on an automatic instrument, it should have a minimum of investigation by photographic methods.

There are three cases where a diffractometer is essential.

1. Where speed of collection of data is all important, because the specimen is deteriorating (although multiple layer photographic recording may be more efficient for crystals with very large unit cells).

2. Where a survey of a number of related structures is required to a medium degree of accuracy, and automatic data collection is the only way of getting the results in a reasonable time. Slight malfunctioning of the apparatus can be tolerated, and the proportion of time 'on the air' can be left high. The main effort is devoted to the use of the data in structure determination.

3. In those rare cases when crystal errors are below photographic recording errors, and it is desired to take advantage of this to achieve high accuracy in structure analysis. Counter recording will be particularly desirable if there are a large number of weak reflections, as is the case if MoKα radiation is used to explore outside the CuKα limiting sphere. The apparatus must be maintained at maximum efficiency; and the effort involved in the collection of data will be considerable, with a large proportion of time 'off the air' for adjustments.

But there will continue to be many problems where these considerations do not apply. In particular, accurate structure determinations of all kinds— to establish key structures beyond all reasonable doubt, to obtain accurate bond lengths and angles and electron densities, to study thermal vibration effects—will require certainty and accuracy rather than speed. It is true that counter recording is inherently more accurate than photographic methods; but if the errors from absorption in the crystal are greater than the recording errors in both cases, there is nothing to choose between them on the score of accuracy. The comparatively effortless certainty and the easy checking of suspect results which photographic recording provides, together with the comparative simplicity, reliability and low cost of the apparatus, may then well outweigh the greater speed and the discrimination of counter methods.

Where the whole of reciprocal space must be investigated, rather than just the relps, photographic recording has considerable advantages.

5. The Uses of CuKα and MoKα Radiation with Photographic and Counter recording

5.1. PHOTOGRAPHIC RECORDING

CuKα radiation has many advantages—greater intensity and higher film sensitivity leading to shorter exposure times; reflections more widely spaced; far less background of Laue streaks; less liability to produce simultaneous reflections—and it will normally be the radiation of choice for cameras able to record high angle reflections. For the precession camera MoKα radiation is essential, if the RL is to be photographed out to the equivalent of the CuKα limiting sphere; but even here for crystals with large unit cells, which do not normally have detectable CuKα reflections with $2\theta > 60°$, CuKα radiation will anyway be essential in order to resolve the reflections.

MoKα radiation reduces absorption errors, usually to a considerable extent, but only to a very limited extent when elements from nickel to yttrium or holmium to uranium are responsible for most of the absorption. (The higher values of μ in the MoKα column of Fig. 165 correspond to such cases). This reduction may, if the crystal cannot be ground into a sphere or cylinder (or measured accurately for the computation of individual absorption correction factors) offset the disadvantages of longer exposure times and higher background errors; but the advantage is not likely to be large. The only overriding reasons for the use of MoKα are (i) the reduction in the size of the RL for the precession camera; and (ii) the need to measure outside the CuKα sphere. In the latter case counters have many advantages over photographic recording.

5.2. COUNTER RECORDING

If one substitutes the linear diffractometer for the precession camera, much of the discussion on photographic recording applies to counters, except for two major considerations. Counters, especially scintillation counters, are almost 100% efficient as detectors for both radiations; and discrimination does away with the background errors of photographic recording. It is likely therefore, that MoKα radiation will more often have overall advantages for counter recording, especially if there is significant information to be obtained outside the CuKα limiting sphere.

6. Reduction to Relative or Absolute F_0^2 Values

6.1. GENERAL

The table of measured intensities for the various indexed reflections will normally be in sections, one for each crystal or each axis of a remounted crystal, with sub-sections, one for each layer. Before the sub-sections can be correlated, they must be multiplied by the velocity (Lorentz) and polarisation factor $(Lp)^{-1}$, and if possible by the absorption factor A^* for each reflection. For most purposes all that is necessary is to use the correct formulae (Appendix V, p. 381) or tables of correction values, in processing the data by electronic computer. But there are occasions when it is desirable to do some at least of the calculations using desk methods (for students to obtain an understanding of the processes being carried out by the computer, or as a check on the computer calculations). The sections below give details of the aids available.

6.2. CORRELATION OF INTENSITIES

When the corrected intensities are available the symmetry related reflections in each layer are averaged, in order to provide the maximum comparisons in subsequent layer correlation.

If complex atomic scattering factors are to be used, the point group symmetry, and not the Laue symmetry (where these are different) must be the criterion in deciding which reflections are symmetry related. If absolute configuration is required from the structure analysis, great care is required in indexing reflections from different crystals, or from different settings of one crystal.

A correlation coefficient K_a^n (for layer n and rotation axis a) is established for each layer, by comparison of a few reflections on each layer with the equivalent ones on the zero layer about another axis (or the K's are all given an initial value of unity). The observational equations

$$K_a^n I_{hkl} - K_b^m I_{hkl} = \Delta \quad \text{are set up.}$$

If symmetry related reflections occur on different layers about the same axis, they should also be included in the observational equations. $\Sigma \Delta^2$ is minimised by L.S. refinement of the values of K; the intensities for each layer are multiplied by the final value of K for that layer; and symmetry related intensities are averaged. This produces a final list of relative F_0^2 values.

If absolute intensities are required, the total energy E, and the incident intensity I_0, must be measured in the same units and E/I_0 processed as above, except that correlation is not required and the symmetry related intensities can be averaged directly. The final values of E/I_0 are then converted to absolute F_0^2 values by the factor derived from 114(1).

All these calculations can be done by computer, but it is always advisable to check a few results by hand. The computer should print out the final values of K and some indication of the spread of values in an averaged group of related reflections, which should be checked to make sure they are reasonable.

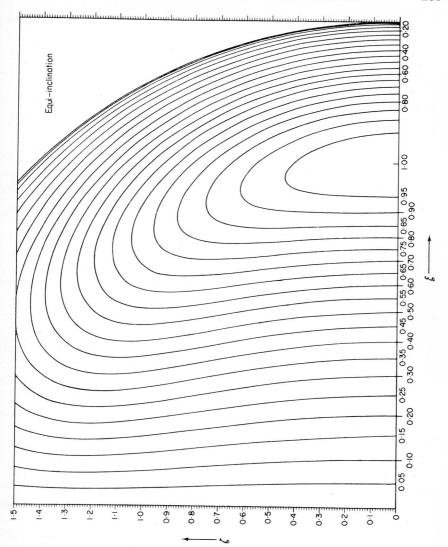

FIG. 172. The Cochrane chart of $(Lp)^{-1}$ correction factors for equi-inclination geometry, and various layer heights from 0 to 1·5 R.U. The reproduction is two-thirds full size.

The $(Lp)^{-1}$ factors are given along the base line where the curves of constant $(Lp)^{-1}$ values cut it. The correction factors have been halved for convenience. Only if absolute intensities are being measured will the factor of 2 have to be taken into account.

This chart is no longer available from the Institute of Physics (p. xxviii), but the reproduction above can be used by reducing the scale of the RL net to 2 R.U. = 134 mm (the length of the base line) or by enlarging the Figure photographically to give a base line of 200 mm.

6.3. Velocity (Lorentz) and Polarisation Correction Factor (Lp)$^{-1}$

For unpolarised incident radiation

$$p^{-1} = \frac{2}{1 + \cos^2 2\theta}.$$

(Section 9.4.1, p. 117). L^{-1} depends on the geometry of the apparatus (Appendix V, p. 381). Desk calculation is likely to be mainly concerned with equi-inclination Weissenberg or precession photographs. For equi-inclination Weissenberg recording a chart has been constructed (Fig. 172) which enables a scale of (Lp)$^{-1}$ corrections to be marked out for any value of the layer height ζ. A strip of thin perspex, with a line scribed down the centre and a small hole near one end marking the origin, is laid across the chart parallel to the base, with the hole on the scale at the left hand side, at the ζ value concerned. The points where the curves cross the scribed line are marked, and labelled with the values for the curves where they cut the base line. A RL net is constructed to the normal scale (100 mm = 1 R.U.) and a drawing pin put upwards through the origin *of construction*. The hole in the scale is placed over the drawing pin and the scribed line swung over a relp. The interpolated reading of the scale gives the (Lp)$^{-1}$ correction for the corresponding reflection. (**Example 43**, p. 321).

For zero layer precession photographs a contour chart is available which can be placed over the photograph, so that the (Lp)$^{-1}$ factors can be read directly from the positions of the reflections. This tends to be rather less accurate because of the smaller scale. Charts are not available for non-zero layers, because a different chart would be required for every layer height. The lack of exact cylindrical symmetry means that the chart has to be two-dimensional.

6.4. Absorption Correction Factor (A*)

For spheres and the zero layers of cylinders, the tables [6(295)] are used for the particular value of μr. The list of values is read in to the computer, and the program interpolates for the value of θ for a particular reflection. Since for a sphere or the zero layer of a cylinder the only variable is θ, it is possible to construct a scale for use with the RL net, as for the (Lp)$^{-1}$ correction. The two can be incorporated into a single scale if the number of reflections on one layer warrants it.

For cylinders and equi-inclination recording, all paths in the cylinder will be increased in the ratio $1/\cos \mu$, so that the A* corrections will be equal to those of the zero layer for a cylinder of radius $r/\cos \mu$. However, half the projected angle of deviation ($\psi/2$, Section 15.2, p. 190) must be used in place of θ in the tables.

For prismatic crystals with the prism axis along the rotation axis, the method of Albrecht enables an approximate value of

$$A = \frac{1}{V} \int e^{-\mu L} \, dV = 1/A*$$

to be evaluated. The elements of volume δV can be taken as elementary prisms of length g, and cross section δC. The path length L in the crystal, for all rays reflected in this small prism, will be constant along its length. δV can therefore be replaced by $g\delta C$ and V by gC, where C is the area of cross section of the whole prism. The integral is reduced to a two-dimensional integration over the cross section:—

$$A = \frac{1}{C} \int e^{-\mu L} \, dC.$$

By dividing up the cross section into n small equal areas, drawing the incident and reflected directions to the centre of each area and measuring the resulting path lengths L_n, the integral can be approximated by the summation,

$$A = \frac{1}{C} \sum e^{-\mu L_n} \frac{C}{n} = \frac{1}{n} \sum e^{-\mu L_n}$$

over the n path lengths. This was the method used to find A* for a square prism in Section 2.1.2 (p. 264). By using ψ instead of 2θ and multiplying the lengths L_n by $1/\cos \mu$, the same method can be used for non-zero layers in orthogonal equi-inclination geometry.

The method is tedious and unlikely to challenge computer methods for dealing with a large number of reflections. But its use can provide a check on the functioning of a computer, and also furnish a reasonably good estimate of the effects of absorption in any given case, merely by the evaluation of A for the same θ, but for two different aspects of the crystal. (**Example 44**, p. 321).

The exact expressions for the absorption factors of flat plate specimens, which are of importance in powder diffraction, are given in Answer 44 (p. 547).

Examples

(Starred Examples should be omitted at a first reading.)

1. The Abbe Theory of the Microscope (p. 4)

(i) How many orders of diffracted beams can enter an objective lens of N.A. 0·85 when focussed on a grating of 1,000 lines mm^{-1}, illuminated normally with parallel light of wavelength 6,000 Å? (ii) What wavelength would be necessary to resolve the lines on a grating with 2,000 lines mm^{-1}? (The Numerical Aperture (N.A.) of an objective used in air is $\sin \psi$, where ψ is the angle of deviation from the axis, of a ray which is just accepted by the objective).

The Ångstrom unit (1 Å $= 10^{-10}$m) and the micro-metre (1 μm $= 10^{-6}$m) are units frequently encountered in X-ray crystallography. It is useful to memorise the relations 1 Å $= 10^{-7}$ mm, 1 μm $= 10^{-3}$mm, 1 μm $= 10^4$ Å.

2. The Laue Conditions Applied to a Rotation Photograph (p. 21)

A crystal is rotated about the **b** axis, and X-rays of wavelength 1·542 Å are incident on it perpendicular to the rotation axis. The photograph (Fig. 4, p. 5) shows the layer lines produced on a cylindrical film of radius 30·0 mm.

Measure the heights of the various layer lines above the zero layer line. (Measure across a corresponding pair above and below and divide by two. (i) Why? (ii) What is the Laue relation for the **b** axis in this case? (iii) Does this relation alter with rotation of the crystal? (iv) Calculate the angles of the cones about the **b** axis, on which the reflections lie as a result of the **b** axis Laue relation. (v) Hence calculate b.

3. The Width of the Diffraction Peaks (p. 21)

For the same arrangement as in Example 2, if v_k is the cone angle given by the Laue condition for the kth order, what is the relative phase angle ϕ_m, of the mth lattice point from the origin along the **b** axis, (i) for a cone angle of v_k, (ii) for $v_k + \Delta v$? (Find the path difference for the rays scattered from the 0 and mth points, and for (ii) expand $\cos (v_k + \Delta v)$). (iii) Using the results of Example 2, plot the Argand polygons (Section 1.3.1, p. 10 and Appendix I, p. 335) for the first order cone given by 5 scattering points along **b** (including

285

the origin at one end), for values of Δv of $0 \cdot 01n$ radians ($n = 1, 2, ...$), until the first zero of the resultant amplitude is reached (or passed). Take the length of the amplitude vector for each point as 50 mm. (iv) What would be the angular spread in degrees of the reflections on this cone, from the maximum to the first zero point on this side? (v) What difference would it make if the position of the first minimum were found on the other side of the maximum? (vi) What is the angular spread between the minima on either side of the maximum? (vii) What would this be for 100 points? (viii) for a 1000? (ix) In the latter case how large (in μm) would the crystal be in the **b** direction?

4. Lattice Planes (p. 23)

Draw the base of a monoclinic cell with **b** vertical and $a = 15$, $c = 10$ Å, $\beta = 110°$, to a scale of 10 mm $\equiv$ 1 Å. Take the origin 10 mm down from the top edge, with **a** horizontally to the right and **c** towards you. Figure 173 shows the minimum space requirements.

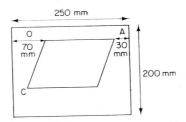

FIG. 173. Sketch showing space required for diagram.

Draw the trace on this (010) plane of the nearest plane to the origin of each of the sets ($h0l$) for all values of h and l from 0–3. Label each trace at one end. This drawing will be required for Examples 6 and 7, and unless the accuracy of measurement and drawing is within the thickness of a fine pencil line (i.e. $0 \cdot 1$ mm or $0 \cdot 1°$) the subsequent exercise will result in failure. This level of accuracy is the normal requirement for the large number of geometrical constructions used in this subject.

Draw the trace of $\overline{3}01$. (A negative index gives a negative intercept, i.e. the axis is cut on the negative side of the origin.)

5. Reciprocal Points and the Sphere of Reflection Construction (p. 28)

The normal to a set of planes (100), ($d_{100} = 5 \cdot 0$ Å) lies along the X-ray beam. (i) Construct the reciprocal points 100, 200, 300 for CuKα ($\lambda = 1 \cdot 54$ Å) and MoKα ($\lambda = 0 \cdot 71$ Å) to a scale of 1 R.U. = 100 mm. (ii) Use the

Sphere of Reflection construction to find how many degrees of rotation about an axis, perpendicular to the X-ray beam, would be required to bring the crystal into the reflecting position for 300 in each case.

(For the purposes of geometrical construction a two-dimensional section of the Sphere of Reflection must be taken. Normally this is a section perpendicular to the rotation axis, since any point in the plane of the section remains in the plane on rotation. In the present case the section passes through the origin, at the end of the diameter representing the X-ray beam, and contains the X-ray beam. It is therefore the equatorial section of the sphere, cutting it in a 'Circle of Reflection' which, in this case, is a great circle, i.e. the plane of the section passes through the centre of the sphere. The reciprocal vector d^*_{h00} rotates about the origin in the plane of the diagram, and the problem is to find what angle of rotation will bring the 300 reciprocal point on to the circle. When the 300 reciprocal point is on the circle the (300) set of planes is in the reflecting position, and d^*_{300} is the normal to these planes. The angle between the X-ray beam and d^*_{300} is therefore the complement of the Bragg angle.) (iii) Use Bragg's Law to calculate the angles of rotation and compare with those obtained by construction.

Keep the diagram for Example 10.

6. The Construction of Reciprocal Points from a Series of Lattice Planes (p. 28)

Start with the diagram from Example 4, and preferably using a different coloured pencil or ink, draw accurate normals from the origin to each of the planes and extend to the edge of the paper. (This is most easily and accurately done by putting a ruler along the trace of the plane, and a set square on the ruler with the perpendicular edge just over the origin. Hold the set square and transfer the ruler to this edge for drawing the normal). Label the end of each normal with the indices of the plane.

Measure the interplanar spacings d_{h0l} (i.e. the distance from the parallel plane through the origin to the intercept plane) for all the planes and convert to Å. Using $\lambda = 4.5$ Å calculate the values of d^*_{h0l}. Using a scale of 100 mm = 1 R.U. mark the reciprocal points and label them with the indices of the corresponding planes. Inspect the arrangement of reciprocal points and describe the basic pattern.

7. The Reciprocal Lattice (RL) Axes of a Monoclinic Crystal (p. 30)

Calculate the RL axes of a monoclinic crystal (with unit cell: $a = 15.0$, $b = 12.0$, $c = 10.0$ Å, $\beta = 110°$) for a wavelength $\lambda = 4.5$ Å, and compare with the diagram of Example 6.

8. Vector Algebra—I (p. 31)

(i) $r_1 = i + 2j + 3k$; $r_2 = 2i - 3j + k$; $r_3 = -i - j + k$.
What are the length and direction cosines of $r = r_1 + r_2 + r_3$?

(ii) $a^* = 0.3$, $b^* = 0.2$, $c^* = 0.4$ R.U.; $\alpha^* = \gamma^* = 90$, $\beta^* = 60°$.
Find the angle between the normals to (101) and (111).

(iii) Derive the relation between the indices (hkl) of the faces of a zone and the indices of the zone axis $[HKL]$. (Find the RL vectors which are perpendicular to $[HKL]$.)

(iv) Find the angle between the normals to (321) and (123) for a cubic crystal.

(v) $r = xa + yb + zc$ is the position vector of the centre of an atom. Find the value of $r \cdot d_{hkl}^*$ if $x = 0.5$, $y = 0.7$, $z = 0.2$, $\lambda = 2.0$ Å and $h = 3$, $k = 2$, $l = 1$.

(vi) Two adjacent sides of a regular octagon are determined by the vectors **a** and **b**. What are the vectors determined by the other sides, taken in order.

(vii) A point describes a circle uniformly in the **i, j** plane, taking 1 ns to complete one revolution. If its initial position vector is **i**, relative to the centre as origin, and the rotation is from **i** to **j**, find the position vectors after 1/12, 3/12, 5/12, 7/12, 1/8 and 3/8 ns.

(viii) Use scalar products to define the reciprocal lattice axes, a^*, b^*, c^* in terms of the crystal lattice axes, a, b, c. State the meaning of the defining relations.

(ix) If the position vectors of P and Q are $i + 3j - 7k$ and $5i - 2j + 4k$ respectively, find $\overrightarrow{PQ}$ and determine its direction cosines.

(x) A monoclinic crystal has $a = 5$, $b = 6$, $c = 7.5$ Å; $\beta (>90) = \sin^{-1} 0.8$. One of a set of lattice planes cuts off intercepts of 2.5 Å on Ox, 1.5 Å on Oy and 2.5 Å on Oz. What are the indices of this plane? Is this the nearest plane of the set from the origin? (Give reasons). What is the shortest distance from the origin to this plane? What are the angles between the normal to these planes and the three axes, Ox, Oy, Oz?

(xi) The velocity of a boat relative to the water is represented by $3i + 4j$, and that of the water relative to the earth by $i - 3j$. What are the speed and direction of the boat's motion relative to the earth, if **i** and **j** represent velocities of one mile an hour E. and N. respectively?

(xii) Express $nd_{h_1 k_1 l_1}^* + md_{h_2 k_2 l_2}^*$ as a single RL vector.

*(xiii) In a central RL plane, any two non-colinear RL vectors define a unit cell (parallelogram). If the cell is primitive it will have the minimum possible area, and all primitive cells will have the same area, equal to that of the polygon whose sides are the perpendicular bisectors of the lines joining

a point to its nearest neighbours. If the cell is multiple (centred) the number
(L) of lattice points included will be given by the multiple cell area, divided
by the primitive cell area.

Show that if the multiple cell sides are the vectors $\mathbf{a}^{*\prime} + \mathbf{b}^{*\prime}$, any lattice
point, including the centering points, can be defined by a position vector
$\mathbf{l}^* = p\dfrac{\mathbf{a}^{*\prime}}{L} + q\dfrac{\mathbf{b}^{*\prime}}{L}$, where p and q are integers.

*(xiv) Show that the *direction* of any RL vector can be expressed by an
integral linear combination of two given non-colinear RL vectors with
which the first is coplanar.

9. Indexing an Almost Undistorted Photograph of a RL Layer (p. 36)

(i) Make an enlarged drawing of the central area in Fig. 26 (b) (p. 35)
and index the relps, given that the crystal is monoclinic, $\mathbf{c}$ is along the
rotation axis (choose $+\mathbf{c}$ upwards), $\mathbf{b}$ along the X-ray beam at the beginning
of the oscillation and $\beta > 90$ and $< 110°$.

Since these relps are projected on to the film as reflections, this process is
equivalent to indexing the reflections on the photograph.

(ii) What is the positive direction of $\mathbf{b}^*$? Axes must be right-handed. (Many
students have difficulty in deciding whether axes are L or R handed. The
easiest rule to remember and use is the corkscrew rule. If the (positive) axes
are taken in cyclic order $(\mathbf{a}, \mathbf{b}, \mathbf{c}, \mathbf{a}, \mathbf{b})$, then starting with any axis, and
rotating it towards the direction of the next in order, produces a corkscrew
motion in the positive direction of the third axis. The rotation must of
course be through the smaller (non-reflex) of the two possible angles).

10. The Sphere of Reflection Construction and the Weighted Rods in the RL due to the White Radiation (p. 57)

(i) If the crystal in Example 5 is rotated 70° from the initial position and
then kept stationary, with a peak voltage on the X-ray tube of 50 kV, what
wavelengths from the white radiation would be reflected and in what direction?
Measure the distance of the point of intersection from the origin and compare with the distance of the relp for $\lambda_{CuK\alpha}$. Repeat for all the orders of $h00$
which contribute to the reflection.) (ii) How could the wavelengths be found
using Bragg's Law directly?

(iii) If the crystal is rotated further, at what angle of deviation of the
reflected beam would reflections cease? (Use the short wavelength cut-off
1(52)] calculated above to find the point at which the radial rod becomes
weightless, and measure the angle of deviation from the diagram. Check
using Bragg's Law.)

11. The Sphere of Reflection and the RL for White Radiation (p. 57)

Plot out the a^*b^* network to $\pm 3a^*$, $+ 3b^*$ for CuKα radiation, $\lambda = 1\cdot54$ Å, and a crystal with an orthogonal lattice: $a = 5\cdot13$, $b = 7\cdot70$ Å. Use a scale of 100 mm = 1 R.U. On the same diagram plot the points for $\lambda = 1\cdot54/4 = 0\cdot385$, and for $\lambda = 1\cdot54 \times \frac{7}{4} = 2\cdot695$ Å (to $\pm 2a^*$, $+2b^*$). Join up the three points for 120 with a heavy line. The complete set of weighted radial lines similar to this constitutes the reciprocal lattice for the case of white radiation with a spectrum from $\lambda = 0\cdot385$ to $2\cdot695$ Å. At the short wavelength end it stops abruptly. (i) If this "cut-off" is at $\lambda = 0\cdot385$ Å, calculate the peak voltage applied to the tube.

The crystal is oscillated about **c** so that the whole reciprocal lattice line for 120 passes through the sphere of reflection. (ii) Use the sphere of reflection construction to measure the angles involved, and draw approximately to scale the effect you would expect to get on the unrolled flat film from this particular weighted line, on a photograph taken with a cylindrical camera of radius 30 mm using CuK radiation ($\lambda_{K\alpha} = 1\cdot54$, $\lambda_{K\beta} = 1\cdot39$ Å). (iii) What characteristics must the crystal have if this is to show up on an actual photograph?

Keep the diagrams for reference in Example 20.

*12. Asterism and Diffuse Reflections (p. 58)

Figure 43 (j) (p. 62) shows radially elongated spots (an effect known as asterism). This is due to the joint effect of the spread of orientation of the mosaic blocks about the mean position, the divergence of the incident X-ray beam and the obliquity of the reflected beam. Assume that the 'spread' is the same in all directions, that the extreme orientations make an angle α with the mean orientation and that the effective X-ray source is much smaller than the spherical crystal (radius $0\cdot1$ mm), i.e. it can be treated as a point source. Crystal–source distance = 100 mm. (i) On this basis explain the formation of the elongated spots, and obtain an estimate of α. (Use the fact that Laue reflections, because they are not limited by Bragg's Law, can be treated as specular reflections from the lattice planes. Different wavelengths will be reflected at different angles of incidence, but over a small range the intensity will only vary slightly. The sensitivity of the photographic emulsion also varies very little over a small range of wavelengths).

Diffuse reflections, reduced to round spots in the reproduction, occur near some of the more intense Laue spots. (ii) Explain the occurrence and positions of these diffuse reflections with the aid of a diagram of the sphere of reflection and use the most favourable case to obtain an estimate of the extent of the effective volume (assumed spherical) of scattering power due to thermal vibrations. Give the result in R.U. for the characteristic wavelength. (CuKα

$\lambda = 1.542$ Å). What fraction of the distance between relps does this represent? (Sucrose parameters, Example 29, p. 305).

*13. Indexing a Laue Photograph (p. 73)

(i) Index the reflections on the Laue photograph (Fig. 174). Construct a rule as in Fig. 49 (b) (p. 70), at 1° intervals, but decide first what range the rule should cover. (ii) Use Section **6.5.3.1** (p. 66) to deduce a minimum value for b, and compare with the value found in Answer 30 (p. 512). (iii) Do the same for **c** (which is along the X-ray beam) in Fig. 43c (p. 59). Original specimen-film distance $= 30$ mm, equivalent to 22 mm for the reproduction. Tube voltage $= 50$ kV.

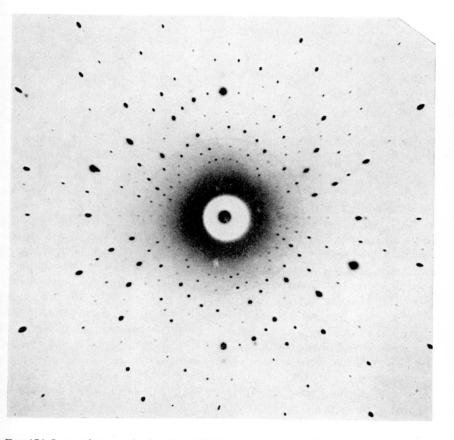

FIG. 174. Laue photograph showing 2-fold symmetry, implying that RL planes are perpendicular to the X-ray beam. Crystal-film distance $= 50.5$ mm.

14. The Transform (Scattering Diagram) of the Benzene Molecule (p. 94)

The expression for calculating the scattering diagram (Fourier transform) is $R = aF = a\sqrt{A^2 + B^2}$ where

$$A = \sum_m f_m \cos \phi_m; \qquad B = \sum_m f_m \sin \phi_m; \qquad \phi_m = \frac{2\pi \mathbf{r}_m \cdot \mathbf{D}^*}{\lambda}.$$

($\mathbf{D}^* = 2 \sin \theta \mathbf{n}$) (Section 9.1, p. 93). If all the atoms are of one kind (or the same to a first approximation as for N, C and O) we can take f_m outside the summation.

$$A = f \sum \cos \phi_m = f A'; \qquad B = f \sum \sin \phi_m = f B';$$

$$R = aF = af\sqrt{A'^2 + B'^2} \qquad \text{or} \qquad F = f\sqrt{A'^2 + B'^2}.$$

If we now take $f = 1$, i.e. assume point atoms of the scattering power of 1 electron, we can write:

$$R' = \sqrt{A'^2 + B'^2}, \qquad F = fR' \quad \text{and} \quad R = afR'.$$

If the scattering diagram is calculated for R', it is only necessary to multiply the scattering value of a particular point $\mathbf{D}^*$, by af, in order to get the actual scattering. The scattering diagram for R' (because of the high symmetry of the benzene molecule) has repeat properties not possessed by the diagram for R, which falls off with f.

Since $\qquad A' = \sum \cos \phi_m, \qquad B' = \sum \sin \phi_m \quad \text{and} \quad \phi_m = \frac{2\pi \mathbf{r}_m \cdot \mathbf{D}^*}{\lambda},$

all points for which $\mathbf{D}^*/\lambda$ is the same, have the same value of R'. Thus if λ is increased, D^* must be increased in proportion; or in other words, the whole diagram is simply enlarged in proportion. Alternatively one can keep the same diagram, but decrease the scale of the drawing and therefore of the sphere of reflection. In Fig. 63 (p. 95) the scale is given for the upper diagram as the length of $0 \cdot 5$ R.U. for $\lambda = 1 \cdot 0$ Å, from which the scale for 1 R.U. in mm can be obtained. 1 R.U. for any other wavelength is therefore obtained by multiplying this length by $1/\lambda$. The scale may also be given as that for $(2 \sin \theta)/\lambda$, since for $\lambda = 1$ this is equivalent to D^* in R.U. For any other wavelength the scale readings must be multiplied by λ to give D^* in R.U., i.e. 1 R.U. in mm is proportional to $1/\lambda$. The scale of the lower diagram is $1/3$ that of the upper.

Take a tracing of Fig. 63 on which to draw the construction required for this and the following Example (15). Use the lower diagram in the first instance, and the upper one where appropriate to obtain more accurate results. The diagram is calculated for the benzene ring as shown, with $x' = +1 \cdot 21, +1 \cdot 21, 0, 0, -1 \cdot 21, -1 \cdot 21$; $y' = 0 \cdot 70, -0 \cdot 70, -1 \cdot 40, 0 \cdot 70,$ $-0 \cdot 70$ Å as the coordinates of the centres of the atoms. The origin of co-ordinates is at the centre of the ring, and the origin of the scattering diagram at the centre of one of the sets of circles (in the lower diagram, take the central set). For convenience the origins of the molecule and the transform have been separated, but all corresponding directions on the two diagrams remain parallel to one another.

Draw in the X-ray beam coming towards the origin of the scattering diagram in a direction perpendicular to the left-hand top side of the hexagon of carbon atoms, as shown. Draw the circles of reflection for CuKα ($\lambda = 1 \cdot 54$) and MoKα ($\lambda = 0 \cdot 71$ Å) radiation. The intersection of the circle of reflection with the diagram gives, at each point, the scattering amplitude in a direction from the centre of the circle to the point on the circumference being considered.

(i) In what directions would scattering be of zero intensity? (Give angles of deviation in the plane of the ring). (ii) For CuKα what would be the scattering amplitude in a direction making an angle of deviation of 60°? (Take the direction to the left of the diagram, and then obtain as accurate a value as possible from the upper diagram). (iii) What is the length and direction of $\mathbf{D^*}$ for the corresponding point on the diagram? For this particular $\mathbf{D^*}$, calculation of R' is particularly simple. (iv) Check the value of R' obtained from the diagram by calculation from the expression given earlier. (v) Repeat this with an incident direction parallel to a side of the benzene ring.

15. The Transform and the RL (p. 98)

(i) If the benzene ring of Example 14 formed a crystal lattice by repeating at intervals of 5·4 Å in the x-direction, and 5·8 Å in the y-direction, what would be the effect on the scattering diagram? (ii) Draw in the reciprocal lattice net on the lower diagram, and label some of the points (320 particu-larly) with the true relative scattering amplitude (the Structure Factor, F) for CuKα radiation ($\lambda = 1 \cdot 54$ Å), given the following values for the atomic scattering factor for carbon. (Refer to the text of Example 14 and notice that f_c is given against $\sin \theta / \lambda$ not $2 \sin \theta / \lambda$).

$\dfrac{\sin \theta}{\lambda}$	0·0	0·1	0·2	0·3	0·4	0·5	0·6	0·7	0·8	0·9	1·0	1·1
f_c	6·0	4·6	3·0	2·2	1·9	1·7	1·6	1·4	1·3	1·16	1·0	0·9

(iii) Use the Structure Factor formula (Section 9.1.4, p. 99) to calculate F_{320}, and compare with the value obtained above.

*16. Orienting a Molecule in the Unit Cell by Packing Considerations and Comparison of the Transform of the Molecule and the Weighted RL (p. 99)

Flavanthrone, $C_{28}H_{12}O_2N_2$, has the structural formula shown in Fig. 175 and crystallises in the space group $P2_1/a$ with cell parameters, $a = 27.92$, $b = 3.80$, $c = 8.10$ Å, $\beta = 95°$, $Z = 2$. (i) What deductions can be drawn about the symmetry of the molecule from this information? Draw to scale ($10\,mm \equiv 1$ Å) the **ac** base of the unit cell; put in the symmetry elements and mark the points where the centres of the molecules must lie. (ii) Make use of the fact that similar large flat molecules lie parallel at a distance apart of about 3.4 Å, to calculate the maximum angle of tilt of the molecule from the **ac** plane.

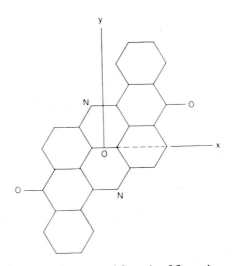

FIG. 175. The structural formula of flavanthrone.

Draw the molecule to the same scale by plotting the calculated positions of the corners of the hexagons on graph paper, taking all bond lengths as 1.40 Å (except $C - 0 = 1.2$ Å), and allowing about 1.7 Å from the centres of the outer atoms to the 'boundary' of the molecule. Cut out four such shapes (marked with the directions of the x and y axes) and place them on the unit cell, in the orientation allowed by the space group which gives the best 'packing', taking account of any differences in height above the base as well as the possibility of 'tilting'. (iii) Mark the direction of the $\mathbf{a}^*$ axis on one of the paper 'molecules'. (iv) If the structure is projected along the **b** axis on to the (010) plane, what is the projected unit cell? (v) What effect will this have on the $h0l$ reflections?

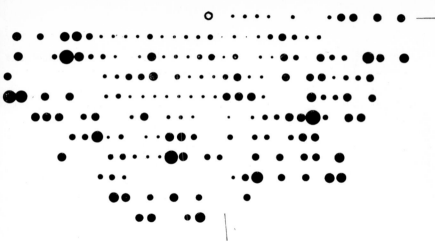

FIG. 176. The weighted *h0l* RL plane of flavanthrone.

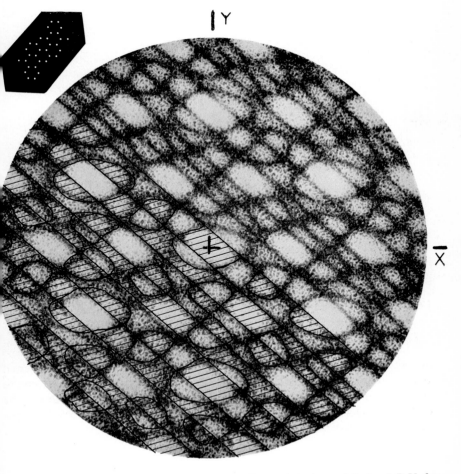

FIG. 177. The optical transform of flavanthrone. The mask is shown full size. 1 R.U. for $\lambda = 1{\cdot}542 \equiv 29{\cdot}2$ mm on the transform.

Figure 176 shows the $h0l$ RL plane to a scale of 1·0 R.U. (for $\lambda = 1·542$ Å)
$\equiv 29·2$ mm. The directions of the axes are marked. (vi) Find which is $\mathbf{a}^*$ and
which is $\mathbf{c}^*$ and the positive direction $\mathbf{b}^*$. The relps are weighted with the
corrected intensities, for comparison with the scattering diagram of the
molecule (Fig. 177), which has been produced optically from the mask
shown and enlarged to the same scale as the weighted RL. The axes OX and
OY on the transform are parallel to the axes Ox and Oy of Fig. 175.
(Ignore shading on Fig. 177 at this stage.) (vii) Why can the transform of a
single molecule be compared directly, in this case, with the weighted RL layer
for a crystal with two molecules in the unit cell? (viii) Would you expect the
comparison to be exact?

Copy the weighted RL layer on to clear plastic sheet or tracing paper. Use
it, together with the packing considerations, (ix) to determine whether the
positive z axis of the molecule is nearly parallel to $+\mathbf{b}$ or $-\mathbf{b}$; and (x) to find
the orientation of the molecule which gives, within the packing limits, the
best 'fit' between the weighted RL and the scattering diagram. Give the
results of at least 3 independent determinations of the angles which Ox and Oy
make with $+\mathbf{a}^*$. (xi) What general features of the transform prevent a direct
comparison with the weighted RL being used, without consideration of
molecular packing? (xii) What do these general features derive from?

17. The Intersections of Related Sets of Lattice Planes (p. 102)

Draw the outline of the unit cell base given in Example 4, and put in the
traces of a number of the planes of the sets (302), ($10\bar{4}$) and $(3 + 1, 0 + 0,
2 + \bar{4}) \equiv (40\bar{2})$. Verify that the planes of the ($40\bar{2}$) set pass through the
intersections of (302) and ($10\bar{4}$).

18. The Indices of Double Reflections (p. 122)

(i) In Fig. 78 (p. 121) what are the indices of the second reflection in the
same direction as primary 101? If the two relps $h_1 k_1 l_1$ and $h_2 k_2 l_2$ are both
in the sphere together, what are the indices HKL of the second reflection
in the same direction as (ii) $h_1 k_1 l_1$, (iii) $h_2 k_2 l_2$? (Obtain $\mathbf{d}^*_{HKL}$ in terms of
$\mathbf{d}^*_{h_1 k_1 l_1}$ and $\mathbf{d}^*_{h_2 k_2 l_2}$ and equate components).

**19. The Construction of the Cubic Stereogram and its Use in Determining
     Multiplicity (p. 130)**

Draw on a stereogram the symmetry elements of class $\dfrac{4}{m} 3 \dfrac{2}{m}$, putting in
first the 4-fold axes and the planes perpendicular to them, in the standard
orientation, then the 2–fold axes (halfway between the 4-fold) and the
corresponding planes. Finally, locate and mark the trigonal axes. Put in the
direction of $\mathbf{d}^*_{h00}$. (Since only directions are involved, the indices of the first
relp from the origin will be used. These are the indices of the lattice planes,

which have no common factor; but the relations found for the hkl relps will also apply to nh, nk, nl). $\mathbf{d}^*_{100}$ is, of course, the direction of the normal to the (100) planes, and the point representing this direction on the stereogram is called the 'pole' of the (100) planes (or crystal face). In the case of a cubic crystal, the RL axes are in the same directions as those of the crystal lattice, i.e. along the 4-fold symmetry axes; and since the lengths of the three axes are equal in each of the lattices, the *direction* (but not the length) of $l\mathbf{a} + m\mathbf{b} + n\mathbf{c}$ is the same as that of $l\mathbf{a}^* + m\mathbf{b}^* + n\mathbf{c}^*$. This is not true for any other system but the cubic. In non-orthogonal systems the two sets of axes are not even in the same directions.

(i) Find and label all the $\mathbf{d}^*$ directions related by symmetry to $\mathbf{d}^*_{100}$. These (or strictly the crystal planes or faces they represent) constitute the 'form' $\{100\}$. Do the same for $\mathbf{d}^*_{110}$ and $\mathbf{d}^*_{111}$.

Any two non-colinear RL vectors, $\mathbf{d}^*_{h_1k_1l_1}$ and $\mathbf{d}^*_{h_2k_2l_2}$, define a plane. Since neither has any component perpendicular to the plane, any linear combination of these two vectors must also lie in the plane, i.e. $n\mathbf{d}^*_{h_1k_1l_1} + m\mathbf{d}^*_{h_2k_2l_2} = \mathbf{d}^*_{(nh_1 + mh_2),\,(nk_1 + mk_2),\,(nl_1 + ml_2)}$ is coplanar with $\mathbf{d}^*_1$ and $\mathbf{d}^*_2$. Such a plane of RL vectors represents a 'zone' of crystal lattice planes or faces, all of which are parallel to a 'zone axis'. This zone axis must be the normal to the plane of RL vectors.

It follows immediately that any pole (RL vector) hkl, whose indices can be formed from those of two other non-colinear vectors by this process, i.e. $h = nh_1 + mh_2$, $k = nk_1 + mk_2$, $l = nl_1 + ml_2$, where n and m are positive or negative integers, lies in the same zone as $h_1 \, k_1 \, l_1$ and $h_2 \, k_2 \, l_2$. Common factors can be ignored, since $\mathbf{d}^*_{ph,\,pk,\,pl}$ lies in the same direction as $\mathbf{d}^*_{hkl}$.

It can be shown (Answer 8 (xiv), p. 462) that any RL direction in the zone can be equated by an integral linear combination of any other two vectors, but the relationship of the indices can only be easily obtained (guessed and verified) for simple indices. (For a more systematic method see Appendix II (5.8), p. 366).

Finally, any RL vector which lies at the intersection of two zones must have indices which can be derived from both zones by this process. This provides a simple trial and error method of finding (a) the indices of such a direction or (b) the zones whose intersections will define a specified direction. (ii) Use this method, together with any symmetry relations between the vectors, to find the positions of the poles of (102) and (112). Put in the symmetry related poles in the positive quadrant only, drawing whatever additional zones are necessary. (iii) Do the same for (123). (iv) Tabulate the number of symmetry related poles in all the above cases, and also the symmetry in the direction of each RL vector. (v) Deduce general rules for the multiplicity of any X-ray powder reflection in the cubic case. (v)i What differences would

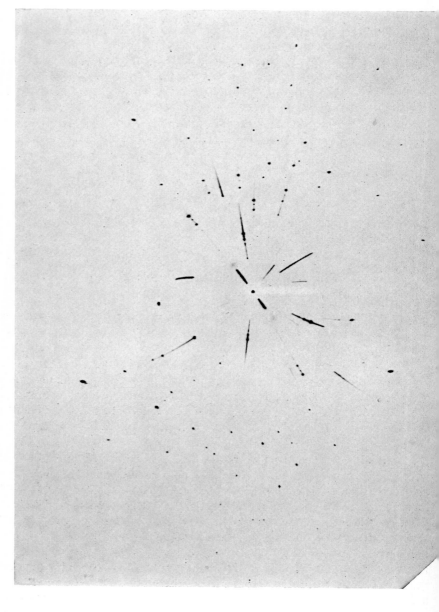

FIG. 178. A random 15° oscillation photograph of hexamine, $(CH_2)_6N_4$.

there be if the crystal class were (a) 23, (b) $\dfrac{2}{m}$ 3? (vii) Deduce from Figure 113 (p. 186) what is the lattice type of cuprite. (Add 50% to any length measured on the photograph before transferring to a standard chart.)

20. The Use of the ρ, ϕ Chart (p. 152)

(i) What form does the reciprocal lattice take when the incident X-rays have a continuous range of wavelengths? The direction of a weighted vector d^*_{hkl} for any particular position of the crystal is given by the angles ρ, ϕ, which are obtained from the position of the corresponding Laue reflection on the ρ, ϕ chart. (ii) Define the angles ρ and ϕ. (Note that the ϕ angles as printed on some charts are the complements of those given by the usual definition. In that case list the angles read from the chart as ϕ' and thence obtain ϕ). If the oscillation described in Example 11 (p. 290) is restricted to 15°, the white radiation streak for 120 has its inner end at $\phi = 81\cdot5°$. (iii) Find, by constructing the circle of reflection, what causes the streak to terminate. (iv) What will be the ϕ value for the outer end of the streak? (v) What is the value of ρ for d^*_{120}?

In Fig. 44 (p. 63) the goniometer reading ω is 93·5° at the beginning of the clockwise oscillation. (vi) What is the reading at the end? (vii) At what value of ω would the RL plane responsible for the crescent of streaks be perpendicular to the X-ray beam?

*21. The Determination of the Orientation of a Cubic Crystal by a General and a Special Method (p. 153)

A spherical crystal of hexamine ($I\bar{4}3m$, $a = 7\cdot03$ Å) is picked up at random on a glass fibre, and the X-ray photograph (Fig. 178) obtained, using CuKα radiation and an oscillation from $\omega = 30\text{--}45°$ on the goniometer scale (readings increase for clockwise rotation viewed from above). (i) Identify the low angle Kα reflections using a table of calculated θ-values (up to $\theta = 20°$); and list the ρ, ϕ coodinates of the corresponding RL vectors when $\omega = 30°$.

Use a stereographic projection (diameter 200 or 300 mm, if such a large scale Wulff net is available) to locate as many [111] directions as possible when $\omega = 30°$: (ii) by calculating the angles between [111] and at least two directions obtained from the photograph; drawing small circles with these radii about the corresponding poles on the stereogram; and identifying the points of intersection which are [111] directions, by measuring the angles from the remaining poles; and (iii) by rotating the experimentally determined poles on the stereogram into the standard cubic orientation, marking in the [111] poles and rotating back to the original orientation. (iv) If the plane of the bottom arc is parallel to the X-ray beam (scale left) when the goniometer scale reads 6°15' and the scale for the top arc is facing the observer, determine the simplest means of bringing a [111] direction into the equatorial plane.

(v) Determine the reading of the goniometer scale which would then bring this direction perpendicular to the X-ray beam. (The arcs were both set at zero when the photograph was taken).

22. The Preliminary Measurement of the Arc Correction and Adjustment of the Crystal by Moving the Supporting Plasticene (p. 153)

A setting photograph (Fig. 179) was taken with both arcs at zero, and the bottom arc perpendicular to the beam at the mid-point of the 15° oscillation. The scale of the bottom arc was towards the observer. Figure 179 shows that a large adjustment is needed. (i) Measure the correction required for the bottom arc (Section 13.2.1.1, p. 149) and give the arc reading after correction (Section 11.2 (iii), p. 135). Since it is required to keep the arc readings small so that the arcs will not foul the layer line screen in subsequent use on a Weissenberg camera, the fibre must be pushed over on the plasticene instead of setting the arc over by a large angle. The fibre is brought upright by moving the bottom arc 5°L and the top arc 7°R. (ii) Describe how you would proceed to adjust the fibre by finger pressure on the plasticene if the required adjustment were 20° clockwise.

23. Final Crystal Setting by Double Oscillation (p. 156)

Figure 179 (see Example 22) was taken with the goniometer reading $\omega = 150\cdot5°$ at the mid-point of the oscillation. The scale of the top arc was to the left. After correcting the orientation about the bottom arc axis, a second oscillation photograph was taken at right angles to the first (mid point, $\omega = 60\cdot5°$) and a correction made about the axis of the top arc. Both corrections were made by pushing the fibre over in the supporting plasticene, so the residual error was larger than would normally be the case if arc movements had been used to correct the orientation.

An oscillation photograph with 45 min exposure was taken with the goniometer scale reading 98·0°, when the cam follower was on the point of the 15° cam. The oscillation was therefore from 98·0 to 113·0°. The motor was stopped with the cam follower again on the point of the cam, and the oscillation reset to 278·0 + 15° for a further 15 min exposure.

Figure 180 is the resulting double oscillation photograph. (i) Draw sketches to show the positions of the arcs, including the arc scales, for the four oscillations. (ii) Measure the zero layer line separations and calculate the arc corrections. (iii) Which arc is corrected from the R.H. side of the photograph? (iv) What are the arc readings after correction?

24. Setting for the Precession Camera (p. 156)

A triclinic crystal with $a^* = 0\cdot25$, $b^* = 0\cdot20$, $c^* = 0\cdot30$ R.U., $\alpha^* = 75°$, $\beta^* = 70°$, $\gamma^* = 80°$, is to be set for transfer to a precession camera. It is oriented on a rotation camera, with $+\mathbf{a}$ upward along the rotation axis and $+\mathbf{b}^*$

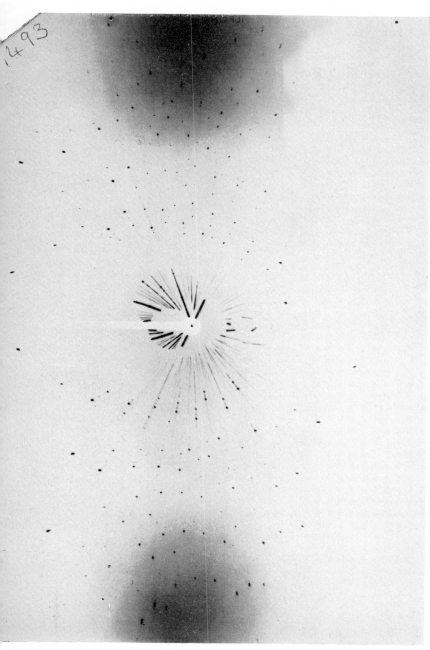

FIG. 179. An oscillation photograph showing the tilt of the Laue streak of the zero layer curve at the origin. This is a measure of the mis-setting in a plane perpendicular to the X-ray beam.

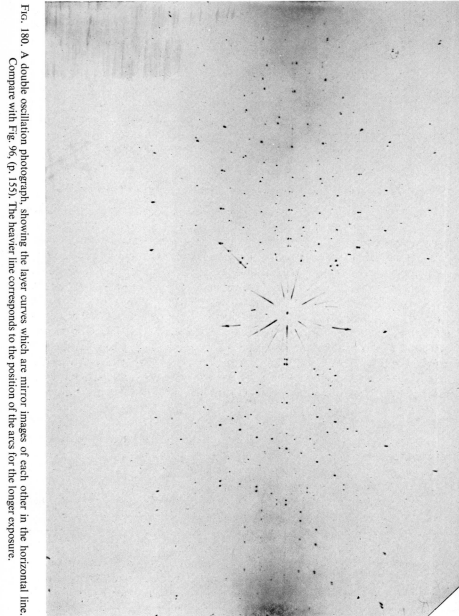

FIG. 180. A double oscillation photograph, showing the layer curves which are mirror images of each other in the horizontal line. Compare with Fig. 96, (p. 155). The heavier line corresponds to the position of the arcs for the longer exposure.

along the negative direction of the X-ray beam (i.e. $\overrightarrow{OC}$ Fig. 105, p. 176) for the goniometer scale reading, $\omega = 75°$. The plane of the bottom arc is parallel to the beam for $\omega = 97°$, with the scale to the right. The scale of the top arc is towards the observer (away from the X-ray tube). The arc readings are: Top, 12·7°R; Bottom, 24·3°L.

(i) Use a stereographic projection to find the arc readings required to bring $\mathbf{a}^*$ along the rotation axis, and the angle which the $\mathbf{a}^*\mathbf{b}^*$ plane then makes with the plane of the bottom arc. (ii) Find also the arc corrections which would be required if the axis of the top arc were horizontal, and compare with those obtained previously. (iii) If the arc readings were T. 12·7°L and B. 24.3°R, what further steps would have to be taken to get $\mathbf{a}^*$ along the rotation axis?

25. The Use of a Geometrical Symmetry Plane in Crystal Setting (p. 163)

In Fig. 43(i) (p. 61) the second photograph is from a crystal which was slightly mis-set about the vertical axis. The camera spindle was vertical and the scale reading 215·6°. The first photograph shows that the crystal is trigonal and has a vertical geometrical symmetry plane, although the lack of fnll symmetry in the intensities of the reflections makes it difficult to determine which are corresponding reflections on either side of the geometrical symmetry plane. (i) Use the method of Section **13**.4, (p. 160) to calculate the correction required to set the crystal for taking the first photograph. (ii) What will the reading of the scale be after applying the correction? (Clockwise rotation increases the scale reading). (iii) How did you decide which were corresponding reflections? (iv) Comment on any differences, other than mis-setting and the appearance of $K\alpha$ reflections, between the two photographs, and suggest reasons for the two photographs being taken under different conditions.

26. The Use of a Greninger Chart (p. 169)

Figure 101 (p. 165) is a back reflection Laue photograph from a grain of aluminium. The Space Group of Aluminium is $Fm3m$. (i) Which of the reflections 100, 110, 111 can occur in this case? (ii) Will the three zone axes [100], [110] and [111] appear as reflections on the Laue photograph? Give reasons. (iii) Find and tabulate the angles ($\leqslant 90°$) which can occur between $\langle 100 \rangle$, $\langle 110 \rangle$ and $\langle 111 \rangle$. ($\langle pqr \rangle$ denotes all the zone axes related by symmetry to $[pqr]$). (iv) How can important zone axes be recognised on the photograph? (v) Which of the marked points on Fig. 101 derive from such axes? (vi) Use the Greninger chart to measure the ε, γ and ψ angles of these points and tabulate the results. (vii) Measure the angles between the various pairs of important zone axes, and index them be comparison with the angles from (iii) (or the larger table in [6(120)]). (viii) Plot the directions on a stereogram of the RL vectors for all the marked points on Fig. 101. By bringing one of the

$\langle 100 \rangle$ poles to the centre, verify the indexing of the zone axes by comparison with the standard cubic stereogram. (ix) Index each of the remaining marked reflections from the zonal relationships (see Example 19, p. 296) and the table of angles [6(120)]. (x) If the specimen is mounted on arcs, with the top arc 12·5° clockwise from the position parallel to the beam and with its scale to the right (reading 5·3°L) and the scale of the bottom arc (reading 20·7°L) towards the observer (away from the X-ray tube) and if $\omega = 259\cdot5°$, what adjustments would be required in order to produce Fig. 45 (p. 64)? (xi) Measure the distance from the centre to the $\gamma = 30°$ hyperbola, along the $\psi = 0$ line in Fig. 103 (p. 168), and calculate the specimen to film distance for which the chart in Fig. 103 would be suitable.

*27. The Determination of Accurate Lattice Parameters (p. 173)

Figure 181 is the full size central portion of the photograph obtained from four 5° oscillations, designed to bring in reflections $10,0\bar{8}$ and $\overline{10},08$ on both sides of the collimator hole (Section 13.6.2.1, p. 170). The approximate value of $\mathbf{d}^*_{1\,0,0\,\bar{8}}$, from the RL net used in designing the oscillation (Section 14.1.4, p. 183) is 1·925 R.U., and the reflections can be identified by means of the

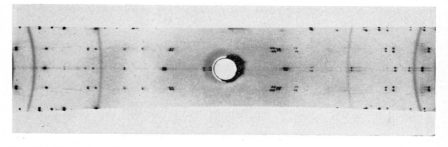

FIG. 181. Four 5° oscillations on one film, to give back reflections for accurate parameter measurement. The shadows of the spring clips used to hold the film firmly against the cassette can be seen, top and bottom. The dark patch at the edge of the hole is light fogging where the collar and collet were not fully effective. The powder rings are from the collimator and would normally be stopped by the guard slit. Only the central portion of the film is shown (full size).

Bernal chart. The measured diameter of the cassette was 60·65 ± 2 mm and the thickness of film and 2 layers of black paper was 0·45 ± 2 mm. From a separate measurement the shrinkage during processing was found to be 0·01 ± 1 %.

$\mathbf{d}^*_{13,00} = 1\cdot8924 \pm 4;\qquad \mathbf{d}^*_{009} = 1\cdot8336 \pm 5;\quad \text{for}\quad \mathrm{CuK}\alpha_1,\ \lambda = 1\cdot5405\ \text{Å},$

had been obtained from earlier back reflection photographs of the same kind.

(i) Measure the photograph, preferably with a travelling microscope, and find $d^*_{10,0\bar{8}}$ for CuKα_1. (ii) Calculate your probable errors, assuming that there has been no measurable change in reproduction of the original photograph. (iii) Find an accurate value of β^* and its probable error.

28. The Relations Between the Bernal (ξ, ζ); ρ, ϕ and θ Charts (p. 177)

On a cylindrical film (radius 30 mm) oscillation photograph the outer end of a Laue streak has the coordinates $x = 20$; $y = 25$ mm (Section **14**.1.2, p. 175). (i) Find the cylindrical coordinates ξ and ζ of the point on the weighted vector which is in the sphere of reflection when the crystal is at the corresponding end of the oscillation. (ii) Derive an expression for cos ϕ and hence find the third coordinate. (iii) Calculate the spherical angular coordinate ρ and the radius of the sphere d^*. (iv) Hence calculate θ. (v) Draw to full scale the position of the outer end of the Laue streak; measure the values of ξ, ζ, ϕ, ρ and θ using the charts, and compare with the calculated values.

29. The Interpretation of an Oscillation Photograph with the Aid of a Cylindrical Laue Photograph (p. 183)

Sucrose crystallises in the monoclinic system with $a = 10\cdot865$, $b = 8\cdot747$, $c = 7\cdot763$ Å, $\beta = 103\cdot0°$. Figure 109 (p. 180), is from a 15° oscillation. Figure 52 (p. 74) is a cylindrical Laue photograph. (i) Determine the crystal lattice direction along the (vertical) rotation axis by inspection of the photographs, giving reasons. (ii) Check your deduction by calculating the lattice spacing along the rotation axis. (CuKα radiation, $\lambda = 1\cdot542$ Å).

ω is the reading of the goniometer scale, which increases for clockwise rotation, looked at from above. (iii) Use Fig. 52, taken when $\omega = 78°$, to deduce the positions of the RL axes in the zero layer for this value of ω. The oscillation of Fig. 109 is from $\omega = 73°$ to 88°. (iv) Draw the **a*c*** net on tracing paper, and determine which of the RL axes whose positions were obtained from Fig. 52 is **a*** and find which is the positive direction of **b**. (v) Then index the zero layer and (vi) the fourth layer above zero on the photograph.

(When the clipped corner of either photograph is right-hand top the beam is coming through the photograph towards the observer, and the top edge in the cassette is uppermost.)

30. The Investigation of an Unknown Crystal (p. 185)

A crystal without recognisable faces, which gave an optically biaxial interference figure, was set by trial oscillation photographs about an important crystal lattice direction. A cylindrical Laue photograph was taken with the goniometer reading $\omega = 0$ (Fig. 182). The goniometer setting was

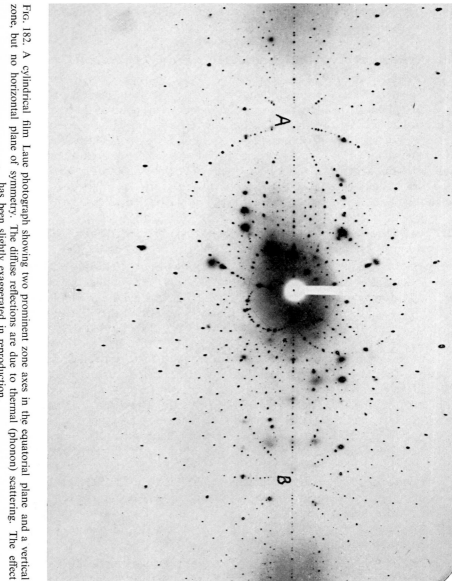

FIG. 182. A cylindrical film Laue photograph showing two prominent zone axes in the equatorial plane and a vertical zone, but no horizontal plane of symmetry. The diffuse reflections are due to thermal (phonon) scattering. The effect has been slightly exaggerated in reproduction.

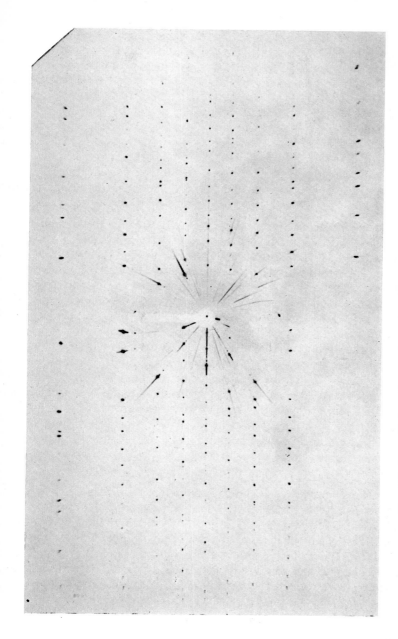

Fig. 183. An oscillation photograph starting with the 2-fold axis along the beam.

changed to $360 - 43·5 = 316·5°$, and a flat plate Laue photograph taken. After a correction to $317·0°$ Fig. 50 (p. 72) was obtained. (i) Give the steps in the reasoning which led to setting the goniometer at $316·5°$ and to the determination of the crystal system. (ii) Would a photograph at $90°$ to the orientation shown in Fig. 50 have given as much information? Figure 183 was taken with an oscillation of $317 + 15°$ and Fig. 184 with an oscillation of $317 + 72·5(\equiv 29·5) + 15°$. CuK$\alpha$ radiation, $\lambda = 1·542$ Å. (iii) Determine the RL parameters as accurately as possible, giving the steps by which you do so. (Take **c** along the rotation axis). (iv) Draw the RL net for the zero layer and index the zero layer of Fig. 183, using any systematic discrepancies that arise with higher angle reflections to obtain more accurate values of the RL parameters. (v) If necessary, redraw the net and use it to index the fourth layer above zero on Fig. 184. (vi) Comment on any pecularities of Fig. 184.

*31. Indexing the Zero Layer (or Row Lines) of a Rotation Photograph (p. 187)

(i) Index the zero layer reflections of Fig. 4 (p. 5) up to $\xi = 0·55$, using Ito's method [1(132)]. If the reflections lie on row lines (Section **14**.2.2(b), p. 186) use the whole row line to obtain the ξ values of the zero layer reflections, by fitting the chart so that reflections top and bottom and left and right all give the same value. ξ must be estimated to the nearest $0·005$, and the upper layers used to resolve reflections which cannot be separated on the zero layer. The success of the method depends on having accurate ξ values.

What is the cause of special difficulty in this case, leading to ambiguity in the indexing even of some of these low angle reflections? (ii) Index the row lines of Fig. 133 (p. 211) i.e. find the indices of the relp where the row line cuts the zero layer, even though the corresponding reflection may be absent. Use the difference in character of the two sets of row lines to help in the indexing, together with the knowledge that the crystal has a rhombohedral lattice. (iii) Explain why a normal rotation photograph necessarily has a horizontal symmetry line through the origin.

32. Vector Algebra—II (p. 187)

(i) Draw a diagram showing how the directions of incidence and scattering are defined by the vector **D***. If the position vector for an atom (scattering factor f_n) is $\mathbf{r}_n = 0·1\mathbf{a} + 0·2\mathbf{b} + 0·3\mathbf{c}$ and the scattering vector **D*** is $3·5\mathbf{a}^* + 2·5\mathbf{b}^* + 1·5\mathbf{c}^*$, what is the phase angle $\phi_n = (2\pi/\lambda)\mathbf{r} \cdot \mathbf{D}^*$ for the wave scattered from this atom under these conditions (a) in radians, (b) in degrees? What is $\cos \phi_n$? If the atom is part of a centrosymmetric set, what is its contribution to the total scattering amplitude F, if $f_n = 6$.

(ii) Use the scalar product to find the angle between **b** and the normal to (111) in a monoclinic crystal with $a^* = 0·1$, $b^* = 0·2$, $c^* = 0·3$ R.U., $\cos \beta^* = 0·4$.

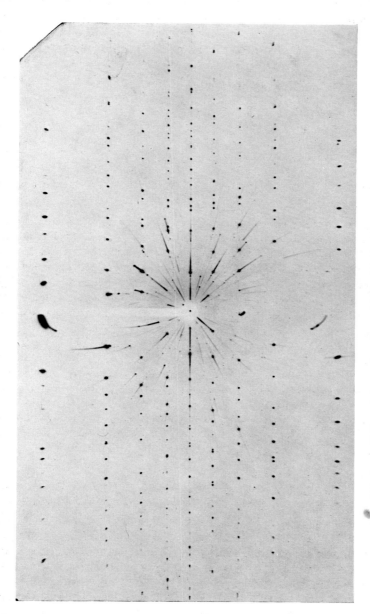

FIG. 184. An oscillation photograph ending with the 2-fold axis nearly at 90° to the beam.

(iii) The unit cell parameters of a crystal are: $a = 5.0$, $b = 6.0$, $c = 7.0$ Å, $\alpha = \gamma = 90$, $\beta = 110°$. Find a^*, b^*, c^*, α^*, β^*, γ^* for CuKα radiation, $\lambda = 1.542$ Å. Calculate the angle between the normals to the faces (210) and (111).

(iv) In a tetragonal crystal the angle between the normals to the faces (130) and (112) is $44.0°$. Find c/a.

(v) The RL for a crystal has parameters, $a^* = 0.2$, $b^* = 0.25$, $c^* = 0.15$ R.U., $\alpha^* = 80$, $\beta^* = 75$, $\gamma^* = 110°$. The crystal is mounted about **a**. Find the position of the origin of construction for the $+4$ layer.

(vi) In the structure of flavanthrone (Example 16, p. 294) two of the carbon atoms C_8 and C_{12} are related by a 2_1 axis at $\frac{1}{4}$, y, $\frac{1}{2}$ to C'_8 and C'_{12}. The co-ordinates of C_8 and C_{12} are 0.215, 0.668, 0.309 and 0.180, 0.674, 0.433 respectively. Find the position vectors, in terms of **a**, **b** and **c**, of C_8 and C'_{12}. Determine the components of the vector $\overrightarrow{C_8 C'_{12}}$ and calculate its length (i.e. the interatomic distance, in Å), and the angle the vector makes with **a**.

*(vii) Give a vectorial solution of the following: the vector area of the triangle whose vertices are the points **a**, **b**, **c** is $\frac{1}{2}(\mathbf{b} \times \mathbf{c} + \mathbf{c} \times \mathbf{a} + \mathbf{a} \times \mathbf{b})$. (Take account of the fact that the origin is not, in general, in the plane of the triangle and **a**, **b** and **c** are therefore not coplanar.)

*(viii) What is the unit vector perpendicular to each of the vectors $2\mathbf{i} - \mathbf{j} + \mathbf{k}$ and $3\mathbf{i} + 4\mathbf{j} - \mathbf{k}$? Calculate the sine of the angle between these two vectors.

(ix) Find the perpendicular distance of a corner of a unit cube from a diagonal not passing through it.

(x) Draw diagrams similar to Figs 130(a) and (b) (pp. 207 and 208) but for the reverse setting of the rhombohedron relative to the hexagonal axes. (a) Express $\mathbf{a}_R$, $\mathbf{b}_R$, $\mathbf{c}_R$ in terms of $\mathbf{a}_H$, $\mathbf{b}_H$, $\mathbf{c}_H$, and (b) use 360(1) to find $\mathbf{a}_R^*$, $\mathbf{b}_R^*$, $\mathbf{c}_R^*$ in terms of $\mathbf{a}_H^*$, $\mathbf{b}_H^*$, $\mathbf{c}_H^*$. Check the result from your second diagram. (c) Use 360(1) to find $\mathbf{a}_H$, $\mathbf{b}_H$, $\mathbf{c}_H$ in terms of $\mathbf{a}_R$, $\mathbf{b}_R$, $\mathbf{c}_R$ and (d) use 359(2) to find $\mathbf{a}_H^*$, $\mathbf{b}_H^*$, $\mathbf{c}_H^*$ in terms of $\mathbf{a}_R^*$, $\mathbf{b}_R^*$, $\mathbf{c}_R^*$. Check from the diagram.

33. The Interpretation of a Zero Layer Weissenberg Photograph (p. 198)

Figure 185 is a simulated Weissenberg photograph of the zero layer about **b**. (i) What features of an actual Weissenberg photograph taken under normal conditions are not represented on the diagram? (ii) A Laue photograph along **b** shows 2-fold symmetry. Choose axes on the Weissenberg diagram consistent with this fact and the normal conventions, given that $a > c$. Draw on tracing paper the curves corresponding to the RL net and label them, so that any spot can be indexed on inspection. (iii) List the absent reflections. What is the cause of these absences? (iv) What is the symmetry

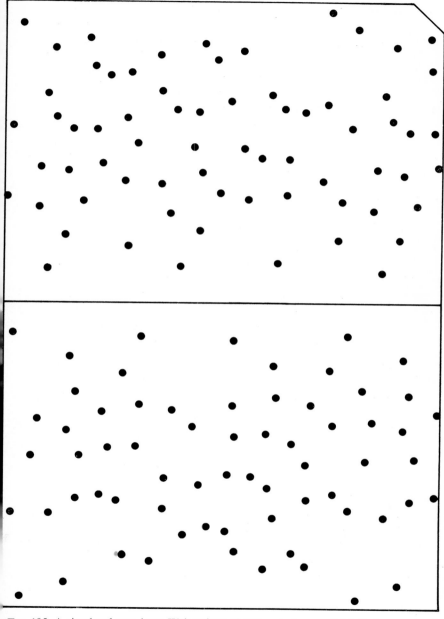

FIG. 185. A simulated zero layer Weissenberg photograph. Ignore slight displacements of spots and allow for a reduction of 1 per cent in the reproduction.

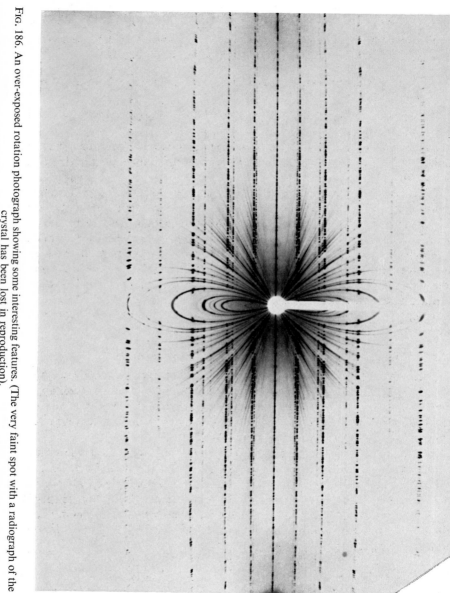

FIG. 186. An over-exposed rotation photograph showing some interesting features. (The very faint spot with a radiograph of the crystal has been lost in reproduction).

of the zero layer RL net? Is this symmetry due to the 2-fold axis along **b** or is it inherent in all zero-layer photographs? Give reasons. (v) Are there any reflections which are inconsistent with the layer symmetry and the absences listed above? (vi) What would be the symmetry of the first ($h1l$) layer? Is this symmetry inherent in all non-zero layer nets and photographs?

34. Setting an Equi-inclination Weissenberg Camera from a Rotation Photograph (p. 200)

The rotation photograph (Fig. 186) was taken with an X-ray tube which had a nominally Cu target.

The crystal is to be transferred to an equi-inclination Weissenberg camera and the second Kα layer *below* the zero layer photographed, using the same radiation.

(i) Draw the diagram showing the arrangement of the X-ray beam and rotation axis for this purpose. (ii) Calculate the equi-inclination angle and the distance that the layer line screen (diameter 50 mm) should be moved from its zero layer position. (iii) What simple checks can be used to make sure that the inclination of the rotation axis and the movement of the layer line screen have been made in the right directions? (iv) If the maximum equi-inclination angle is 30°, what is the highest layer that can be photographed about this axis? (v) If the radiation is CuKα ($\lambda = 1.542$ Å) what is the length of the crystal axis which is along the rotation axis? (vi) Comment on any unusual features of this rotation photograph.

*35. The Effect of Centred Cells on X-ray Photographs (p. 201)

Figures 89 (p. 140) and 120 (p. 196) are copies of a zero layer Weissenberg photograph, and Fig. 187 is the corresponding first layer photograph about the **c** axis of a crystal whose Laue photograph along **c** is shown in Fig. 43(c) (p. 59). CuKα radiation ($\lambda = 1.542$ Å) was used and a rotation photograph about **c** gave a layer separation of 0.353 R.U.

(i) Explain why there are no reflections along the directions of the **a*** and **b*** axes on the first layer photograph. (ii) Given that the second layer photograph is similar to the zero layer, and that from morphological evidence the class is $\bar{4}$, suggest the space group referred to these axes. (iii) In Fig. 28 (p. 37) determine the possible crystal lattice types, given that a Laue photograph along **b** shows only 2-fold symmetry, and that the crystal gives a biaxial optical interference figure.

*36. The Choice of RL Axes (p. 201)

(i) Do the axes chosen in Fig. 120 (p. 196) and Fig. 187 give the smallest crystal cell which conforms to the symmetry of the crystal? (ii) If not, choose other RL axes to give a smaller crystal cell, and the positive

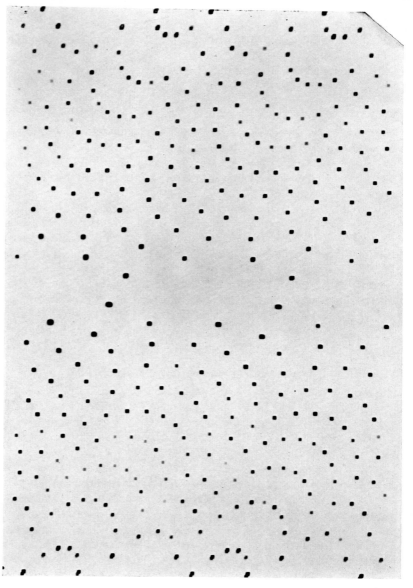

FIG. 187. A first layer Weissenberg photograph corresponding to the zero layer of Figure 120 (p. 196). The extension of low angle spots on the bottom half and contraction on the top half is already evident (Section **18**.2.2.2, p. 272). This does not affect the optical density of the centre of the spot if correct integration limits have been set. (Allow for a 1 per cent reduction from the original).

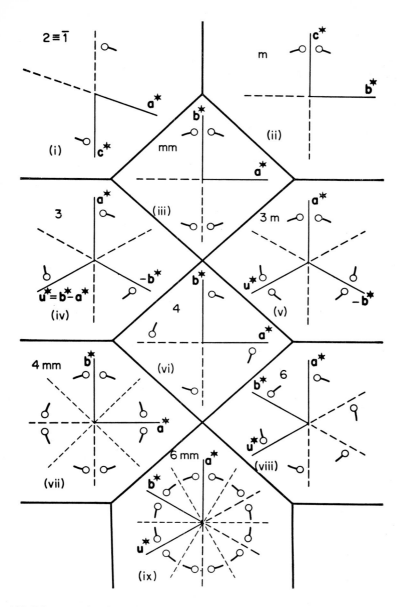

Fig. 188. Diagrams showing the possible two-dimensional symmetries of a RL layer. The tails (arcs of circles centred on the origin) indicate identical sequences of weighted relps as in Figure 127 (p. 204). The symmetry symbol is given for each diagram.

directions of **a*** and **b*** so that the intensity of the 310 reflection is greater than that of 130. (iii) Use direct measurement in mm on the film to get as accurate a value of a as possible.

37. The Symmetry of Weissenberg Photographs (p. 205)

The diagrams in Fig. 188 represent the various possible symmetries of RL planes (layers). Only one independent relp is shown, together with the symmetrically related points. The tails to the spots represent the directions of similar sequences of weighted points in the RL.

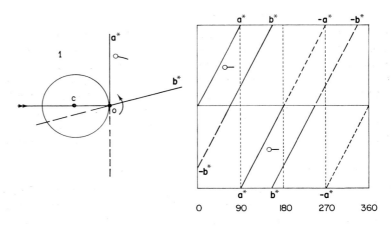

FIG. 189. Diagram of a RL layer with no symmetry (1) and reduced scale outline Weissenberg photograph covering 360° rotation of the layer. Dashed lines mark every 90° rotation. An actual photograph would cover just over half the length of this diagram, i.e. about 200° instead of 360°.

(i) Show on outline diagrams similar to Fig. 189 how the corresponding reflections would appear on a Weissenberg photograph, starting in each case with the vertical axis perpendicular to the beam as in Fig. 189, which shows the case of no symmetry. Put in and label the axes, using dashed lines for the 'negative' ends. Where the symmetry element appears directly on the photograph, put it in on your diagram. If the symmetry element appears indirectly, state how you would recognise it. The slope of a central lattice line on the photograph is $\tan^{-1} 2 = 63° 26'$, and goes from the centre to the edge in 90° rotation.

The n-fold rotors and symmetry lines in the diagrams, which are regarded as operating only in the plane, can be replaced by 3-dimensional symmetry elements as follows:

n-fold rotor: n-fold axis perpendicular to the plane.

2-fold rotor: inversion centre in plane.

Symmetry line: reflection plane normal to plane *or*

2-fold rotation axis in plane.

These equivalences are important when relating the symmetry of a plane of the reciprocal lattice to its point group symmetry as a whole.

(ii) Which of the nine diagrams could not be zero layers?

38. Calculations for a Precession Camera (p. 228)

An orthorhombic crystal is mounted on a precession camera with **c** along the beam when $\mu = 0$. The crystal parameters are: $a = 5\cdot0$, $b = 6\cdot0$, $c = 4\cdot5$ Å. MoKα radiation, $(\lambda = 0\cdot710$ Å$)$ is to be used to photograph the second layer, with $\mu = 30°$, $X = 60$ mm. (i) How far must the film holder be advanced? (ii) Layer screens of diameters 30, 40, 50 and 60 mm are available. Decide which screen to use and how far from the crystal it must be set. (iii) At what distance from the centre will the cones for the first and third layers hit the screen? (iv) Which reflections near the centre will not be recorded?

39. Final Setting on a Precession Camera (p. 234)

Figure 25 (p. 34) was taken after setting a spherical crystal of sucrose $(a = 10\cdot86,\ b = 8\cdot75,\ c = 7\cdot76$ Å$,\ \beta = 103°;\ $S.G. P2$_1$) on the precession camera. It was first set on a rotation camera with **c** along the rotation axis, and adjusted to bring **c*** along the axis. This was combined with reducing the arc readings to smaller values, by pushing the fibre over in the plasticene (to reduce the possibility of the arcs fouling when transferred to the precession camera). Therefore the setting was not very accurate. It was decided to obtain the ϕ setting on the precession camera, and initially it was only known that **c*** was nearly along the rotation axis.

A setting photograph was taken on a camera similar to Fig. 141 (p. 220) under the following conditions:

$X = 60$ mm; $\mu = 10°$; setting screen with $r = 10$ mm, $s = 30$ mm; MoK radiation, $\lambda_{K\alpha} = 0\cdot710$ Å; 16 ma, 40 kV (reduced to 2 ma, 10 kV to register the direct beam, with $\mu = 0$, for 1s and obtain a radiograph of the crystal to act as origin). The top arc axis was vertical (plane horizontal) with scale downwards reading $14\cdot75°$L (Section **11**.2 (iii), p. 135). The dial axis reading was $89\cdot5°$. Figure 148 (p. 233), shows the setting photograph obtained. The nearly horizontal streak is clearly c^* (confirmed by measurement of the spacings of the reflections, allowing for the mis-setting) but no obvious RL plane is present. Corrections were calculated from the c^* streak for the two arcs (the scale of the bottom arc was towards the X-ray tube and reading $5.6°L$).

(i) Deduce the readings of the arcs after the corrections have been applied.

Since the correction for the bottom arc is less accurate than that for the top arc ((ii) Why?), the dial axis was turned through 90° (scale reading 359·5°) so that the scale of the bottom arc was upwards and Fig. 145(b) (p. 229) was obtained. By a lucky chance an important RL plane was nearly perpendicular to the precessing axis, so the dial axis was adjusted to set this plane correctly, and the small correction applied to the bottom arc.

(iii) Deduce the bottom arc and dial axis scale readings after the corrections.

The corrected photograph (Fig. 190) was used to identify the RL plane.

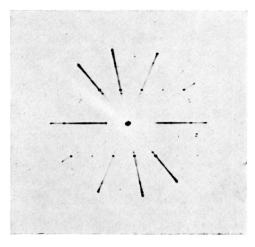

FIG. 190. Central portion of a setting photograph showing an unknown zero RL plane containing c^* along the dial axis, i.e. horizontal. The plane has been set perpendicular to the precessing axis. Since the lines of relps parallel to c^* are rather widely spaced, the plane is not very densely populated and the neighbouring planes are therefore not far away. Some split reflections from these neighbouring planes can be seen in the central area defined by the zero layer.

(iv) Deduce the RL vector d^*_{hk0} which, with d^*_{001} (i.e. c^*) defines the plane of Fig. 190.

(v) Calculate the angle through which the dial axis should be turned to bring the a^*c^* plane perpendicular to the precessing axis. Does this give an unambiguous dial axis scale reading? What reading should the scale be set at?

Figure 191 is the photograph obtained after adding the dial axis correction. After a final correction of dial and arcs a check photograph was taken (Fig. 145 (a), p. 229) followed by a precession photograph of the **a*c*** layer (Fig. 25, p. 34).

(vi) Deduce the final arc and dial readings.

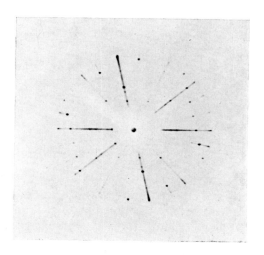

Fig. 191. The setting photograph obtained by adding the angle, calculated from Fig. 190, required to bring the **a*c*** plane perpendicular to the precessing axis.

40. The Measurement of Preferred Orientation in Rolled Sheet (p. 239)

Figures 55 (p. 80) and 192 (a) and (b) are photographs of Mo foil in which the continual reduction by rolling has induced the originally randomly oriented grains to line up, relative to the rolling direction and the normal to the sheet. By convention these directions and the transverse direction perpendicular to both, are given the symbols W (Waltzrichtung), N (Normalrichtung) and Q (Querrichtung).

Photographs with cylindrical film of unoriented polycrystalline Mo give uniform powder rings. When $d^{*2} = 4 \sin^2 \theta$ is calculated for each of the rings, it is found that d^{*2} for the first line is equal to the difference in d^{*2} values for all pairs of adjacent lines (with one exception), and that only the 14th line is missing. (i) Deduce the lattice type. (ii) Index by inspection of Fig. 192 the powder rings on which the heavy arcs due to preferred orientation lie,

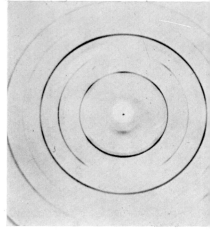

(a)   (b)

FIG. 192. Preferred orientation photographs of Mo foil, taken with MoKα radiation and specimen-film distance 35 mm. (a) W vertical, Q parallel to X-rays. (b) Q vertical, W parallel to X-rays.

and find the θ-values for 110, 200 and 220. (iii) Measure with a protractor the angles ψ, between the ends of the strong parts of the arcs and the horizontal, for each of these reflections. As a first approximation we can assume that the three directions W, N and Q define three planes of symmetry in the specimen, and therefore only the values of ψ between 0–90 (i.e. in one quadrant) need be recorded, although all four quadrants should be measured and the averages taken. Tabulate the values of $(90 - \theta)$ and ψ for each of these three lines. (iv) Construct two stereograms, one for the 200 reflection, form {100}, and the other for 110 and 220, form {110}, both with N in the centre and W from north to south. Draw lines on the stereograms representing the theoretically possible range of orientation of the normals to these planes, for the three photographs. Using the ψ values, mark on these lines the portions corresponding to the arcs observed on the photographs.

Draw round and shade in those areas of the stereograms which contain the marked portions, and which represent the limits of variation of position of the normal to the plane considered. The result is the pole figure for this plane. Mark the centres of these shaded areas as the idealized positions of the normals, and from this determine which crystallographic directions are predominantly parallel to N, W and Q respectively.

A larger number of photographs with various orientations of the foil is needed to define the shaded areas more precisely.

(v) From consideration of the data given below, give two reasons why it would be impracticable to use CuKα radiation for this experiment.

λ CuKα = 1·54; λ MoKα = 0·71 Å.

μ_m of Mo for CuKα = 164, for MoKα = 20·2 cm^{-1}.

Density of Mo = 10·1 gm/cc.

Thickness of foil = 0·12 mm.

41. Preferred (Fibre) Orientation in a Wire (p. 249)

Figure 154 (p. 248) shows the effects of preferred orientation in a copper wire. Use the methods of Section **17**.3 (p. 247) to decide the type of preferred orientation.

42. Formulae and Calculations for X-ray Filters (p. 258)

(i) Prove the formula given in Section **18**.1.1.2 (p. 257) for the optimum thicknesses of a pair of balanced filters. Given the data below, and assuming that, between absorption edges, $\mu_m \propto \lambda^3$, (ii) find the optimum thicknesses for Y and Zr as filters for MoKα, $\lambda = 0·710$ Å. (iii) What proportion is the difference intensity of the incident intensity? (iv) What should be the thickness of a Zr β filter, in order to cut down the relative Kβ intensity ($\lambda = 0·632$ Å) from 1/6 to 1/600 of the Kα? (Note the units for μ_m)

	μ_m cm^{-1}			
λ	0·632	0·710	0·75	D_m
Y	75	100	17·2	4·57
Zr	79	15·9	18·6	6·50

43. The Use of the Cochrane Chart (p. 282)

A monoclinic crystal has $a^* = 0·3$, $b^* = 0·35$, $c^* = 0·25$, $\beta^* = 70°$. Use the Cochrane chart to find the (Lp)$^{-1}$ factors for all the independent reflections of the + 3 layer about **c**, taken on an equi-inclination Weissenberg camera. The length of the base of the chart is 2 R.U. If Fig. 172 (p. 281) is used the RL net must be drawn to the same scale.

44. The Calculation of A* (p. 283)

A single crystal oscillation photograph of a cylindrical crystal of Co(CNS)$_4$Hg was taken, with the axis of the cylinder parallel to the rotation axis. The diameter of the crystal was 0·2 mm. (i) Calculate the absorption correction factor A* for a CuKα reflection on the zero layer at a θ angle of 22·5°, given that:

Atom	Co	C	N	S	Hg
Atomic weight	58·9	12·0	14·0	32·1	200·6
$\mu_m\,\mathrm{cm}^{-1}$	313	4·6	7·5	89·1	216

The density of the crystal $D_m = 3\cdot02\ \mathrm{gm\ cm}^{-3}$.

Method: Represent the cross-section of the crystal by a circle of 100 mm radius, and draw a grid of lines parallel to the incident and reflected directions spaced 30 mm apart. Draw the first pair of lines through the centre of the circle. Measure the path length $l_1 + l_2 = L_n$ for each of the n intersections within the boundary, and evaluate $A = (1/n)\,\Sigma\,e^{-\mu L_n}$, making use of any symmetry in the diagram. Hence find A* and compare with the value given in the tables 6(295).

Derive exact expressions for the absorption factor of (ii) a flat plate specimen in a transmission Guinier powder camera, and (iii) a thick plate powder specimen used in the reflection mode. In each case assume that the normal to the plate is in the plane of reflection and that the incident beam makes an angle ϕ with the normal. (iv) What form does the factor take in the case of a powder diffractometer, when the incident and reflected beams make equal angles with the normal?

Appendices

Notes on Vectors

The following results of vector algebra are relevant to the text. Sections 1–9 and 15 must be fully mastered since these results will be in constant use. Other sections can be referred to as required.

1. Conventions

A line representing the direction as well as the magnitude of a vector quantity will be denoted by a letter in bold type, viz. **a**. The length of the line, corresponding to the magnitude of the vector quantity, will be denoted by the same letter in *italics*. In certain cases, where a vector is denoted by the points at each end of a line, e.g. *AB*, the direction will be given by an arrow—$\overrightarrow{AB}$. Where necessary the magnitude of a vector (or vector expression) will be denoted by vertical lines, e.g. $|\mathbf{a}| = a$.

a + **b** = **r** means that **r** is the vector represented by the third side of a triangle whose other two sides are **a** and **b** (Fig. 193). Vector addition is *defined* to take place according to the triangle law, but the usefulness of the definition depends on the fact that many physical quantities are combined in this way. The result of repeated application of this law gives the result of adding any number of vectors: **r** = **a** + **b** + **c** .... **r** is the closing side of a polygon of vectors whose other sides are **a**, **b**, **c**, etc. In general such a polygon is non-planar. It is easily shown that **a** + **b** = **b** + **a**, that **a** + (**b** + **c**) = (**a** + **b**) + **c**, etc., i.e. the commutative and associative laws of ordinary

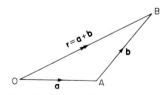

Fig. 193. The addition of vectors.

324

algebra also hold for vector algebra. The sum is independent of the order and the grouping of terms.

ma, where m is a scalar factor, is defined as meaning a vector in the direction of a and of length ma. $-$a is defined as a vector in the opposite direction to a, but of the same length. a $-$ b means 'reverse the direction of b and add to a'.

$$-m\mathbf{a} = m(-\mathbf{a}); \qquad m(n\mathbf{a}) = (mn)\,\mathbf{a} = n(m\mathbf{a}); \qquad (n+m)\,\mathbf{a} = n\mathbf{a} + m\mathbf{a}.$$

It can be easily shown that $m(\mathbf{a} + \mathbf{b}) = m\mathbf{a} + m\mathbf{b}$.

Brackets have the same meanings as in normal algebra.

2. Position Vectors

The position vector of a point P, relative to the origin O, is simply the line joining OP, i.e. $\overrightarrow{OP}$. Three or more vectors are said to be coplanar if a plane can be drawn parallel to all of them. If they are position vectors the plane through the origin will contain them all. If such a plane cannot be drawn the vectors are said to be non-coplanar.

3. Components of a Vector

If a, b, c are three non-coplanar vectors, giving three axial directions from the origin, and r is the position vector of the point P, then $\mathbf{r} = x\mathbf{a} + y\mathbf{b} + z\mathbf{c}$. xa, yb and zc are the *components* of the vector r in the directions of the three axes. In general they are *not* the resolved parts of r in the directions of the axes. This is only true for orthogonal axes. In the general case, planes are drawn through P parallel to the planes defined by ab, bc and ca, which, together with the parallel planes through the origin, form a parallelepiped (Fig. 194). The edges of this parallelepiped are the components xa, yb and

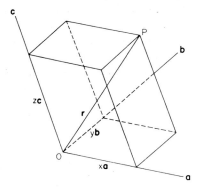

FIG. 194. The components of a vector relative to axes a, b and c. $r = x\mathbf{a} + y\mathbf{b} + z\mathbf{c}$.

z**c**. One can get from O to P along the edges of the parallelepiped in six ways, corresponding to the six ways of arranging the components in order, but the parallelepiped can only be constructed in one way, i.e. it is unique. Therefore the components of a vector for a particular set of axes **a**, **b**, **c** are unique; and it follows that *two vectors which have the same components must be equal*. If **a**, **b** and **c** form a right-hand set of orthogonal axes and are *unit vectors*, i.e. vectors of unit length, then x, y and z are the Cartesian coordinates of the point P. In this very important case the symbols **i**, **j**, **k** are always used for the axial vectors.

Position vectors are attached to an origin. Other vectors may be used merely to indicate direction and can be attached to different points as required. Their components relative to a particular set of axes remain the same.

4. The Vector $\overrightarrow{AB}$

There is one minor problem which occurs so frequently that it is worth remembering a rule for it. In Fig. 193, if $\overrightarrow{OA}$ and $\overrightarrow{OB}$ are the position vectors of the points A and B, what is the vector $\overrightarrow{AB}$? From the Figure

$$\overrightarrow{OA} + \overrightarrow{AB} = \overrightarrow{OB}; \therefore \ \overrightarrow{AB} = \overrightarrow{OB} - \overrightarrow{OA}, \tag{1}$$

i.e. the 'end' vector minus the 'beginning' vector.

5. Scalar Product

Many scalar quantities depend upon two vector quantities in such a way as to be proportional to the product of the magnitudes of the vectors and the cosine of the angle between them. Work done during the displacement of a body by a force acting on it is a typical example. It is therefore convenient to *define* the scalar product of two vectors, **a**·**b**, as a *scalar* quantity, equal to the product $ab \cos \psi$ where ψ is the angle between **a** and **b**. It is obvious from the definition that **a**·**b** = **b**·**a**, i.e. scalar multiplication is commutative. From the definition, **a**·**b** is the length of the resolved part of **b** in the direction **a** multiplied by a. If **a** is a unit vector, **a**·**b** is simply the length of the resolved part of **b** in the direction of **a**. This particular use of scalar products is constantly occurring, and should be thoroughly grasped.

If **a**·**b** $= 0$, $\cos \psi = 0$, i.e. the vectors are at right angles. **a**·**a** $= a^2$. (This relation is the basis for finding the length of a vector. 'Squaring' a vector means taking the dot product with itself). If **i**, **j**, **k** are unit right-hand orthogonal vectors (i.e. Cartesian axes)

$$\mathbf{i}^2 = \mathbf{j}^2 = \mathbf{k}^2 = 1, \qquad \mathbf{i} \cdot \mathbf{j} = \mathbf{j} \cdot \mathbf{k} = \mathbf{k} \cdot \mathbf{i} = 0. \tag{2}$$

$\mathbf{a} \cdot \mathbf{b} \, \mathbf{c} = (\mathbf{a} \cdot \mathbf{b}) \, \mathbf{c}$ is a vector in the direction of $\mathbf{c}$ of magnitude $(\mathbf{a} \cdot \mathbf{b}) \, c$. Similarly $\mathbf{a} \cdot \mathbf{b} \, \mathbf{c} \cdot \mathbf{d}$ is simply the product of two scalar quantities $\mathbf{a} \cdot \mathbf{b}$ and $\mathbf{c} \cdot \mathbf{d}$.

Since $\mathbf{r} \cdot \mathbf{i}$ is the resolved part of $\mathbf{r}$ in the direction of $\mathbf{i}$, it is also the length of the component of $\mathbf{r}$ in the direction of $\mathbf{i}$.

$$\therefore \ \mathbf{r} = \mathbf{r} \cdot \mathbf{i} \, \mathbf{i} + \mathbf{r} \cdot \mathbf{j} \, \mathbf{j} + \mathbf{r} \cdot \mathbf{k} \, \mathbf{k}. \tag{1}$$

The direction cosines of $\mathbf{r}$ are:

$$\mathbf{r} \cdot \mathbf{i}/r = l, \qquad \mathbf{r} \cdot \mathbf{j}/r = m, \qquad \mathbf{r} \cdot \mathbf{k}/r = n. \tag{2}$$

These results are only true for orthogonal axes. (The use of $\mathbf{i}$, $\mathbf{j}$ and $\mathbf{k}$ implies unit orthogonal right hand axes.)

The meaning of the scalar product $\mathbf{a} \cdot (\mathbf{b} + \mathbf{c})$ is not immediately obvious. In Fig. 195, OA and OB represent the vectors $\mathbf{a}$ and $\mathbf{b}$, which can always be drawn in the plane of the diagram. However BC, representing $\mathbf{c}$, and $\overrightarrow{OC}$ will, in general, be out of the plane. Draw planes through B and C perpendicular to OA, cutting it in D and E. However far out of the paper C is, OEC

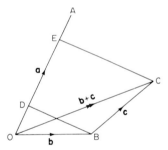

FIG. 195. The distributive law for scalar multiplication:— $\mathbf{a} \cdot (\mathbf{b} + \mathbf{c}) = \mathbf{a} \cdot \mathbf{b} + \mathbf{a} \cdot \mathbf{c}$. In general, C is *not* in the plane defined by $\mathbf{a}$ and $\mathbf{b}$.

is a right angle and OE is the resolved part of OC in the direction of OA, i.e. $(\mathbf{b} + \mathbf{c}) \cdot \mathbf{a} = OEa$. Similarly, if B is transferred to D, C will remain on the plane through E, and DE is the resolved part of BC in the direction of OA, i.e. $\mathbf{c} \cdot \mathbf{a} = DEa$. $\mathbf{b} \cdot \mathbf{a} = ODa$ in the same way.

$$\therefore \ (\mathbf{b} + \mathbf{c}) \cdot \mathbf{a} = OEa = ODa + DEa$$

$$= \mathbf{b} \cdot \mathbf{a} + \mathbf{c} \cdot \mathbf{a}$$

$$\text{and } \mathbf{a} \cdot (\mathbf{b} + \mathbf{c}) = \mathbf{a} \cdot \mathbf{b} + \mathbf{a} \cdot \mathbf{c},$$

i.e. the ordinary distributive law is obeyed. It is extremely important to be convinced of the validity of this law for vectors, including the general case obtained by repeated application:

$$(\mathbf{a} + \mathbf{b} + \mathbf{c} + \mathbf{d} \ldots) \cdot (\mathbf{l} + \mathbf{m} + \mathbf{n} \ldots) = \mathbf{a} \cdot \mathbf{l} + \mathbf{a} \cdot \mathbf{m} + \mathbf{a} \cdot \mathbf{n} \ldots$$
$$+ \mathbf{b} \cdot \mathbf{l} + \mathbf{b} \cdot \mathbf{m} + \mathbf{b} \cdot \mathbf{n} \ldots$$
$$+ \ldots .$$

Previous results either derive directly from the definitions or are intuitively easy to grasp. This is not the case for the distributive law for scalar multiplication of vectors; but a great deal of the power of vector algebra to simplify 3-dimensional geometry depends on the use of this law.

Since the angle between $-\mathbf{a}$ and $-\mathbf{b}$ is the same as between $\mathbf{a}$ and $\mathbf{b}$ (vertically opposite angles), $-\mathbf{a} \cdot -\mathbf{b} = \mathbf{a} \cdot \mathbf{b}$.

6. The Equation of a Straight Line

A straight line is defined by a point P through which it passes, and a vector defining its direction. Let $\mathbf{a}$ be the position vector defining P, $\mathbf{b}$ the vector defining the direction of the line and $\mathbf{r}$ the position vector of any point on the line. Then

$$\mathbf{r} = \mathbf{a} + t\mathbf{b} \tag{1}$$

where t is a scalar variable. The line through two points defined by $\mathbf{a}$ and $\mathbf{b}$ is

$$\mathbf{r} = \mathbf{a} + t(\mathbf{b} - \mathbf{a}) = \mathbf{a}(1 - t) + t\mathbf{b} \tag{2}$$

and if three points whose position vectors are $\mathbf{r}$, $\mathbf{a}$ and $\mathbf{b}$ satisfy (2), then they are colinear. (If the expression relating $\mathbf{r}$, $\mathbf{a}$ and $\mathbf{b}$ can be put in the form

$$L\mathbf{r} + M\mathbf{a} + N\mathbf{b} = 0 \quad \text{with} \quad L + M + N = 0,$$

then the expression can be put in the form (2) and the three points are colinear.)

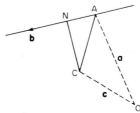

FIG. 196. The perpendicular distance from a point to a line. O is not in the plane of the diagram.

It is sometimes necessary to find the perpendicular distance p, from a point to a line $\mathbf{r} = \mathbf{a} + t\mathbf{b}$.

Let the point be C, with position vector $\mathbf{c}$ and CN the perpendicular from C to the line (Fig. 196). Then $\overrightarrow{AC} = \mathbf{c} - \mathbf{a}$ and $AN = (\mathbf{b}/b)\cdot(\mathbf{c} - \mathbf{a})$.

$$\therefore\ CN^2 = (\mathbf{c} - \mathbf{a})^2 - (1/b)^2(\mathbf{b}\cdot(\mathbf{c} - \mathbf{a}))^2. \tag{1}$$

7. The Normal Equation of a Plane

Let $\mathbf{n}$ be a vector perpendicular to the plane and p the perpendicular distance from the origin to the plane. If X, with position vector $\mathbf{r}$, lies in the plane then

$$\mathbf{r}\cdot(\mathbf{n}/n) = p \quad \text{or} \quad \mathbf{r}\cdot\mathbf{n} = pn = q. \tag{2}$$

For a plane normal to $\mathbf{n}$ and passing through a point P with position vector $\mathbf{c}$; if $\mathbf{r}$ is the position vector of any point in the plane, $\mathbf{r} - \mathbf{c}$ is a vector in the plane and therefore perpendicular to $\mathbf{n}$.

$$\therefore\ (\mathbf{r} - \mathbf{c})\cdot\mathbf{n} = 0 \quad \text{or} \quad \mathbf{r}\cdot\mathbf{n} = \mathbf{c}\cdot\mathbf{n} = q,$$

which is of the same form.

The distance p, of a point P, position vector $\mathbf{c}$, from the plane $\mathbf{r}\cdot\mathbf{n} = q$ can be found as follows:

$\mathbf{r}$ is the position vector of any point in the plane, therefore

$(\mathbf{r} - \mathbf{c})\cdot(\mathbf{n}/n)$ is the required distance, and

$$p = (1/n)(\mathbf{r}\cdot\mathbf{n} - \mathbf{c}\cdot\mathbf{n}) = (1/n)(q - \mathbf{c}\cdot\mathbf{n}) \tag{3}$$

8. The Intercept Equation of a Plane

If a plane cuts three axes, given by the vectors $\overrightarrow{OA} = \mathbf{a}$, $\overrightarrow{OB} = \mathbf{b}$, $\overrightarrow{OC} = \mathbf{c}$ (Fig. 197), making three intercepts, $\overrightarrow{OX} = x\mathbf{a}$; $\overrightarrow{OY} = y\mathbf{b}$; $\overrightarrow{OZ} = z\mathbf{c}$; and $\mathbf{n}$

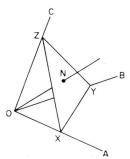

FIG. 197. The intercepts cut off on the three axes by a plane whose normal is ON.

is a unit vector along ON, normal to the plane, then $x\mathbf{a} \cdot \mathbf{n} = y\mathbf{b} \cdot \mathbf{n} = z\mathbf{c} \cdot \mathbf{n} = ON =$ the perpendicular distance from the origin to the plane.

Conversely, if

$$x\mathbf{a} \cdot \mathbf{n} = y\mathbf{b} \cdot \mathbf{n} = z\mathbf{c} \cdot \mathbf{n} = d, \tag{1}$$

then the unit vector $\mathbf{n}$ is normal to a plane which cuts the axes at $x\mathbf{a}$, $y\mathbf{b}$, $z\mathbf{c}$, and the perpendicular distance from the origin to the plane is d.

9. To Find the Angle, ψ Between Two Vectors

Let the reference axes be defined by 3 non-coplanar vectors, $\mathbf{a}$, $\mathbf{b}$ and $\mathbf{c}$ with angles α, β and γ between $\mathbf{bc}$, $\mathbf{ca}$ and $\mathbf{ab}$ respectively. Let the two vectors be

$$\mathbf{p} = x_1\mathbf{a} + y_1\mathbf{b} + z_1\mathbf{c}$$

and
$$\mathbf{r} = x_2\mathbf{a} + y_2\mathbf{b} + z_2\mathbf{c}.$$

$$p^2 = x_1^2 a^2 + y_1^2 b^2 + z_1^2 c^2 + 2x_1 y_1 \mathbf{a} \cdot \mathbf{b} + 2y_1 z_1 \mathbf{b} \cdot \mathbf{c} + 2z_1 x_1 \mathbf{c} \cdot \mathbf{a}$$

$$= x_1^2 a^2 + y_1^2 b^2 + z_1^2 c^2 + 2x_1 y_1 ab \cos \gamma + 2y_1 z_1 bc \cos \alpha + 2z_1 x_1 ca \cos \beta. \tag{2}$$

Similarly

$$r^2 = x_2^2 a^2 + y_2^2 b^2 + z_2^2 c^2 + 2x_2 y_2 ab \cos \gamma + 2y_2 z_2 bc \cos \alpha + 2z_2 x_2 ca \cos \beta. \tag{3}$$

$$\mathbf{p} \cdot \mathbf{r} = x_1 x_2 a^2 + y_1 y_2 b^2 + z_1 z_2 c^2 + (x_1 y_2 + x_2 y_1) ab \cos \gamma$$

$$+ (y_1 z_2 + y_2 z_1) bc \cos \alpha + (z_1 x_2 + z_2 x_1) ca \cos \beta. \tag{4}$$

$$\mathbf{p} \cdot \mathbf{r} = pr \cos \psi, \quad \therefore \cos \psi = \mathbf{p} \cdot \mathbf{r}/(pr). \tag{5}$$

Substitute the values of $\mathbf{p} \cdot \mathbf{r}$, p and r from (3), (1) and (2) and evaluate $\cos \psi$.

If the axes are orthogonal, all cross products are zero and

$$\cos \psi = \frac{x_1 x_2 a^2 + y_1 y_2 b^2 + z_1 z_2 c^2}{\sqrt{(x_1^2 a^2 + y_1^2 b^2 + z_1^2 c^2)(x_2^2 a^2 + y_2^2 b^2 + z_2^2 c^2)}}. \tag{6}$$

If, in addition, $a = b = c$

$$\cos \psi = \frac{x_1 x_2 + y_1 y_2 + z_1 z_2}{\sqrt{(x_1^2 + y_1^2 + z_1^2)(x_2^2 + y_2^2 + z_2^2)}}. \tag{7}$$

The angle between the normals to two sets of planes, $h_1k_1l_1$ and $h_2k_2l_2$ can be found using the reciprocal lattice axes, $\mathbf{a}^*, \mathbf{b}^*, \mathbf{c}^*$ (Section **3.2** p. 28) and

$$\mathbf{p}' = \mathbf{d}^*_{h_1k_1l_1} = h_1\mathbf{a}^* + k_1\mathbf{b}^* + l_1\mathbf{c}^*$$

$$\mathbf{r}' = \mathbf{d}^*_{h_2k_2l_2} = h_2\mathbf{a}^* + k_2\mathbf{b}^* + l_2\mathbf{c}^*$$

in exactly the same way.

10. Vector Multiplication

$\mathbf{a} \times \mathbf{b}$ is *defined* as equal to $ab \sin \psi\, \mathbf{n}$, where $\mathbf{n}$ is a unit vector perpendicular to $\mathbf{a}$ and $\mathbf{b}$ and in the direction given by the corkscrew rule (**Example 9**, p. 289) on rotating from $\mathbf{a} \to \mathbf{b}$. $\psi(<180°)$ is the angle of rotation from $\mathbf{a}$ to $\mathbf{b}$. $\mathbf{b} \times \mathbf{a}$ *is therefore equal to* $-\mathbf{a} \times \mathbf{b}$. The order of vectors in a vector product is not commutative. The distributive law holds provided the order of the vectors is maintained, i.e. $\mathbf{a} \times (\mathbf{b} + \mathbf{c}) = \mathbf{a} \times \mathbf{b} + \mathbf{a} \times \mathbf{c}$. [16(40)]. If $\mathbf{a}$ and $\mathbf{b}$ are parallel $\mathbf{a} \times \mathbf{b} = 0$. In particular $\mathbf{a} \times \mathbf{a} = 0$. If they are perpendicular, then $|\mathbf{a} \times \mathbf{b}| = ab$. If $\mathbf{i}, \mathbf{j}, \mathbf{k}$ are a right-handed set of orthogonal unit vectors, then

$$\mathbf{i} \times \mathbf{i} = \mathbf{j} \times \mathbf{j} = \mathbf{k} \times \mathbf{k} = 0 \quad \text{and} \quad \mathbf{i} \times \mathbf{j} = \mathbf{k}; \quad \mathbf{j} \times \mathbf{k} = \mathbf{i} \quad \text{and} \quad \mathbf{k} \times \mathbf{i} = \mathbf{j}. \quad (1)$$

If $\mathbf{a}$ and $\mathbf{b}$ are expressed in terms of their components for the axes $\mathbf{i}, \mathbf{j}$ and $\mathbf{k}$, then

$$\mathbf{a} \times \mathbf{b} = (a_1\mathbf{i} + a_2\mathbf{j} + a_3\mathbf{k}) \times (b_1\mathbf{i} + b_2\mathbf{j} + b_3\mathbf{k})$$

$$= a_1b_2\mathbf{i} \times \mathbf{j} + a_1b_3\mathbf{i} \times \mathbf{k} + a_2b_1\mathbf{j} \times \mathbf{i} + a_2b_3\mathbf{j} \times \mathbf{k} + a_3b_1\mathbf{k} \times \mathbf{i}$$

$$+ a_3b_2\mathbf{k} \times \mathbf{j}$$

$$= \mathbf{i}(a_2b_3 - a_3b_2) + \mathbf{j}(a_3b_1 - a_1b_3) + \mathbf{k}(a_1b_2 - a_2b_1). \quad (2)$$

This is usefully written in the form of a determinant

$$\mathbf{a} \times \mathbf{b} = \begin{vmatrix} \mathbf{i} & \mathbf{j} & \mathbf{k} \\ a_1 & a_2 & a_3 \\ b_1 & b_2 & b_3 \end{vmatrix}. \quad (3)$$

Since $(\mathbf{a}/a)$ and $(\mathbf{b}/b)$ are unit vectors $(|\mathbf{a} \times \mathbf{b}|/ab) = \sin \psi$, and this expression can be used for finding the angle between two vectors, in a somewhat similar way to the scalar expression, 330(5) for $\cos \psi$. In particular, by squaring (2) and dividing by a^2b^2, we have

$$\sin^2 \psi = \frac{(a_2b_3 - a_3b_2)^2 + (a_3b_1 - a_1b_3)^2 + (a_1b_2 - a_2b_1)^2}{(a_1^2 + a_2^2 + a_3^2)(b_1^2 + b_2^2 + b_3^2)}. \quad (4)$$

It must be emphasised that these expressions 331(2, 3 and 4) only apply when the orthogonal axes $\mathbf{i}$, $\mathbf{j}$, $\mathbf{k}$ are being used.

The vector product is useful in finding areas and volumes. A parallelepiped, whose edges are the right-hand set of vectors $\mathbf{a}$, $\mathbf{b}$ and $\mathbf{c}$, has the area of its base given by $\mathbf{a} \times \mathbf{b} = A\mathbf{n}$, where $\mathbf{n}$ is a unit vector perpendicular to the base. $\mathbf{n} \cdot \mathbf{c}$ is the resolved part of $\mathbf{c}$ along $\mathbf{n}$ and is the height of the parallelepiped. Its volume is therefore:

$$V = A\mathbf{n} \cdot \mathbf{c} = \mathbf{a} \times \mathbf{b} \cdot \mathbf{c} = \mathbf{b} \times \mathbf{c} \cdot \mathbf{a} = \mathbf{c} \times \mathbf{a} \cdot \mathbf{b} = [\mathbf{abc}] = -[\mathbf{bac}] \quad (1)$$

where $[\mathbf{abc}]$ stands for any cyclical permutation of the vectors with the $\cdot$ and $\times$ either way round. One form of the scalar expression for V is

$$V = abc(1 + 2\cos\alpha \cos\beta \cos\gamma - \cos^2\alpha - \cos^2\beta - \cos^2\gamma)^{\frac{1}{2}}$$

where α, β and γ are the angles between $\mathbf{b}$ and $\mathbf{c}$, $\mathbf{c}$ and $\mathbf{a}$, $\mathbf{a}$ and $\mathbf{b}$. This is a striking and useful example of the simplification achieved by the use of vector algebra.

It follows that if three vectors are co-planar, $[\mathbf{abc}] = 0$ and in particular if one of the vectors is repeated, e.g. $[\mathbf{aac}]$, the triple product is zero. Conversely, if $[\mathbf{abc}] = 0$ and none of the vectors is zero, $\mathbf{a}$, $\mathbf{b}$ and $\mathbf{c}$ are co-planar.

In any mixed product the $\times$ multiplication must be performed first.

$\mathbf{a} \times \mathbf{b} \cdot \mathbf{c}$ can be found in terms of Cartesian components by multiplying the expression 331(2) for $\mathbf{a} \times \mathbf{b}$ by $\mathbf{c} = c_1\mathbf{i} + c_2\mathbf{j} + c_3\mathbf{k}$,

$$\therefore \mathbf{a} \times \mathbf{b} \cdot \mathbf{c} = c_1(a_2b_3 - a_3b_2) + c_2(a_3b_1 - a_1b_3) + c_3(a_1b_2 - a_2b_1)$$

$$= \begin{vmatrix} a_1 & a_2 & a_3 \\ b_1 & b_2 & b_3 \\ c_1 & c_2 & c_3 \end{vmatrix} \quad (2)$$

Therefore, if $\mathbf{a}$, $\mathbf{b}$ and $\mathbf{c}$ are coplanar the determinant of the Cartesian coefficients is zero and vice versa.

11. The Equation of a Plane through Three Points

The points A, B, C, are defined by position vectors $\mathbf{a}$, $\mathbf{b}$, $\mathbf{c}$. Let $\mathbf{r}$ be any point on the plane. Then the three vectors $(\mathbf{r} - \mathbf{a})$, $(\mathbf{a} - \mathbf{b})$, $(\mathbf{b} - \mathbf{c})$ are co-planar

$$\therefore (\mathbf{r} - \mathbf{a}) \cdot (\mathbf{a} - \mathbf{b}) \times (\mathbf{b} - \mathbf{c}) = 0.$$

$$\therefore (\mathbf{r} - \mathbf{a}) \cdot (\mathbf{a} \times \mathbf{b} + \mathbf{b} \times \mathbf{c} + \mathbf{c} \times \mathbf{a}) = 0.$$

$$\therefore \mathbf{r} \cdot (\mathbf{a} \times \mathbf{b} + \mathbf{b} \times \mathbf{c} + \mathbf{c} \times \mathbf{a}) = [\mathbf{abc}]. \quad (3)$$

This is the normal equation of a plane, with

$$\mathbf{n} = \mathbf{a} \times \mathbf{b} + \mathbf{b} \times \mathbf{c} + \mathbf{c} \times \mathbf{a} \quad \text{and} \quad q = [\mathbf{abc}].$$

12. Double Vector Products

To find $(\mathbf{a} \times \mathbf{b}) \times (\mathbf{c} \times \mathbf{d})$. Let $(\mathbf{a} \times \mathbf{b}) = \mathbf{r}$. $\mathbf{r} \times (\mathbf{c} \times \mathbf{d}) = l\mathbf{c} + m\mathbf{d}$ (perpendicular to the normal to the $\mathbf{cd}$ plane, $\therefore$ in the plane). Define orthogonal axes $\mathbf{i}$ and $\mathbf{j}$ in the $\mathbf{cd}$ plane, such that $\mathbf{c} = c_1\mathbf{i}$. Let

$$\mathbf{d} = d_1\mathbf{i} + d_2\mathbf{j} \text{ and } \mathbf{r} = r_1\mathbf{i} + r_2\mathbf{j} + r_3\mathbf{k}, \text{ where } \mathbf{k} = \mathbf{i} \times \mathbf{j}$$

$$\therefore \mathbf{c} \times \mathbf{d} = c_1 d_2 \mathbf{k} \text{ and } \mathbf{r} \times (\mathbf{c} \times \mathbf{d}) = (r_1\mathbf{i} + r_2\mathbf{j} + r_3\mathbf{k}) \times c_1 d_2 \mathbf{k}$$

$$= -r_1 c_1 d_2 \mathbf{j} + r_2 c_1 d_2 \mathbf{i}$$

$$= r_2 d_2 c_1 \mathbf{i} - r_1 c_1 (d_1 \mathbf{i} + d_2 \mathbf{j}) + r_1 d_1 c_1 \mathbf{i}$$

$$= r_2 d_2 \mathbf{c} - r_1 c_1 \mathbf{d} + r_1 d_1 \mathbf{c}$$

$$= \mathbf{c}(r_1 d_1 + r_2 d_2) - r_1 c_1 \mathbf{d}.$$

$$\therefore \mathbf{r} \times (\mathbf{c} \times \mathbf{d}) = \mathbf{c}(\mathbf{r} \cdot \mathbf{d}) - \mathbf{d}(\mathbf{r} \cdot \mathbf{c}). \tag{1}$$

Substitute $\mathbf{a} \times \mathbf{b} = \mathbf{r}$

$$\begin{aligned}
(\mathbf{a} \times \mathbf{b}) \times (\mathbf{c} \times \mathbf{d}) &= \mathbf{c}(\mathbf{a} \times \mathbf{b} \cdot \mathbf{d}) - \mathbf{d}(\mathbf{a} \times \mathbf{b} \cdot \mathbf{c}) \\
&= (\mathbf{d} \times \mathbf{c}) \times (\mathbf{a} \times \mathbf{b}) = \mathbf{a}(\mathbf{d} \times \mathbf{c} \cdot \mathbf{b}) - \mathbf{b}(\mathbf{d} \times \mathbf{c} \cdot \mathbf{a})
\end{aligned} \tag{2}$$

To find $\mathbf{a} \times \mathbf{b} \cdot \mathbf{c} \times \mathbf{d}$.

From 332 (1)

$$\mathbf{a} \times \mathbf{b} \cdot \mathbf{c} \times \mathbf{d} = \mathbf{a} \cdot \mathbf{b} \times (\mathbf{c} \times \mathbf{d}).$$

From (1)

$$\mathbf{a} \cdot \mathbf{b} \times (\mathbf{c} \times \mathbf{d}) = \mathbf{a} \cdot (\mathbf{b} \cdot \mathbf{d} \, \mathbf{c} - \mathbf{b} \cdot \mathbf{c} \, \mathbf{d}) = \mathbf{b} \cdot \mathbf{d} \, \mathbf{a} \cdot \mathbf{c} - \mathbf{b} \cdot \mathbf{c} \, \mathbf{a} \cdot \mathbf{d}.$$

$$\therefore \quad \mathbf{a} \times \mathbf{b} \cdot \mathbf{c} \times \mathbf{d} = \mathbf{a} \cdot \mathbf{c} \, \mathbf{b} \cdot \mathbf{d} - \mathbf{a} \cdot \mathbf{d} \, \mathbf{b} \cdot \mathbf{c}. \tag{3}$$

13. Conversion to Cartesian Coordinate Axes

To convert from fractional coordinates in terms of general (right-hand) axes $\mathbf{a}$, $\mathbf{b}$ and $\mathbf{c}$ ($a \neq b \neq c$) $\widehat{\mathbf{a}\,\mathbf{b}} = \gamma$, $\widehat{\mathbf{b}\,\mathbf{c}} = \alpha$, $\widehat{\mathbf{c}\,\mathbf{a}} = \beta$, to coordinates in terms of orthogonal unit vectors with $\mathbf{i}$ in the direction of $\mathbf{a}$, $\mathbf{j}$ in the $\mathbf{ab}$ plane with $\mathbf{j} \cdot \mathbf{b}$ positive and $\mathbf{k} = \mathbf{i} \times \mathbf{j}$. If $\mathbf{a}$, $\mathbf{b}$ and $\mathbf{c}$ are right handed, $\mathbf{k} \cdot \mathbf{c}$ (or $[\mathbf{abc}]$) will be positive.

Let $r = x\mathbf{a} + y\mathbf{b} + z\mathbf{c} = X\mathbf{i} + Y\mathbf{j} + Z\mathbf{k}$. To find X, Y and Z in terms of x, y and z,

$$\mathbf{a} = a\mathbf{i}.$$

$$\mathbf{b} = (\mathbf{b}\cdot\mathbf{i})\,\mathbf{i} + (\mathbf{b}\cdot\mathbf{j})\,\mathbf{j} \quad (327\,(1)).$$

$$\therefore\ \mathbf{b} = b\cos\gamma\mathbf{i} + b\sin\gamma\mathbf{j}.$$

$$\mathbf{c} = (\mathbf{c}\cdot\mathbf{i})\,\mathbf{i} + (\mathbf{c}\cdot\mathbf{j})\,\mathbf{j} + (\mathbf{c}\cdot\mathbf{k})\,\mathbf{k}; \quad \mathbf{c}\cdot\mathbf{i} = c\cos\beta.$$

$$\therefore\ \mathbf{c}\cdot\mathbf{b} = bc\cos\beta\cos\gamma + (\mathbf{c}\cdot\mathbf{j})\,b\sin\gamma = bc\cos\alpha$$

$$\therefore\ \mathbf{c}\cdot\mathbf{j} = \frac{c(\cos\alpha - \cos\beta\cos\gamma)}{\sin\gamma} = cA.$$

$$\mathbf{c}\cdot\mathbf{c} = c^2 = c^2\cos^2\beta + (cA)^2 + (\mathbf{c}.\mathbf{k})^2.$$

$$\therefore\ (\mathbf{c}\cdot\mathbf{k})^2 = c^2 - c^2\cos^2\beta - c^2\,\frac{(\cos^2\alpha + \cos^2\beta\cos^2\gamma - 2\cos\alpha\cos\beta\cos\gamma)}{\sin^2\gamma}$$

$$= \frac{c^2}{\sin^2\gamma}\,(\sin^2\gamma - \cos^2\beta\sin^2\gamma - \cos^2\alpha - \cos^2\beta\cos^2\gamma$$
$$+ 2\cos\alpha\cos\beta\cos\gamma)$$

$$= \frac{c^2}{\sin^2\gamma}\,(\sin^2\alpha + \sin^2\beta + \sin^2\gamma + 2\cos\alpha\cos\beta\cos\gamma - 2)$$

$$= c^2 B^2.$$

$$\therefore\ x\mathbf{a} + y\mathbf{b} + z\mathbf{c} = xa\mathbf{i} + yb(\cos\gamma\mathbf{i} + \sin\gamma\mathbf{j}) + zc\cos\beta\mathbf{i} + zcA\mathbf{j} + zcB\mathbf{k}$$

$$= \mathbf{i}(xa + yb\cos\gamma + zc\cos\beta)$$
$$+ \mathbf{j}(yb\sin\gamma + zcA)$$
$$+ \mathbf{k}(zcB).$$

$$\therefore\ X = xa + yb\cos\gamma + zc\cos\beta$$
$$Y = yb\sin\gamma + zcA$$
$$Z = zcB$$
$$A = \frac{\cos\alpha - \cos\beta\cos\gamma}{\sin\gamma}$$
$$B = \frac{(\sin^2\alpha + \sin^2\beta + \sin^2\gamma + 2\cos\alpha\cos\beta\cos\gamma - 2)^{\frac{1}{2}}}{\sin\gamma}.$$

$$(1)$$

14. Volume of the Parallelepiped Defined by a, b, c

$\mathbf{a} = a\mathbf{i}; \ \mathbf{b} = b\cos\gamma\mathbf{i} + b\sin\gamma\mathbf{j}; \ \mathbf{c} = c\cos\beta\mathbf{i} + cA\mathbf{j} + cB\mathbf{k}.$

$$
\begin{aligned}
V = \mathbf{a}\cdot\mathbf{b}\times\mathbf{c} = \mathbf{a}\times\mathbf{b}\cdot\mathbf{c} &= a\mathbf{i}\times(b\cos\gamma\mathbf{i}\sin + b\sin\gamma\mathbf{j})\cdot\mathbf{c} \\
&= ab\sin\gamma\mathbf{k}\cdot(c\cos\beta\mathbf{i} + cA\mathbf{j} + cB\mathbf{k}) \\
&= abc\sin\gamma B \\
&= abc(\sin^2\alpha + \sin^2\beta + \sin^2\gamma + 2\cos\alpha\cos\beta\cos\gamma - 2)^{\frac{1}{2}}.
\end{aligned} \tag{1}
$$

15. Vectors on the Argand Diagram

The amplitude and relative phase of a wave front in diffraction theory can be represented by a vector in the $\mathbf{i}\,\mathbf{j}$ plane, whose anti-clockwise angle to a fixed zero line, OX (in the direction of unit vector $\mathbf{i}$), is the phase angle of the

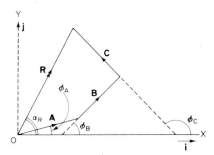

FIG. 198. The vector polygon on the two-dimensional Argand diagram.

wave front (Section 1.3.1 p. 10). The resultant amplitude and phase of a combination of wave fronts, arriving at a point by different paths, is obtained by finding the vector sum of all the individual amplitude vectors. Let the (two-dimensional) amplitude vectors $\mathbf{A}, \mathbf{B}, \mathbf{C}\ldots$ (Fig. 198) make angles ϕ_A, ϕ_B, $\phi_C\ldots$ with the OX axis defined by $\mathbf{i}$. Then $\mathbf{R} = \mathbf{A} + \mathbf{B} + \mathbf{C} + \ldots$ and the component of $\mathbf{R}$ along OX equals $\mathbf{R}\cdot\mathbf{i} = X_R = \mathbf{A}\cdot\mathbf{i} + \mathbf{B}\cdot\mathbf{i} + \mathbf{C}\cdot\mathbf{i} + \ldots =$ the sum of the components of $\mathbf{A}, \mathbf{B}, \mathbf{C}\ldots$ along OX.

$$
\left.
\begin{aligned}
&\therefore \ X_R = A\cos\phi_A + B\cos\phi_B + C\cos\phi_C + \ldots . \\
&\text{Similarly the component of } \mathbf{R} \text{ perpendicular to } OX \text{ (in the direction defined by } \mathbf{j}) \\
&= Y_R = \mathbf{R}\cdot\mathbf{j} \\
&= A\sin\phi_A + B\sin\phi_B + C\sin\phi_C\ldots \\
&R = (X_R^2 + Y_R^2)^{\frac{1}{2}} \text{ and the phase angle } \alpha_R = \tan^{-1}\frac{Y_R}{X_R}
\end{aligned}
\right\} \tag{2}
$$

16. Intensity for Constructive Interference and for Scattering when Phase Relationships are Random

For scattering from n atoms all of the same kind, with atomic scattering factor f, 17(2), we have

$$I = R^2 = a^2 \left[\left(\sum_n f \cos \phi_n \right)^2 + \left(\sum_n f \sin \phi_n \right)^2 \right]$$

$$= \left[a^2 \sum_n f^2 \cos^2 \phi_n + \sum_n f^2 \sin^2 \phi_n + \sum_{n \neq m} f^2 \cos \phi_n \cos \phi_m \right.$$

$$\left. + \sum_{n \neq m} f^2 \sin \phi_n \sin \phi_m \right]$$

$$= a^2 \left[f^2 \sum_n (\cos^2 \phi_n + \sin^2 \phi_n) + f^2 \sum_{n \neq m} (\cos \phi_n \cos \phi_m \right.$$

$$\left. + \sin \phi_n \sin \phi_m) \right]$$

$$= a^2 \left[nf^2 + f^2 \sum_{n \neq m} \cos(\phi_n - \phi_m) \right].$$

For constructive interference

$$\phi_n - \phi_m = 2p\pi \quad \therefore \quad I = a^2 \left[nf^2 + (n^2 - n) f^2 \right] = (naf)^2 \qquad (1)$$

If the $(\phi_n - \phi_m)$ are random, $I = n(af)^2$,

$$\text{since } \Sigma \cos (\phi_n - \phi_m) \text{ tends to zero.} \qquad (2)$$

If the f's are not the same (1) becomes

$$I = a^2 \left(\sum_m nf_m \right)^2 \qquad (3)$$

(2) becomes

$$I = a^2 \left(\sum_m nf_m{}^2 \right). \qquad (4)$$

In other words, for constructive interference, amplitudes are added; for random phase relationships, intensities are added.

If the scattering units are not spherically symmetrical, e.g. 'molecules' instead of atoms, they must all be parallel in the first case (as in a crystal) and parallel or randomly oriented in the second. For random orientation the 'scattering factor' corresponding to f will be that calculated for a spherically 'averaged' molecule.

17. The Use of the Operator $i = \sqrt{-1}$

Since multiplying a vector by -1 turns the vector through 180°, the operation of turning the vector through 90° can be represented by multiplication with $i = \sqrt{-1}$. A second multiplication by i completes the rotation of 180° which corresponds to multiplying by $i^2 = -1$. The operator -1 can be used three dimensionally, but for rotation through 90° one must define the plane in which rotation takes place and the sense of rotation. The operator i can therefore only be used in two-dimensional geometry. The plane used can be defined as the XY plane, with OX to the right and OY upwards, as for normal 2-dimensional Cartesian axes. The sense of rotation is anticlockwise, from OX to OY. 'Real' numbers represent vectors in the OX direction; 'pure imaginary' numbers, i.e. real numbers multiplied by i, represent vectors in the OY direction. Mixed numbers represent vectors making some general angle ϕ with the OX axis.

Thus any vector in the plane can be represented by the sum of the magnitudes of its components in the OX and OY directions, if the second component is multiplied by i. $\mathbf{A} = X + iY$ is a vector on the Argand diagram with components X and Y, making an angle $\phi = \tan^{-1}(Y/X)$ with the OX axis. Since $X = A\cos\phi$ and $Y = A\sin\phi$, $\mathbf{A} = A(\cos\phi + i\sin\phi)$.

It is possible to sum the Argand diagram in terms of the components of the amplitude vectors as in I(15) (p. 335), but it is far more convenient for many purposes to use the operator i, especially in the exponent of an exponential expression since this effectively converts two-dimensional vector algebra into normal scalar algebra.

Defining $e^{i\phi}$ in terms of the expansion of e^x, it is shown in any standard textbook that the exponential with imaginary exponent obeys the same rules as for a real exponent. In addition, from the expansions of $\cos x$ and $\sin x$ in powers of x, it can be shown that $e^{i\phi} = \cos\phi + i\sin\phi$. A vector $\mathbf{A}$ on the Argand diagram, with phase angle ϕ (i.e. making an angle ϕ with the X-axis) has components $A\cos\phi$ and $A\sin\phi$.

$$\therefore \mathbf{A} = A\cos\phi + iA\sin\phi = Ae^{i\phi}. \tag{1}$$

It is thus possible to represent a vector on the Argand diagram by a single term, and the sum of two vectors,

$$\mathbf{A} + \mathbf{B} = (A\cos\phi_1 + B\cos\phi_2) + i(A\sin\phi_1 + B\sin\phi_2)$$

$$= Ae^{i\phi_1} + Be^{i\phi_2}. \tag{2}$$

If we have a vector $\mathbf{A} = A(\cos\psi + i\sin\psi) = Ae^{i\psi}$ and multiply it by $e^{i\phi} = \cos\phi + i\sin\phi$, we obtain

$$Ae^{i\phi} = Ae^{i\psi}e^{i\phi} = Ae^{i(\psi+\phi)}.$$

Therefore multiplying by $e^{i\phi}$ turns a vector through the angle ϕ. (Multiplication of the two component forms shows that $(\cos\psi + i\sin\psi)(\cos\phi + i\sin\phi) = \cos(\psi + \phi) + i\sin(\psi + \phi)$). Finally, multiplying a vector by its complex conjugate (the image of the vector in the X-axis, with the same X-components, but all Y-components, i.e. the i terms, of opposite sign), gives the square of the length of the vector,

$$\mathbf{A}\mathbf{A}^* = Ae^{i\phi} \times Ae^{-i\phi} = A^2e^0 = A^2. \tag{1}$$

(This is also easily proved in terms of the components).

(Note that if ϕ is a linear function of time, $Ae^{i\phi}$ represents a uniformly rotating vector; if it is a linear function of distance from an origin it will repeat regularly as the distance from the origin is increased (and can therefor be used to represent a Fourier term (Section 9.1.4.3, p. 105, and Appendix III, p. 372)). It can also be used to represent the electric field of an E.M. wave varying both with time and distance, i.e. the analytical equivalent of Fig. 6 (p. 10).

18. The Determination of the Coefficients of a Quadratic Equation after a Change of Axes

In Appendix VI (p. 390) the formula $B' = L^*BL$ is given with a reference to [18(135)]. Unfortunately the derivation given in that reference is of doubtful validity. The form actually derived is incorrect for axes which are not orthonormal, although the final form, obtained by a substitution which is only valid for orthonormal axes, does have general application! The following derivation, in terms of the multiplication of square matrices, and various other results derived generally in [18], is not restricted in any way.

The equation of a quadric surface in terms of general axes $\mathbf{e}_1$, $\mathbf{e}_2$, $\mathbf{e}_3$ but with origin at the centre, has coefficients B_{11}, B_{22}, B_{33}, $2B_{12}$, $2B_{23}$, $2B_{31}$, which can be represented by a symmetric matrix (i.e. $B_{ij} = B_{ji}$):

$$B = \begin{bmatrix} B_{11} & B_{12} & B_{13} \\ B_{21} & B_{22} & B_{23} \\ B_{31} & B_{32} & B_{33} \end{bmatrix}$$

Define the matrix

$$X^* = \begin{bmatrix} x & y & z \\ 0 & 0 & 0 \\ 0 & 0 & 0 \end{bmatrix}$$

X, the transpose of X^*, is given by:

$$X = \begin{bmatrix} x & 0 & 0 \\ y & 0 & 0 \\ z & 0 & 0 \end{bmatrix}$$

X and X^* are equivalent to column and row matrices respectively. Simple multiplication of matrices shows that the quadratic form of the equation is given by the single non-zero term of the matrix:

$$X^*BX, \tag{1}$$

i.e. $B_{11}x^2 + B_{22}y^2 + B_{33}z^2 + 2B_{12}xy + 2B_{23}yz + 2B_{31}zx$.

The axes are changed (without change of origin) to $\tilde{e}_1$, $\tilde{e}_2$, $\tilde{e}_3$, where

$$\tilde{e}_1 = l_{11}e_1 + l_{21}e_2 + l_{31}e_3$$
$$\tilde{e}_2 = l_{12}e_1 + l_{22}e_2 + l_{32}e_3$$
$$\tilde{e}_3 = l_{13}e_1 + l_{23}e_2 + l_{33}e_3$$

with transformation matrix L^*.

Then the old coordinates x, y, z are obtained from the new ones $\tilde{x}$, $\tilde{y}$, $\tilde{z}$, by the matrix L, the transpose of L^*.

In matrix symbols:

$$X = L\tilde{X}. \tag{2}$$

Taking the transpose of both sides of (2)

$$X^* = (L\tilde{X})^* = \tilde{X}^*L^* \quad [18(79)]. \tag{3}$$

Substituting for X and X^* from (2) and (3) in (1) gives the matrix of the quadratic form in terms of the new coordinates as:

$$\tilde{X}^*L^*BL\tilde{X}. \tag{4}$$

This again is a matrix with a single non-zero term, since $\tilde{X}$ and $\tilde{X}^*$ are also equivalent to column and row matrices respectively. Therefore the new matrix of coefficients $\tilde{B}$, for the new coordinates, i.e. the matrix of the new quadratic form $\tilde{X}^*\tilde{B}\tilde{X}$, is obtained by comparison of (1) and (4) to give

$$\tilde{B} = L^*BL. \tag{5}$$

Since $X^*\tilde{B}X = \tilde{X}^*\tilde{B}\tilde{X}$, the two non-zero terms must be equal and therefore both equal to H, the constant term of the quadratic equation.

Take the transpose of both sides of (5).

$$\tilde{B}^* = (L^*BL)^* = (BL)^*L = L^*B^*L = L^*BL = \tilde{B}.$$

since $B^* = B$ for a symmetric matrix.

Since $\tilde{B}^* = \tilde{B}$, $\tilde{B}$ is also a symmetric matrix.

Lattice Theory

1. The Properties of a Point Lattice and of Lattice Planes

We define a lattice as an infinite set of points whose position vectors, from one of the points as origin, are given by $\mathbf{g} = p\mathbf{a} + q\mathbf{b} + r\mathbf{c}$ where p, q and r are all the $+ve$ and $-ve$ integers, and $\mathbf{a}$, $\mathbf{b}$ and $\mathbf{c}$ are the basis vectors of the lattice. Any other point $P\mathbf{a} + Q\mathbf{b} + R\mathbf{c}$ can be chosen as origin, and then the position vectors from the new origin will be

$$(p - P)\mathbf{a} + (q - Q)\mathbf{b} + (r - R)\mathbf{c} \equiv p'\mathbf{a} + q'\mathbf{b} + r'\mathbf{c},$$

where $p'q'r'$ are all the $+ve$ and $-ve$ integers. *Therefore any point has the same surroundings as any other point.*

The lattice can be built up as a line of points given by $\mathbf{g} = p\mathbf{a}$ and parallel lines of points given $\mathbf{g} = \mathbf{b} + p\mathbf{a}$; $2\mathbf{b} + p\mathbf{a}$; $...q\mathbf{b} + p\mathbf{a}$ to form a two-dimensional net. Parallel $2 - D$ nets given by $\mathbf{g} = \mathbf{c} + p\mathbf{a} + q\mathbf{b}$; $2\mathbf{c} + p\mathbf{a} + q\mathbf{b}$; $... r\mathbf{c} + p\mathbf{a} + q\mathbf{b}$ build up the $3 - D$ lattice. The lines and nets are obviously equally spaced, even though $\mathbf{a}$, $\mathbf{b}$ and $\mathbf{c}$ are non-orthogonal.

Any of an infinite number of position vectors $U\mathbf{a} + V\mathbf{b} + W\mathbf{c}$ defining a point A can be chosen as the first vector $\mathbf{A}$ of a new basis, providing the integers U, V and W have no common factor, so that A is the nearest point to the origin in the direction $\mathbf{A}$. A second vector $\mathbf{B}' = U'\mathbf{a} + V'\mathbf{b} + W'\mathbf{c}$ defines a lattice plane; and if the parallelogram defined by $\mathbf{A}$ and $\mathbf{B}'$ contains lattice points, another vector $\mathbf{B}$ which defines a primitive parallelogram can be chosen in that plane.† The lines defined by $\mathbf{g}' = p\mathbf{A}$; $\mathbf{B} + p\mathbf{A}$; $...q\mathbf{B} + p\mathbf{A}$ are equally spaced in the $\mathbf{AB}$ plane and include all the lattice points in the plane, since if there were any points not included they would be inside a parallelogram with sides parallel to $\mathbf{A}$ and $\mathbf{B}$. The points of the corners of the parallelogram would not then have the same surroundings as those of the origin parallelogram. Any third non-coplanar vector $\mathbf{C} = U''\mathbf{a} + V''\mathbf{b} + W''\mathbf{c}$, which defines a primitive parallelepiped‡ (unit cell), gives a series of

---

† Specifically, if the line through the origin and A is moved perpendicular to its length in the plane until it coincides with the nearest parallel line of lattice points, a vector $\mathbf{B}$ to any of these points will give a primitive parallelogram.

‡ Similarly, if the $\mathbf{AB}$ plane is moved parallel to itself until it coincides with the nearest plane of points, a vector $\mathbf{C}$ to any of these points will give a primitive unit cell.

equally spaced layers of points making up the $3-D$ lattice. These layers must all be similar, and no points can lie between them since otherwise there would be points of the lattice with different surroundings.

There are thus an infinite number of ways in which the $3-D$ lattice can be described in terms of equally spaced lattice planes. Any such plane can be defined by 3 non-collinear lattice points. Planes of the set will pass through all the lattice points, in particular through the lattice point at the origin and through the lattice point at **a**. Let there be $(H-1)$ equally spaced lattice planes between these two. Then the intercept cut off by the nearest plane to the origin is a/H. Similarly the intercepts on **b** and **c** will be b/K, c/L where H, K and L are integers. These are intercepts cut by the nearest plane to the origin of the set. If the plane cuts the axis between the origin and $-\mathbf{b}$, then K will be a negative integer, and similarly for L. If one chooses to start with the plane going through $-\mathbf{a}$, then H will be a negative integer.

Reference to Section 3.3 (p. 30) shows that $\mathbf{d}^*_{HKL} = H\mathbf{a}^* + K\mathbf{b}^* + L\mathbf{c}^*$ is normal to the plane cutting off intercepts a/H, b/K, c/L on the crystal lattice axes, and that $d^*_{HKL} = \lambda/d_{HKL}$ where d_{HKL} is the perpendicular distance from the origin to the plane.

Project any crystal lattice vector $\mathbf{g}_{pqr}$ on to the unit vector $\mathbf{d}^*_{HKL}/d^*_{HKL}$.

$$\frac{\mathbf{d}^*_{HKL} \cdot \mathbf{g}_{pqr}}{d^*_{HKL}} = \frac{d_{HKL}}{\lambda}(H\mathbf{a}^* + K\mathbf{b}^* + L\mathbf{c}^*) \cdot (p\mathbf{a} + q\mathbf{b} + r\mathbf{c})$$

$$= \frac{d_{HKL}}{\lambda}(Hp + Kq + Lr)\lambda$$

$$= d_{HKL}(Hp + Kq + Lr).$$

For all position vectors $\mathbf{g}_{pqr}$ which define points lying on the nearest plane to the origin of the set HKL, the projection is $\pm d_{HKL}$, i.e. for these points

$$Hp + Kq + Lr = \pm 1.$$

Suppose H, K and L have a common factor m, so that $H = mh$, $K = mk$, $L = ml$.

$$\therefore \quad Hp + Kq + Lr = m(hp + kq + lr) = \pm 1.$$

Since the smallest absolute value which the sum of products of integers can have (except zero) is 1, m must also equal 1 if HK and L define the intercepts of the nearest plane of lattice points to the origin. Thus a set of lattice planes can only be described by a triplet of integers with no common factor. A triplet of integers with a common factor m describes a set of equidistant planes, only $1/m$ of which contain lattice points and whose spacing $d_{hm, km, lm}$ is $(1/m) \times d_{hkl}$, i.e. $1/m$, of the spacing of the lattice planes.

2. Systematic Absences due to Centred Cells and Translation Symmetry Elements

2.1 CENTRED CELLS

The symmetry of a crystal lattice is that of the highest class (or point group) of the system to which the crystal belongs. However, in order to achieve this symmetry in the unit cell it is frequently necessary to choose axes which define a centred cell. There are 7 different arrangements of centring (including no-centring or Primitive unit cells and the doubly centred hexagonal cell) which may be required to satisfy the lattice symmetry [4(221)] and obtain convenient axes. Cell centring causes systematic absences of reflections which would be present in the case of a primitive (P) unit cell (Section 10.1, p. 127). The derivation of the relations between the indices of the systematically absent reflections for the various types of centring can be simply achieved with the aid of the RL.

2.2 SYSTEMATIC ABSENCES DUE TO CENTRED CELLS

The crystal can be treated as a lattice of point scatterers, each point having the same amplitude of scattering (Section 1.4.2, p. 16).

(a) Face centred cells. In the case of C centring, as well as the origin lattice point, there is an exactly equivalent point at $\mathbf{r} = \frac{1}{2}(\mathbf{a} + \mathbf{b})$. By projecting $\mathbf{r}$ on to the unit vector $(\mathbf{d}_{hkl}^*/d_{hkl}^*)$,

$$\frac{\mathbf{r} \cdot \mathbf{d}_{hkl}^*}{d_{hkl}^*} = \frac{\frac{1}{2}(h + k)\,\lambda}{d_{hkl}^*} = \frac{h + k}{2} \cdot d_{hkl},$$

we find that the centring point is either on a lattice plane if $h + k$ is even, or half way between planes if $h + k$ is odd. In the latter case its (equal) contribution to the reflection hkl will be exactly out of phase with that from the origin point, and the resultant will be zero. Reflections will therefore be systematically absent for $h + k$ odd. Similarly, for A and B face centring (which can become C by rearranging the axes) reflections will be absent for $k + l$ odd and $h + l$ odd respectively.

(b) Body centred cells (I). $\mathbf{r} = \frac{1}{2}(\mathbf{a} + \mathbf{b} + \mathbf{c})$ for the body centring point, and the projected distance from the origin is

$$\frac{h + k + l}{2} \cdot d_{hkl}.$$

Therefore reflections are absent for $h + k + l$ odd.

(c) All face centred cells (F). Here there are 3 points additional to the origin point in the unit cell, at

$$\mathbf{r}_1 = \frac{\mathbf{a} + \mathbf{b}}{2}\; ;\; \mathbf{r}_2 = \frac{\mathbf{b} + \mathbf{c}}{2}\; ;\; \mathbf{r}_3 = \frac{\mathbf{c} + \mathbf{a}}{2}.$$

In this case it is necessary to sum the Argand diagram vectors (I(15), p. 335) for the four points. Since a uniformly weighted point lattice is necessarily centrosymmetrical, the resultant is given by $F_{hkl} = \Sigma F'_{hkl} \cos \phi$ (Answer 14 (iv), p. 475), where $\phi = (2\pi/\lambda)\,\mathbf{r}\cdot\mathbf{d}^*_{hkl}$ (99(1)) and F'_{hkl} is the scattering amplitude from a single lattice point.

$$\therefore\ \phi_0 = 0; \qquad \phi_1 = 2\pi\frac{h+k}{2}; \qquad \phi_2 = 2\pi\frac{k+l}{2}; \qquad \phi_3 = 2\pi\frac{l+h}{2}.$$

$$\therefore\ F_{hkl} = F'_{hkl}\left(\cos 0 + \cos 2\pi\frac{h+k}{2} + \cos 2\pi\frac{k+l}{2} + \cos 2\pi\frac{l+h}{2}\right).$$

Since in this case ϕ can only be 0 or π, F_{hkl} can only be zero if one of ϕ_1, ϕ_2, $\phi_3 = 0$ and the other two $= \pi$. Therefore at least one index must be odd and one even to get the sum odd. The third can be either odd or even, since two of the pairs will be odd and the third even in either case. Therefore if there are mixed even and odd indices the corresponding reflection will be absent. If the indices are all odd or all even, the sums in pairs will all be even and $F_{hkl} = 4F'_{hkl}$, the maximum possible resultant.

Therefore for an F-centred lattice, reflections are only present if the indices are all even or all odd.

(d) For an R-centred hexagonal cell (R_{HEX}—Fig. 130(a) p. 207). the centring points (for the obverse setting) are at:

$$\mathbf{r}_1 = \tfrac{2}{3}\mathbf{a} + \tfrac{1}{3}\mathbf{b} + \tfrac{1}{3}\mathbf{c}; \qquad \mathbf{r}_2 = \tfrac{1}{3}\mathbf{a} + \tfrac{2}{3}\mathbf{b} + \tfrac{2}{3}\mathbf{c}.$$

$$\therefore\ F_{hkl} = F'_{hkl}\left(\cos 0 + \cos 2\pi(\tfrac{2}{3}h + \tfrac{1}{3}k + \tfrac{1}{3}l) + \cos 2\pi(\tfrac{1}{3}h + \tfrac{2}{3}k + \tfrac{2}{3}l)\right).$$

Since 3 equal vectors on the Argand diagram can only have zero resultant if they form a closed equilateral triangle, either below or above the OX axis, we have:

$$\tfrac{2}{3}h + \tfrac{1}{3}k + \tfrac{1}{3}l = \tfrac{1}{3} + m \qquad \tfrac{2}{3}h + \tfrac{1}{3}k + \tfrac{1}{3}l = -\tfrac{1}{3} + m$$

or

$$\tfrac{1}{3}h + \tfrac{2}{3}k + \tfrac{2}{3}l = \tfrac{2}{3} + m' \qquad \tfrac{1}{3}h + \tfrac{2}{3}k + \tfrac{2}{3}l = -\tfrac{2}{3} + m'.$$

Multiply by 3 and subtract

$$-h + k + l = 1 + 3(m' - m) \text{ or } -h + k + l = -1 + 3(m' - m).$$

Therefore the resultant will be zero if

$$-h + k + l = 3n \pm 1.$$

If $-h + k + l = 3n$, add $3h$ to each side, and divide by 3:

$$2h + k + l = 3(n + h),$$
$$\tfrac{2}{3}h + \tfrac{1}{3}k + \tfrac{1}{3}l = n + h.$$

For the point at $\mathbf{r}_1$ the phase angle equals $(n + h)\,2\pi \equiv 0$. Similarly, by multiplying both sides by -1 and adding $3(k + l)$ to each side:

$$h + 2k + 2l = 3(k + l - n)$$

and the third phase angle is also zero, so that the resultant, $F_{hkl} = 3F'_{hkl}$. Therefore reflection only occurs in an R-centred hexagonal lattice when $-h + k + l = 3n$, for the obverse setting. For the reverse setting

$$\mathbf{r}_1 = \tfrac{2}{3}\mathbf{a} + \tfrac{1}{3}\mathbf{b} + \tfrac{2}{3}\mathbf{c} \quad \text{and} \quad \mathbf{r}_2 = \tfrac{1}{3}\mathbf{a} + \tfrac{2}{3}\mathbf{b} + \tfrac{1}{3}\mathbf{c}$$

and a similar analysis shows that reflection only occurs for $h - k + l = 3n$. It is thus possible, from the systematic absences, to detect not only the presence of a rhombohedral lattice referred to hexagonal axes, but also which of the two possible orientations of the rhombohedron relative to the hexagonal axes, has been chosen.

2.3 ABSENCES DUE TO TRANSLATION SYMMETRY ELEMENTS

Space group symmetry elements may include a translation of $(1/n\text{th})$ of the distance between lattice points in the direction of translation, where n is 2, 3, 4 or 6. [4(228)].

In Section 9.1.4 (p. 99) it is shown that only the position of the atom, in the direction of the normal to the set of planes giving rise to a reflection, affects the intensity of the reflection. It is therefore possible to project the atoms on to the normal to the planes or on to any plane containing that normal, i.e. on to $\mathbf{d}^*_{hkl}$ or a plane containing it, in order to simplify consideration of the effect of translation symmetry by reducing the problem to 1 or 2 dimensions.

(a) *Screw Axes*

For screw axes the symbol is T_m, which means that the structure does not simply repeat after a rotation of $2\pi/T$, but only after a rotation of $2\pi/T$ followed by a translation of m/T of the distance between lattice points in the direction of the rotation axis. The possible values of T_m are:

$$2_1; \quad 3_1(3_2); \quad 4_1(4_3), 4_2; \quad 6_1(6_5), 6_2(6_4), 6_3.$$

The symbols in brackets are left handed versions of the right hand ones preceding them.

In the case of screw axes the problem can be reduced to one dimension, since only in a projection on the axis will the translation effect result in an alteration in the periodicity. In any other direction the differing positions of the atoms in the planes perpendicular to the rotation axis, whether related by symmetry or not, will destroy the effect of the translation in the axial direction. This can be shown analytically, and the systematic absences found for a 2_1 axis as follows.

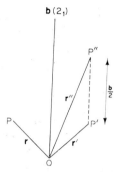

FIG. 199. The 2_1 symmetry axis along $\mathbf{b}$ takes P to P' by a rotation of π and then to P'' by a translation of $\mathbf{b}/2$.

In Fig. 199 O is an origin on the $2_1(\mathbf{b})$ axis which has $\mathbf{a}$ and $\mathbf{c}$ necessarily perpendicular to it. Let $\mathbf{r} = x\mathbf{a} + y\mathbf{b} + z\mathbf{c}$ be the position vector of an atom P (taken for simplicity in the plane of the paper). The position P', related by a rotation of π, is given by $\mathbf{r}' = -x\mathbf{a} + y\mathbf{b} - z\mathbf{c}$; and after a translation of $(\mathbf{b}/2)$ the position P'', related by 2_1 to P, is given by $\mathbf{r}'' = -x\mathbf{a} + (y+\tfrac{1}{2})\,\mathbf{b} - z\mathbf{c}$. Consider any reflection hkl, and project $\mathbf{r}$ and $\mathbf{r}''$ on to the unit vector $(\mathbf{d}_{hkl}^{*}/d_{hkl}^{*})$. The distance between the projected points is:

$$\frac{\mathbf{d}_{hkl}^{*}}{d_{hkl}^{*}} \cdot (\mathbf{r}'' - \mathbf{r}) = \mathbf{d}_{hkl}^{*} \cdot (-2x\mathbf{a} + \tfrac{1}{2}\mathbf{b} - 2z\mathbf{c}) \frac{d_{hkl}}{\lambda}$$

$$= \lambda\!\left(-2hx + \frac{k}{2} - 2lz\right)\frac{d_{hkl}}{\lambda}$$

$$= \left(-2hx + \frac{k}{2} - 2lz\right)d_{hkl}.$$

Since positional relations with other atoms are not fixed by symmetry, if the resultant amplitude for the reflection hkl is to be zero, the contribution of each pair related by the 2_1 axis must be zero. Their phases must therefore differ by π; or, since d_{hkl} corresponds to a phase difference of 2π, the distance between them in the direction of $\mathbf{d}_{hkl}^*$ must be $(n + \tfrac{1}{2}) d_{hkl}$.

$$\therefore \quad -2hx + k/2 - 2lz = (n + \tfrac{1}{2})$$

for all possible values of x and z. This is clearly only possible if $h = l = 0$ and $k = 2n + 1$. The effect is therefore restricted to reflections $0k0$, for which $\mathbf{d}_{hkl}^*$ lies along the 2_1 axis, and these have zero intensity, i.e. are absent, for k odd.

It is clear that similar considerations will limit the effects of other screw axes to reflections from planes normal to the axes, i.e. to $00l$ reflections, since for 3, 4 and 6, the axis of symmetry is also the $\mathbf{c}$ axis of the lattice. Consider the 6_1 axis and take the atoms in the bottom $1/6$ of the unit cell. After a rotation of $2\pi/6$ they are translated upwards to the second $1/6$ of the cell. When both sections are projected on to the 6_1 axis they are indistinguishable and so are the upper 4 sections; so that for these $00l$ reflections the periodicity is $1/6$ of the non-projected periodicity in this direction, which is $\mathbf{c}$. The effective lattice spacing in the direction of $\mathbf{d}_{00l}^*$ is therefore only $1/6$ of the spacing on which the RL is based; the RL vector will consequently be 6 times as long; and only every 6th point of the RL in this direction will actually correspond to a reflection. Therefore for $6_1(6_5)$ the $00l$ reflections only occur for $l = 6m$. Similarly for $6_2(6_4)$ the translation is $2/6$ and $00l$ reflections are only present for $l = 3m$. For 6_3 the translation is $3/6$ and $00l$ only present for $l = 2m$. For $4_1(4_3)$, $00l$ is only present for $l = 4m$; 4_2 only for $l = 2m$. For $3_1(3_2)$ $00l$ is only present for $l = 3m$.

(b) *Glide planes*

Consider a glide plane perpendicular to $\mathbf{c}$ and gliding in the direction of $\mathbf{b}$. (Fig. 200). (The glide direction must be in the plane). If we take the left hand half of the unit cell, reflect it across the plane and move all the atoms by $\mathbf{b}/2$, we obtain the right hand half of the unit cell. If all the atoms are projected on to the glide plane in the direction of its normal, i.e. $\mathbf{c}$, the projected periodicity is halved in the direction of $\mathbf{b}$. Only reflections from planes which are parallel to $\mathbf{c}$, the line of projection, will be unaffected by the movement of the atoms.

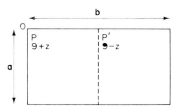

FIG. 200. The b glide plane through the origin perpendicular to $\mathbf{c}$ reflects a group of atoms at $P(+z)$ to $-z$, below the plane and translates them by $\mathbf{b}/2$ to P'.

Projection in any other direction, moreover, will not produce the increased periodicity. Therefore only $hk0$ planes parallel to **c** will have their periodicity affected by the glide. For these planes **b*** is twice as long as for the rest of the RL and therefore only relps with $h = 2m$ will give rise to reflections. In the case of an n (diagonal) glide plane in which the glide direction is $(\mathbf{a}/2) + (\mathbf{b}/2)$, the projected cell is face-centred and $hk0$ reflections will only occur for $h + k = 2m$. Other reflections hkl (with $l \neq 0$) will not be affected by systematic absences due to the glide.

The same principles operate even when the glide plane is not normal to an axis and if its glide is $\frac{1}{4}$ of the cell repeat distance. (Since this is only possible along the body diagonal of an I-lattice or the face diagonal of an F-lattice, it is still only half the distance between lattice points). In the case of $Ia3d$, the d, or diamond, glide plane is perpendicular to a [110] direction and the glide is $\frac{1}{4}$ of the body diagonal. If we take, as one of the many symmetry related combinations, the glide planes parallel to $(1\bar{1}0)$ (there are always 2 parallel planes, one gliding along each of the two diagonals), the glides will be

$$\frac{\mathbf{a} + \mathbf{b} + \mathbf{c}}{4} \quad \text{and} \quad \frac{-\mathbf{a} - \mathbf{b} + \mathbf{c}}{4}.$$

We first find the lattice planes which are perpendicular to $(1\bar{1}0)$, i.e. those whose normals $\mathbf{d}^*_{hkl}$ are perpendicular to $\mathbf{d}^*_{1\bar{1}0}$.
We have:

$$\mathbf{d}^*_{hkl} \cdot \mathbf{d}^*_{1\bar{1}0} = 0. \quad \therefore \quad h - k = 0 \text{ (cubic lattice)}.$$

Therefore the planes concerned are hkl.

The I-lattice means that $h + h + l = 2n$. There are therefore no restrictions on h, but l must be even for reflection to occur.

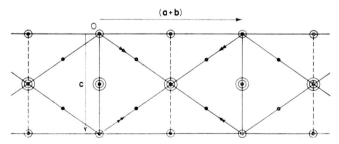

FIG. 201. The projection of the $Ia3d$ lattice on the d glide plane $(1\bar{1}0)$, including the additional lattice points (plain dots) produced in the 2-dimensional projected lattice by the d glides. The points from the cell corners are ringed once and the body-centring points twice. The projection includes all points in a band parallel to $(1\bar{1}0)$, of depth equal to the repeat distance perpendicular to $(1\bar{1}0)$, i.e. $(\mathbf{a} - \mathbf{b})$ in this case. The directions of the d glides are shown by double arrows.

Projecting on to the $(1\bar{1}0)$ plane we obtain Fig. 201. The lattice points at $\mathbf{c}/2$ are the projections of the body centring points and halve the c axis in this projection. The corresponding hhl RL plane has axes $\mathbf{c}^{*\prime} = 2\mathbf{c}^*$ and $\mathbf{a}^{*\prime} = \mathbf{a}^* + \mathbf{b}^*$. (It might be thought that as the projected lattice has a repeat of $(\mathbf{a}+\mathbf{b})/2$, this would give a doubling of the RL axis in this direction. But the halving in the projected crystal lattice in this direction would occur with a primitive unit cell, and is already accounted for in the hhl relps for a P cell. This can be seen from first principles by defining a projected axis $\mathbf{a}' = (\mathbf{a}+\mathbf{b})/2$ in the [110] direction. Then

$$ a' = \frac{(a^2+b^2)^{\frac{1}{2}}}{2} = \frac{a}{\sqrt{2}} = d'_{100}. \quad \therefore \ a^{*\prime} = \lambda/d'_{100} = \frac{\lambda}{a/\sqrt{2}} = \sqrt{2}a^*. $$

Since the first relp in the [110] direction is $\mathbf{d}^*_{110} = \mathbf{a}^* + \mathbf{b}^*$, which is $\sqrt{2}a^*$ from the origin, this corresponds to $\mathbf{a}^{*\prime}$. There is therefore no doubling of the RL in this direction for the projected unit cell).

Due to the operation of the diamond glides, the ringed projected lattice points in Fig. 201 are accompanied by face-centring points which are equivalent in projection to lattice points. The indices $h'0l'$, referred to the axes $\mathbf{a}^{*\prime}$ and $\mathbf{c}^{*\prime}$, therefore have the face centring restriction that reflections are only present for $h' + l' = 2n$.

Since for the same relp:

$$ h'\mathbf{a}^{*\prime} + l'\mathbf{c}^{*\prime} = h'(\mathbf{a}^* + \mathbf{b}^*) + l'2\mathbf{c}^* = h\mathbf{a}^* + h\mathbf{b}^* + l\mathbf{c}^* $$

we have $h' = h$, $l' = l/2$.

$$ \therefore \ h' + l' = h + l/2 = 2n \quad \text{or} \quad 2h + l = 4n. $$

The effect of the d glide is therefore that for reflections hhl to be present, $2h + l = 4n$. The symmetry related sets can be obtained by cyclic permutation of the indices.

2.4 Systematic Determination of Centred Cells, Glide Planes and Screw Axes

(a) *The determination of the type of centring*

In 2.2 (p. 342) the conditions for reflections to be present in the case of centred cells were derived by considering any RL vector $\mathbf{d}^*_{hkl}$. The conditions therefore apply to all reflections. In 2.3 (p. 344), only certain classes of reflections were affected, corresponding to relps along the axis in the case of a screw axis, and relps in the plane in the case of a glide plane. We therefore inspect all the reflections first to see if there are conditions for systematic absences which apply throughout, and if so which centred cell is present. Before proceeding further, the original choice of axes should be checked. If a non-standard setting has resulted it may be necessary, or at least worthwhile, changing the order or positions of the axes so that they accord with the presentation in the International Tables [5]. If, for instance, it turns out that an F centred cell has been chosen for a tetragonal crystal, then new **a** and **b** axes should be chosen at 45° to the original ones. This produces an I centred cell of half the size, in the standard orientation. In an orthorhombic crystal a single face centring should in most cases be C, and the axis per-perpendicular to the centred face should be chosen as **c**. The indices of the reflections must be transformed to conform to the new axes (II (4), p. 357), although this should be delayed until after the next stage if the selection of axes is likely to be influenced by the direction of a glide. For example, if axes have been chosen which produce a B face centred monoclinic cell with an a glide, then new axes should be chosen to give a primitive (P) cell with a c glide to conform to the standard setting. This means changing the original **a** axis to the new **c'** axis and choosing a new **a'** axis to define a primitive cell (Fig. 202).

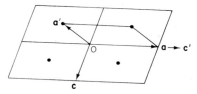

Fig. 202. Change of axes in a B-centred monoclinic cell with an a glide, to a Primitive cell with a c glide.

The transformation is given by $\mathbf{a}' = -\frac{1}{2}(\mathbf{a}+\mathbf{c})$; $\mathbf{c}' = \mathbf{a}$. If the necessity of transforming to a P cell had not demanded a change of axes, it would hardly have been worthwhile making a change just to get a c glide rather than an a glide, since the figures and tables for the standard c glide setting can be easily adapted for an a glide.

(b) *The determination of glide planes*

For a P cell the procedure is straightforward. All the possible types of mirror plane which are present in the point group [5(22)] of the crystal (or the highest possible point group symmetry if the precise point group has not been determined) are investigated in turn. The reflections from sets of planes perpendicular to the mirror plane are inspected for systematic absences. The effect of a glide is either to halve the projected two-dimensional unit cell in the direction of the glide, or to centre it, as in the case of the diagonal glide. The corresponding RL plane either has one axis effectively doubled by the glide, so that the relps with odd indices in that direction are 'absent'; or it is itself face centred, so that the sum of the two indices referred to the projected cell axes must be even. By finding the projected cell without a glide, defining new axes if necessary, and then putting in the effect of the glide, one can determine the 'absent' relps for the various possible cases, and compare with the absences for the corresponding set of reflections. In most cases the projected axes are the same as the crystal lattice axes, but this is not the case for body diagonal glide planes. In the tetragonal system ($1\bar{1}0$) is a possible glide plane, with the (hhl) lattice planes perpendicular to it. A glide in the $\mathbf{a} + \mathbf{b}$ direction would merely reproduce the halving already there from the projection of lattice points (Fig. 201, p. 347), and would produce no absences. Such a glide plane may or may not be present in the space group, but its presence or absence cannot be inferred from systematic absences.

A glide in the $\mathbf{c}$ direction on the other hand halves the $\mathbf{c}$ axis in projection and doubles the $\mathbf{c}^*$ axis, so that hhl reflections only occur for l even. An inspection of Figure 201 (p. 347) shows that an n glide, i.e. $(\mathbf{a} + \mathbf{b})/2 + \mathbf{c}/2$, produces the same projected cell as a c glide. If present an n glide will therefore produce no further absences. Since one might start by inferring an n-glide with the c glide as a possible non-essential companion, it follows that if the systematic absences are to give a non-equivocal result, both glides must be present, and this is in fact, the case. In the conventional order of precedence c comes before n, so that conventionally a c glide is inferred and the n glide becomes a consequence of the presence of the c glide. It could however, apart from convention, be the other way round. The possibilities in the ($1\bar{1}0$) plane are therefore (i) no symmetry plane; (ii) an m plane; (iii) a c glide Positive evidence from the occurrence of systematic absences can only be obtained for the c glide.

For possible point group mirror planes containing two axes the analysis is simpler. Consider the (001) plane containing the $\mathbf{a}$ and $\mathbf{b}$ axes. The reflections which may be affected have indices $hk0$. An a glide halves the projected cell and doubles the $\mathbf{a}^*$ axis, so that reflections $hk0$ only appear for h even. Similarly for a b glide only k even reflections occur. An n glide produces a face centred projected cell, so that reflections $hk0$ only occur for $h + k = 2n$.

Thus for a P cell the systematic absences, which occur only in central RL planes parallel to possible symmetry planes in the crystal (or in reflections from a zone of lattice planes perpendicular to the symmetry plane, which is the equivalent description) can be readily analysed to give the positions of glide planes and the direction of glide. The relevant letter can then be inserted in the space group symbol [4(105, 228) and 5(47)].

For a centred cell the situation is complicated by the fact that the general absences due to centring may include absences in a zone corresponding to a glide plane. For example, in a C centred orthorhombic cell reflections only occur for $h + k$ even. Because of this $h0l$ reflections only occur for h even and $0kl$ reflections only for k even. In a primitive unit cell these conditions would be proof of the existence of an a glide plane perpendicular to **b**, and a b glide plane perpendicular to **a**. In a C centred cell they are merely part of the general condition arising from the C centring. a and b glides may or may not be present in the space group, but this cannot be deduced from systematic absences since no additional absences are caused. However, if $h0l$ and $0kl$ reflections only occur for l even, then c glides exist as essential symmetry elements perpendicular to **a** and **b**, since this condition is not part of the general C centring condition.

The procedure then is the same as for the primitive cell, except that any condition in a zone which arises from the cell centring must be disregarded.

In other words, any translation symmetry element which is essential to the specification of the space group will cause absences additional to those caused by the cell centring.

(c) *The determination of screw axes*

The procedure is similar to that for glide planes. All the independent symmetry axes possible in the point group must be investigated. Any absences in the relps along the direction of a symmetry axis *may* determine a screw axis in this direction. If, however, the condition is already part of a more general condition due to cell centring *or the presence of glide planes,* then it must be disregarded. Only those conditions causing *additional* absences, over and above those due to cell centring or glide planes, determine the presence of a screw axis as an essential element of symmetry. Screw axes may or may not be present, corresponding to conditions for reflection already produced by cell centring or glide planes; but this cannot be deduced from systematic absences since no additional absences will be caused.

For example, in a P orthorhombic cell, if the only condition is that $h00$ is absent for h odd, then there is a 2_1 axis in the **a** direction. But if a single a glide is present, giving absences in $h0l$ reflections for h odd, this includes the $h00$ absences, and a 2_1 axis may or may not be present. There are three space groups which would have these systematic absences—$Pma2$, $P2_1am$ and $Pmam$. The last two (which are in non-standard orientation—the standard symbols are $Pmc2_1$ and $Pmma$) have 2_1 axes in the glide direction; the first has not.

(d) *The strategy of the determination of space group symmetry elements from systematic absences*

The symmetry should be determined first, using optical and morphological as well as X-ray evidence. If there is any doubt, work to the higher point group symmetry.

The determination of cell centring should be accomplished as early as possible, in order that the correct choice of axes can be made. This should be possible from photographs of the 0, 1 and 2 RL layers (Section **15**.4, p. 205).

Any critical glide directions should next be investigated, and the remainder of the translational symmetry elements left until the whole of the data has, been collected. The possible symmetry planes and axes should then be tried, testing for the independence of any conditions found each time.

Finally, the space groups compatible with all the possible combinations of Laue symmetry and systematic absences have been tabulated [5(349)] and this table enables the possible space groups for a particular case to be listed. If more than one space group is possible, any common symmetry elements will be determined. But the complete specification of the symmetry, including the presence or not of symmetry elements with systematic absences which are part of more general conditions, can only be accomplished by finding which of these possible space groups is actually present in the crystal. This requires evidence outside Laue symmetry and systematic absences [5(41, 539)]

3. Choice of Axes

3.1 AXIAL LENGTHS, ANGLES AND INTENSITIES KNOWN

It is assumed that a crystal is to be set up in accordance with a previous investigation, either using the original axes or more convenient ones derived from them, so that the axial lengths and angles are known. The problem is to find the axial directions in the crystal. In addition, where intensity differences between related reflections are used as a criterion, it is assumed that sufficient information about the intensities is known from a previous investigation. If departures from Friedel's law are to be taken into account, corresponding information is assumed to be available about 'non-Friedel' pairs.

These assumed conditions will most often arise when a second crystal is being set up to extend an existing investigation, but they may arise in checking or improving the results of earlier work. The problem of starting with an unknown crystal is dealt with later (II 3.2(b), p. 355).

(a) *Triclinic*

$+\mathbf{a}^*$, $+\mathbf{b}^*$ and $+\mathbf{c}^*$ are found from their known lengths and angles, and if the resulting set of axes is right hand, then the correct orientation has been found. If the axes are left hand, reverse the directions of all three axes to produce a right hand set with the same lengths and (vertically opposite) angles. There is only one correct set of axes for a triclinic crystal, given the axial lengths and angles.

(b) *Monoclinic*

Symmetry determines the $\mathbf{b}^*$ direction, and $+\mathbf{a}^*$ and $+\mathbf{c}^*$ are found from their known lengths and β^* angle. The direction of either positive $\mathbf{a}^*$ or positive c^* can be chosen arbitrarily as long as Friedel's law is obeyed, or in any case if there is a two fold symmetry axis in the space group. Once this choice has been made, the positive direction of the second axis is fixed by the value of β^*, and the direction of $+\mathbf{b}^*$ (which is independent of the previous choice) by right hand axes. For a space group without a two fold axis, anomalous scattering can give rise to differences in intensity between hkl and $\bar{h}k\bar{l}$, and if this difference is known, there is only one correct choice of axes.

(c) *Orthorhombic*

The axes are determined by their lengths and symmetry. As long as Friedel's law is obeyed an arbitrary choice can be made of the positive directions of any two of the axes, and the third is then fixed by right hand axes. When anomalous scattering is being taken into account for space groups without a centre of symmetry, the positive direction of $\mathbf{c}^*$ (the direction of the 2-fold axis in the mm point group) will have to be designated in terms of the differing intensities of hkl and $hk\bar{l}$. The positive direction of $\mathbf{a}^*$ or $\mathbf{b}^*$ is a matter of choice, and the positive direction of the third axis then comes from right hand axes. In the 222 point group a consistent choice of right hand axes is all that is required for dealing with anomalous scattering.

(d) *Tetragonal*

Symmetry determines the direction of $\mathbf{c}^*$, and the length of $\mathbf{a}^*$ determines the $\mathbf{a}^*(\mathbf{b}^*)$ directions. If departures from Friedel's law are ignored, any right hand set of axes can be chosen in the Laue group [5(30)] $4/mmm$, i.e. any of the four possible directions can be chosen for $+\mathbf{a}^*$, either of the remaining two for $+\mathbf{b}^*$, and $+\mathbf{c}^*$ is determined by right hand axes. In the Laue group $4/m$, positive $\mathbf{a}^*$ and $\mathbf{b}^*$ must be chosen to give the correct relation between the intensities of hkl and khl reflections, i.e. $+\mathbf{a}^*$ may be chosen arbitrarily from the four possible directions, but $+\mathbf{b}^*$ and $+\mathbf{c}^*$ are then determined. If departures from Friedel's law are to be taken into account, the non-centro-symmetric space groups have the following *additional* restrictions on choice of positive axial directions. In point groups $4mm$ and 42, $+\mathbf{c}^*$ must take account of differences in hkl and $hk\bar{l}$ intensities. In point groups $\bar{4}$ and $\bar{4}2m$, the choice of $+\mathbf{a}^*$ and $+\mathbf{b}^*$ must be in accord with the differences in intensity of the hkl and $k\bar{h}l$ reflections (or any equivalent pairs), i.e. the first choice of $+\mathbf{a}^*$ is restricted to two directions at $180°$ to one another. For $\bar{4}2m$ the second choice of two possible directions for $+\mathbf{b}^*$ remains, since the difference is equivalent to rotation about one of the horizontal 2-fold symmetry axes.

(e) *Trigonal*

Laue group $\bar{3}$. In rhombohedral space groups the axial directions are found from symmetry and the length and direction of $\mathbf{a}^*_R$. $+\mathbf{a}^*_R$ can be chosen in 3 ways, but $+\mathbf{b}^*_R$ and $+\mathbf{c}^*_R$ are then fixed by symmetry and right hand axes.

By starting with one of the $-\mathbf{a}^*_R$ directions as $+\mathbf{a}^{*'}_R$, another three sets of right hand axes can be set up, but these would have the wrong relative intensities for $hk0$, $h0k$ and similar pairs of reflections. In terms of the centred hexagonal cell (Fig. 130, p. 207) $+\mathbf{a}^*_H$ and $+\mathbf{b}^*_H$ can be chosen in three ways, determined by the relative intensities of hkl and $h+k$, $\bar{k}l$ reflections, with $+\mathbf{c}^*_H$ fixed by right hand axes. In space groups based on a primitive hexagonal cell the same considerations apply.

Laue group $\bar{3}m (\equiv \bar{3}(2/m))$. In a rhombohedral lattice the 2-fold axis and mirror plane makes $hk0 = h0k$, and the two sets of rhombohedral axes are therefore equivalent. There are therefore 6 choices for $+\mathbf{a}^*_R$ (or $\mathbf{b}^*_R$ or $\mathbf{c}^*_R$) and the other two axes are then fixed by symmetry and right hand axes. In terms of hexagonal axes there is a second set of three $+\mathbf{a}^*_H$, $+\mathbf{b}^*_H$ directions with $+\mathbf{c}^*_H$ given by right hand axes, which are related to the first set by the 2-fold axes. hkl and $h+k$, $\bar{k}l$ reflections are equal by the operation of the mirror plane, and it is no longer possible to distinguish between the two sets.

For the non-centrosymmetrical space groups, if departures from Friedel's law are to be taken into account no further restrictions operate for Laue group $\bar{3}$; but for point groups $3m$ and 32 the two sets of rhombohedral axes will no longer be equivalent, since $hkl \neq \bar{h}\bar{k}\bar{l}$, and for these two point groups the restrictions become the same as for $\bar{3}$.

(f) *Hexagonal*

Laue groups $6/m$ and $6/mmm$. $+\mathbf{a}^*$ (or $+\mathbf{b}^*$) can be chosen in 6 ways, with two choices for the positive direction of the second axis and $+\mathbf{c}^*$ determined by right hand axes. There are thus twelve equivalent sets of axes.

Departures from Friedel's law restrict the choice of $+\mathbf{a}^*$ to three directions for point groups $\bar{6}$ and $\bar{6}m2$, since $h+k$, $\bar{h}l$ is no longer equal to hkl, but still allow two choices for $+\mathbf{b}^*$. In point groups 6, $6mm$, 62, there are six choices for $+\mathbf{a}^*$, but only one for $+\mathbf{b}^*$, since hkl is no longer equivalent to $hk\bar{l}$.

(g) *Cubic*

Symmetry fixes the axial directions. For Laue group $4/m\,\bar{3}m$ there are six ways of choosing $+\mathbf{a}^*$, (or $\mathbf{b}^*$ or $\mathbf{c}^*$), four ways of choosing the positive direction of the second axis, and the third is then fixed by right hand axes. There are thus 24 ways of choosing a set of axes and these sets are indistinguishable as long as Friedel's law is obeyed. For Laue group $2/m\,\bar{3}$ the first axis can be chosen in 6 ways as before, but the choice of second axis is limited to the two opposite directions which give the right relative intensities to hkl and khl or any symmetry equivalent pair of reflections. (A Laue photograph taken with the beam along the chosen $\mathbf{a}^*$ axis shows characteristic mm symmetry, and can be used to define the two possible directions of the $\mathbf{b}^*$ axis if a photograph from the original setting is available for comparison.) The third direction is fixed by right hand axes.

If departures from Friedel's law are taken into account, then in point group 23 the (6-fold) choice of the first axis fixes the other two, because the differences between hkl and $hk\bar{l}$ (and their symmetry equivalent pairs) introduce a further restriction. In point group 432 the choice of axes is not restricted, since any right hand set can be transformed into any other of the 24 possible combinations by operation of the 4-fold and 3-fold axes. However, hkl and $hk\bar{l}$ (etc.) reflections will no longer be equal in intensity. In point group $\bar{4}\,3m$ the choice of second axis is restricted to the two opposite directions which make hkl and $hk\bar{l}$ (etc.) have the correct relative intensities.

3.2 CHOOSING AXES FOR A PREVIOUSLY UNKNOWN CRYSTAL

(a) *Axes defined but no intensities measured*

In those cases where a choice was made in II (3.1) on the basis of the known differences between certain reflections, additional freedom of choice is available when first setting the crystal up to measure intensities. If the differences are due to the breakdown of Friedel's law, the additional choice is available until the differences due to this breakdown have been measured.

(b) *Completely unknown crystal*

In the higher symmetry systems the directions and lengths of the axes are determined by the symmetry and the lattice. Only the choices in 3.2 (a) are available.

For orthorhombic space groups with P I or F cells the six right hand permutations of the axes are available initially. For single face centring C should be chosen, except for four mm space groups, where c is the unique direction, and A should be chosen to maintain this [4(244)]. In general the orientation should be that adopted as standard in the International Tables, but this cannot be found until the space group is known. In 15 cases the evidence from Laue symmetry and systematic absences determines the orthorhombic space group uniquely, and these cases can be oriented to conform with the International Tables convention as soon as this evidence has been collected. In 24 cases, however, (12 pairs) the diffraction evidence is compatible with two space groups which differ only by the presence of a centre of symmetry in one but not in the other. If positive evidence of the absence of a centre of symmetry (e.g. a positive test for piezo-electricity or sufficient morphological evidence or definite statistical evidence) can be obtained, the space group will be determined and orientation can proceed. If the tests are negative it may be necessary to try both space groups in the structure analysis. Unfortunately in 7 of these pairs the orientation of the axes is different in the two space groups. A further 20 space groups have groups of 3 or 4 which are compatible with the diffraction evidence, and of these a further 6 cases occur in which axes, which would be correct for one of the space groups, might be incorrect (i.e. not conforming to the International Tables convention) for another in the group. It is thus not always possible to know which is the correct orientation until the structure analysis is almost complete.

In the case of a monoclinic crystal, **b*** must be along the symmetry axis, and the other two axes should give a primitive or a C-centred cell, with **c** and **a** the two shortest translations perpendicular to **b** and $\beta > 90°$. Problems of orientation similar to those for orthorhombic crystals do not arise, since any alternative space groups have the same orientation of the axes.

In the case of a triclinic crystal, in the absence of other criteria, the cell with the three shortest axes should be chosen, with α and $\beta \geqslant 90°$.

When the first three layers of the RL about an important crystal lattice direction have been obtained, a few units of the net should be drawn out on perspex to a fairly large scale, and the 3 sheets stacked in correct height and orientation so that the possible cells can be marked out for comparison and final choice. Delaunay and other reductions of the initial choice give an analytical method of dealing with the problem [1(135), 5(530)].

3.3 Non-Standard Choice of Axes (Including Left Hand Axes)

For some purposes it may be desirable to adopt non-standard axes, such as a B-centred monoclinic cell, a C or I-centred tetragonal cell or a centred triclinic cell, in order to compare one crystal structure with another of different symmetry and therefore axial directions. If a triclinic crystal is one of a long chain series it may be better to have a non-standard cell, in order to establish a relationship with a monoclinic member of the series. There may be structural reasons for not choosing the three shortest translations. If the substance dimerises, the best choice of axes might be one that emphasises the relationship with the dimer. If there are good reasons for not employing conventional axes, then unconventional ones should be chosen. But the penalties may be severe, and if there is no reason against it, the International Tables conventions should be followed. For some purposes it may be worth transforming to conventional axes, even if the final result is given in terms of unconventional ones.

If left hand axes have been inadvertently chosen, they need not be transformed if one of the axes is perpendicular to a mirror plane in the point group (or the Laue group, if departures from Friedel's law are being ignored) since, in the case of the **c** axis for example, $hkl = hk\bar{l}$, and it is immaterial whether one measures the intensities of the $+l$ or the $-l$ reflections. However, if the corresponding plane in the space group is a glide plane or does not pass through the origin, the phase of hkl will not be the same as that of $hk\bar{l}$, so that it is wise even in such cases to work with right hand axes all the time.

4. Transformation of Axes and Consequent Transformation of Co-ordinates and Indices of Planes

If axes have to be changed for any reason, the consequent changes of indices and coordinates should be dealt with systematically, otherwise it is extremely easy for mistakes to occur.

Let $\mathbf{a}$, $\mathbf{b}$ and $\mathbf{c}$ be the original axes and $\mathbf{a}'$, $\mathbf{b}'$ and $\mathbf{c}'$ the new ones, given in terms of $\mathbf{a}$, $\mathbf{b}$, $\mathbf{c}$ by the equations:

$$\mathbf{a}' = X_1\mathbf{a} + Y_1\mathbf{b} + Z_1\mathbf{c}$$

$$\mathbf{b}' = X_2\mathbf{a} + Y_2\mathbf{b} + Z_2\mathbf{c} \tag{1}$$

$$\mathbf{c}' = X_3\mathbf{a} + Y_3\mathbf{b} + Z_3\mathbf{c}.$$

A simple example is the reorientation of a monoclinic B face-centred cell to give a primitive cell with $\mathbf{c}'$ in the glide direction, as in Figure 202 (p. 349). The transformation is given by:

$$\mathbf{a}' = -\tfrac{1}{2}\mathbf{a} - \tfrac{1}{2}\mathbf{c}$$

$$\mathbf{b}' = \mathbf{b}$$

$$\mathbf{c}' = \mathbf{a}.$$

Take a position vector

$$\mathbf{g} = x\mathbf{a} + y\mathbf{b} + z\mathbf{c} = x'\mathbf{a}' + y'\mathbf{b}' + z'\mathbf{c}' \tag{2}$$

defining the point P. It is required to find the new coordinates, x', y', z' of P in terms of the old coordinates, x, y, z. If we substitute for $\mathbf{a}$, $\mathbf{b}$ and $\mathbf{c}$ in the left-hand side of (2) we can compare coefficients of $\mathbf{a}'$ on the two sides to find x', y' and z'. In the example given $\mathbf{a} = \mathbf{c}'$; $\mathbf{b} = \mathbf{b}'$; $\mathbf{c} = -2\mathbf{a}' - \mathbf{c}'$.

In general the inverse matrix of coefficients given by the equations:

$$\mathbf{a} = X_1'\mathbf{a}' + Y_1'\mathbf{b}' + Z_1'\mathbf{c}'$$

$$\mathbf{b} = X_2'\mathbf{a}' + Y_2'\mathbf{b}' + Z_2'\mathbf{c}' \tag{3}$$

$$\mathbf{c} = X_3'\mathbf{a}' + Y_3'\mathbf{b}' + Z_3'\mathbf{c}'$$

can be obtained by inspection of the diagram showing the relation between the new and old axes as above, but if necessary it can be obtained by matrix algebra (or very laborious normal algebra) in terms of the undashed matrix, as follows:

$$
\begin{bmatrix} X'_1 & Y'_1 & Z'_1 \\ X'_2 & Y'_2 & Z'_2 \\ X'_3 & Y'_3 & Z'_3 \end{bmatrix} = \begin{bmatrix} \dfrac{\Delta X_1}{\Delta} & \dfrac{-\Delta X_2}{\Delta} & \dfrac{\Delta X_3}{\Delta} \\[2ex] \dfrac{-\Delta Y_1}{\Delta} & \dfrac{\Delta Y_2}{\Delta} & \dfrac{-\Delta Y_3}{\Delta} \\[2ex] \dfrac{\Delta Z_1}{\Delta} & \dfrac{-\Delta Z_2}{\Delta} & \dfrac{\Delta Z_3}{\Delta} \end{bmatrix} \qquad (1)
$$

where Δ is the determinant

$$
\begin{bmatrix} X_1 & Y_1 & Z_1 \\ X_2 & Y_2 & Z_2 \\ X_3 & Y_3 & Z_3 \end{bmatrix}
$$

and ΔX_1 is the minor of X_1, i.e. it is the determinant obtained by striking out the row and column on which X_1 stands.

For example,

$$
\Delta Y_2 = \begin{bmatrix} X_1 & Z_1 \\ X_3 & Z_3 \end{bmatrix}
$$

Using equations 357(3), however obtained, and substituting in 357(2) we have:

$$
\mathbf{g} = x(X'_1\mathbf{a}' = Y'_1\mathbf{b}' + Z'_1\mathbf{c}') + y(X'_2\mathbf{a}' + Y'_2\mathbf{b}' + Z'_2\mathbf{c}'
$$
$$
+ z(X'_3\mathbf{a}' + Y'_3\mathbf{b}' + Z'_3\mathbf{c}')
$$
$$
= \mathbf{a}'(xX'_1 + yX'_2 + zX'_3) + \mathbf{b}'(xY'_1 + yY'_2 + zY'_3) + \mathbf{c}'(xZ'_1 + yZ'_2 + zZ'_3)
$$
$$
= \mathbf{a}'x' \quad + \quad \mathbf{b}'y' \quad + \quad \mathbf{c}'z'.
$$
$$
\therefore \; x' = xX'_1 + yX'_2 + zX'_3; \; y' = xY'_1 + yY'_2 + zY'_3; \; z' = xZ'_1 + yZ'_2 + zZ'_3.
$$

This is conveniently expressed in terms of a matrix, as:

		a'	b'	c'
		x'	y'	z'
a	x	X'_1	Y'_1	Z'_1
b	y	X'_2	Y'_2	Z'_2
c	z	X'_3	Y'_3	Z'_3

The arrow denotes the direction in which the matrix is read to produce x' (or $\mathbf{a}$) etc. in terms of x, y, z (or $\mathbf{a}'$, $\mathbf{b}'$, $\mathbf{c}'$).

To obtain the transformation of indices we require the corresponding relations in the reciprocal lattice.

$$\mathbf{d}^*_{hkl} = h\mathbf{a}^* + k\mathbf{b}^* + l\mathbf{c}^* = \mathbf{d}^*_{h'k'l'} = h'\mathbf{a}^{*\prime} + k'\mathbf{b}^{*\prime} + l'\mathbf{c}^{*\prime}.$$

We require to find $h'k'l'$ in terms of $h\,k$ and l.

Let

$$\mathbf{a}^* = U'_1\mathbf{a}^{*\prime} + V'_1\mathbf{b}^{*\prime} + W'_1\mathbf{c}^{*\prime}$$

$$\mathbf{b}^* = U'_2\mathbf{a}^{*\prime} + V'_2\mathbf{b}^{*\prime} + W'_2\mathbf{c}^{*\prime} \tag{1}$$

$$\mathbf{c}^* = U'_3\mathbf{a}^{*\prime} + V'_3\mathbf{b}^{*\prime} + W'_3\mathbf{c}^{*\prime}$$

We have $\mathbf{a}' = X_1\mathbf{a} + Y_1\mathbf{b} + Z_1\mathbf{c}$ etc., from 357(1). Multiply the first equation of (1) by $\mathbf{a}'$ and the first equation of 357(1) by $\mathbf{a}^*$.

$$\therefore \ \mathbf{a}^* \cdot \mathbf{a}' = U'_1\lambda = \mathbf{a}' \cdot \mathbf{a}^* = X_1\lambda, \quad \text{since} \quad \mathbf{a}' \cdot \mathbf{b}^{*\prime} = \mathbf{a}' \cdot \mathbf{c}^{*\prime} = \mathbf{a}^* \cdot \mathbf{b} = \mathbf{a}^* \cdot \mathbf{c} = 0.$$

Similarly, by multiplying the first equation of (1) by $\mathbf{b}'$ and the second equation of 357(1) by $\mathbf{a}^*$ it can be shown that $V_1' = X_2$ and, by extension, that the inverse matrix for the reciprocal lattice is the transpose (rows and columns interchanged) of the direct matrix of the crystal lattice—one more example of the reciprocal relationship between the two lattices. Therefore the transformation of the indices—the co-ordinates in the RL—and the RL axes is given by a similar matrix as follows.

$$(2)$$

It can easily be shown by substituting 357(1) instead of 357(3) in the equation 357(2) for $\mathbf{g}$, that

$$x = X_1x' + X_2y' + X_3z' \text{ etc.} \tag{3}$$

i.e. the transformation matrix is the transpose of 357(1) for $\mathbf{a}'$, etc. in terms of $\mathbf{a}$, $\mathbf{b}$, $\mathbf{c}$.

Equations 359(1) and 359(3) can therefore be added to the matrix representation as shown in 359(2).

The corresponding transformation of $\mathbf{a}^*$, $\mathbf{b}^*$, $\mathbf{c}^*$ and h, k, l can be added to the first matrix, to give:

$$
\begin{array}{cccc|ccc}
 & & & \rightarrow h' & k' & l' \\
 & & \rightarrow \mathbf{a}^{*\prime} & \mathbf{b}^{*\prime} & \mathbf{c}^{*\prime} \\
 & & \rightarrow \mathbf{a}' & \mathbf{b}' & \mathbf{c}' \\
 & & x' & y' & z' \\
h & \mathbf{a}^* & \mathbf{a} & x & X'_1 & Y'_1 & Z'_1 \\
k & \mathbf{b}^* & \mathbf{b} & y & X'_2 & Y'_2 & Z'_2 \\
l & \mathbf{c}^* & \mathbf{c} & z & X'_3 & Y'_3 & Z'_3
\end{array}
\qquad (1)
$$

Any other co-ordinates in the RL will transform in the same way as indices, and Zone Axis Symbols (which are the co-ordinates defining a lattice direction in the crystal lattice) will transform in the same way as cell co-ordinates.

5. The Vector Algebra of the Crystal Lattice and RL.

5.1 GENERAL

Because of the complete symmetry of the relationship between the two lattices, any expression which is true for one lattice is true for the other merely by adding or taking away the star superscripts. Relationships established between the two lattices are also true if inverted, i.e. the corresponding symbols for the two lattices are interchanged. It is therefore not necessary to give all the relevant expressions both ways round. (This does not apply to relationships between *alternative* sets of axes, e.g. rhombohedral and hexagonal.)

5.2 TRICLINIC LATTICES

The relations between the two lattices are adequately expressed by 30 (1 and 2), but for some purposes a single vector equation giving an axis in one lattice in terms of those of the other, is more convenient.

Since $\mathbf{a}$ is perpendicular to $\mathbf{b}^*$ and $\mathbf{c}^*$ we have:

$$\mathbf{a} = K\,\mathbf{b}^* \times \mathbf{c}^*.$$

Take the dot product with $\mathbf{a}^*$:

$$\mathbf{a} \cdot \mathbf{a}^* = \lambda = K\,\mathbf{b}^* \times \mathbf{c}^* \cdot \mathbf{a}^*$$
$$= KV^*,\ 332(1).$$

$$\therefore \ K = \lambda/V^* \text{ and } \mathbf{a} = \frac{\lambda}{V^*} \mathbf{b}^* \times \mathbf{c}^*.$$

Similarly

$$\mathbf{b} = \frac{\lambda}{V^*} \mathbf{c}^* \times \mathbf{a}^* \text{ and } \mathbf{c} = \frac{\lambda}{V^*} \mathbf{a}^* \times \mathbf{b}^*. \tag{1}$$

The lengths of **a**, **b** and **c** are given by

$$|\mathbf{a}| = a = \frac{\lambda}{V^*} |\mathbf{b}^* \times \mathbf{c}^*| = \frac{\lambda}{V^*} b^* c^* \sin \alpha^* \tag{2}$$

and two similar equations.

Three similar equations give **a***, **b***, **c*** in terms of **a**, **b**, **c** and V.

The relation between V and V^* can be shown very simply using (1).

$$V^* = \mathbf{a}^* \times \mathbf{b}^* \cdot \mathbf{c}^* = \frac{\lambda^3}{V^3} (\mathbf{b} \times \mathbf{c}) \times (\mathbf{c} \times \mathbf{a}) \cdot (\mathbf{a} \times \mathbf{b})$$

$$= \frac{\lambda^3}{V^3} \left((\mathbf{b} \times \mathbf{c} \cdot \mathbf{a}) \mathbf{c} - (\mathbf{b} \times \mathbf{c} \cdot \mathbf{c}) \mathbf{a} \right) \cdot (\mathbf{a} \times \mathbf{b}), \quad 333\,(2)$$

$$= \frac{\lambda^3}{V^3} V\mathbf{c} \cdot \mathbf{a} \times \mathbf{b} = \frac{\lambda^3}{V^3} \times V^2 = \frac{\lambda^3}{V}. \tag{3}$$

V and V^* can be calculated from 335(1) and (3). The relations between a^*, b^*, c^* and a, b, c can then be calculated from (2).

α, β and γ can be obtained in two ways, (i) by taking the vector products and (ii) the scalar products, of **a**, **b** and **c** in pairs.

$$|\mathbf{b} \times \mathbf{c}| = bc \sin \alpha: \quad \mathbf{b} \times \mathbf{c} = \frac{\lambda^2}{V^{*2}} (\mathbf{c}^* \times \mathbf{a}^*) \times (\mathbf{a}^* \times \mathbf{b}^*)$$

$$= \frac{\lambda^2}{V^{*2}} [(\mathbf{c}^* \times \mathbf{a}^* \cdot \mathbf{b}^*) \mathbf{a}^* - (\mathbf{c}^* \times \mathbf{a}^* \cdot \mathbf{a}^*) \mathbf{b}^*)], \quad 333\,(2)$$

$$= \frac{\lambda^2}{V^{*2}} V^* \mathbf{a}^*$$

$$\therefore \ \sin \alpha = \frac{\lambda^2}{V^*} a^*/bc.$$

Similarly

$$\sin \beta = \frac{\lambda^2}{V^*} b^*/ca$$

$$\sin \gamma = \frac{\lambda^2}{V^*} c^*/ab$$

$$\tag{4}$$

$$\mathbf{b} \cdot \mathbf{c} = \frac{\lambda^2}{V^{*2}} \mathbf{c}^* \times \mathbf{a}^* \cdot \mathbf{a}^* \times \mathbf{b}^* = \frac{\lambda^2}{V^{*2}} c^* a^* \sin \beta^* a^* b^* \sin \gamma^* \cos \alpha.$$

$$\therefore \ a^{*2} b^* c^* \sin \beta^* \sin \gamma^* \cos \alpha = \mathbf{c}^* \times \mathbf{a}^* \cdot \mathbf{a}^* \times \mathbf{b}^*$$

$$= \mathbf{c}^* \cdot \mathbf{a}^* \mathbf{a}^* \cdot \mathbf{b}^* - \mathbf{c}^* \cdot \mathbf{b}^* \mathbf{a}^* \cdot \mathbf{a}^*, \quad 333(3)$$

$$= c^* a^* \cos \beta^* a^* b^* \cos \gamma^* - c^* b^* \cos \alpha^* a^{*2}$$

$$= a^{*2} b^* c^* (\cos \beta^* \cos \gamma^* - \cos \alpha^*).$$

$$\therefore \ \cos \alpha = \frac{\cos \beta^* \cos \gamma^* - \cos \alpha^*}{\sin \beta^* \sin \gamma^*} \left.\begin{array}{l} \\ \\ \\ \\ \\ \end{array}\right\}$$

Similarly
$$\cos \beta = \frac{\cos \alpha^* \cos \gamma^* - \cos \beta^*}{\sin \alpha^* \sin \gamma^*} \qquad (1)$$

$$\cos \gamma = \frac{\cos \alpha^* \cos \beta^* - \cos \gamma^*}{\sin \alpha^* \sin \beta^*}.$$

All these expressions can be inverted to give the RL in terms of the crystal lattice.

It may sometimes happen that the RL has been interpreted on one zero layer to give a^*, b^*, γ^*; c has been found from the layer line spacing; and α and β from Laue photographs along $\mathbf{a}^*$ and $\mathbf{b}^*$ (Section 13.6.1, p. 170). The values of α^*, β^* and c^* have to be found to complete the description of the RL.

By inverting the expression for $\cos \gamma$ in (1) and re-arranging, we have:

$$\cos \gamma = \cos \alpha \cos \beta - \sin \alpha \sin \beta \cos \gamma^*.$$

Thus γ can be found from α, β and γ^*, and α^* and β^* can then be calculated from the inverted expressions (1). c^* can be calculated as follows,

$$\mathbf{c} \cdot \mathbf{c}^* = \lambda = cc^* \cos \psi \qquad \therefore \ c^* = \lambda/(c \cos \psi) \qquad (2)$$

where ψ is the angle between $\mathbf{c}$ and $\mathbf{c}^*$ and therefore between $\mathbf{a}^* \times \mathbf{b}^*$ and $\mathbf{c}^*$.

$$\therefore \ \mathbf{a}^* \times \mathbf{b}^* \cdot \mathbf{c}^* = V^* = a^* b^* \sin \gamma^* c^* \cos \psi.$$

$$\therefore \ \cos \psi = \frac{V^*}{a^* b^* c^* \sin \gamma^*}$$

$$= \frac{(\sin^2 \alpha^* + \sin^2 \beta^* + \sin^2 \gamma^* + 2 \cos \alpha^* \cos \beta^* \cos \gamma^* - 2)^{\frac{1}{2}}}{\sin \gamma^*} \qquad (3)$$

from 335(1). c^* can be obtained by substituting this value of $\cos \psi$ into (2).

.3 To Find $d^{*2}_{hkl} = 4\sin^2\theta_{hkl}$ **in Terms of** $a^*b^*c^*$ **and** $\alpha^*\beta^*\gamma^*$.

$$d^*_{hkl} = h\mathbf{a}^* + k\mathbf{b}^* + l\mathbf{c}^*.$$

$$d^{*2} = (h\mathbf{a}^* + k\mathbf{b}^* + l\mathbf{c}^*)^2 = h^2a^{*2} + k^2b^{*2} + l^2c^{*2} + 2hka^*b^* \cos\gamma^*$$
$$+ 2klb^*c^* \cos\alpha^* + 2lhc^*a^* \cos\beta^*. \tag{1}$$

.4 To Find the Distance between Two Points A, B.

Let the position vectors be $\mathbf{r}_1 = x_1\mathbf{a} + y_1\mathbf{b} + z_1\mathbf{c}$ and $\mathbf{r}_2 = x_2\mathbf{a} + y_2\mathbf{b} + z_2\mathbf{c}$.

$$\overrightarrow{AB} = \mathbf{r}_2 - \mathbf{r}_1 = (x_2 - x_1)\,\mathbf{a} + (y_2 - y_1)\,\mathbf{b} + (z_2 - z_1)\mathbf{c}$$

$$\therefore\ AB^2 = (x_2 - x_1)^2a^2 + (y_2 - y_1)^2b^2 + (z_2 - z_1)c^2$$
$$+ 2(x_2 - x_1)(y_2 - y_1)ab \cos\gamma + 2(y_2 - y_1)(z_2 - z_1)bc \cos\alpha$$
$$+ 2(z_2 - z_1)(x_2 - x_1)ca \cos\beta. \tag{2}$$

If, however, $\mathbf{r}_1$ and $\mathbf{r}_2$ have been converted to Cartesian components, 334(1), $\mathbf{r}_1 = X_1\mathbf{i} + Y_1\mathbf{j} + Z_1\mathbf{k}$ and $\mathbf{r}_2 = X_2\mathbf{i} + Y_2\mathbf{j} + Z_2\mathbf{k}$, then $AB^2 = (X_2 - X_1)^2 + (Y_2 - Y_1)^2 + (Z_2 - Z_1)^2$.

.5 The Effect of Symmetry on the General Relations

These relations are very much simplified for systems other than triclinic, and in many cases it is easier to obtain the simpler expressions from first principles, rather than deriving them from the general expressions by putting in the special conditions.

In the monoclinic system for example $\alpha = \gamma = \alpha^* = \gamma^* = 90°$ and $\beta^* = 180 - \beta$. By putting $\sin 90° = 1$ and $\cos 90° = 0$ in the general expressions, those for the monoclinic case can be found; but it is easier to find a^* in terms of a and β from first principles as in Answer 7 (p. 459) and this applies to most if not all the relations deduced above. These expressions can however be used to check any derived directly for more symmetrical crystals.

In the hexagonal and trigonal systems $a = b$ and $a^* = b^*$. $\gamma = 120°$ and $\gamma^* = 180 - \gamma = 60°$. The rhombohedral lattice has equal axes and angles, but calculations are usually carried out in the doubly centred hexagonal cell.

The relations between $\mathbf{a}_R$, $\mathbf{b}_R$, $\mathbf{c}_R$ and $\mathbf{a}_H$, $\mathbf{b}_H$, $\mathbf{c}_H$ can be deduced from Figure 130 (a) (p. 207). For the standard, obverse setting:

$$\mathbf{a}_R = \tfrac{2}{3}\mathbf{a}_H + \tfrac{1}{3}\mathbf{b}_H + \tfrac{1}{3}\mathbf{c}_H; \qquad \mathbf{a}_H = \mathbf{a}_R - \mathbf{b}_R;$$
$$\mathbf{b}_R = \tfrac{1}{3}\mathbf{a}_H + \tfrac{1}{3}\mathbf{b}_H + \tfrac{1}{3}\mathbf{c}_H; \qquad \mathbf{b}_H = \mathbf{b}_R - \mathbf{c}_R; \tag{3}$$
$$\mathbf{c}_R = -\mathbf{b}_H^* + \mathbf{c}_H^*; \qquad \mathbf{c}_H = \mathbf{a}_R + \mathbf{b}_R + \mathbf{c}_R.$$

The corresponding relations for the RLs, from Fig. 130(b) (p. 208) are:

$$\mathbf{a}_R^* = \mathbf{a}_H^* + \mathbf{c}_H^* ; \qquad\qquad \mathbf{a}_H^* = \tfrac{2}{3}\mathbf{a}_R^* + \tfrac{1}{3}\mathbf{b}_R^* - \tfrac{1}{3}\mathbf{c}_R^*;$$

$$\mathbf{b}_R^* = -\mathbf{a}_H^* + \mathbf{b}_H^* + \mathbf{c}_H^* ; \qquad \mathbf{b}_H^* = \tfrac{1}{3}\mathbf{a}_R^* + \tfrac{1}{3}\mathbf{b}_R^* - \tfrac{2}{3}\mathbf{c}_R^*; \qquad (1$$

$$\mathbf{c}^*{}_R = -\mathbf{b}_H^* + \mathbf{c}_H^* ; \qquad\qquad \mathbf{c}_H^* = \tfrac{1}{3}\mathbf{a}_R^* + \tfrac{1}{3}\mathbf{b}_R^* + \tfrac{1}{3}\mathbf{c}_R^*.$$

The expressions for the reverse setting are given in Answer 32(x) (p. 526)

The other three systems are orthogonal, with $a = b$ and $a^* = b^*$ for th$\cdot$ tetragonal and $a = b = c$, $a^* = b^* = c^*$ for the cubic. All cross terms in th$\cdot$ expressions are zero for the orthogonal lattices.

5.6 d_{hkl}^{*2} FOR CUBIC LATTICES

The cubic system is unique in having all its axes equal and orthogonal. I$\bullet$ consequence:

$$d_{hkl}^{*2} = (h^2 + k^2 + l^2)\, a^{*2} = N a^{*2},$$

where N is an integer. This has important consequences, especially fo$\cdot$ powder photographs which have a distinctive regular appearance, wit$\blacksquare$ characteristic differences between the patterns for P, I and F lattices [1(123)] For P lattices $N = h^2 + k^2 + l^2$ can have all integral values except fo$\cdot$ $N = m^2(8n - 1)$ where m and n are integers. For $m = 1$, the missing d^{*2} value (and therefore missing powder lines) have $N = 7, 15, 23, 31, 39, 47, 55, 63.$. and for $m = 2$, $N = 28, 60 \ldots$. For most values of N at any rate up to 30, ther$\cdot$ is one characteristic triplet of indices (representing the symmetry related set) e.g. 211 gives $N = 6$, as do 121, 112, $\bar{1}12$, etc. In some cases more than on$\cdot$ representative triplet of indices give the same value of N. 300 and 221 is th$\cdot$ first example, with $N = 9$; then $N = 17, 18, 25, 26, 27, 29, 33$. The first tripl$\cdot$ set is for $N = 41$ (621, 540 and 443).

For I lattices $h + k + l = 2n$ and therefore $h^2 + k^2 + l^2 = 2n$. Only eve$\blacksquare$ values of N are possible, so the first 'missing' value is $N = 28$, correspondin$\cdot$ to the 14th line on a powder photograph. Powder photographs correspondin$\cdot$ to P and I lattices are very similar; but whereas the missing lines for P ar$\cdot$ 7 and 15, for I 7 and 15 are present and the 14th absent. Since the first lin$\cdot$ may be absent for space group reasons and others accidentally absent be cause of very weak intensities, it is necessary to check thoroughly on an$\cdot$ preliminary identification.

The pattern for an F lattice is very distinctive. The requirement that hk should be all even or all odd, reduces the possible values of N to 3, 4, 8, 11 12, 16, 19, 20, 24, 27, (28) 32, 35, 36 with 28 (and 60) the only values whic$\blacksquare$ are missing from the 1, 2, 1, 2 pattern of N values, and consequently from th similar pattern of lines on a powder photograph.

If the measurements on a cubic powder photograph are presented as a set of d values in descending order, it is a simple matter to verify the values of N with a slide rule which has an upper x^2 scale. If the length of the bottom scale from 1 to x represents $\log x$, that from 1 to x on the top scale represents $\log x$.

$$d_N^2 = \lambda^2/d_N^{*2}; \qquad \therefore \ 2\log d_N = 2\log \lambda - \log Na^{*2}.$$

$$\therefore \ \log d_N + \tfrac{1}{2}\log N = \log \lambda/a^* = \log d_1, \quad (d_1 = d \text{ for } N = 1; \qquad \therefore \ d_1 = a)$$

Reverse the centre slide, so that the $\log x$ and $\tfrac{1}{2}\log x$ scales are adjacent, but with opposite senses. Figure 203 shows the way in which integers on the $\log x$ scale (i.e. $\tfrac{1}{2}\log N$) can be lined up with the d values on the $\log x$ scale. If 1 on the $\tfrac{1}{2}\log x$ (top) scale is opposite the first d value on the bottom scale, then 2 (top) should be opposite d_2 (bottom) and so on. If the first line is not d_1 a match will not be obtained. 2 on the top scale must then be put against the first observed d value, and if necessary, 3, 4, etc. until a match is obtained within the limits of error in the d values. If the first d value is greater than 10 Å, take the first value less than ten and start with the appropriate integer on the top scale.

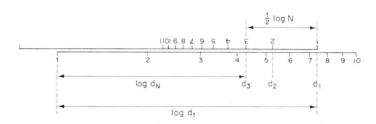

FIG. 203. The use of a slide rule with reversed slide to find the N values of powder lines of a cubic material from the listed d values of the lines.

The value of $d_1 = a$ obtained from the slide rule may be too small by a factor of $\sqrt{2}$ if an I lattice has not been recognised; but the slide rule will enable a certain check on N' (the values first found) to be made. If an I lattice is present, then the values of N' must be doubled to give N.

.7 TO FIND THE RL VECTORS WHICH ARE PERPENDICULAR TO A GIVEN CRYSTAL LATTICE LINE

A lattice line in the crystal lattice is defined by $\mathbf{g} = p\mathbf{a} + q\mathbf{b} + r\mathbf{c}$ (p, q, r integers). To find the lattice planes which are parallel to this line (i.e. the planes for which it is the zone axis) we find which planes have their normals perpendicular to **g**. The direction of the normal to the set of planes (hkl) i the reciprocal lattice vector $h\mathbf{a}^* + k\mathbf{b}^* + l\mathbf{c}^*$. If this is perpendicular to

$$(h\mathbf{a}^* + k\mathbf{b}^* + l\mathbf{c}^*) \cdot (p\mathbf{a} + q\mathbf{b} + r\mathbf{c}) = 0.$$

$$\therefore \ hp + kq + lr = 0. \tag{1}$$

This is the Weiss Zone Law, usually expressed in terms of a zone axis $[HKL$ which is the crystal lattice vector $H\mathbf{a} + K\mathbf{b} + L\mathbf{c}$. Therefore

$$hH + kK + lL = 0, \tag{2}$$

the usual form of the Weiss Law.

5.8 To Find the Zone Axis $[HKL]$ of Two Sets of Planes, $(h_1 k_1 l_1)$ AN $(h_2 k_2 l_2)$

From (2)

$$h_1 H + k_1 K + l_1 L = 0$$

$$h_2 H + k_2 K + l_2 L = 0.$$

$$\therefore \ \frac{H}{L} = \frac{l_2 k_1 - l_1 k_2}{h_1 k_2 - h_2 k_1} \ ; \ \frac{K}{L} = \frac{l_1 h_2 - l_2 h_1}{k_2 h_1 - k_1 h_2} \ .$$

The simplest solution of these equations is:

$$H = k_1 l_2 - k_2 l_1 ; \quad K = l_1 h_2 - l_2 h_1 ; \quad L = h_1 k_2 - h_2 k_1.$$

A simple mnemonic gives these three expressions:

$$
\begin{array}{ccccccc}
h_1 & | & k_1 & l_1 & h_1 & k_1 & | & l_1 \\
 & & & \times & \times & \times & & \\
h_2 & | & k_2 & l_2 & h_2 & k_2 & | & l_2 \\
 & & \parallel & & \parallel & & \parallel & \\
 & & H & & K & & L &
\end{array}
\tag{3}
$$

The three determinants of order 2 shown by the crosses give H, K and L. I these have a common factor this should be extracted to give the simples form of the direction $[HKL]$.

The derivation of the zone axis symbols is a more systematic (if less gen erally useful) way than the linear combination of indices in Example 1 (p. 296), of testing whether a third set of planes (hkl) is in the same zone a the first two. If HKL are derived from $h_1 k_1 l_1$, $h_2 k_2 l_2$ and $hH + kK + lL = 0$ then (hkl) is in the same zone as $(h_1 k_1 l_1)$, $(h_2 k_2 l_2)$.

5.9 To Find the Indices of Reflections in the nTH Layer Line of a Rotation Photograph when the Crystal is Rotated about a General Lattice Line $\mathbf{g}_{pqr} = p\mathbf{a} + q\mathbf{b} + r\mathbf{c}$.

This is equivalent to finding the RL coordinates in the nth layer of the RL perpendicular to $\mathbf{g}$. The interplanar spacing, ζ (Fig. 204) is given by

$$\zeta = \frac{\lambda}{g}. \qquad \therefore \ g = \frac{\lambda}{\zeta}.$$

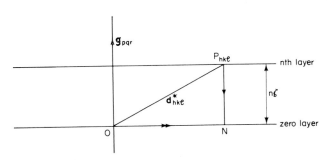

FIG. 204. The relation of a relp P_{hkl} on the nth RL layer above zero, to the crystal lattice vector $\mathbf{g}_{pqr}$ perpendicular to the RL planes.

$\therefore \ \mathbf{g}/(\lambda/\zeta)$ is a unit vector in the direction of $\mathbf{g}$.

$\therefore$ The vector $\overrightarrow{ON} = \overrightarrow{OP} + \overrightarrow{PN} = \mathbf{d}_{hkl}^* - n\zeta\mathbf{g}/(\lambda/\zeta)$.

Since ON is perpendicular to $\mathbf{g}$
$$\mathbf{g} \cdot (\mathbf{d}_{hkl}^* - n\zeta^2\mathbf{g}/\lambda) = 0$$

$\therefore \ \mathbf{g} \cdot \mathbf{d}_{hkl}^* - n\zeta^2 \ (\lambda^2/\zeta^2)/\lambda = 0.$

$\therefore \ (p\mathbf{a} + q\mathbf{b} + r\mathbf{c}) \cdot (h\mathbf{a}^* + k\mathbf{b}^* + l\mathbf{c}^*) - n\lambda = 0$

$\therefore \ \lambda(ph + qk + rl) - n\lambda = 0$

$\therefore \ ph + qk + rl = n.$

6. The Calculation of the Position of the Origin of Construction in Non-orthogonal Lattices

The RL axes are parallel to the crystal lattice axes in all orthogonal lattices; the $\mathbf{c}^*$ axis is parallel to $\mathbf{c}$ in the hexagonal case; and the $\mathbf{b}^*$ axis is parallel to $\mathbf{b}$ for monoclinic crystals. If any of these crystal lattice axes have been aligned along a rotation axis or the dial axis of a precession camera, the corresponding RL axis is automatically in the same direction.

The axes of the two lattices diverge in the case of the hexagonal **a** (or **b** axis, any of the three equivalent rhombohedral axes, the **a** or **c** monoclinic axes, or any of the triclinic axes. In these cases a further setting operation i required for the precession camera (Answer 24, p. 499). For rotation camera the origin of the non-zero layer RL net is no longer the same as the origin o construction (the point where the rotation axis cuts the layer, Section **14**.1.3 p. 181). The position of the origin of construction O_n has to be found relative to the RL net origin L, for each layer.

(a) *Triclinic about* **a**, **b** *or* **c**

The case of $+$**a** upwards will be considered; the other cases can be deal with in the same way. In Figure 205, L is the origin of the **b*****c*** net for the nth layer about **a**. $OL = n$**a***. $s = O_nL$ is the displacement of the net origin from the rotation axis. O_n is the origin of construction which we wish to find

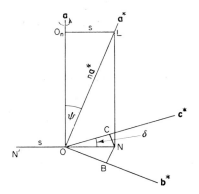

FIG. 205. The relation between the origin of construction O_n (where the rotation axis intersects the nth RL layer), the origin of the net L, and the axes and angles of a triclinic lattice rotating about **a**.

From L drop a perpendicular LN on to the **b*****c*** zero layer. Then ON is parallel and equal to O_nL. Draw NB and NC at right angles to **b*** and **c***. Since the plane LNB is perpendicular to **b***, OLB is a right angled triangle and $OB = OL \cos LOB = n$**a*** $\cos \gamma$*. Similarly $OC = n$**a*** $\cos \beta$*.

To find N calculate the lengths of OB and OC and mark them off along **b*** and **c***. Draw perpendiculars to **b*** and **c*** through B and C in the **b*****c*** plane. The intersection of these lines gives N. Produce NO backwards to N' with $ON' = ON$. N' is the position of the origin of construction, O_n, for the nth layer. The length of ON and $\widehat{CON} = \delta$ can be calculated as follows:

$$\frac{OC}{ON} = \cos \delta; \quad \frac{OB}{ON} = \cos(\alpha^* - \delta)$$

$$\therefore \quad \frac{OB}{OC} = \frac{\cos(\alpha^* - \delta)}{\cos \delta} = \frac{\cos \alpha^* \cos \delta + \sin \alpha^* \sin \delta}{\cos \delta}$$

$$= \cos \alpha^* + \sin \alpha^* \tan \delta.$$

$$\therefore \quad \tan \delta = \left(\frac{OB}{OC} - \cos \alpha^*\right) \bigg/ \sin \alpha^* = \left(\frac{na^* \cos \gamma^*}{na^* \cos \beta^*} - \cos \alpha^*\right) \bigg/ \sin \alpha^*$$

$$= \left(\frac{\cos \gamma^*}{\cos \beta^*} - \cos \alpha^*\right) \bigg/ \sin \alpha^*. \tag{1}$$

$$ON = OC \sec \delta = na^* \cos \beta^* \sec \delta. \tag{2}$$

Alternatively, if $\psi = O_n OL$ has been evaluated from 362(3),

$$s = na^* \cos \psi \text{ and } \cos \delta = \frac{na^* \cos \beta^*}{na^* \cos \psi} = \frac{\cos \beta^*}{\cos \psi}. \tag{3}$$

(b) *Rhombohedral about* **a** (**b** *or* **c**)

This is a special case of the triclinic lattice with

$$a^* = b^* = c^*, \quad \alpha^* = \beta^* = \gamma^*; \quad \cos \alpha^* = \frac{\cos^2 \alpha - \cos \alpha}{\sin^2 \alpha} = \frac{-\cos \alpha}{1 + \cos \alpha}, \quad 362(1)$$

If $+$**a** is upwards along the rotation axis, O_n is in the direction of $-(\mathbf{b}^* + \mathbf{c}^*)$ from L, and $s = na^* \sin \psi$, where $90 - \psi$ is the angle between $(\mathbf{b}^* + \mathbf{c}^*)$ and $\mathbf{a}^*$. (Fig. 206).

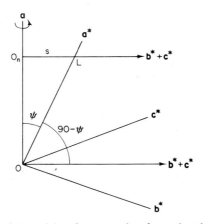

FIG. 206. The position of the origin of construction for a rhombohedral lattice rotating about a rhombohedral axis.

If $\alpha^* < 90$ ($\alpha > 90$), $\psi > 0$ and s is positive. If $\alpha^* > 90$ ($\alpha < 90$), $\psi < 0$, and s is negative, i.e. O_n is in the direction of $+(\mathbf{b}^* + \mathbf{c}^*)$ from L.

From 330(5), $\cos(90 - \psi) =$

$$\sin\psi = \frac{\mathbf{a}^* \cdot (\mathbf{b}^* + \mathbf{c}^*)}{a^* |\mathbf{b}^* + \mathbf{c}^*|} = \frac{\mathbf{a}^* \cdot \mathbf{b}^* + \mathbf{a}^* \cdot \mathbf{c}^*}{a^* (b^{*2} + c^{*2} + 2\mathbf{b}^* \cdot \mathbf{c}^*)^{\frac{1}{2}}} = \frac{2a^{*2} \cos\alpha^*}{a^* (2a^{*2} + 2a^{*2} \cos\alpha^*)^{\frac{1}{2}}}$$

$$= \frac{\sqrt{2} \cos\alpha^*}{(1 + \cos\alpha^*)^{\frac{1}{2}}}.$$

Alternatively, the procedure for the triclinic case can be followed.

(c) *Monoclinic about* **a** (*or* **c**)

If $+\mathbf{a}$ is upwards, and the plane containing **a**, **a***, **c***, **c** (all perpendicular to **b**) is the plane of the diagram (Fig. 207), then $s = na^* \cos\beta^*$. For n positive, O_n is in the direction of $-\mathbf{c}^*$ from the origin of the net L. For $+\mathbf{c}$ upwards interchange **a*** and **c***.

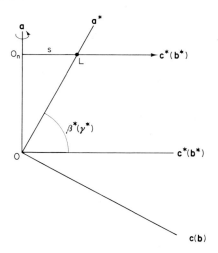

FIG. 207. The a^*c^* plane of a monoclinic lattice (or the a^*b^* plane of a hexagonal lattice) showing the position of the origin of construction O_n, for rotation about **a**.

Because the net origin L is moving away from the rotation axis, as we go to higher layers in the direction of positive $\mathbf{c}^*$, there is considerable danger of putting O_n on the positive side of L, and this tendency must be consciously countered. O_n is *left behind* on the negative side of the net origin, L. If n is negative, the positive is reversed.

(d) *Hexagonal about* **a**

This is a special case of the monoclinic lattice with $\mathbf{b}^*$ and $\mathbf{c}^*$, β^* and γ^* interchanged. $a^* = b^*$ and $\gamma^* = 60°$. If $+\mathbf{a}$ is upwards, and n is positive, O_n (Fig. 207) will be in the direction of $-\mathbf{b}^*$ from the net origin, L, and $O_n L = s = na^* \cos 60° = \frac{1}{2}n\,a^*$.

O_n therefore falls on a lattice point for n even and half way between two lattice points for n odd.

The Electron Density Expressed as a Fourier Series

1. The Fourier Coefficients

The most general form of the Fourier series for the electron density ρ at the point $\mathbf{r}_m = x\mathbf{a} + y\mathbf{b} + z\mathbf{c}$, is

$$\rho_{xyz} = \sum_{hkl} A_{hkl} \exp\left\{-i\left[\frac{2\pi}{\lambda}\mathbf{r}_m \cdot \mathbf{d}_{hkl}^* - \alpha_{hkl}'\right]\right\} \tag{1}$$

with values of h, k and l from $-\infty$ to $+\infty$. The expression for the structure factor $\mathbf{F}_{HKL}$, in terms of the electron density, is

$$\mathbf{F}_{HKL} = \int_0^{abc} \rho_{xyz}\, dv_m \exp\left[i\left(\frac{2\pi}{\lambda}\mathbf{r}_m \cdot \mathbf{d}_{HKL}^*\right)\right]. \tag{2}$$

Substituting for ρ and taking the integral inside the summation:

$$\mathbf{F}_{HKL} = \sum_{hkl} \int_0^{abc} A_{hkl} \exp\left\{i\left[\frac{2\pi}{\lambda}\mathbf{r}_m \cdot (\mathbf{d}_{HKL}^* - \mathbf{d}_{hkl}^*) + \alpha'\right]\right\} dv_m$$

Putting $\mathbf{d}_{HKL}^* - \mathbf{d}_{hkl}^* = \mathbf{d}_{h_1 k_1 l_1}^*$ and considering one term at a time, we reduce the integral to one dimension. Take new axes $\mathbf{a}' = \mathbf{a}$; $\mathbf{b}' = h_1\mathbf{b} - k_1\mathbf{a}$; $\mathbf{c}' = h_1\mathbf{c} - l_1\mathbf{a}$ (and corresponding RL axes $\mathbf{a}^{*\prime}$, $\mathbf{b}^{*\prime}$, $\mathbf{c}^{*\prime}$). $\mathbf{b}'$ and $\mathbf{c}'$ are perpendicular to $\mathbf{d}_{h_1 k_1 l_1}^*$ ($\mathbf{b}' \cdot \mathbf{d}^* = \mathbf{c}' \cdot \mathbf{d}^* = 0$) and therefore parallel to the planes $h_1 k_1 l_1$. The new unit cell is multiple, $V' = h_1^2[\mathbf{abc}] = h_1^2 V$, so that for the $h_1 k_1 l_1$ term the integral over the new cell will be h_1^2 times the required integral. For this term the integral

$$= \frac{1}{h_1^2} \int_0^{a'b'c'} A_{hkl} \exp\left[i\left(\frac{2\pi}{\lambda}\mathbf{r}_m \cdot \mathbf{d}_{h_1 k_1 l_1}^* + \alpha'\right)\right] dv_m.$$

Let $\mathbf{r}_m = x'\mathbf{a}' + y'\mathbf{b}' + z'\mathbf{c}'$.

Since the set of planes $h_1 k_1 l_1$ (referred to $\mathbf{a}$, $\mathbf{b}$ and $\mathbf{c}$) are parallel to $\mathbf{b}'$ and $\mathbf{c}'$ and cut off $\mathbf{a}'/h_1$ on $\mathbf{a}'$, the new indices $h_1' k_1' l_1'$ relative to the new axes $\mathbf{a}'$, $\mathbf{b}'$, $\mathbf{c}'$ are $h_1 00$.

$$\therefore \quad \mathbf{r}_m \cdot \mathbf{d}_{h_1 00}^* = h_1 \mathbf{r}_m \cdot \mathbf{a}^{*\prime} = h_1 x'\lambda$$

$\therefore$ Integral for the $h_1 k_1 l_1$ term

$$= \frac{1}{h_1^2} \int_0^{a'b'c'} A_{hkl} \exp\left[i\left(2\pi h_1 x' + \alpha'\right)\right] \mathrm{d}v.$$

Since the exponential term is constant for a given x', $\mathrm{d}v$ can be put in the form

$$\mathrm{d}v = \mathbf{b'} \times \mathbf{c'} \cdot \mathbf{da'} = \mathbf{b'} \times \mathbf{c'} \cdot \mathbf{a'}\mathrm{d}x' = V'\mathrm{d}x' = h_1^2 V \mathrm{d}x'.$$

$$\therefore \text{ Integral} = A_{hkl} V \int_0^1 \exp\left[i(2\pi h_1 x' + \alpha')\right] \mathrm{d}x'.$$

To see the meaning of this integral, we put it in the approximate form of a summation, taking equal increments

$$\mathrm{d}x' = \frac{1}{N}, \text{ where } N = h_1 n.$$

$$\therefore h_1 x' = h_1 m/N = m/n$$

$$\therefore \text{ Integral} \approx A_{hkl} \frac{V}{N} \sum_{m=0}^{h_1 n} \exp\left[i\left(2\pi \frac{m}{n} + \alpha'\right)\right].$$

This is a sum of vectors on the Argand diagram (I(15), p. 334), all of the same amplitude $\left(A_{hkl}\,(V/N)\right)$ and each making an angle ϕ with the X-axis which increases regularly by $(2\pi/n)$ until the $(n-1)$th vector closes the polygon. When $m = n$ the vector is parallel to that for $m = 0$. This closing of the polygon is repeated h_1 times, the final resultant being zero (or, strictly, one amplitude). As $N \to \infty$ the polygon becomes a circle, the summation becomes an integral and the resultant $\to 0$. (See Fig. 208).

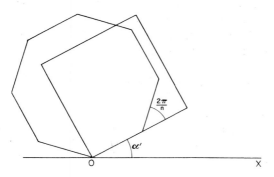

FIG. 208. The closed Argand polygons for $n = 4$ and 8.

This can be repeated for every term of the original summation, all the integrals being zero, except for the term in which $hkl = HKL$. In this case $\mathbf{d}^*_{h_1k_1l_1}$ is zero and the integral becomes

$$\int_0^{abc} A_{HKL}\, e^{i\alpha'} dv = A_{HKL}\, e^{i\alpha'} V.$$

$\therefore$ Since all other terms of the summation are zero

$$\mathbf{F}_{HKL} = A_{HKL}\, e^{i\alpha'} V$$

or since

$$\mathbf{F}_{hkl} = F_{hkl}\, e^{i\alpha}$$

$$A_{hkl} = \frac{F_{hkl}}{V}$$

$$\alpha' = \alpha.$$

2. Friedel's Law and the Form of the Fourier Series

If Friedel's Law is obeyed,

$$F_{\bar{h}\bar{k}\bar{l}} = F_{hkl} \quad \text{and} \quad \alpha_{\bar{h}\bar{k}\bar{l}} = -\alpha_{hkl}.$$

Since

$$\mathbf{d}^*_{\bar{h}\bar{k}\bar{l}} = -\mathbf{d}^*_{hkl}$$

$$\rho = \frac{1}{2V}\sum_{hkl} F_{hkl} \exp\left[-i\left(\frac{2\pi}{\lambda}\mathbf{r}\cdot\mathbf{d}^*_{hkl} - \alpha_{hkl}\right)\right] + F_{\bar{h}\bar{k}\bar{l}} \exp\left[-i\left(\frac{2\pi}{\lambda}\mathbf{r}\cdot\mathbf{d}^*_{\bar{h}\bar{k}\bar{l}} - \alpha_{\bar{h}\bar{k}\bar{l}}\right)\right]$$

$$= \frac{1}{2V}\sum_{hkl} F_{hkl} \left\{\exp\left[-i\left(\frac{2\pi}{\lambda}\mathbf{r}\cdot\mathbf{d}^*_{hkl} - \alpha_{hkl}\right)\right] + \exp\left[i\left(\frac{2\pi}{\lambda}\mathbf{r}\cdot\mathbf{d}^*_{hkl} - \alpha_{hkl}\right)\right]\right\}$$

$$= \frac{1}{V}\sum_{hkl} F_{hkl} \cos\left[\frac{2\pi}{\lambda}\mathbf{r}\cdot\mathbf{d}^*_{hkl} - \alpha_{hkl}\right].$$

The Integrated Reflection
and the Shape Transform

1. The Integrated Reflection

The total energy E', reflected as a mosiac block rotates with angular velocity ω through the reflecting position, can be obtained from a more detailed consideration of the transform of a crystal (Sec. **9**.1.3, p. 97). In that section the transform was treated as a point function, which is a good enough approximation for most purposes, but is only accurate for an infinite crystal. For a finite crystal the relps are surrounded by small regions in which the transform has finite values (the 'shape transform'—Ans. 15, p. 477). However, these regions are so small that we can neglect any variation of the transform of the unit cell contents and within one small region take the amplitude and phase of scattering from a single lattice point of the crystal lattice (i.e. the resultant scattering of the unit cell contents) as equal to the value, $\mathbf{F}_{hkl}$, at the relp. In Fig. 209 the position of a small area on the surface of the Sphere of Reflection, within the region surrounding a relp, hkl, for which $d^*_{hkl} = \sqrt{2}$ $(2\theta = 90°)$, is given by $\mathbf{D}^* = \mathbf{d}^*_{hkl} + \Delta\mathbf{d}^*$. Since the point at the end of $\mathbf{D}^*$ is in the Sphere of Reflection, $\mathbf{D}^* = \mathbf{s} - \mathbf{i}$, where $\mathbf{s}$ and $\mathbf{i}$ have their usual meanings (Sec. **1**.3.1, p. 12). However, since the point is not a relp, the phase angles between waves scattered from different crystal lattice points are no longer multiples of 2π, and the resultant amplitude of scattering in this direction must be obtained by calculating the phase angles, and finding the closing side of the Argand polygon for all the lattice points in the crystal block. The energy s^{-1} in the solid angle defined by the small area round $\mathbf{D}^*$, is obtained and summed (integrated) for all such areas over the surface of the sphere within the region of finite amplitude. This is in turn summed for the different positions of the surface of intersection as the crystal rotates through the reflecting position, and the relp and its surrounding region move through the surface of the sphere. For simplicity we take a crystal block in the form of a rectangular parallelepiped, with edges $N_1\mathbf{a}$, $N_2\mathbf{b}$, $N_3\mathbf{c}$. We use the exponential with imaginary index to represent the amplitude vector

(I(17), p. 337). We must first find the phase angles of the waves scattered from all the lattice points in the crystal. The amplitude is the same for all points, and can be taken as constant and equal to F_{hkl} over the small range of angles for which significant resultant scattering occurs. We take the origin for the crystal at a lattice point such that the coordinates of all other

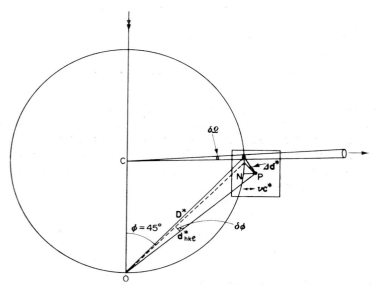

FIG. 209. A greatly exaggerated diagram of the volume of diffracting power round a relp P with $d^*_{hkl} = \sqrt{2}$ R.U. When P is in the sphere $2\theta = 90°$ and $\phi = 45°$. In the diagram P has moved through $\delta\phi$ from the reflecting position and the diffraction from a small area whose position is defined by $\Delta\mathbf{d}^*$, is pictured.

$$PN = vc^* = d^*_{hkl}\, \delta\phi \cos\phi = \delta\phi \text{ for } \phi = 45°.$$

lattice points are positive. To find the phase difference for neighbouring lattice points in the three axial directions we use, in a more general way, the expression for the path difference derived in Section 2.1.1 (p. 18),

i.e. path difference $= (\mathbf{s} - \mathbf{i}) \cdot \mathbf{a} = \mathbf{D}^* \cdot \mathbf{a}$.

$\therefore$ Phase difference, $\phi_a = 2\pi \dfrac{\mathbf{D}^* \cdot \mathbf{a}}{\lambda}$.

$\mathbf{D}^* = \mathbf{d}^*_{hkl} + \Delta\mathbf{d}^*$. Let $\Delta\mathbf{d}^* = \xi\mathbf{a}^* + \zeta\mathbf{b}^* + v\mathbf{c}^*$ where ξ, ζ, v are small fractions.

Then

$$\phi_a = \frac{2\pi}{\lambda}\, \mathbf{D}^* \cdot \mathbf{a} = \frac{2\pi}{\lambda}\,(h + \xi)\lambda = 2\pi(h + \xi) \equiv 2\pi\xi$$

since $\mathbf{a} \cdot \mathbf{a^*} = \lambda$ and $\mathbf{a} \cdot \mathbf{b^*} = \mathbf{a} \cdot \mathbf{c^*} = 0$ and we are only concerned with ϕ (mod 2π), i.e. we can ignore integral multiples of 2π.

Similarly $\phi_b = 2\pi\zeta$ and $\phi_c = 2\pi v$.

We first sum the amplitude vectors for the line of points through the origin in the $\mathbf{a}$ direction.

$$\mathbf{L} = \mathbf{F}_{hkl} + \mathbf{F}_{hkl}\, e^{i\phi_a} + \mathbf{F}_{hkl}\, e^{i2\phi_a} \ldots + \mathbf{F}_{hkl}\, e^{i(N_1 - 1)\phi_a}$$

(the Nth lattice point is not the origin of a unit cell)

$$= \mathbf{F}_{hkl}\, \frac{1 - e^{iN_1\phi_a}}{1 - e^{i\phi_a}} \quad \text{(sum of a geometric series)}.$$

Since each point in the line parallel to $\mathbf{a}$ and passing through the point $\mathbf{b}$ (i.e. the first line away from the origin in the $\mathbf{ab}$ plane) has a phase difference ϕ_b from the corresponding point in the origin line, the Argand diagram will be the same, except that all vectors, including the resultant, will be turned through the angle ϕ_b. The resultant is therefore $\mathbf{L}e^{i\phi_b}$. Summing over all the lines of points parallel to $\mathbf{a}$ in the $\mathbf{ab}$ plane, gives:

$$\mathbf{A} = \mathbf{L}(1 + e^{i\phi_b} + e^{i2\phi_b} + \ldots e^{i(N_2 - 1)\phi_b})$$

$$= \mathbf{L}\, \frac{1 - e^{iN_2\phi_b}}{1 - e^{i\phi_b}}.$$

Similarly, summing the resultants for all the $\mathbf{ab}$ planes gives

$$\mathbf{V} = \mathbf{A}\, \frac{1 - e^{iN_3\phi_c}}{1 - e^{i\phi_c}} = \mathbf{F}_{hkl}\, \frac{1 - e^{iN_1\phi_a}}{1 - e^{i\phi_a}} \cdot \frac{1 - e^{iN_2\phi_b}}{1 - e^{i\phi_b}} \cdot \frac{1 - e^{iN_3\phi_c}}{1 - e^{i\phi_c}}.$$

In Answer 3 (p. 455) it is shown that, for a single line of points, the intensity I', in terms of the energy scattered by 1 electron, is given by:

$$I' = \mathbf{L}\mathbf{L^*} = F_{hkl}^2 \left(\sin^2 \frac{N_1\phi_a}{2} \Big/ \sin^2 \frac{\phi_a}{2} \right).$$

By a simple extension of this procedure

$$I' = \mathbf{V}\mathbf{V^*} = F_{hkl}^2\, \frac{\sin^2 2\pi N_1\xi/2}{\sin^2 2\pi\xi/2} \cdot \frac{\sin^2 2\pi N_2\zeta/2}{\sin^2 2\pi\zeta/2} \cdot \frac{\sin^2 2\pi N_3 v/2}{\sin^2 2\pi v/2}.$$

The trigonometrical part of this expression is known as the 'interference function'. I' is therefore a function of ξ, ζ and v and is the flux of energy in the direction and position of the crystal defined by $\mathbf{d}_{hkl}^*$ and ξ, ζ and v.

The intensity in J $mm^{-2}\,s^{-1}$ at a distance R mm from the crystal is given by

$$I(\xi, \zeta, v) = F_{hkl}^2\, I_0 \left(\frac{e^2}{mc^2}\right)^2 \cdot \frac{1}{R^2} \cdot \frac{\sin^2 N_1 \pi \xi}{\sin^2 \pi \xi} \cdot \frac{\sin^2 N_2 \pi \zeta}{\sin^2 \pi \zeta} \cdot \frac{\sin^2 N_3 \pi v}{\sin^2 \pi v}.$$

We now assume that $\mathbf{a}^*$ and $\mathbf{b}^*$ lie in the tangent plane to the sphere and integrate over ξ and ζ with constant v, i.e. over the (approximate) surface of intersection with the sphere. Consider a small solid angle, $\delta\Omega$, subtended by an area $a^*\delta\xi\, b^*\delta\zeta$ ($\delta\Omega = a^*b^*\delta\xi\delta\zeta/1^2$) for the value of v given by the tangent plane of the sphere in Fig. 209. The area defined by $\delta\Omega$ at a distance R mm from the crystal $= R^2\delta\Omega$. Therefore the energy s^{-1} in the solid angle $\delta\Omega = IR^2\delta\Omega = IR^2 a^* b^* \delta\xi\delta\zeta$.

$\therefore$ Energy s^{-1} over all directions giving significant diffraction for this value of v

$$= I(v) = R^2 a^* b^* \int_{-\xi}^{\xi} \int_{-\zeta}^{\zeta} I(\xi, \zeta, v)\, d\xi\, d\zeta.$$

$\therefore$ Total energy, E', for rotation through the reflecting position

$$E' = \int_{t_1}^{t_2} I(v)\, dt.$$

From Fig. 209 $c^*\delta v = d^*\delta\phi \cos 45° = \delta\phi$. ($\phi$ is here the angular coordinate of $\mathbf{d}_{hkl}^*$ and $\delta\phi$ the displacement from the position where the relp hkl is in the sphere. $\mathbf{d}_{hkl}^*$ for $2\theta = 90$ is $\sqrt{2}$ R.U.)

$$\therefore \quad \frac{dv}{dt} = \frac{1}{c^*}\frac{d\phi}{dt} = \frac{\omega}{c^*},$$

where $\omega = d\phi/dt$ is the angular velocity of the crystal rotating through the reflecting position.

$$\therefore \quad E' = \int_{-v}^{v} I(v)\frac{dt}{dv}\cdot dv = \frac{c^*}{\omega}\int_{-v}^{v} I(v)\, dv$$

$$= \frac{R^2 a^* b^* c^*}{\omega}\int_{-\xi}^{\xi}\int_{-\zeta}^{\zeta}\int_{-v}^{v} I(\xi, \zeta, v)\, d\xi\, d\zeta\, dv.$$

To perform the integration we separate the integrals and use the fact that ξ, ζ and v are all small to put $\sin \pi\xi = \pi\xi$, etc.

Since the function only has appreciable values near the relp, we can also extend the limits to infinity.

The integral then reduces to three integrals of the form

$$\int_{-\infty}^{\infty} \frac{\sin^2 N_1 \pi \xi}{(\pi \xi)^2} \, d\xi.$$

Put $u = N_1 \pi \xi$. $\therefore du = N_1 \pi d\xi$.
  The integral then becomes

$$\int_{-\infty}^{\infty} \frac{N_1^2 \sin^2 u}{u^2} \cdot \frac{du}{N_1 \pi} = \frac{N_1}{\pi} \int_{-\infty}^{\infty} \frac{\sin^2 u}{u^2} \, du.$$

This is a standard integral and with infinite limits is equal to π.

$$\therefore E' = \frac{R^2 a^* b^* c^*}{\omega} I_0 F_{hkl}^2 \left(\frac{e^2}{mc^2} \right)^2 \cdot \frac{1}{R^2} \cdot N_1 N_2 N_3$$

$$a^* b^* c^* = V^* = \frac{\lambda^3}{V}; \quad N_1 N_2 N_3 = M; \quad MV = \delta v,$$

the volume of a mosiac block

$$\therefore E' = \frac{I_0}{\omega} \cdot \frac{\lambda^3}{V^2} \cdot F_{hkl}^2 \left(\frac{e^2}{mc^2} \right)^2 \cdot \delta v.$$

Although this has, for simplicity, been derived for an orthogonal lattice, it can be modified with the use of more vector algebra to cover the general case and produce the same result.

2. The Size of the Shape Transform

We can get an estimate of the size of the volume round the relps in which significant diffracting power exists by assuming that it extends to the points at which the first zero on one side of the Bragg peak occurs. The interference function is zero if $2\pi N_1 \xi / 2 = \pi$,

$$\text{i.e. } \xi = 1/N_1 \quad \text{or} \quad \xi a^* = a^*/N_1 = \frac{\lambda}{N_1 a} = \frac{\lambda}{t} \text{ R.U.}$$

where t is the thickness of the crystal. For a mosaic block of 0.1 μm and $a = 10$ Å, $N_1 = 1000$ and the volumes only extend from the relps 1/1000 of the distance, a^*, between relps. The dimension across the volume,

$$2\Delta d^* = 2\lambda/t. \tag{1}$$

The spread of the reflection, $\Delta\theta$, corresponding to these volumes of diffracting power is given by:

$$d^* = 2\sin\theta; \quad \Delta d^* = 2\cos\theta\,\Delta\theta = \lambda/t.$$

$$\therefore \; 2\Delta\theta = \frac{\lambda}{t\,\cos\theta}.$$

For the case above, and CuKα radiation

$$2\Delta\theta \approx 10^{-2}/\cos\theta \text{ degrees.}$$

Even at $\theta = 85°$, this gives a spread of less than 1/10 of a degree.

The Polarisation and Velocity (Lorentz) Factors

1. The Polarization Factor

The polarization factor for unpolarized incident radiation has been dealt with in Section **9.4.1** (p. 117). In the case of incident radiation which has been monochromatised by crystal reflection, the factor depends on the Bragg angle for the monochromator and the geometry of the apparatus. When the plane of reflection for the monochromator is also the equatorial (zero layer) plane for the crystal under investigation, the form of the correction is relatively simple. This is also the normal geometry for powder photographs taken with curved crystal monochromators.

Refer to Section **9.4.1** and Fig. 77 (p. 117). Let the Bragg angle for the monochromator be α. Then:

$$\overline{A_{M\perp}^2} = \tfrac{1}{2}K_M^2 A_0^2 \text{ and } \overline{A_{M\|}^2} = \tfrac{1}{2}K_M^2 A_0^2 \cos^2 2\alpha$$

where K_M and K are the reflection factors for the monochromator and crystal planes respectively.

$$\therefore A_R^2 = K_1^2 \left(\frac{A_0^2}{2} + \frac{A_0^2}{2} \cos^2 2\alpha \cos^2 2\theta \right), \text{ where } K_1 = K_M K.$$

For X-rays polarised perpendicular to the plane of reflection, $A_R^2 = K_1^2 A_0^2$

$$\therefore p = \frac{1 + \cos^2 2\alpha \cos^2 2\theta}{2} . \tag{1}$$

For non-zero layer reflections the expression is more complicated [17]. One of the advantages of balanced filters for monochromatisation (Sec. **18.1.1.2**, p. 257) is that the normal polarisation factor is unaffected.

2. The Velocity Factor (Refer to Section 9.4.2 p. 118)

2.1 *Rotation Geometry*, (a) The general case. This is only normally required for the Rimsky retigraph, but special cases can be deduced from it. Figure

210 shows the geometry of the general case. The expression is most easily obtained in terms of μ, v, and ξ. The full velocity of P is perpendicular to O_nP, but we require the velocity along the radius of the sphere, CP. We first resolve along and perpendicular to the radius of the circle, C_nP, and then along CP. The resolved part perpendicular to C_nP is tangential to the sphere and has no component along CP.

$$\text{Velocity along } C_nP = \omega\xi \sin\delta.$$

In O_nC_nP, $\xi/\sin\psi = C_nO_n/\sin\delta = CO\cos\mu/\sin\delta.$ $\therefore \sin\delta = \cos\mu \sin\psi/\xi.$

$\therefore$ Velocity along $CP = V = \omega\xi \sin\delta \cos v = \omega \cos\mu \cos v \sin\psi.$

$\therefore$ the correction factor, $L^{-1} = V/\omega = \cos\mu \cos v \sin\psi.$ (1)

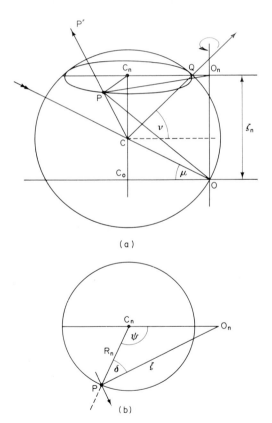

(a)

(b)

FIG. 210. (a) The velocity factor diagram for the general case, which may be met with n using a Rimsky retigraph. The velocity of P along the sphere radius PC has to be found. (b) The plane containing the nth RL layer, looking down the rotation axis.

In $\Delta O_n C_n P$, $\xi^2 = \cos^2 \mu + \cos^2 v - 2 \cos \mu \cos v \cos \psi$.

$$\therefore \cos \psi = \frac{\cos^2 \mu + \cos^2 v - \xi^2}{2 \cos \mu \cos v}. \tag{1}$$

In the Rimsky retigraph, ζ_n and μ are set manually, but v is automatically set by the linkage. ξ^2 varies with the reflection, but is usually calculated from d_{hkl}^{*2}.

In Fig. 210(a), $OO_n = \zeta_n = \sin \mu + \sin v$. $\therefore \sin v = \zeta_n - \sin \mu$. $\qquad(2)$

$$OP^2 = d^{*2} = \zeta^2 + \xi^2. \quad \therefore \xi^2 = d^{*2} - \zeta^2. \tag{3}$$

By putting these values into (1) and 382(1), L^{-1} can be expressed in terms of μ, ζ_n and d_{hkl}^*. The full expression is very clumsy and it is better to take the calculation in stages.

(b) *The normal beam case.*

$$\mu = 0 \text{ and } \sin v = \zeta_n; \quad \cos v = (1 - \zeta_n^2)^{\frac{1}{2}}.$$

$$\cos \psi = \frac{1 + 1 - \zeta_n^2 - \xi^2}{2 \cos v} = \frac{2 - d^{*2}}{2 \cos v}.$$

$$\therefore \sin \psi = \left(1 - \frac{(2 - d^{*2})^2}{4 \cos^2 v}\right)^{\frac{1}{2}} = \frac{(4 - 4\zeta_n^2 - 4 + 4d^{*2} - d^{*4})^{\frac{1}{2}}}{2 \cos v}$$

$$= \frac{(4d^{*2} - d^{*4} - 4\zeta_n^2)^{\frac{1}{2}}}{2 \cos v}.$$

$$\therefore L^{-1} = \tfrac{1}{2}(4d^{*2} - d^{*4} - 4\zeta_n^2)^{\frac{1}{2}}. \tag{4}$$

For the zero layer, $\zeta_n = 0$ and $L^{-1} = \tfrac{1}{2}d^*(4 - d^{*2})^{\frac{1}{2}} = \sin \theta (4 - 4 \sin^2 \theta)^{\frac{1}{2}}$

$$= 2 \sin \theta \cos \theta = \sin 2\theta. \tag{5}$$

(c) *Equi-inclination.*

$$v = \mu; \quad \sin \mu = \zeta_n / 2 \text{ from (2) and } \cos \mu = \left(1 - \frac{\zeta_n^2}{4}\right)^{\frac{1}{2}}$$

$$\cos \psi = \frac{2 \cos^2 \mu - \xi^2}{2 \cos^2 \mu} = 1 - \frac{\xi^2}{2 \cos^2 \mu};$$

$$\therefore \sin \psi = \left(1 - 1 + \frac{\xi^2}{\cos^2 \mu} - \frac{\xi^4}{4 \cos^4 \mu} \right)^{\frac{1}{2}}$$

$$= \frac{\xi}{2 \cos^2 \mu} (4 - \zeta_n^2 - \xi^2)^{\frac{1}{2}} = \frac{\xi}{2 \cos^2 \mu} (4 - d^{*2})^{\frac{1}{2}}.$$

$$\therefore L^{-1} = \frac{\cos^2 \mu \, \xi}{2 \cos^2 \mu} (4 - d^{*2})^{\frac{1}{2}} = \tfrac{1}{2} (d^{*2} - \zeta_n^2)^{\frac{1}{2}} (4 - d^{*2})^{\frac{1}{2}}. \tag{1}$$

The three expressions 382(1), 383(4) and (1) can be combined with the polarisation factor. For unpolarised X-rays this can be expressed in terms of d^* as follows:

$$p = \frac{1 + \cos^2 2\theta}{2} = \frac{2 - \sin^2 2\theta}{2} = 1 - 2 \sin^2 \theta \cos^2 \theta$$

$$= 1 - 2 \sin^2 \theta \, (1 - \sin^2 \theta) = 1 - 2 \sin^2 \theta + 2 \sin^4 \theta$$

$$= 1 - \frac{d^{*2}}{2} + \frac{d^{*4}}{8} = \tfrac{1}{2} (2 - d^{*2} + 0 \cdot 25 d^{*4}). \tag{2}$$

Combined with (1) to give $(Lp)^{-1}$ for equi-inclination geometry, we have:

$$(Lp)^{-1} = \frac{(d^{*2} - \zeta_n^2)^{\frac{1}{2}} (4 - d^{*2})^{\frac{1}{2}}}{(2 - d^{*2} + 0 \cdot 25 d^{*4})} . \tag{3}$$

2.2 *Precession Geometry* (Refer to Section **16**.3.4 and Fig. 144, p. 225).
   In Fig. 144

$$\delta = X_1 X_2 = KX_2 - KX_1 = KX_2 - KH' = KX_2 - (90 - \psi).$$

In right-angled $\triangle KH'X_2$, $\cos \mu = \tan (90 - \psi) \cot KX_2$.

$$\therefore \cot KX_2 = \cos \mu \tan \psi.$$

$$\therefore \quad \delta = \cot^{-1} (\cos \mu \tan \psi) - (90 - \psi).$$

$$\therefore \frac{d\delta}{dt} = \frac{d\delta}{d\psi} \cdot \frac{d\psi}{dt} = \Omega \frac{d\delta}{d\psi} \qquad \text{(clockwise rotation positive for an observer at } N)$$

$$\frac{d\delta}{d\psi} = - \frac{\sec^2 \psi \cos \mu}{1 + \tan^2 \psi \cos^2 \mu} + 1.$$

$$\therefore \frac{d\delta}{dt} = \Omega \left(1 - \frac{\sec^2 \psi \cos \mu}{1 + \tan^2 \psi \cos^2 \mu} \right) .$$

The circles of reflection are rotating with angular velocity Ω about O_nON (Fig. 142(a), p. 221), clockwise looked at along NO. This is equivalent to a stationary circle and an anti-clockwise rotation of O_nP_n (Fig. 142(b)) with angular velocity Ω. From this must be subtracted the clockwise rotation $d\delta/dt$.

∴. Resultant angular velocity of O_nP_n relative to the circle of reflection

$$= \omega = \Omega - \Omega \left(1 - \frac{\sec^2 \psi \cos \mu}{1 + \tan^2 \psi \cos^2 \mu} \right)$$

$$= \frac{\Omega \sec^2 \psi \cos \mu}{1 + \tan^2 \psi \cos^2 \mu} .$$

In Fig. 144, ψ is the angle between the plane containing ON and CO (Fig. 142(b), p. 222) and the plane containing HH' and CO. The angle between O_nN_n and $O_nH_n = N_0OH = \phi_0 = QOH' = QX_2$ on the stereogram. (X_2 is the new direction of H' and Q is the direction in the RL plane cut by the plane containing ON and CO).

In right-angled ΔOQX_2, $\sin (90 - \mu) = \tan \phi_0 \tan (90 - \psi)$

$$\therefore \tan \psi = \tan \phi_0 / \cos \mu$$

$$\therefore \frac{\sec^2 \psi}{1 + \tan^2 \psi \cos^2 \mu} = \frac{1 + \tan^2 \phi_0 / \cos^2 \mu}{1 + \tan^2 \phi_0}$$

$$= \cos^2 \phi_0 + \sin^2 \phi_0 / \cos^2 \mu$$

$$= 1 + \sin^2 \phi_0 (\sec^2 \mu - 1)$$

$$= 1 + \sin^2 \phi_0 \tan^2 \mu$$

$$\therefore \omega = \Omega \cos \mu (1 + \sin 2\phi_0 \tan^2 \mu).$$

Velocity of P_n perpendicular to $O_nP_n = \xi\omega$ (Fig. 211) where $\xi = O_nP_n$ is the cylindrical coordinate of P.

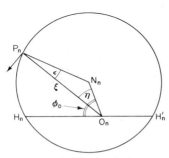

FIG. 211. The section containing the nth RL layer and a relp P_n, in the sphere of reflection.

∴ Velocity of P_n along $P_nN_n = \xi\omega \sin \varepsilon$. But

$$\frac{O_nN_n}{\sin \varepsilon} = \frac{R_n}{\sin \eta} \quad \therefore \sin \varepsilon = \frac{\sin \mu \sin \eta}{\sin \nu} \; ;$$

P_nN_n makes an angle of $90 - \nu$ with P_nC (Fig. 142(a));

∴ Velocity of P_n along $P_nC = \xi\omega \sin \varepsilon \sin \nu$

$$= V = \xi\omega \frac{\sin \mu \sin \eta}{\sin \nu} \cdot \sin \nu$$

$$= \xi\omega \sin \mu \sin \eta.$$

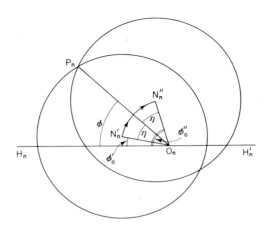

FIG. 212. The two positions during a complete precession in which P_n is on the sphere of reflection. The same section as Fig. 211.

For the two penetrations of the sphere taking place in a complete precession (Fig. 212),

$KF_{hkl}^2 L' = E'$ and $KF_{hkl}^2 L'' = E''$ where L', L'', E', E'' are the Lorentz (velocity) factors and the integrated intensities for the two penetrations. Total energy $E = E' + E'' = KF_{hkl}^2(L' + L'') = KF_{hkl}^2 L$.

∴ Velocity factor for a complete precession $= L = L' + L''$.

(With the use of a 'Zoltai' screen only one penetration is recorded, and the velocity factor must be amended accordingly.)

$L' = \dfrac{\Omega}{V'}$ and $L'' = \dfrac{\Omega}{V''}$ (if Ω is used in the expression for the integrated reflection).

$$. L = \frac{1}{\xi \sin \mu \sin \eta \cos \mu} \left\{ \frac{1}{1 + \tan^2 \mu \sin^2 (\phi + \eta)} + \frac{1}{1 + \tan^2 \mu \sin^2 (\phi - \eta)} \right\}$$

where ϕ is the angle $O_n P_n$ makes with $O_n H_n$ (Fig. 212) and $\phi_0' = \phi - \eta, \phi_0'' = \phi + \eta$.

In $\Delta O_n N_n P_n$ (Fig. 211)

$$R_n^2 = O_n N_n^2 + O_n P_n^2 - 2 O_n N_n \cdot O_n P_n \cos \eta.$$

$$\therefore \cos \eta = \frac{\sin^2 \mu + \xi^2 - \sin^2 v}{2\xi \sin \mu}.$$

If the coordinates of P_n relative to axes in the zero RL plane ($a*b*$ say) with origin at O, Ox along OH and $Oy \perp$ to OH, and a third (ζ) axis along OO_n are $x\ y\ \zeta$, then:

$$\zeta = OO_n = \lambda/c = \cos \mu - \cos v.$$

$$\therefore \cos v = \cos \mu - \zeta.$$

$$\xi = (x^2 + y^2)^{\frac{1}{2}}; \quad \phi = \tan^{-1} \frac{y}{x}; \quad \zeta = \frac{\lambda}{c}.$$

Using these relations it is possible to express L as a function of μ, λ, (c), x and y. The polarisation factor 384(2) can then be incorporated in the same terms since $d*^2 = \xi^2 + \zeta^2$.

3. The Geometrical Factor for Powder Photographs

Since a powder sphere uniformly weighted with relps is not altered by rotation, we can develop the expressions for a powder from consideration of rotating crystals. The powder reflection is continuous, so that we measure the power (i.e. energy s^{-1}) in a small length Δl mm of the powder ring. Let this arc of the powder ring lie symmetrically about the equatorial plane relative to the rotation axis, and let the plane of reflection containing the direct beam and the reflection at one end of the arc make an angle ψ with the equatorial plane. Let this plane cut the circle of intersection of the powder sphere and the sphere of reflection at $\pm \Delta \zeta/2$. From Fig. 213:

$\psi = \frac{1}{2}(\Delta l)/P'N' = \Delta l/2r \sin 2\theta$, where r mm is the radius of the film,

and $\psi = \frac{1}{2}(\Delta \zeta)/PN = \Delta \zeta/(2d_{hkl}^* \cos \theta)$.

$\therefore \Delta \zeta = \Delta l\, d_{hkl}^*/2r \sin \theta.$

In one revolution (time T s) the proportion of crystallites reflecting

$$= \frac{\text{Area of band of width } \Delta \zeta}{\text{Area of powder sphere}} = \frac{2\pi d^* \Delta \zeta}{4\pi d^{*2}} = \frac{\Delta \zeta}{2d^*} = \frac{\Delta l}{4r \sin \theta}.$$

Each crystallite, of volume Δv, contributes energy,

$$\Delta E = \frac{I_0}{\omega} Q' \Delta v \text{ (Lp), } 114(1). \quad \therefore \text{ total energy E in one revolution is given by}$$

$$E = \frac{I_0}{\omega} Q' \text{ (Lp) } \Sigma \Delta v$$

$$= \frac{I_0}{\omega} Q' \text{ (Lp) } \frac{v\Delta l}{4\,r\sin\theta} = \frac{I_0 T}{2\pi} \cdot \frac{Q'(\text{Lp})v\Delta l}{4\,r\sin\theta}$$

where v = total irradiated volume of the specimen.

$$\therefore \text{ Energy s}^{-1} = P'_{\Delta l} = \frac{I_0 v\Delta l}{8\pi r\sin\theta} \cdot Q'(\text{Lp}).$$

This is for one relp, hkl, per crystal. If the multiplicity (Section **9.4.4**, p. 118 is p'', the total power

$$P_{\Delta l} = P'_{\Delta l} p'' = \frac{I_0 v p'' \Delta l}{8\pi r\sin\theta} \cdot Q'(\text{Lp}).$$

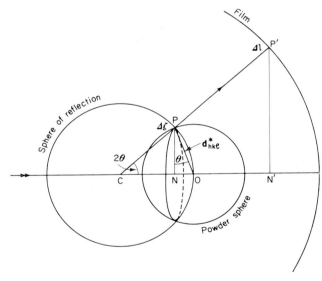

FIG. 213. The sphere of reflection and the hkl powder sphere. The powder sphere is rotated about an axis perpendicular to the plane of the diagram. If the diagram is oscillated through a small angle ψ about $CNON'$, P will define a small arc $PN\psi$, and P' a small arc $P'N'\psi$ All reflections arising from the arc on either side of P will be registered in the arc on either side of P'.

Since the powder ring is uniform,† the power is proportional to the length of arc.

$$\therefore P_l = \frac{I_0 lvp''}{8\pi r \sin\theta} \cdot Q'(Lp).$$

The (Lp) factor is that for the zero layer of a single crystal.

$$\therefore P_l = \frac{I_0 lvp'' \cdot Q'}{8\pi r \sin\theta} \cdot \frac{1 + \cos^2 2\theta}{2\sin 2\theta}$$

$$= \frac{I_0 lvp''}{32\pi r} \cdot \frac{Q'(1 + \cos^2 2\theta)}{\sin^2\theta \cos\theta}.$$

Although this expression has been derived for an equatorial arc, it is clearly independent of position round the ring.† The expression is the same for a transmission specimen, which can be divided into a large number of small rods perpendicular to the plane of reflection. In the case of a Guinier camera the polarisation factor for crystal monochromatised radiation, 381(1), must be used.

For completeness the necessity for correcting for absorption (Section **18**.6.4, p. 282) must be mentioned.

† This assumes cylindrical symmetry about the incident beam. For a beam with vertical divergence the expression will only hold for a small equatorial arc.

The Anisotropic Temperature Factor

1. The Meaning of the Coefficients B_{ij}

Refer to Section 9.5.3(p. 119). It is required to find the relation between the anisotropic temperature factor coefficients B_{ij} and the assumed ellipsoidal distribution of $\bar{\mathbf{u}}$, the r.m.s. displacement of the atom in any direction. Define $\mathbf{w} = \bar{\mathbf{u}}/\bar{u}^2$ (i.e. $w = 1/\bar{u}$).

Since the ends of the displacement vectors lie on an ellipsoid, the inverse vectors $\mathbf{w} = x\mathbf{a}^* + y\mathbf{b}^* + z\mathbf{c}^*$ will vary with direction in such a way that the end of the vector $\mathbf{w}$ lies on the inverse ellipsoid, given by

$$B_{11}x^2 + B_{22}y^2 + B_{33}z^2 + B_{12}xy + B_{23}yz + B_{31}zx = 2\pi^2/\lambda^2 \qquad (1)$$

based on the axes $\mathbf{a}^*$, $\mathbf{b}^*$, $\mathbf{c}^*$. It is required to show that the coefficients of (1) are the same as those of the anisotropic temperature factor.

In the direction of $\mathbf{d}^*_{hkl}$ the point on the ellipsoid is at

$$\frac{h\mathbf{a}^* + k\mathbf{b}^* + l\mathbf{c}^*}{d^*_{hkl}} \, w,$$

where w is the length of the radius vector of the ellipsoid in this direction i.e. at

$$\frac{hw}{d^*}, \frac{kw}{d^*}, \frac{lw}{d^*}.$$

Substituting these co-ordinates in (1) we get:

$$\frac{w^2}{d^{*2}}(B_{11}h^2 + B_{22}k^2 + B_{33}l^2 + B_{12}hk + B_{23}kl + B_{31}lh) = 2\pi^2/\lambda^2.$$

$$\therefore \quad B_{11}h^2 + B_{22}k^2 + B_{33}l^2 + B_{12}hk + B_{23}kl + B_{31}lh = \frac{2\pi^2 d^{*2}}{w^2\lambda^2}$$

$$= \frac{2\pi^2 4 \sin^2\theta \bar{u}^2}{\lambda^2} = \frac{\sin^2\theta}{\lambda^2} \cdot 8\pi^2\bar{u}^2 \qquad (2)$$

This (times -1) is the exponent of the exponential, which is the temperature factor of the particular atom for the reflection hkl (since $\bar{\mathbf{u}}$ is in the direction normal to the planes (hkl)). Thus the temperature factor for any reflection is given by

$$q(hkl) = \exp - (B_{11}h^2 + B_{22}k^2 + B_{33}l^2 + B_{12}hk + B_{23}kl + B_{31}lh)\dagger \quad (1)$$

where the B_{ij} are independent (within limits) and vary from atom to atom.

The ellipsoid, whose coefficients are B_{ij}, is reciprocal to the vibration ellipsoid whose radii give the r.m.s. displacement of the atom; but its principle axes are in the same direction as those of the vibration ellipsoid, and their lengths are the reciprocals of the vibration ellipsoid axes.

2. The Lengths and Directions of the Vibration Ellipsoid Axes

To obtain the lengths and directions of the vibration ellipsoid axes and to find the restrictions on and relation between the B_{ij}'s when atoms lie on special positions, it is first of all necessary to refer the ellipsoid to orthonormal axes—the unit vectors $\mathbf{i}, \mathbf{j}, \mathbf{k}$, with $\mathbf{i}$ in the direction of $\mathbf{a}$, $\mathbf{j}$ in the $\mathbf{a}, \mathbf{b}$ plane with $\mathbf{j} \cdot \mathbf{b}$ positive and $\mathbf{k} = \mathbf{i} \times \mathbf{j}$. (It is assumed that the original axes $\mathbf{a}^*, \mathbf{b}^*$, $\mathbf{c}^*$ form a R.H. set, i.e. $[\mathbf{a}^*\mathbf{b}^*\mathbf{c}^*]$ is positive).

The equations defining the new axes in terms of the old are:

$$\mathbf{i} = l'_{11}\mathbf{a}^* + l'_{21}\mathbf{b}^* + l'_{31}\mathbf{c}^*$$
$$\mathbf{j} = l'_{12}\mathbf{a}^* + l'_{22}\mathbf{b}^* + l'_{32}\mathbf{c}^* \quad (2)$$
$$\mathbf{k} = l'_{13}\mathbf{a}^* + l'_{23}\mathbf{b}^* + l'_{33}\mathbf{c}^*$$

† For some purposes it is convenient to define $q(hkl)$ in terms of U_{ij} as follows:

$$q(hkl) = \exp\left[-\frac{2\pi^2}{\lambda^2}(h^2a^{*2}U_{11} + k^{*2}b^{*2}U_{22} + l^2c^{*2}U_{33} + \right.$$

$$\left. 2hka^*b^*U_{12} + 2klb^*c^*U_{23} + 2lhc^*a^*U_{31} \right],$$

i.e. $U_{11} = \dfrac{\lambda^2}{2\pi^2 a^{*2}} B_{11}$, etc; $U_{12} = \dfrac{\lambda^2}{4\pi^2 a^*b^*} B_{12}$, etc.

Notice the factor 2 in the cross terms.

Normally λ is taken as $1\cdot0$ A, since the temperature factor is a function of $\bar{u}_{hkl}/d_{hkl}$ and is independent of wavelength. By substituting 100 for hkl and a^* for d^*_{hkl} in 390(2) it is easily shown that $U_{11} = \bar{u}^2_{100}$, i.e. it is the mean square of the amplitude of vibration in the direction of $\mathbf{a}^*$. Similarly $U_{22} = \bar{u}^2_{010}$ and $U_{33} = \bar{u}^2_{001}$.

However, in finding the lengths and directions of the axes of the ellipsoid, i.e. the latent roots and vectors, the values of B_{ij} must be used, not U_{ij}.

The restrictions on the values of U_{ij} are the same as those for B_{ij}, if the factor of 2 noted above is taken into account, i.e. $U_{ii} > 0$; $U_{ii}U_{jj} - U_{ij}^2 > 0$ and $\Delta U > 0$. The restrictions due to symmetry are exactly the same in the two cases.

It should be noted that the expression for $q(hkl)$ in terms of U_{ij} and a^*, b^* and c^* has been quoted incorrectly by a number of authors. The cosines of α^*, β^* and γ^* have been included in the cross terms, e.g. $2hka^*b^* \cos \gamma U_{12}$. This has the effect of making $U_{ij} = 0$, leaving only U_{ii}, for all orthogonal lattices, and is clearly incorrect.

Denote the matrix of coefficients, l'_{ij}, by L'^*. Multiply the equations of (1)
by $\mathbf{a}$, $\mathbf{b}$, $\mathbf{c}$ in turn, to find the expressions for l'_{ij}

$$\mathbf{a} \cdot \mathbf{i} = l'_{11}\lambda + 0 + 0. \qquad \therefore l'_{11} = \mathbf{a} \cdot \mathbf{i}/\lambda.$$

$$\mathbf{b} \cdot \mathbf{i} = l'_{21}\lambda \quad \text{etc.}$$

$$\therefore L'^* = \frac{1}{\lambda} \begin{bmatrix} \mathbf{a} \cdot \mathbf{i} & \mathbf{b} \cdot \mathbf{i} & \mathbf{c} \cdot \mathbf{i} \\ \mathbf{a} \cdot \mathbf{j} & \mathbf{b} \cdot \mathbf{j} & \mathbf{c} \cdot \mathbf{j} \\ \mathbf{a} \cdot \mathbf{k} & \mathbf{b} \cdot \mathbf{k} & \mathbf{c} \cdot \mathbf{k} \end{bmatrix}$$

$$\therefore L'^* = \frac{1}{\lambda} \begin{bmatrix} a & b\cos\gamma & c\cos\beta \\ 0 & b\sin\gamma & cA \\ 0 & 0 & cB \end{bmatrix}$$

where $A = \dfrac{\cos\alpha - \cos\beta\cos\gamma}{\sin\gamma}$

$$B = \frac{(\sin^2\alpha + \sin^2\beta + \sin^2\gamma + 2\cos\alpha\cos\beta\cos\gamma - 2)^{\frac{1}{2}}}{\sin\gamma},$$

(See 334(1)).

To enable the ellipsoid coefficients to be represented by a symmetric matrix
it is necessary to divide the cross terms into equal parts, i.e.

$$B_{12}xy = \frac{B_{12}}{2}xy + \frac{B_{21}}{2}yx, \quad \text{where } B_{ji} = B_{ij}.$$

The matrix of coefficients of the quadratic form is then given by

$$2B = \begin{bmatrix} 2B_{11} & B_{12} & B_{13} \\ B_{21} & 2B_{22} & B_{23} \\ B_{31} & B_{32} & 2B_{33} \end{bmatrix}$$

We require to find the coefficients B'_{ij} of the equation of the ellipsoid relative
to the new axes. The new symmetric matrix $2B'$ is given by L'^*2BL', where
L' is the transpose of L'^* (columns and rows of L'^* transposed to give L')
[18(135)], (I(18), p. 338). For the general case this multiplication of three
matrices is somewhat tedious and the result, given below, difficult to interpret.

$$\lambda^2 B'_{11} = a^2 B_{11} + b^2 \cos^2 \gamma B_{22} + c^2 \cos^2 \beta B_{33} + ab \cos \gamma B_{12}$$
$$+ bc \cos \gamma \cos \beta B_{23} + ca \cos \beta B_{31}$$
$$\lambda^2 B'_{22} = b^2 \sin^2 \gamma B_{22} + c^2 A^2 B_{33} + bcA \sin \gamma B_{23}$$
$$\lambda^2 B'_{33} = c^2 B^2 B_{33}$$
$$\lambda^2 B'_{12} = 2b^2 \cos \gamma \sin \gamma B_{22} + 2c^2 A \cos \beta B_{33} + ab \sin \gamma B_{12}$$
$$+ bc(A \cos \gamma + \sin \gamma \cos \beta)B_{23} + caAB_{31}$$
$$\lambda^2 B'_{23} = 2c^2 ABB_{33} + bcB \sin \gamma B_{23}$$
$$\lambda^2 B'_{31} = 2c^2 B \cos \beta B_{33} + bc \cos \gamma B_{23} + caBB_{31}.$$

It can be simplified somewhat for monoclinic and hexagonal lattices, and for orthogonal lattices has the simple form

$$2B' = \frac{1}{\lambda^2} \begin{bmatrix} 2a^2 B_{11} & abB_{12} & acB_{13} \\ baB_{21} & 2b^2 B_{22} & bcB_{23} \\ caB_{31} & cbB_{32} & 2c^2 B_{33} \end{bmatrix}.$$

The matrix B' gives the form of the equation of the ellipsoid for arbitrary (relative to the ellipsoid) orthonormal axes, $\mathbf{i}, \mathbf{j}, \mathbf{k}$. It is then required to find the directions of a new set of orthonormal axes, $\mathbf{i}', \mathbf{j}', \mathbf{k}'$, for which the equation takes the form $\mu_1 x^2 + \mu_2 y^2 + \mu_3 z^2 = 2\pi^2/\lambda^2$. The new axes $\mathbf{i}', \mathbf{j}'$, $\mathbf{k}'$ are then along the axes of the ellipsoid. $\mathbf{i}', \mathbf{j}', \mathbf{k}'$ can be found by setting up the *characteristic equation* of the quadratic form B' and solving for the three roots, μ_1, μ_2 and μ_3. For each of these roots in turn we can then obtain the direction cosines of the new axes, $\mathbf{i}', \mathbf{j}', \mathbf{k}'$ relative to $\mathbf{i}, \mathbf{j}, \mathbf{k}$. In matrix algebra this can be done by finding the *latent roots* μ_1, μ_2, μ_3 and the *latent vectors*—unit vectors in the direction of the ellipsoid axes—in terms of their components, which are also their direction cosines relative to $\mathbf{i}, \mathbf{j}, \mathbf{k}$. Since the latent roots and vectors will probably have been found for the matrix $2\lambda^2 B'$, the roots μ'_1, μ'_2, μ'_3, must be divided by $2\lambda^2$ to give μ_1, μ_2, μ_3.

$$\therefore \frac{1}{2\lambda^2} (\mu'_1 x^2 + \mu'_2 y^2 + \mu'_3 z^2) = 2\pi^2/\lambda^2.$$

The length of the x axis of this $w = 1/\bar{u}$ ellipsoid is given by the value of x when $y = z = 0$.

$$\therefore w_x^2 = 1/\bar{u}_x^2 = \frac{4\pi^2}{\mu'_1}. \qquad \therefore \bar{u}_x = \frac{\sqrt{\mu'_1}}{2\pi}.$$

Similarly,

$$\bar{u}_y = \frac{\sqrt{\mu'_2}}{2\pi}; \qquad \bar{u}_z = \frac{\sqrt{\mu'_3}}{2\pi}.$$

The direction cosines are unaffected by multiplying the matrix by a constant.

So far we have assumed that the original B_{ij}'s from the least squares analysis are accurate. But they may very well have 'soaked up' some of the inevitable experimental errors in the intensities, as well as inadequacies in the atomic scattering factors, and have produced an inconsistent set of coefficients.

3. The General Restrictions on the Values of B_{ij}

If the vibration ellipsoid is a reasonable approximation to the actual motion of the atom, then the latent roots of B' must all be positive, or the matrix must be 'positive definite'. The matrix of the quadratic form for the axes, $\mathbf{i}', \mathbf{j}', \mathbf{k}'$ (B'') is given by:

$$B'' = \begin{bmatrix} \mu_1 & 0 & 0 \\ 0 & \mu_2 & 0 \\ 0 & 0 & \mu_3 \end{bmatrix}.$$

with μ_1, μ_2 and μ_3 all positive. To find what restriction this places on the original coefficients B_{ij}, we transform back from the final axes, $\mathbf{i}'$, $\mathbf{j}'$, $\mathbf{k}'$, to the original axes $\mathbf{a}^*$, $\mathbf{b}^*$ and $\mathbf{c}^*$.

Let
$$\mathbf{a}^* = l_{11}\mathbf{i}' + l_{21}\mathbf{j}' + l_{31}\mathbf{k}'$$

$$\mathbf{b}^* = l_{12}\mathbf{i}' + l_{22}\mathbf{j}' + l_{32}\mathbf{k}' \tag{1}$$

$$\mathbf{c}^* = l_{13}\mathbf{i}' + l_{23}\mathbf{j}' + l_{33}\mathbf{k}'$$

and
$$L^* = \begin{bmatrix} l_{11} & l_{21} & l_{31} \\ l_{12} & l_{22} & l_{32} \\ l_{13} & l_{23} & l_{33} \end{bmatrix}.$$

Then $B = L^*B''L$ (L is the transpose of L^*).

$$B''L = \begin{bmatrix} \mu_1 l_{11} & \mu_1 l_{12} & \mu_1 l_{13} \\ \mu_2 l_{21} & \mu_2 l_{22} & \mu_2 l_{23} \\ \mu_3 l_{31} & \mu_3 l_{32} & \mu_3 l_{33} \end{bmatrix}$$

$$L^*B''L = B = \begin{bmatrix} \mu_1 l_{11}^2 + \mu_2 l_{21}^2 + \mu_3 l_{31}^2 \\ \mu_1 l_{11} l_{12} + \mu_2 l_{21} l_{22} + \mu_3 l_{31} l_{32} \\ \mu_1 l_{11} l_{13} + \mu_2 l_{21} l_{23} + \mu_3 l_{31} l_{33} \\ \\ \mu_1 l_{12} l_{11} + \mu_2 l_{22} l_{21} + \mu_3 l_{32} l_{31} \\ \mu_1 l_{12}^2 + \mu_2 l_{22}^2 + \mu_3 l_{32}^2 \\ \mu_1 l_{12} l_{13} + \mu_2 l_{22} l_{23} + \mu_3 l_{32} l_{33} \\ \\ \mu_1 l_{13} l_{11} + \mu_2 l_{23} l_{21} + \mu_3 l_{33} l_{31} \\ \mu_1 l_{13} l_{12} + \mu_2 l_{23} l_{22} + \mu_3 l_{33} l_{32} \\ \mu_1 l_{13}^2 + \mu_2 l_{23}^2 + \mu_3 l_{33}^2 \end{bmatrix} \tag{1}$$

$$= \tfrac{1}{2} \begin{bmatrix} 2B_{11} & B_{12} & B_{13} \\ B_{21} & 2B_{22} & B_{23} \\ B_{31} & B_{32} & 2B_{33} \end{bmatrix}.$$

It is immediately clear that $B_{ij} = B_{ji}$ and that B_{11}, B_{22} and B_{33} must all be positive.

If we calculate $\dfrac{2B_{11} 2B_{22} - B_{12}^2}{4}$ we find:

$$B_{11} B_{22} = (\mu_1 l_{11}^2 + \mu_2 l_{21}^2 + \mu_3 l_{31}^2)(\mu_1 l_{12}^2 + \mu_2 l_{22}^2 + \mu_3 l_{32}^2)$$

$$= \mu_1^2 l_{11}^2 l_{12}^2 + \mu_2^2 l_{21}^2 l_{22}^2 + \mu_3^2 l_{31}^2 l_{32}^2 + \mu_1 \mu_2 (l_{11}^2 l_{11}^2 + l_{21}^2 l_{12}^2)$$

$$+ \mu_2 \mu_3 (l_{21}^2 l_{32}^2 + l_{31}^2 l_{22}^2) + \mu_3 \mu_1 (l_{31}^2 l_{12}^2 + l_{11}^2 l_{32}^2).$$

$$\frac{B_{12}^2}{4} = (\mu_1 l_{12} l_{11} + \mu_2 l_{22} l_{21} + \mu_3 l_{32} l_{31})^2$$

$$= \mu_1^2 l_{12}^2 l_{11}^2 + \mu_2^2 l_{22}^2 l_{21}^2 + \mu_3^2 l_{32}^2 l_{31}^2 + 2\mu_1 \mu_2 l_{12} l_{11} l_{22} l_{21}$$

$$+ 2\mu_2 \mu_3 l_{22} l_{21} l_{32} l_{31} + 2\mu_3 \mu_1 l_{32} l_{31} l_{12} l_{11}.$$

$$\therefore \frac{4B_{11} B_{22} - B_{12}^2}{4} = \mu_1^2 (l_{11}^2 l_{12}^2 - l_{12}^2 l_{11}^2) + \mu_2^2 (l_{21}^2 l_{22}^2 - l_{22}^2 l_{21}^2)$$

$$+ \mu_3^2 (l_{31}^2 l_{32}^2 - l_{32}^2 l_{31}^2) + \mu_1 \mu_2 (l_{11} l_{22} - l_{21} l_{12})^2$$

$$+ \mu_2 \mu_3 (l_{21} l_{32} - l_{31} l_{22})^2 + \mu_3 \mu_1 (l_{31} l_{12} - l_{11} l_{32})^2.$$

$\therefore 4B_{11} B_{22} - B_{12}^2 > 0$ and similar calculations show that $4B_{ii} B_{jj} - B_{ij}^2 > 0$.

From the theorem that the determinant of the product of two matrices is equal to the product of the determinants of the matrices [18(135)], we have

$$\Delta B = \Delta L^* . \Delta B'' . \Delta L$$

Since the value of the determinant is not altered by interchanging its rows and columns

$$\Delta B = (\Delta L^*)^2 \, \mu_1 \mu_2 \mu_3 > 0.$$

(ΔL^* cannot be zero since this would make $\mathbf{a}^*$, $\mathbf{b}^*$, $\mathbf{c}^*$ co-planar, 332(1)).

$$\therefore \quad \Delta B = \tfrac{1}{4}(4B_{11}B_{22}B_{33} + B_{12}B_{23}B_{31} - B_{11}B_{23}^2 - B_{22}B_{31}^2 - B_{33}B_{12}^2) > 0.$$

Therefore the general restrictions on the possible values of the B_{ij}'s are that:

$$B_{ii} > 0; \qquad 4B_{ii}B_{jj} - B_{ij}^2 > 0; \qquad \Delta B > 0.$$

If the B_{ij}'s produced by least squares refinement violate these conditions, this implies that the atomic vibrations are described by an imaginary ellipsoid or a paraboloid or hyperboloid, and probably means that the data are not accurate enough to justify using anisotropic temperature factors.

4. Particular Restrictions on the Values of B_{ij} due to Symmetry

If an atom lies at a centre of symmetry the vibration ellipsoid, having itself a centre, conforms to the symmetry and no additional condition arises. If the atom lies on a 2–fold axis (including $\bar{2} = m$), then one of the three axes of the ellipsoid (which are also 2–fold axes of symmetry) must lie along this axis. If the symmetry of the atomic position is $2/m$ no further conditions arise, since the symmetry of the ellipsoid is $2/m, 2/m, 2/m$. Let the symmetry axis be $\mathbf{b}^*$. Then $\mathbf{a}^*$ and $\mathbf{c}^*$ will be perpendicular to it, $\mathbf{j}$ and $\mathbf{j}'$ will both lie in the direction of $\mathbf{b}^*$, and $\mathbf{i}'$ and $\mathbf{k}'$ will be in the $\mathbf{a}^*\mathbf{c}^*$ plane. The matrix L^* (from 394(1)) will be given by

$$L^* = \begin{bmatrix} l_{11} & 0 & l_{31} \\ 0 & l_{22} & 0 \\ l_{13} & 0 & l_{33} \end{bmatrix}$$

From 395(1), $B = \begin{bmatrix} \mu_1 l_{11}^2 + \mu_3 l_{31}^2 & 0 & \mu_1 l_{13} l_{11} + \mu_3 l_{33} l_{31} \\ 0 & \mu_2 l_{22}^2 & 0 \\ \mu_1 l_{11} l_{13} + \mu_3 l_{31} l_{33} & 0 & \mu_1 l_{13}^2 + \mu_3 l_{33}^2 \end{bmatrix}$

In this case B_{12} and B_{23} are zero.

For an atom at a position with point symmetry $2/m\,2/m\,2/m$, the ellipsoid axes lie in the direction of the RL axes and

$$L^* = \begin{bmatrix} l_{11} & 0 & 0 \\ 0 & l_{22} & 0 \\ 0 & 0 & l_{33} \end{bmatrix} \qquad B = \begin{bmatrix} \mu_1 l_{11}^2 & 0 & 0 \\ 0 & \mu_2 l_{22}^2 & 0 \\ 0 & 0 & \mu_3 l_{33}^2 \end{bmatrix}$$

and $B_{11} \neq B_{22} \neq B_{33}$; $\quad B_{ij} = 0.$

For an atom at 4, $\bar{4}$ or $4/m$, $\mathbf{i}'$ and $\mathbf{j}'$ are any two directions at right angles in the plane perpendicular to the axis (which is $\mathbf{k}'$). Also $\mu_1 = \mu_2$ and

$$L^* = \begin{bmatrix} l_{11} & -l_{12} & 0 \\ l_{12} & l_{11} & 0 \\ 0 & 0 & l_{33} \end{bmatrix}$$

$$B = \begin{bmatrix} \mu_1(l_{11}^2 + l_{12}^2) & 0 & 0 \\ 0 & \mu_1(l_{12}^2 + l_{11}^2) & 0 \\ 0 & 0 & \mu_3 l_{33}^2 \end{bmatrix}$$

$B_{11} = B_{22} \neq B_{33}$; $\quad B_{ij} = 0.$

For 3 or 6–fold symmetry (with or without planes) $\mu_1 = \mu_2$

$$L^* = \begin{bmatrix} l_{11} & l_{21} & 0 \\ \dfrac{l_{11} - \sqrt{3}l_{21}}{2} & \dfrac{\sqrt{3}l_{11} + l_{21}}{2} & 0 \\ 0 & 0 & l_{33} \end{bmatrix}$$

$$B = \begin{bmatrix} \mu_1(l_{11}^2 + l_{21}^2) & \dfrac{\mu_1}{2}(l_{11}^2 + l_{21}^2) & 0 \\ \dfrac{\mu_1}{2}(l_{11}^2 + l_{21}^2) & \mu_1(l_{11}^2 + l_{21}^2) & 0 \\ 0 & 0 & \mu_3 l_{33}^2 \end{bmatrix}$$

$\therefore\, B_{11} = B_{22} \neq B_{33}$; $\quad B_{12} = \tfrac{1}{2}B_{11}$; $\quad B_{13} = B_{23} = 0.$

Appendix VII

Mechanical Integration of X-ray Reflections

1. The Integrating Mechanism for a Weissenberg Camera

A 14-notch star wheel A, (Fig. 214(a) & (b)) with locating roller and spring B is turned by means of pins on its front face which interact with adjustable stops on the camera base at each end of the oscillation. There are 7 pins, and 7 double oscillations are required for a complete revolution of the star wheel in 14 steps. Attached to A is a step cam C, which raises a rod D and lever E about its fixed pivot F. The movement of the application pad G is controlled from zero to about 4 mm by the position of its carrier block on the arm E. G supports a weighted flap H on the side of the cylindrical film holder I, which rests on rollers J. One pair of rollers is plain, the other V-shaped for location in a V-groove on the film holder. The peripheral movement of the film in mm is given by the reading of the scale on E. When the step of the cam is reached the mechanism returns to its starting point under its own weight, providing the bearing surfaces are adequately lubricated.

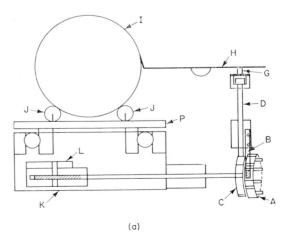

(a)

FIG. 214. (a) A section of the carriage perpendicular to the rotation axis.

398

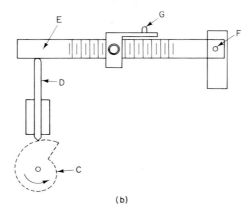

(b)

FIG. 214(b) The step cam operating lever, with its adjustable pad, G.

The spindle of the star wheel and cam is extended into the main base of the carriage K, and a worm at its far end rotates a worm wheel carrying a heart-shaped cam L (Figs. 214(c) & (d). This cam operates a rod which pushes against a slotted lever M, pivoted in a block N, fixed to the main base K. The slot extends above the pivot. The lever M is held against the rod by a spring. The pointer O indicates the number of revolutions of the star wheel from the maximum position of the cam L (30 revolutions from point to heel).

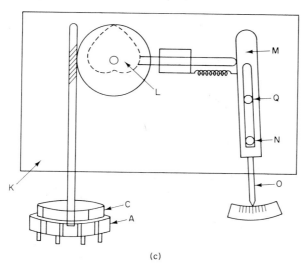

(c)

FIG. 214. (c) A plan view of the main carriage, showing the operation of the heart shaped cam and lever.

The movement of the lever M is used to give longitudinal motion to the subsidiary carriage base P. This is carried in the main base by sprung V-grooves and ball bearings. It is connected to the slotted lever M by a movable pivot Q, which is clamped to the subsidiary base P. The relative movement of the main and subsidiary base is thus equal to the relative movement of N and Q. This movement can be adjusted from 0 to 6 mm by altering the position of the clamped pivot Q. The amount of movement is given by the reading of the scale R.

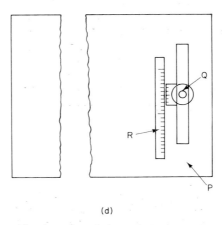

(d)

FIG. 214. (d) The movable pivot Q, and the scale R, for setting the longitudinal range.

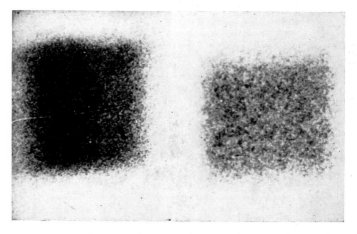

FIG. 215. Two enlargements of integrating Weissenberg reflections, one showing the effect of a sticking cam-follower rod. The graininess has been enhanced by reproduction.

Any sticking of the integrating movement will normally show up as irregularities in the spots on the photograph, Fig. 215, but two points should be particularly inspected for wear. These are the points (one inside the platform) where the linear movement of a plunger is turned into the radial movement of an arm, and where there is therefore sliding movement between plunger and arm. The arm is liable to wear into pits and give rise to sticking on the spring or gravity loaded return movement. If the arm is not already of hardened steel, an insert should be let in where the plunger bears against it. Adequate lubrication must be provided to prevent sticking.

The effect of the integration is to divide up the area of any given reflection by a slightly distorted rectangular grid, whose unit is 1/14 of the peripheral movement and 1/30 of the longitudinal movement. These grid units will normally be of the order of 0·1 mm. Over these areas the intensity is assumed to be uniform, and therefore to produce a uniform increase in optical density on the film proportional to this uniform intensity. The sum of the densities due to all the grid areas is taken as a measure of the integrated intensity.

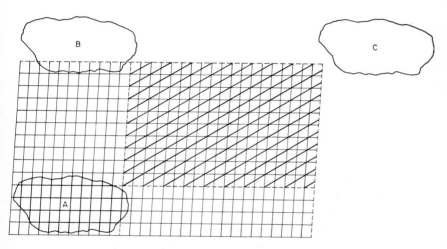

Fig. 216. The movement of the reflection relative to the film in the course of an integration cycle.

At each oscillation the effect is to move the spot relative to the film from its initial position A (Fig. 216), to a position one grid distance higher, until it reaches position B. It then drops back to position A at the 14th oscillation, but one step to the right, and repeats the 13 steps to the top. This process continues until at the end of the 30th movement from bottom to top it arrives at position C. Any of the grid areas in the shaded centre portion has had all the areas outlined in A fall on it once, so that the final density will be the sum of the densities due to all the areas, which is the integrated density. The central shaded area will therefore be of uniform density equal to the integrated density for the reflection. The size of this area is slightly larger than the difference between the integrating range and the size of the reflection in the two directions. If the integrating range is made 0·6 mm larger than the size of the largest spot to be measured, then there should be an area about 0·7 mm square available for photometry. If a photometer beam of 0·5 mm diameter is used, errors of placing the spot over the photometer beam will normally be avoided.

The integrating effect is enhanced by the fact that the grid slopes in the opposite direction when the heart-shaped cam is making its return movement.

2. Integrating Mechanism for a Precession Camera

The film holder is located in a secondary frame which has linear movements in two directions at right angles similar to those of a Weissenberg cassette. These movements are operated by the rotation of the precession arm, and variable pivots enable the extent of the movement to be set as in the Weissenberg carriage.

3. Setting the Integration Limits

3.1 LOW AND MEDIUM ANGLE REFLECTIONS

Integration limits should in general be 0·6 mm greater than the largest dimension of any spot on a non-integrated film, in there levant direction. It is important to note that an oscillation spot (e.g. on a layer line screen checking photo) will not be as long as the corresponding Weissenberg spot even on the zero layer. If such spots are used to find the integration limits, the central plateau of the integrated spot will be too small for accurate photometry. It is better to have the integration limits slightly too large than too small.

On non-zero equi-inclination Weissenberg photographs the lengthening of spots from relps near the rotation axis (Sec. **18**.2.2.2 p. 272) could lead to impossibly large integration limits, especially if the effect is enhanced by deliberate missetting of μ to avoid double reflections (IX 1.3(c), p. 433). Since such reflections will also have large errors in the calculated Lp correction factors, they should be ignored and the intensities measured about another axis.

3.2 HIGH ANGLE REFLECTIONS WITH $\alpha_1 \alpha_2$ SEPARATION

This is a problem which only occurs with Weissenberg cameras, since a precession camera can only record up to $\theta = 30°$. For integrated Weissenberg films, errors can arise in the photometry of high angle reflections, due to α_1, α_2 separation, unless this factor is taken into account in setting the integrating ranges. Of course, if the doublet is treated as a single reflection, and the greatest measurement across any doublet on a non-integrating photograph is added to the diameter, b, of the photometer beam to give the integrating range in that direction, then there will be an area of diameter b, in the centre of the resulting integrated spot, whose uniform density will be proportional to the total energy in $\alpha_1 + \alpha_2$. However, this may result in unnecessarily large integration ranges, with consequent long exposures, and possibly in overlapping of neighbouring reflections. What is required are integrating ranges which enable $\alpha_1 + \alpha_2$ to be measured as a single reflection just up to the point where α_1 and α_2 can be measured separately.

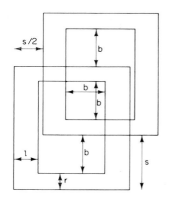

Fig. 217. Diagram of an integrated $\alpha_1\alpha_2$ doublet at the 'changeover point' between making a single measurement and treating α_1 and α_2 separately.

The situation at the changeover point is illustrated in Fig. 217. l is the longitudinal non-integrated spot extension (assumed equal for α_1 and α_2) and r the extension in the perpendicular (circumferential) direction. The inner rectangles represent the area of uniform density for the individual members of the doublet; and the overlap of these two and the areas clear of overlap top and bottom must just be large enough to take the photometer beam. If the α_1 α_2 separation, s, is less than the critical value (i.e. for smaller θ) the central, overlap, area will be larger and can be photometered; if s is larger, the two individual areas will be larger and can be photometered separately. For a standard Weissenberg camera the sideways (longitudinal) separation will be $s/2$, as shown for zero layers, and not very different for non-zero layers.

For the critical separation, $s = b + r$, the circumferential integration range $= 2(b + r)$, and the longitudinal range $= s/2 + l + b = 3/2b + l + r/2$. The critical value of θ can be calculated as follows:

$$s = 2R\Delta\theta \text{ where } R \text{ is the camera radius.}$$

$$\Delta\lambda = 2d \cos \theta\Delta\theta = \lambda \cot \theta\Delta\theta.$$

$$\therefore \Delta\theta = \frac{\Delta\lambda}{\lambda} \tan \theta = \frac{s}{2R}.$$

$$\therefore \tan \theta = \frac{s}{2R} \cdot \frac{\lambda}{\Delta\lambda} = \frac{b+r}{2R} \cdot \frac{\lambda}{\Delta\lambda}. \tag{1}$$

Because of the back reflection focusing effect, r for good crystals will be smaller than the corresponding measurement r' for low angle reflections. If we take $r = r'/2 = 0.2$, $b = 0.5$ and $2R = 57.3$ mm, for CuKα radiation we have

$$\tan \theta = \frac{0.7}{57.3} \cdot \frac{1.54}{3.82 \times 10^{-3}} = 4.93.$$

$$\therefore \theta = 78.5°.$$

If the integration range required for low angle reflections $(r' + b)$ were to be chosen, it would still be possible to measure all reflections with $\theta > 78.5°$ accurately, by taking α_1 and α_2 separately; but there would be a large number with $\theta < 78.5°$ which would not have sufficient uniform areas to measure either separately or overlapping. When θ becomes small enough for adequate overlap area to occur for this integration range, we have:

$$r' + b - (s + r) = b \quad \therefore s = r' - r = r$$

and this is assuming that the focusing effect is still keeping $r = r'/2$.

$$\therefore \tan \theta = \frac{r}{2R} \cdot \frac{\lambda}{\Delta\lambda} = 1.41$$

$$\therefore \theta = 54.5°.$$

Some 40% of the observable reflections would be affected, and since r would have increased the number would be even greater. Most would only be slightly affected, because the plateau of density at the centre of an integrated spot slopes very gradually at the edges. However, the problem is obviously a serious one and the correct integrating limits should be applied whenever possible. Figure 218 gives the integrating limits for $b = 0.5$ mm and two sets of values of r, r' and l. l can be taken as the same for high and low angle reflections. The limits are given (a) as derived for low angle reflections, (b) the correct values for all reflections as derived above and (c) treating all reflections as single (i.e. producing an adequate overlapping uniform area for all reflections) for θ up to (i) 80°, (ii) 85°.

				Integration limits (mm)			
Spot size				Case	Case	Case (c)	
r	r'	l		(a)	(b)	(i)	(ii)
0·2	0·4	0·6	Circ.	0·9	1·4	1·5	2·3
			Long.	1·1	1·45	1·5	1·9
0·5	1·0	1·5	Circ.	1·5	2·0	1·85	2·7
			Long.	2·0	2·5	2·45	2·85

FIG. 218. Spot size and integration limits for selected conditions. r = circumferential spot size for high angle reflexions; r' = the corresponding measurement for low angles; l = spot size in the axial direction. Limits for case (a) are calculated for low angle reflexions; case (b) gives the optimum limits; and case (c) the limits required treating $\alpha_1\alpha_2$ doublets as single reflexions for maximum θ (i) 80°, (ii) 85°.

4. Photometry of Integrated Reflections

Graininess and slight undulations of density across the central plateau of an integrated reflection are averaged by the employment of a large area photometer beam (0·5 mm in diameter). It is this large beam (compared with a 'flying spot' integrating photometer) together with a 500 mm scale on which the photo cell output is measured, which is largely responsible for the higher accuracy of intensity measurement attainable with mechanically integrated reflections. A simple photometer suitable for this purpose is described in Appendix X, p. 466).

Appendix VIII

A Safe X-ray Tube Shutter and Universal Mounting and Alignment System for X-ray Goniometers

1. General

This appendix describes an existing arrangement for the safe operation of most X-ray cameras, with a note on methods of dealing with exceptional problems. The mounting and alignment system is a slight elaboration of the Manchester design described by Hughes and Taylor in 1961, and the shutter is based on a design produced by the National Physical Laboratory.

2. The Shutter

This was designed to be attached to a standard sealed tube shield by four waisted 2BA screws at the corners of a 47·5 mm square. The principle is very simple. Two lugs riveted and soldered to the springs E (Fig. 219) fit into recesses in the shutter plate D. Either is sufficient to prevent the shutter being opened and both must be pushed down by the projecting hooks on the cap F, before the shutter can be moved. The cap F is fastened to the camera and forms part of the two labyrinths connecting the camera collimator to the tube (Figs 220, 221 and 222). The labyrinth in the claw cap allows 3 mm relative movement in any direction and also takes up angular misalignment. When the shutter is opened the inner edges of plate D engage in the hooks of cap F, locking the camera to the tube. The plates at the two ends of the shutter body C trap any slight radiation leakage along the shutter when it is open. This leakage is extremely small and cannot be detected when the plates are attached; but it might be a slight improvement in the design to extend downwards the recess in the base plate A, so that a second shouldered hard lead plug, similar to that which comes over the window in the 'closed' position of the shutter, but with a central hole in it, could be incorporated to come opposite the window in the 'open' position.

406

The base plate of the shutter is wedge shaped, so that the front face is 'square' to the apparatus for a target take off angle of 6° [1(82)]. (This fore-shortens a 10 × 1 mm focus to 1 × 1 mm). However the linkage will allow the angle to be varied from 0 to 10° if required, although variations of more than 1 or 2° will require repositioning of the shutter (VIII (4.1), p. 413). A simple gravity shutter on the camera (Figs 221, 222) stops the beam if the collima-tor is accidentally omitted.

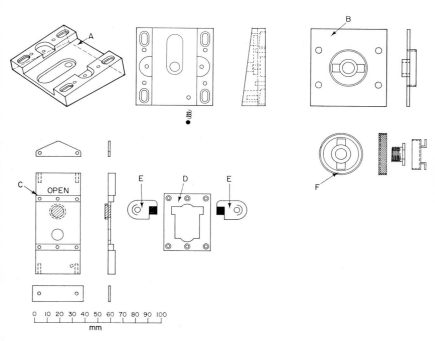

FIG. 219. Details of the construction of the N.P.L. type safety shutter. The springs E, which are the only critical parts, are made from 0·2 mm hard phosphor bronze sheet (BSS 407/2). The components of the cap F are soldered together, giving a strength ap-proximately equal to that of the hook lugs. A maniac could break off the lugs or rupture the joint! For vertical tubes the friction generated by the springs E is aided by a spring loaded ball and dimple, to prevent the shutter falling closed.

3. Stands and Alignment

3.1 GENERAL

The object of the system is (a) to allow a satisfactory alignment of a camera against an X-ray tube, using optical methods, and (b) to allow the inter-change of cameras without realignment.

A standard tube window height is chosen which will accommodate the greatest collimator height above base that is likely to be required. 375 mm above the table top has proved to be sufficient.

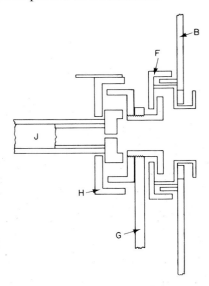

FIG. 220. A sectional diagram showing the labyrinth arrangements connecting the shutter cover plate *B*, to the collimator *J*. *H* is a lead cap which fits over the collimator *J*, and has a brass segment riveted to it to support the gravity shutter (Figure 229, p. 418 and Figure 87(a), p. 136). *G* is the pivoted arm, attached to the camera (Figures 221 and 87(a). *F* is the clawed cap, whose construction allows both lateral and angular movement relative to the camera arm, to take account of slight variations in the geometry and to allow final adjustment using X-rays, in special cases. The inner flanged sleeve is screwed into the arm *G*, and the outer cap screwed on as a lock nut.

3.2 MAIN STAND ALIGNMENT USING A 'DUMMY CAMERA'

An adjustable main stand (400 × 450 mm wide in 12·5 mm Grade A paxolin sheet), with 'groove and plane' mounting, enabling a subsidiary stand to be withdrawn from the tube (Fig. 223), is fitted to the bench in front of each window. These stands are adjusted using a simple 'dummy camera' (Fig. 224) which has a central 0·55 mm pinhole with a standard shutter attachment at the tube end, and a fluorescent screen with central cross lines and a lead glass plate at the other. The pinhole picture of the foreshortened focus is brought on to the cross lines. This automatically aligns the 'groove' parallel to the direction of the X-ray beam, since the feet which fit in the groove are made parallel to the axis of the tube. Each main stand after alignment has exactly the same geometrical relation to its X-ray tube as every other stand, so that a camera on its subsidiary stand (Figs. 225 and 226) which is aligned for one main stand can be transferred to another without requiring any readjustment (other than the stop on the V-groove, VIII 4.3 (b), p. 417).

FIG. 221. Close-up of the camera attachment.

3.3 CAMERA AND SUBSIDIARY STAND ALIGNMENT USING A 'DUMMY X-RAY TUBE'

A subsidiary stand is designed for its particular camera, with three pillars B (Fig. 226) to take the feet of the camera in a ball, slot and plane mounting. The pillars with the conical hole and the V-groove can be adjusted in slots C. The 'hole' pillar adjustment moves the camera collimator across the tube window, and final positioning in this direction can be done by differential adjustment of the camera feet. The 'slot' pillar adjustment is more critical, because there is no equivalent adjustment on the feet. It is used to bring the vertical

FIG. 222. The camera locked to the tube by the open shutter.

FIG. 223. The main stand, with detachable adjusters in position, carrying the dummy camera with plumb line extension for preliminary lining up using the bench mark.

Fig. 224. The dummy camera, locked to the tube for final alignment of the main stand, using X-rays.

Fig. 225. A Nonius integrating, equi-inclination Weissenberg camera, with its subsidiary stand, on the optical aligning jig. The detachable screw adjuster for the 'slot' pillar is shown in position.

plane containing the collimator parallel to the vertical plane through the X-ray beam (or the V-groove on the main stand). A detachable screw adjuster is therefore provided for this pillar (Fig. 225).

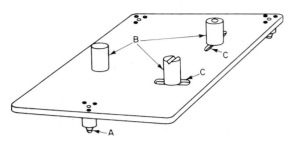

FIG. 226. Diagram of the construction of a subsidiary stand. The positions of the pillar will be determined by the geometry of the feet of the particular camera.

A camera is aligned on its subsidiary stand using an optical dummy X-ray tube (Fig. 225), consisting of two pinholes, the rear one of 1·0 mm diameter illuminated by a projector lamp with integral reflector, and the front one, corresponding to the window of the X-ray tube, either 1·75 mm or 0·33 mm

FIG. 227. The dummy X-ray tube with the dummy camera in position for checking that the original alignment of the stand has not changed. The 6° upward direction of the beam is an 'accident' in carrying out the design. It could equally well have been horizontal.

depending on whether or not the small hole cap is in position. This dummy tube is mounted so that the front pinhole is at the standard tube window height above the table, to which a main stand is fitted as for an X-ray tube. The direction of the beam, so long as it is within the range of the adjustments on the stand, is immaterial. The dummy camera is used to adjust this stand, a cap with a ground lead glass screen replacing the one with a fluorescent screen for this purpose (Fig. 227). The geometrical relation of the light beam to the stand is then exactly the same as that of the X-ray beam to the main tube stands.

A camera on its subsidiary stand can now have the collimator lined up up with the light beam, by adjustment of the pillars B and the camera legs, and is then ready for use against any of the X-ray tube windows.

4. Detailed Procedures

4.1 PROCEDURE FOR SETTING UP A MAIN STAND AGAINST AN X-RAY TUBE

(a) *Safety*

Shut off the power and turn the H.T. and filament controls to zero. Turn off the water supply and **remove the mains plug or fuses.** Turn on and check the portable audible warning monitor and place it on the bench where scatter from the window could reach it.

(b) *Removal of shutter*

Take out the four round headed screws holding the shutter cover plate B (Fig. 219, p. 407) in place and remove the cover. Take out the screws holding the shutter base plate on to the tube face, taking care not to let the shutter mechanism drop. Note that the screws are 'waisted' to allow free adjustment, and that the position of the base cannot be properly adjusted until these screws are almost fully home. **Do not dismantle the shutter further.**

(c) *Marking out the stand position*

Attach the jig provided (Fig. 228) to the tube face, so that the bottom bar rests squarely on the table. Use the bottom plate to mark out the centre line on the table—at right angles for a vertical tube and at a 6° take off angle for a horizontal tube.

(d) *Location of shutter*

Remove the jig, place the appropriate aligning insert in the tube shield window and offer up the shutter base to the tube face, so that the projection on the insert goes in to the hole in the base. Make a pencil mark on the tube

Fig. 228. The marking out jig in position. The vertical and horizontal pieces are separate, with a tongue and groove location.

face to mark the position of the base, remove the insert, and fasten the base to the tube face in this position.†

A certain amount of movement is possible in the direction at right angles to the main slot adjustment. In this direction the base should be in the mid-way position.

† This assumes that the 6° take-off line from the center of the focus comes approximately in the middle of the window. If a new type of shield is to be used the Radiation Protection Officer, or someone designated by him, must try this out, using the full beam of X-rays from the shutter window. For this purpose a large (250 × 250 mm) lead glass-covered fluorescent screen should be placed 100 mm from the shutter and surrounded by overlapping screens. A special cap (without the central disc), for which the R.P.O. must be responsible, is used to open the shutter with the set OFF and the screens put in place. The set is operated at low voltage until the shape of the beam can be seen, and the direction and amount of movement of the shutter base deduced. The set is turned OFF and the shutter moved before a final check. During this process it is essential that an audible warning monitor be in position.

In the case of Nonius square shields the mark for the shutter should be 2·0 mm away from the thick end of the base.

Replace the cover.

The window of the tube itself, inside the shield, is very delicate. On no account must anything be pushed through the shield window while it is exposed during this operation.

(e) *Adjustment of stand*

    (i) Place the main stand centrally on the table by eye, so that the base is just not overlapping the edge of the table (or up to 450 mm from the tube face). Place the dummy camera on the stand and move it up until it is nearly touching the shutter housing.

    (ii) Adjust the front links equally (using the screw attachments, Fig. 223 (p. 410) to bring the dummy camera shutter cap to the same height as the shutter opening, keeping the front edge of the stand roughly parallel to the front face of the shutter.

    (iii) Adjust the rear links equally to give the slope of the dummy camera from the table (a) 6° for a vertical tube; (b) 0° for a horizontal tube, using the wedge and spirit level, as required.

    (iv) Adjust the stand on the table (using the dummy camera extension and plumb line for the rear alignment) and if necessary do a further adjustment on the front links to enable the shutter cap to be pressed into the shutter housing. Lock them together by opening the shutter.

    (v) Do final adjustments to bring the bob over the centre line on the table and the front of the dummy camera central with the cap.

    (vi) Make sure that the cap with the fluorescent screen is on the dummy camera and turn on the tube. **Check with the monitor that no X-rays are coming from the cap window.** A pinhole picture of the focus should be seen on the screen. If not, check the previous procedure and if the focus is still not visible, move the dummy camera slightly by hand, or by raising or lowering the front links, if necessary, until an image is seen. By small movements of the tube end of the stand (using light taps from the flat of the hand) and equal movement of the front links, bring the image roughly on to the cross wires. Screw the base to the table.

    (vii) Using equal and if necessary small differential movements of the front links, bring the image exactly on to the cross wires.

(f) *Checking the shutter position*

    (i) Check the position of the front of the dummy camera relative to the cap. If this is out by more than 0·5 mm in the direction of the length of the tube shield, mark the alteration required in the position of the shutter housing. If it is out in the direction at right angles to the length.

this indicates either a fault in carrying out the procedure or a tube focus which is off centre. In the latter case it may be necessary to alter the angular adjustment

(ii) Close the shutter, withdraw the dummy camera and **SWITCH OFF THE TUBE.** If it is necessary to adjust the shutter, take off front plate B and loosen the fixing screws half a turn only. Make the adjustment tighten the screws and replace the cover B. In the case of an off centre focus, all the adjustment possible at right angles to the slot should be taken up in the appropriate direction, and the angles of 4.1(e) (iii) or (v) altered if further movement is required (1° produces a movement of approximately 1 mm at the cap).

4.2 PROCEDURE FOR ALIGNING A CAMERA AND ITS SUBSIDIARY STAND

If in any doubt check the camera adjustments (App. IX) and correct if necessary.

(a) Rough alignment. Place the camera and stand on the alignment jig (Fig. 225, p. 411) with the small hole cap removed. Adjust the camera legs so that they are approximately at the mid-position of the range. With the claw arm lowered check that the (large) collimator is nearly the same height as the beam from the dummy X-ray tube. If it is too far out the height of the pillars B (Fig. 226, p. 412) will have to be altered. If it is within the range adjust the screws equally until the collimator is the same height as the beam If necessary loosen the fixing nut on the 'hole' pillar, and adjust its position until the beam is vertically above or below the hole in the collimator. Tighten the nut on the 'hole' pillar and loosen that on the 'slot' pillar. Adjust the pillar (keeping the 'slot' perpendicular to the beam) so that the vertical plane containing the collimator is approximately parallel to the beam, and attach the screw adjustment jig. (Fig. 225, p. 411).

(b) Fine adjustment. By equal and if necessary differential adjustments to the legs and by movement of the 'slot' pillar, bring the collimator hole in to the middle of the light beam and the collimator parallel to the beam. Use the microscope to observe the light beam coming through the collimator and the adjustments to get the brightest beam through. Draw the camera stand back, substitute the small collimator and the small-hole cap, raise the claw arm from its vulnerable lowered position and push the stand forward until the cap is nearly touching the tube. Make final adjustments to maximize the brightness of the beam through the small collimator, withdrawing the stand to increase the sensitivity of the adjustment, if required, and lock the 'slot' pillar and adjusting legs. Check that this has not altered the adjustment, and if necessary do a final correction.

4.3 USE OF SAFETY SHUTTERS AND INTERCHANGEABLE STANDS

Experience has shown the necessity for particular care being given to the following points.

(a) *General*

All movements connected with the use of X-ray tubes and cameras must be slow and deliberate. **Haste is absolutely impermissible.** Never use force on the shutters or in moving the sliding stand. **Both should move with light finger pressure, particularly the shutter.** It is also important to be gentle in removing the cylindrical film holder in case the collimator has not been removed, since it is very easy to bend the collimator in these circumstances. In case of difficulty ask for assistance.

(b) *Stands*

Check that the camera and stand numbers are the same. When moving the camera towards or away from the X-ray tube it is essential to hold the stand and **not** the camera. When interchange of cameras (and stands) occurs the only adjustment necessary is to the adjustable stop at the end of the V-groove on the main stand. This should be screwed out to make sure that it does not obstruct the movement of the camera stand towards the tube. The clawed camera attachment must be **gently** pushed into the shutter opening so that the shutter can be pushed up, locking the attachment in place. The camera base is then pushed gently further towards the tube until the movement allowed by the loose coupling is fully taken up. The adjustable stop on the main stand is then screwed in until the camera and stand are pushed back about 0·5 mm. Finally the lock nut is tightened. No further adjustment should be necessary until the camera is replaced by another.

(c) *Camera attachment*

This is the arm which pivots downwards to enable the collimator to be withdrawn (*G*, Fig. 87 (a), p. 136). In this condition it can be easily bent if any weight is accidentally put on the end of the arm. It should therefore not be left in this position for longer than is necessary to withdraw or replace the collimator. Only a quarter turn is needed to secure or release the arm but the securing screw must be tight.

(d) *Lead caps*

These are both visible evidence that an X-ray tube is safe and a second line of defence against any shutter failure (*P*, Fig. 230, p. 420) and Fig. 87(a), p. 136). The following drill should be obligatory on everyone using X-ray cameras.

(i) Bring the camera up to within 25 mm of the tube window.

(ii) Remove the lead cap in such a way that the fingers do not come in front of the window (i.e. treat the tube as though it were on).

(iii) Couple the camera to the window and open the shutter.

(iv) After the exposure, close the shutter and withdraw the camera about 25 mm.

(v) **Replace the lead cap, treating the tube as though it were on and the window open.**

(vi) Withdraw the camera fully for removal of collimator, film holder, etc.

It is essential that this procedure become habitual, and a conscious effort must be made to this end.

(e) *High tension sets*

Before switching on a set, check that the opposite window, as well as your own, is closed with cap in place, or has a camera properly locked into

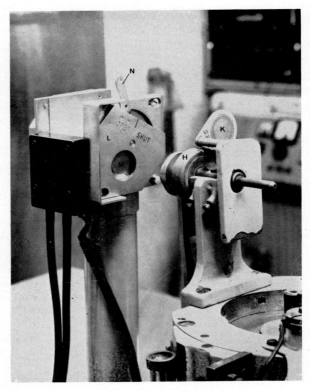

FIG. 229. The Hilger safety shutter. The body L, is screwed to the face of the tube, and slots allow some lateral adjustment. The shouldered disc M is held up by the shutter operating arm N in the 'shut' position, and falls under gravity to the position shown, in front of the tube window, when N is in the 'open' position, unless a nosepiece on the camera is far enough in the shutter opening to support it.

t with a collimator in position. (The gravity shutter will prevent an extremely dangerous large main beam coming through if the collimator has been omitted, but will also produce considerable sideways scatter. A camera should never be left longer than can be helped in this condition.) At the end of an exposure, switch off the high tension unless an exposure is going on at another window. In either case, carefully observe the procedure of paragraph 4.3(d).

5. Non-standard Cameras and Tubes

5.1 THE HILGER FINE FOCUS X-RAY TUBE

There are three problems connected with this tube. (a) The window is too low to use main and subsidiary stands; (b) the face is not suitable for mounting the standard shutter; and (c) the fine focus becomes part of the collimating system, so that final adjustment using X-rays is essential.

(a) A special stand has to be constructed for each type of camera, and adjustments have to be made with reference to the tube window. It is essential to use a safety shutter (see (b)) because the camera is not locked to the tube, and in final adjustments using X-rays the adjustment of the legs may bring the collimator away from the window. The initial part of the X-ray adjustments (getting a beam through the collimator) are made much easier and therefore quicker, if a special large single aperture collimator is used. The large beam from this collimator is roughly centered (as in (c) below) and then the normal collimator put back for final adjustment.

(b) The Hilger safety shutter should be fitted. (Fig. 229). This works by gravity, and for the window to be open at all, a nosepiece on the camera must project into the opening to hold up the disc which otherwise falls down over the window when the lever is in the 'open' position. In Fig. 87(c) (p. 137) the original nosepiece can be seen. (An improved version has the flange brought forward, so that it almost touches the face of the shutter when the camera is in position).

Cleanliness is all-important in the use of a gravity shutter, since the only failure (if the shutter is properly assembled) that can arise, is for the disc to stick open.

(c) The camera is adjusted optically as in Appendix IX, and at the same time a dense, preferably spherical specimen is adjusted to the centre of the light beam coming through the collimator. The camera is transferred to the tube, and the beam coming through the collimator (stopped from going into the room by a microscope or plate camera and an angled table screen round one side and the rear of the camera) is observed on a fluorescent screen. The legs are given slight adjustments until the shadow of the specimen is in the centre of the beam from the smallest collimator available. If it is found that

the nosepiece fouls the hole before this adjustment can be made, two possibilities arise. If it fouls top or bottom, the angle with the table must be altered
(raising or lowering the back legs). If the fouling is on one side the shutter
housing must be moved across or the take-off angle altered.

Although the scatter from the beam is only of the order of 1 mR/hr about
a foot from the specimen, and much less in the room, the apparatus should
always be screened as much as possible when a collimated beam is in the open.
Observation of the fluorescent screen should be kept as short as possible. For
normal use a scatter screen should be fitted to the camera. This must be capable of stopping the direct beam if necessary, and constructed so that it
cannot be removed unless the camera is pulled back from the tube. If such
a shield is always used it becomes impossible for a collimated beam to escape
into the room as a result of a plate holder being removed without the tube
shutter being closed.

5.2 Nonius Guinier Powder Camera

This camera incorporates a curved crystal monochromator which must be
at a specified distance from the tube focus. For CuKα radiation and a normal

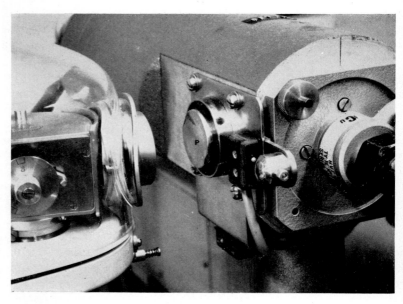

Fig. 230. Close-up of the labryinth and shutter arrangement for a Nonius Guinier camera.
The two micro-switches operated by spring loaded plungers on the shutter knob and camera
base are clearly visible, and the lead cap P is in position as a visible (and essential) second
line of defence against accidental exposure.

sealed tube, this distance is too short to enable an automatic safety shutter to be installed if the plastic cover is to be used. In such a case two electrical switches, operated by the shutter knob and the camera base respectively (Fig. 230) should be incorporated in parallel into the electromagnetic H.T. switch circuit. If both switches are open, i.e. the shutter open and the camera pulled back, the tube cannot be energised. If either is closed the circuit can be made. Since such a system is more liable to failures than a mechanical shutter, the lead cap as second line of defence is even more important. Such a cap can be seen in position in Fig. 230.

5.3 SPECIAL TUBES

The construction of some special tubes, such as rotating anode tubes, may make incorporation of automatic shutters difficult. In such a case the electrical method described above must be used. Magnetic proximity switches can, with advantage, replace micro-switches for the camera base at least. Again, the lead cap should be insisted on as second line of defence.

Checks, Adjustments and Procedures
for Achieving High Accuracy in Measurement
of X-ray Diffraction Geometry and Intensity

1. Cameras, Checks and Adjustments

1.1 GENERAL

The checks and adjustments are described in terms of specific cameras but the principles involved apply generally, so that variations of the method described (which include optical and mechanical lining up procedures) can be applied to other designs. If faults are found the rectification may be a work shop job, if adjustments similar to those described are not provided.

1.2 CHECKING AND ADJUSTING A UNICAM ROTATION CAMERA

(a) *General description.* Refer to Section **11**.2 (p. 135).

(i) *Permanent adjustments.* The axis of rotation of the goniometer spindl is fixed relative to the base. The cylindrical film holder is located by th steel ring at its base, which is coaxial with the body and has three feet restin on the plate holding the spindle bearing. This plate is machined perpendicula to the spindle axis, and forms a circular recess in the top plate of th camera base. Two lugs project from the wall of this recess and ar machined to a cylindrical form, coaxial with the spindle and with a radiu equal to that of the outer surface of the film holder ring. This ring is presse against the lugs by a spring loaded plunger fitting into a groove in th film holder ring. The film holder is thus held coaxial with the spindle and located so that the collimator hole and the beam exit hole are lined u with the collimator axis. Any dirt between the feet and the plate or th ring and the lugs will obviously throw the alignment out, and cleanlines at these points is essential for accurate work. If checks after careful cleanin reveal that the cylinder is not coaxial with the spindle, this can only b rectified by an instrument maker.

The combination microscope–telescope is carried on an arm, the bottom of which is machined with a cylindrical surface to mate with a cylindrical surface on the base, which is itself coaxial with the spindle. The clamping screw holds these two surfaces in contact. The bottom of the arm is pressed sideways against two adjusting screws by a spring, and a third screw controls the height. The two side screws, if turned the same amount, move the microscope round the axis of the spindle. If they are given different adjustments, the axes of the two mating cylindrical surfaces are set out of alignment, and this makes it difficult to replace the microscope in position reproducibly. However, very slight relative adjustment of these two screws is permissible to bring the cross-wires on to the rotation axis. The slide of the microscope tube is also adjustable, either by four screws, or, in older models, by shims, to bring the optical and mechanical axes (which should be coincident) perpendicular to and intersecting the rotation axis.

The X-ray collimator rests on four adjustable screws so that its axis can be made to intersect the rotation axis at right angles. The cylindrical film holder has adjustments for positioning the beam trap.

The plate holder rail fits into the same position as the microscope, and in older models is adjusted by the same two side screws. After this rail is correctly adjusted these two screws must not be altered. Final adjustments on the microscope must be made by the four slide screws (and, of course, the height adjustment screw). In later models the rail has independent adjusting screws incorporated which are used after the microscope has been finally adjusted.

The plate holder has three adjusting screws in the frame to bring the plate perpendicular to the beam.

The optical collimator has adjusting screws to bring its axis perpendicular to the rotation and collimator axes.

All the permanent adjustments described above are made before the camera leaves the manufacturers. It should not be necessary to alter them. However instruments do go out of adjustment, and if accurate work is to be done the camera should at least be checked before use.

(ii) *Running adjustments.* There are two adjustments which are designed to be continually altered. The first is the height of the inner spindle. This slides in the outer spindle, but is constrained to rotate with it. When the adjustment is correct the inner spindle is locked in position by a set screw, turned with a key. The adjustment is made by a screw in the front of the base which pushes against a spring loaded lever through a loose connecting pin. The other end of the lever is connected by a sliding ball joint to the bottom of the spindle. If this adjustment is altered when the spindle has been locked by the set screw: (a) a considerable end thrust will be put on the spindle bearing, which will interfere with its free rotation; and (b) if the knob is unscrewed in an effort to lower the spindle, the loose connecting pin will become free and fall into the bottom of the case. The height adjustment will still work, possibly rather jerkily, but about 10 mm of the range will be lost, and it may well not be possible to bring the crystal high enough to be in the X-ray beam. Therefore the height adjustment of the spindle should never be touched until the set screw has been unlocked.

The second adjustment is to the oscillation arm. This has a removable extension which carries the cam follower (adjustable in height to press against any of the 3 cams) at the end. The oscillation arm is fixed to a loose collar on the spindle, and this collar can be clamped to the spindle in any position by a second set screw turned with the same key as the first. These two set screws are very similar, but the one in the oscillation collar is above the scale (and, when the extension arm is on, always projects towards the back of the camera); whereas the height adjustment screw is below the scale, and the spindle must be rotated until it projects towards the back before the key can be fitted on to it.

When the camera is to be used with a complete rotation of the spindle the oscillation extension arm is withdrawn and the rider pulley allowed to press on the belt. For an oscillation the rider pulley is locked back, the extension arm replaced, the cam follower adjusted to the desired cam, the weight hooked on to the screw at the end of the arm and hung over the adjustable pulley. If the camera is level the pulley is adjusted so that the weight moves freely in the hole. If the camera is tilted it is usually necessary to make a paper tube for the weight to slide in, to prevent it being caught on one or other of the two holes in the top and bottom of the casing.

(b) *Detailed Procedures*

(vii, x—xiii and xv may be omitted if only the cylindrical film holder is to be used.)

(i) The requirements are: a darkened cubicle; Unicam rotation camera, complete with arcs, optical collimator and back reflection camera; two screwdrivers, $\frac{1}{4}''$, $\frac{1}{16}''$; 4 B.A. and 10 B.A. spanners; glass fibres; an auto collimating attachment; two plane mirrors (20×10 mm and $\frac{1}{4}$ plate); a front lens from another microscope; wax; a table lamp; a focussing microscope lamp and diffusing screen.

(ii) Go through the description of the adjustments on the instrument and locate all the parts referred to.

(iii) Place a thin fibre on the arcs and adjust the arcs until the fibre does appear to move in the field of view of the microscope when the spindle is rotated.

(iv) Remove the microscope, adjust the lens of the eyepiece until the cross wires are sharply in focus, swing aside the front lens of the microscope objective and train the resulting telescope on a distant object. Move the eyepiece in and out until the distant object is focussed on the crosswires without parallax. (The locking ring must be loosened to enable the eyepiece to be moved, and all the lenses should be checked and cleaned if necessary.)

(v) Replace the microscope (making sure the front lens is fully down); adjust by eye the bottom of the arm parallel to the plate holding the two side adjusting screws; focus on the fibre and check whether it has moved. If necessary, adjust the four slide screws of the microscope tube to bring the crosswires near to the image of the fibre, and turn the eyepiece (without moving it in or out) to bring one cross wire parallel to the fibre. Lock the eyepiece in position.

(vi) Adjust the fibre until the tip is on the horizontal crosswire, and adjust one of the side screws to bring the vertical crosswire exactly on to the fibre. (Tighten the lock nuts and check that this has not altered the adjustment). Fasten the auxiliary lens (the movable front lens of another similar microscope) on to the front of the microscope objective with wax, so that the tip of the fibre still focusses on the cross wire. Place the cylindrical film holder in position, taking care that the spring loaded plunger is in the V-groove, and focus on the beam exit hole (remove the lead lined cap). Adjust the two side screws at the base equally until the vertical crosswire bisects the image of the hole. Adjust the height screw until the crosswires are centred on the hole. Place the lid and suspended lead stop on the film holder. Observe the stop as the lid is rotated to and fro. Its appearance should be symmetrical for similar rotations on either side of the central position, and the back and front should bisect the horizontal crosswire. If not, adjust the stop by turning, raising or lowering the support. If necessary the support may be slightly bent in order to obtain the correct position of the back stop. Remove the film holder, focus on the fibre, adjust the height of the *fibre*, and one of the side screws, if necessary, to recenter the cross wires on the tip of the fibre. Remove the auxiliary lens and check that the tip of the fibre is still centered on the crosswires.

(vii) Replace the fibre by the mirror with its plane parallel to the plane of one arc. View the edge of this in the microscope and adjust it roughly parallel to the crosswire. Swing back the front lens of the objective. Fit the auto-collimating attachment, taking care not to alter the position of the eyepiece. The attachment is simply a lamp shining through a glass plate coated with a partially reflecting film and set at 45° to the axis. The lamp illuminates the cross wires, and light from the plane of the crosswires, which is parallel on emerging from the telescope, is reflected back from the plane mirror and refocussed in the plane of the crosswires. If the mirror is parallel to the rotation axis, the image of the horizontal crosswire will be in the same position after rotating through 180°. If, in addition, it is coincident with the real crosswire, the optical axis of the telescope is perpendicular to the rotation axis. The real crosswires and the image are both seen by reflection in the 45° plate.

Now adjust the mirror on the arc until the image is in the same position on rotating through 180°. Then adjust the slide screws of the telescope tube until the image is coincident with the cross wires. Adjust one pair of screws at a time (back or front) by equal amounts, so that the telescope moves in a vertical plane. (Note the reading of the goniometer scale when the crosswires are coincident). Move the telescope to and fro on its slide and verify that the position of the image does not alter. It is necessary to get the mechanical axis of the telescope tube perpendicular to the rotation axis, but the mechanical axis may not be parallel to the optical axis. Unscrew the locating bar from the telescope tube. The tube can now be rotated on the screws. Rotate the tube through 180° and check the position of the crosswire image. If it is still coincident with the crosswires, the mechanical axis is also perpendicular to the rotation axis. If not, adjust the slide screws until the errors are equal and opposite for the normal and 180° positions of the telescope tube. The mechanical axis is now perpendicular to the rotation axis. Replace the locating bar.

(viii) Remove the auto-collimating attachment and place the optical collimator in the left hand position on the base (the fitting is similar to that for the microscope). See that the tube adjusting screws are in approximately their middle positions. Fix a piece of white paper with a straight edge on the horizontal diameter of the front lens of the collimator. Mark the center of the lens aperture on the paper. Swing the front lens of the microscope into position and focus on the image of this white paper in the mirror on the arcs. Adjust the height screw on the collimator until the edge of the paper is nearly coincident with the horizontal crosswire. Use the microscope lamp to direct an intense and uniform beam of light on the signal aperture of the collimator and approximately along its axis. The signal will be seen in the **telescope** on rotating the mirror. If it is too bright put a diffusing screen in front of the lamp. Move the front lens of the collimator in or out until the signal is sharply in focus.

Set the goniometer scale exactly at 45° to the reading noted in (vii). The signal should be vertical and bisected by both horizontal and vertical crosswires. If not alter the tube adjusting screws on the collimator (or the side screws on the mounting) and turn the body if necessary, until it is. Replace the front lens of the telescope and focus on the paper stuck on the collimator. The centre of the lens aperture should be at the intersection of the crosswires. This adjustment is not critical, but if it is far out, move the collimator parallel to itself by means of the tube and height adjusting screws until the misadjustment is corrected.

(ix) Remove the mirror. Replace and recenter the fibre, adjusting the tip on to the crosswire. Remove and replace the microscope several times to check the reproducibility of its position. If this is poor, try slightly greasing the mating surfaces.

(x) Place the microscope lamp, adjusted for an approximately parallel beam, so that it shines through the large collimator. Adjust the position of the lamp so that the collimator is filled with light. This position, which is not critical, can be found by catching the beam through the collimator on a piece of white paper. Close the lamp diaphragm down as much as possible.

View the beam through the microscope while racking it in and out. The cross-section of the beam which is focussed at different points by this movement should be centered on the crosswires. If it is not, or if it is centered at one point but moves off as the microscope is racked away, the collimator screws require adjustment. This adjustment must be divided into two parts if it is to be quickly and easily dealt with. (Failure to do this can lead to complete failure to achieve the adjustment at all). Remembering that positioning and movement in the microscope field of view is reversed, adjust the pair of screws (back or front) which will bring the beam parallel to the microscope axis and as near the center as possible. This adjustment is fairly simple. Then, by moving the two screws on one side by equal amounts, move the beam parallel to itself into the central position. Since it is not easy to move the screws exactly equal amounts, repeat the first adjustment when the beam is nearly centered. A repetition of these two methods of adjustment in turn will fairly quickly bring the beam centered on the crosswires for all positions of the microscope. Check the collimator by rotating it on the screws. The cross section of the beam furthest from the collimator should not move by more than 1/10th of its diameter. Replace the large by the small collimator and make final adjustments on the screws. These must be within the limits of error in positioning the large collimator. Check the small collimator by rotating it on the screws.

If the collimators, while not satisfying the rotation test, can be adjusted correctly for a particular position, then the tops of the collimators should be marked and care taken always to replace them with these marks uppermost.

(xi) Remove the black paper from the flat plate camera, and insert the loose frame, a $\frac{1}{4}$-plate glass and a piece of squared paper. Fix the flat-plate camera slide to the goniometer in place of the telescope.

(xii) Adjust the flat-plate camera slide turning both its adjusting screws (or, in older models, the side screws) by equal amounts, until the beam of light through the collimator remains stationary on the squared paper when the film holder is moved from front to back of the slide. Lock the adjusting screws.

(xiii) Using a silvered $\frac{1}{4}$-plate glass, adjust the three small film-holder screws until the beam is reflected back along its own path. Lock these screws, checking the adjustment after locking.

(xiv) Stick fresh black paper in the frame of the camera, and fasten the lead stop in position on the front so that the beam through the collimator hits its center. (Stick fluorescent powder in the stop).

(xv) For older models, remove the slide and replace the microscope. Use the microscope tube adjusting screws to bring the microscope co-axial with the collimator, as shown by the test in (x). The collimator must not be touched.

(xvi) If the back reflection plate camera is to be used, the collimator support must be replaced by that for the plate camera. Since the collimator is machined perpendicular to the plate holder, it is only necessary to ensure that the collimator axis is perpendicular to the rotation axis and intersects it. Proceed as for the normal collimator adjustment, using the three adjusting screws at the back of the support. Use the bottom screw to bring the beam horizontal (i.e. the cross-section remaining at the same **height** in the field of view as the microscope is racked in and out), and adjust the other two screws by equal but opposite amounts to make the beam intersect the rotation axis.

(If possible, the back reflection camera should be permanently fitted to a base, and the collimator lined up not only to intersect the rotation axis but also to be parallel to the microscope axis, so that the optical collimator can be used for accurate alignment of a specimen with flat faces. In the case of a permanent fitment the microscope can be moved round the spindle axis by equal adjustment of the side screws to achieve this, after using the auto-collimator to bring the microscope axis horizontal. The auto-collimator can then be used again to adjust the optical collimator perpendicular to the microscope axis as in (vii) and (viii).)

If the back reflection camera has been used, the normal collimator must be rechecked on replacement.

(xvii) If the camera is to be used for accurate parameter work (Section 13.6.2, p. 170) the cylindrical film holder should be checked for distortion and misalignment. A miniature dial gauge indicator is fastened to the spindle, so that it bears against the wall of the film holder as near the collimator entry hole as possible. A total variation up to 0·1 mm is the maximum which can be tolerated, and if this increases when the test is done near the top of the cylinder, the camera should be sent to the workshop for adjustment.

1.3 CHECKING AND ADJUSTING A NONIUS INTEGRATING WEISSENBERG CAMERA

(a) *General description.* Refer to Section 11.3 (p. 139).

Since the Weissenberg camera is a development of the rotation camera (usually turned on its side, with the rotation axis horizontal) some of the checks, such as that for the coaxiality of the spindle and film holder, can be done by a similar procedure. However, the additional requirements introduced by the third, inclination axis make a mechanical method for lining up the collimator more appropriate. Before carrying out any checks the instrument should be carefully cleaned, and mating surfaces, especially moving ones, lightly oiled. This is especially important for the sliding surfaces of the integrating mechanism.

(b) *Preliminary checking and adjustments.*

(i) *Workshop adjustments.* The machined face of the main base determines the direction of the inclination axis, and its position is located by the pivot pin and its mating hole in the movable base. The housing grooves for the rails are machined parallel to the flat bottom face of the movable base and equi-distant at the two ends. The rails should therefore be parallel and their plane perpendicular to the equi-inclination axis. If warping of or damage to either of the base castings is suspected, the main base top surface and the mating bottom surface of the movable base should be carefully checked with a straight edge, and if plane, adjusted horizontal with a sensitive level. With the movable base in place, use the level at various positions across the rails to check them for parallelism with the machined surfaces of the base. An internal gauge can be used to check that the rails are parallel in the horizontal plane. If the rails are bent by more than 25 μm they should be replaced.

Assuming that these checks are satisfactory, the rotation spindle must next be checked to see that it is (i) parallel to the rails and (ii) intersecting the equi-inclination axis. (i) can be easily checked by fastening a dial gauge to the carriage and running it along the side and top (or bottom) of the spindle. The variation should be less than 1 in 1000, otherwise the instrument should be sent to the workshop for adjustment. Check (ii) has to be done optically, and requires a small (0·1–0·2 mm diameter) spherical specimen, which is centred on the spindle axis and in or near the collimator axis as a starting point. (The longitudinal movement of the spindle, corresponding to the vertical movement

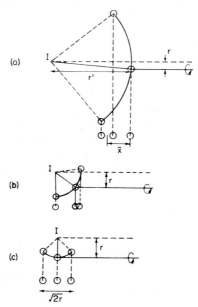

FIG. 231. The movement of the spherical test specimen when the moving base is oscillated from one extreme inclination (45°) to the other. The movement seen in the microscope is shown by the projections on to a line perpendicular to the microscope axis. (a) Large error in 'height' adjustment. (b) Error in height adjustment similar to the distance apart of the two axes. (c) Height adjustment correct. Sphere at the point of closest approach to the inclination axis *I*. The scale of the diagrams is increasing from (a) to (c).

in the rotation camera, is directly adjusted by a knurled nut in the spindle housing, and locked with a thumb screw). The moving base locking nut is loosened and the base rotated about the inclination axis, as far as it will go in both directions. At the same time the sphere is observed in the microscope which is attached to the main base, and has its axis perpendicular to the rotation axis at zero inclination. The displacement of the sphere parallel to the rotation axis is measured on the micrometer eyepiece scale, and the directions of movement from the central position recorded. Large, nearly equal movements of $\bar{x}$ mm in the same direction from the central position (Fig. 231) are due to incorrect adjustment of the spindle 'height'. $\bar{x} = 0.3 \, r'$ (for 45° inclination) where r' is the distance of the sphere from the inclination axis. The direction of movement (allowing for reversal in the microscope) enables a correction of approximately $3\bar{x}$ to be made on the spindle. When the spindle 'height' error is nearly the same as the distance of separation of the spindle and inclination axes, the two movements become very unequal (Fig. 231) and eventually become opposite in direction. When the two movements are equal and opposite the sphere is at the point of nearest approach of the two axes. For 45° inclination the total movement from one extreme to the other is $\sqrt{2}r$, where r is the distance between the two axes (Fig. 231). If this movement is less than 1/10th the diameter of a 150 μm sphere, i.e. <15 μm, the separation of the axes is $15/\sqrt{2} \sim 10$ μm, which is within the acceptable limits. Movement greater than this requires corrections in the workshop.

If the movement at this position is sufficiently small the microscope is focussed sharply on the sphere (with substitution of a higher power objective if necessary to get a depth of focus of 10–20 μ) and the crosswires brought on to the centre of the sphere exactly. In the case of the Nonius camera, large horizontal adjustments are carried out by adjusting the stop in the foot of the microscope, and final adjustments by loosening the Allen set screw under the front lens and moving the two eccentric adjusters. When the microscope has been set in this position *it must not be moved until the collimator adjustments have been finished,* since the arcs carrying the sphere must now be removed and the microscope is effectively marking the point of (near) intersection of the two axes.

(ii) *Adjustment of the collimator and microscope.* Place a pointed uniform steel rod of the same diameter as the body of the collimator in the collimator housing. The fine central point on the axis of the rod is moved into focus in the fixed microscope. The point is made to coincide with the crosswires by adjusting the screws upon which the dummy collimator rests. A ground steel base plate is placed across the rails of the camera and a dial gauge with magnetic base is placed on it.

The reading of the dial gauge at the top of the dummy collimator, as near as possible to the pointed end, is taken as standard and maintained.

Another point, as far from the first as possible, is measured and the screws adjusted to bring this to the same standard reading. The first point is then re-measured and further adjustments made until the same standard reading is obtained at both points. Care must be taken to use equal adjustments on all four screws to raise or lower the collimator, and equal adjustments on the front and back pairs to alter the angle. This operation brings the collimator axis parallel to the plane of the rails and therefore perpendicular to the inclination axis of the camera.

The inclination angle is set accurately to zero and another fine pointed steel rod mounted with plasticine on the crystal rotation axis. This rod is off-set so that the point describes a circle perpendicular to the crystal rotation axis, as shown in Fig. 232.

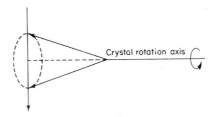

Crystal rotation axis

FIG. 232. The movement of the second rod (3 mm diam, 50 mm long) defining a plane perpendicular to the rotation axis. The collimator axis must be brought parallel to this plane.

By adjusting the 'height' of the crystal rotation axis this circle is made to pass very near (30 μm) to the surface of the dummy collimator, which is now pushed towards the microscope as far as it will go, at one of the two points of closest approach. A feeler gauge is used to measure the gap between the surface of the collimator and the rotation axis point, and the screws adjusted to make these gaps as nearly equal as possible.

The height and angle of the dummy collimator are checked by dial gauge and feeler gauge respectively in turn, and the position of the point of the dummy collimator brought back into focus for checking by the microscope. The necessary adjustments are made of the supporting screws to align the collimator so that its axis is coincident with, and perpendicular to, the other two axes, within acceptable limits.

The difference in the dial gauge readings at points 100 mm apart should be less than 25 μm, and when the rotating rod just touches the collimator at one extreme position, the gap at the other position 80 mm away, should be less than 40 μm. 25 μm difference in the dial gauge readings corresponds to an inclination of ·014° between the plane of the rails and the collimator axis, and a gap of 40 μm to an inclination of 0·03° between the plane perpendicular to the crystal rotation axis and the collimator axis.

Within these and the previous limits the three axes will then be intersecting in a point and mutually perpendicular to each other.

When the collimator alignment is achieved the microscope alignment can be carried out, using the adjusting screws in the slide. The dummy collimator is pushed in and out to line up the microscope slide parallel to the collimator. After the screw in the foot has been adjusted, the final adjustment to bring the crosswires on to the point of the dummy collimator is carried out using the eccentric adjustment of the front lens.

(iii) *Adusting the camera coaxial with the spindle*. The rollers on which the cassette rests (VII(1), p. 398) are carried in frames attached to the carriage. Slotted attachment holes allow horizontal adjustment. Vertical adjustment is carried out using shims. If adjustment is required, this is a workshop job.

(c) *Running adjustments*

(i) *Equi-inclination angle*. Cell parameters must be known to 0·1% in order to calculate equi-inclination settings with sufficient accuracy; and the setting of the reciprocal lattice layer accurately perpendicular to the rotation axis is of even more importance than for the zero layer, because displacement of the reflections for a given setting error is greater. If the angle has been calculated correctly the first problem is which way to set it. If the layer to be photographed is 'above' zero, the upper base must be moved so that the direct beam (i.e. the collimator) enters the cassette or the layer screen above zero and vice versa.

However, if a reciprocal lattice line coincides with the crystal lattice line along the rotation axis, there is a second problem. It is important that the equi-inclination angle should be deliberately misset (*ca.* 0·5°) so that the axial reciprocal lattice point is not in the sphere, since otherwise the condition will be fulfilled for producing multiple reflections for all reflections of the layer (Section 9.5.4, p. 120). To check that this does not occur, it is necessary to modify the layer line screen so that this axial reflection, if it occurs, will register on the center line of the film in a preliminary photograph.

(ii) *Layer screen settings*. This has been dealt with in Section **15**.2.1.2 (p. 198) and the correct direction of movement can be determined in relation to the collimator, as in (i). It is essential to take a check photograph for each layer setting before putting on the Weissenberg photograph.

(iii) *The oscillation angle*. This is in the first place a problem of ϕ-setting (Section **13**.4, p. 160) and designing the oscillation to collect the desired information (Section **15**.1.3, p. 190). When the goniometer scale readings (ω values) for the two ends of the oscillation have been determined for a non-integrating photograph, the goniometer spindle is locked in the two positions in turn, and the appropriate micro-switch actuating stop on the spindle head locked in position against the switch. The split carriage traverse nut is released, the carriage moved to the appropriate end of its travel and the nut closed. When the motor is switched on a compatible movement of crystal and carriage should take place. Always check the setting by running a complete oscillation before proceeding to take a photograph.

If an integrating photograph is to be taken it is necessary to add 7·5° to each end of the designed oscillation, because the film is rotating at each end and this interferes with the integrating process over this interval. Run the cassette to one end of the oscillation, stopping the motor just as the micro-switch operates. Adjust the star wheel in the operating position for that end of the oscillation, push the integrating mechanism stop along its slide until the star wheel just turns and then on for a further 1 mm, and lock in position. Repeat at the other end. Try out a complete double oscillation. The carriage should carry on a small but definite amount after the star wheel operates, before reversing. If the interval is too short (or too long) make a slight adjustment to the stop. Once this setting has been started the traverse nut must not be released.

If the pointer for the longitudinal integration is not at 0 or 30, hold back the spring stop and turn the star wheel by hand to bring the pointer to the nearer end.

(iv) *Integration limits.* The problems of the setting of integration limits are dealt with in VII (3) (p. 402).

(v) *Cassette setting.* The starting point of a photograph (A, Fig. 114, p. 189) moves from the zero layer position by $r \tan \mu$ for a non-zero layer. In non-integrating instruments there is often a slide and scale to enable this movement to be allowed for, so that A comes at the same point on the film for all layers. In an integrating instrument this adjustment is not provided because of mechanical difficulties, and the only way it can be allowed for is by releasing the traverse nut and re-setting the carriage on the lead screw. This involves resetting the integration stops, and should only be done if A (or the far end of the photograph) will otherwise be off the end of the film.

(vi) *Position of the β-filter.* If the β-filter is placed over the layer screen opening it helps to cut down fluorescent and air scattered radiation, thus reducing the background on the film. However, this introduces variation in absorption which is approximately the same as the variation in thickness. In the case of Ni foil, the thickness ariation on 20 μm is of the order of 1 μm in commercially available foil, so hat errors from this source are likely to be greater than those arising from ncreased background when the β-filter is put in the incident beam. If β eflections can be excluded by the layer screen, it is usually better not to employ β-filter.

(d) *Adjustments involving the X-ray tube*

(i) *Test for uniform traverse of the film.* Providing the gearing onnecting the traverse screw with the spindle is in good condition, this est is also a check on uniform rotation of the crystal. A stationary owder specimen (e.g. a wire wound round and projecting in front of the ollimator) is used with a narrow, 0·5 mm gap in the layer screen. traight powder lines are produced along the film, and variations in speed of raverse will produce variations in density along the line. For this and the ollowing test short term variation in beam intensity must be less than 1 %.

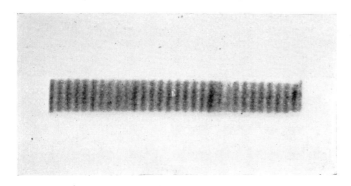

FIG. 233. Direct beam streak, showing non-uniform traverse of the cassette.

In the case of the Integrating Weissenberg a much better check is obtained y placing over the collimator a sleeve with a pinhole (c. 0·2 mm in diameter) the end. The direct beam intensity is reduced by filtering or lowering the ube output so that one vertical cycle of integration, with the maximum pos- ble traverse of the cassette, gives a band with optical density between 0·5 nd 1·0. The uniformity of density of this band along its length is a measure f the uniformity of translation of the cassette. Fig. 233 shows such a band ith a non-uniform density which has the periodicity of the traverse screw. his effect was eventually traced to a faulty bearing.

(ii) *Tests for a uniform incident beam.* Errors will be introduced into the easurement of the intensities of X-ray reflections unless the crystal bathed in a uniform incident beam. Even for a spherical crystal with egligible absorption variations will occur, due to the lateral movements

(a)

(b)

FIG. 234. Photographs of the shutter attachment on a Weissenberg camera. (a) Recording the shadow of the specimen. (b) Photographing the beam as near as possible to the specimen position.

which inevitably accompany rotation in an actual apparatus. In the case of medium or high absorption crystals, reflections on one side will effectively have a higher intensity incident beam than those on the side where the beam is weaker.

One way to avoid this is for all parts of the crystal to receive X-rays from all parts of the focus. However, for most purposes a simple pinhole collimator, large enough to enable all parts of the crystal to 'see' the whole of a normal 1×1 mm focus, would give far too large a beam and/or obstruct too many back reflections. The objective is to get the maximum useful X-rays with the minimum unused radiation in order to minimize unwanted background fogging. Whatever compromise is reached between these two opposing requirements inevitably means that different points in the crystal 'see' different parts of the focus, and reasonable uniformity of the tube focus is therefore a prerequisite for a uniform beam bathing the specimen. Variations on a scale small compared with the area of the focal spot 'seen' by a point in the crystal will not matter, if the average intensity over the focus is uniform.

A really uniform tube focus, even on this 'average' basis, appears to be impossible to achieve in a commercial tube, and one can only reject the worst by taking pinhole photographs of the foci of a number of tubes.

The practical test is to photograph the beam at a point as near the crystal as possible, and to combine this with small (0·1 mm) adjustments of the camera relative to the tube, in the horizontal as well as the vertical directions, to find the position giving the best beam. For this purpose a simple attachment is required consisting of a normal photographic 6 blade shutter with a grooved film holder behind it, which can be mounted vertically in two positions on a stand which fits the carriage of the Weissenberg camera.

In one position the beam can be photographed with a specimen in place (Fig. 234(a)), and in the other the film is as near as possible to the specimen position (Fig. 234(b)). Because of the thickness of the shutter it is not possible to get nearer than 12 mm from the specimen position, so that the beam cross section photographed is slightly larger than that at the specimen; but the intensity variation at the specimen should be rather less than at the film.

The centre of the shutter has 6 overlapping blades in the closed position. It is essential that this part of the shutter is centred on the collimator axis, and a hole in the film holder enables this adjustment to be done using the camera microscope. An auxiliary lead shutter to prevent fogging is required, which can be swung clear just before an exposure.

Photographs are taken on fine grain film (Ilford Industrial C). The exposure times should be made long, 15–20 sec, so as to reduce any effect of the shutter blades moving across the beam at different speeds. These long exposures can be attained by using ten layers of nickel foil of total thickness 0·16 mm. between the tube and the collimator.

Any variation in thickness of the β-filter foil, which is known to be very uniform, will be evened out for the large number employed.

With the shutter in the position nearest the tube, photographs are taken up to 0·3 mm on either side of the position given by optical alignment, using the horizontal traverse of the camera, and in the vertical direction by calibrated movement of the front feet of the camera, starting from the best horizontal position.

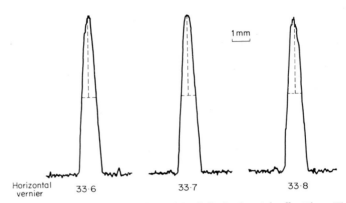

Horizontal vernier   33·6   33·7   33·8

FIG. 235. Variation of beam spot profile with slight horizontal off-setting. The vertical bisector of the beam at half height is included to make the asymmetry more obvious. The scale refers to the film.

Figure 235 gives photometer traces of a beam profile and shows that movements of as little as 0·1 mm can give rise to detectable asymmetry.

An independent check on the alignment can be made by mounting and centering an absorbing specimen such as tinned copper wire 0·25 mm in diameter, and checking photographically that the specimen is in the center of the beam. Figure 236 shows a vertical trace, and corresponds to a vertical missetting of approximately 1/16 mm at the plane of the film and less at the

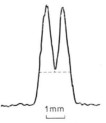

FIG. 236. Beam spot profile with a specimen of 0·25 mm copper wire, showing slight vertical off-setting. The scale refers to the film.

plane of the specimen. The beam profile appears wider in this figure than in Fig. 235 as the recording film is further away from the collimator.

The alignment procedure should be repeated before the investigation of a crystal and whenever a tube insert is changed.

Using this method of alignment it is possible to obtain accurately and fairly quickly a position of the camera so that the collimated beam is symmetrical. The recorded intensity variation over the central part of the beam should then be no worse than that of the background, i.e. it can be attributed to the graininess of the film.

In addition to aligning the camera, the attachment can be used to take beam photographs to test the reproducibility of an aligned camera as follows:

(a) before and after sliding the camera away from the tube to check the reproducibility of its position;

(b) between rotations of the collimator by a quarter of a revolution at a time;

(c) before and after removing and replacing a nickel filter in front of the collimator;

The camera and filter should be satisfactory in all these respects.

By modifying the camera shutter stand it would be possible to use the same technique with any kind of camera, or for checking the beam of a single crystal diffractometer.

1.4 PRECESSION CAMERA   Refer to Section **16.3.2** (p. 218).

The main mechanical requirements are that the two gimbal axes of each set should intersect, and the collimator and rotation axes pass through both points of intersection. In addition the dial axis must be coaxial with the horizontal gimbal axis and the arc radius pass through the gimbal intersection. Usually these requirements are achieved by careful design and machining, so that only damage or wear can destroy the alignment; and in this case replacement of the damaged or worn parts is the only remedy. However, the crystal gimbals can be tested by centering an absorbing spherical specimen on the dial axis and adjusting its 'height' until it is centered in a light beam shone down the collimator. When the specimen is observed in a fixed microscope, it should have no perceptible movement during a complete rotation with the largest possible precession angle. Any appreciable error in the film holder gimbals will show up in double spots on the photograph. When the precession angle is set at zero there should be no movement of the gimbals on rotation of the axis, and the film holder should be perpendicular to the X-ray beam. The latter can be checked by taking a powder photograph, and measuring in various directions the distance of a powder ring from the central spot. In general, the correction of errors is a matter for the manufacturers. See IX 1.3(d) (ii) (p. 435) for tests of the uniformity of the incident beam.

2. Specimen Selection and Preparation

2.1 SELECTION OF CRYSTALS

The final results of a structure analysis cannot be better than the imperfections of the crystals allow. It is therefore worth while spending considerable time and effort in obtaining good, single crystals as starting material. This may involve recrystallization or even a different method of production. Crystals which are only slightly soluble can be recrystallized using a Soxhlet condenser, or grown from materials which react slowly to give the compound concerned. Calcium hydroxide crystals can be grown on Portland cement which is hardening under water, and a generally useful formula is to stick the ingredients in a bucket and forget about them! This emphasizes that haste is the main enemy in producing reasonably perfect crystals. In rare cases too much perfection will produce extinction difficulties (Section 9.5, p. 119); but the effect is likely to be so small that it need not normally be considered until the final stages of structure analysis, when a correction can be applied.

In difficult cases the extensive literature on crystallization, which is an important industrial process, should be consulted.

2.2 TWINNING

One problem in obtaining suitable starting material is directly related to crystallization, i.e. the growth and selection of untwinned individuals.

The relationship between the lattices of twinned crystals is normally that of rotation of 180° about a central lattice line, or reflection across a lattice plane. If the lattice is not geometrically symmetrical about the line or plane, two lattices with differing orientations will be produced. The corresponding RL's will show up on an X-ray photograph as a double lattice, and the fact of twinning is usually obvious. The exact relationship may not be easy to find, but if one is merely concerned to avoid using a twinned crystal, this is not difficult to achieve. Either it is not possible to set the crystal about an axis (or rather, one twin can be set to give straight layer lines, but the other remains mis-set) or, if this setting can be achieved, the RL plane perpendicular to the rotation axis shows two inclined nets. A precession photograph of this plane is particularly useful in sorting out the fact and type of twinning.

However, if the lattice is geometrically symmetrical about the twin axis or plane, so will the RL be. In this case there will not be any obvious geometrical signs of twinning since the relps of the two RL's will coincide. If the twins are present in fairly equal amounts the result will be an apparent mirror plane and perpendicular 2-fold axis in the Laue symmetry. If only one crystal is used in the investigation it will only be the failure of the structure analysis which points to twinning as the cause of the failure; and unless accurate data have been collected, an incorrect structure may be deduced, with a rather large *R*-value which is put down to normal errors in the data. Since accurate data collection is a difficult and time consuming process, the Laue symmetry should be checked on a number of crystals, preferably from different sources (i.e. crystals that have formed in different environmental conditions). Some of these crystals should have the two twins present in sufficiently different proportions to show up the lack of a true symmetry plane. Morphological and optical evidence of twinning should also be looked for; but perfect optical extinction is not positive evidence of lack of twinning, since the geometrical symmetry plane (or axis) on which twinning takes place may be parallel to a symmetry plane (or axis) in the optical properties of the crystal. Partial extinction in two directions not at 90° is, however, positive evidence of the presence of two individuals.

2.3 CRYSTAL SHAPE AND SIZE

The necessity for shaping most crystals into spheres or at least cylinders, and the intensity errors arising from departures from geometrical perfection, have been discussed in Section **18.**2.1.3 (p. 265).

Spheres are best ground in the Schuyff and Hulscher modification of the original Bond gas driven grinder. Figure 237(a) is a workshop drawing of a grinder based on their design and Fig. 237(b) a photograph of the instrument.

Cylinders can be ground on a watchmaker's lathe, using an Airbrasive unit jet as cutting tool. For difficult cases it may be necessary to get an approximation to a cylinder by allowing the edges of an equant prism to sublime or dissolve away, and to coat the crystal with an impervious layer or mount it in a capillary. To enable the specimen to be bathed in a uniform beam, the diameter should not be greater than 0·25 mm and the practical lower limit is about 0·1 mm. A cylinder will give rise to end-effect errors if it is short, or errors due to lack of uniformity of the beam over its length if it is long. Since the size of these errors has not yet been investigated it is worth persevering with sphere grinding. The degree of perfection necessary depends on the final accuracy of atomic coordinates or electron density required. The residual *R*, corresponding to this required accuracy, should be calculated from Cruickshank's approximate relations between *R* and the errors in electon density or coordinates [21]. $\sigma(I)/I$ can now be estimated in terms of *R*, and the corresponding shape error in the crystal $\sigma(r)/r$ obtained from Section **18.**3 (p. 274). For cylinders one can only assume, at present, that the relation is of the same order of magnitude as for a sphere. A rough calculation of this sort will prevent the use of a crystal which cannot produce sufficiently accurate results.

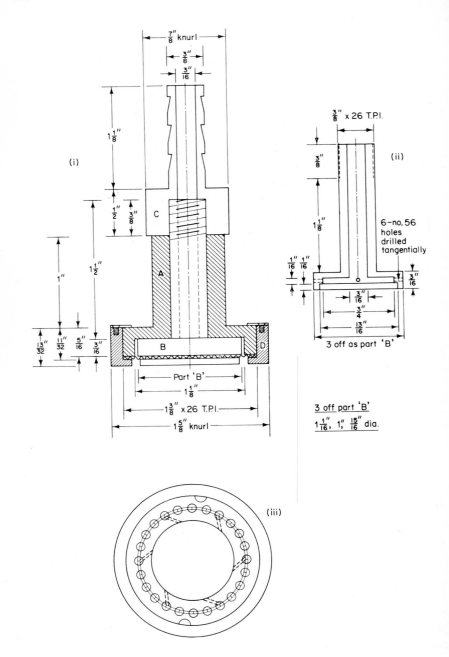

(i)

$\frac{7}{8}''$ knurl

$\frac{3}{8}''$

$\frac{3}{16}''$

$1\frac{1}{8}''$

$\frac{1}{2}''$ $\frac{3}{8}''$

C

$1\frac{1}{2}''$

$1''$

A

$\frac{13}{32}''$ $\frac{11}{32}''$ $\frac{5}{16}''$ $\frac{3}{16}''$

B   D

Part 'B'

$1\frac{1}{8}''$

$1\frac{3}{8}''$ x 26 T.P.I.

$1\frac{5}{8}''$ knurl

(ii)

$\frac{3}{8}''$ x 26 T.P.I.

$\frac{3}{8}''$

$1\frac{1}{8}''$

6 – no. 56
holes
drilled
tangentially

$\frac{1}{16}''$ $\frac{1}{16}''$

$\frac{3}{16}''$

$\frac{3}{16}''$

$\frac{3}{4}''$

$\frac{13}{16}''$

3 off as part 'B'

3 off part 'B'

$1\frac{1}{16}''$, $1''$, $\frac{15}{16}''$ dia.

(iii)

FIG. 237(b). Photograph of the crystal grinder on its stand.

FIG. 237(a). The sphere grinder. (i) Section of the complete assembly. The body A houses the inner gas chamber B which is held in place by the threaded knurled nozzle C. The cap D clamps the nylon gauze over the grinding annulus and has a ring of holes to allow the gas to escape. It is held on by two plates which pass down grooves in the head and engage under the head when the cap is turned. (i) Detail of the gas chamber with angled gas jets which trundle or spin the crystal round the annulus. The top of B is a force fit in the head. (iii) Plan view, showing the holes in the cap and in the inner gas chamber, and the grooves in the head.

The cavity in A is lined with fine emery paper, cut to fit accurately with a butt joint, and put in place before B is inserted. The crystals to be ground are placed in the cavity, the gauze and cap replaced and gas from a reducing valve fed into the nozzle. The pressure required varies between 5 and 60 p.s.i. depending on the hardness and brittleness of the crystal, and the time from a few seconds up to 24 hr or more.

2.4 Crystal Washing

Grinding usually leaves powder on the surface and this requires washing off by a liquid in which the crystal is slightly soluble. For a sphere this is done in a shallow dish before mounting, great care being required not to lose the tiny crystals. A cylinder is best washed on the lathe, using a brush dipped in solvent as a lathe tool (utilizing surface tension to keep the solvent round the cylinder).

2.5 Thermal Shock

To improve the mosaic character of the crystal it should be dipped in liquid air contained in a shallow hemispherical Dewar flask, so that the crystal can be recovered if it comes away from the fibre. For a sphere, this should be done before final mounting.

3. Film Packs and Processing

3.1 Film Packs

The minimum detectable optical density is about 0·005 and the maximum density which is still linearly dependent on X-ray intensity is about 1·0, so

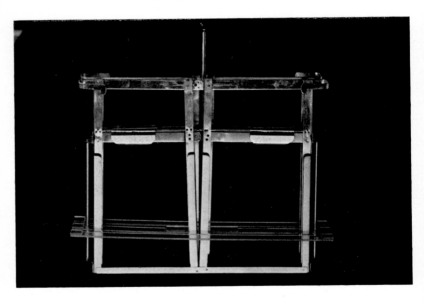

Fig. 238. Photograph of the frame and paddle used to produce controlled agitation during development. The frame is formed from two adapted Kodak No. 16 double film channel hangers joined together to give a 4-film rack. The 3-strip Perspex paddle with 9 mm gaps for the films can be raised by the central rod. The inner film groove supports are a push fit on the top and bottom rails, to allow for adjustment.

that on a single film the measurable range of intensities is 200 : 1. If a double film pack is used with Ilford Industrial G on top and Industrial B underneath, then the ratio of densities on the two films is about 10:1 and an overall range of 2000 :1 can be obtained.

Any departure from linearity below density 1·0 is small and can be ignored except in the most favourable circumstances, when the error will be comparable to or greater than absorption errors. In such cases the densities should be corrected in the first processing of the data.

3.2 PROCESSING FILMS FOR PHOTOMETRY OF INTEGRATED REFLECTIONS

Tank development should be employed, using filtered solutions in a thermostatic bath, with paddle agitation of the developer every 15 seconds. The film holder and paddle are shown in Fig. 238. The use of old developer appreciably increases the background fog, and lack of agitation causes a detectable increase in intensity errors due to processing.

An Accurate Photometer for Measuring the Optical Density of Integrated Reflections

1. Description

1.1 GENERAL LAYOUT

Figure 239 shows the general layout of the cubicle. *A* is the photometer box, *B* the stabilized 24 V d.c. power supply unit, *C* the galvanometer (Cambridge D'Arsonval 470 ω, 1260 mm μA^{-1}), *D* the front aluminized plate-glass mirror, *E* the cylindrical* scale centered on the image of the galvanometer mirror and slung by adjustable rods from a beam *F*, *G* the galvanometer lamp (including a green Wratten filter (No. 59) to reduce eyestrain), *H* the desk lamp (flexible stem with base screwed to the bench *J*) and *K* the switch for the desk lamp. Because of the temperature sensitivity of the photo-

* A square section brass holder was grooved to hold the flexible scale. A brass strip was put in place of the scale and the holder curved to the correct radius in bending rollers. The straight ends were cut off leaving sufficient curved length to take the scale. This form of scale is essential if corrections due to the non-linearity of a straight scale are to be avoided.

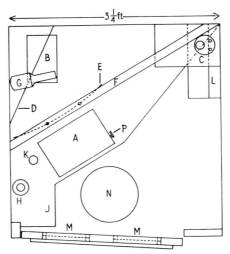

FIG. 239. Scale diagram of the layout of the photometer cubicle.

cell adequate ventilation is essential and dust must be exluded. N is an adjustable chair.

1.2 THE PHOTOMETER BOX AND LIGHT SOURCE

The photometer box, with a base 12 in. × 7·5 in., is 5·5 in. high at the front and 10 in. at the back, giving a sloping top of 12 in. × 9 in. The lamp, mounted on cross slides by means of a tight lamp-holder with contact plungers removed, is a 24 V, 60 W bus headlamp bulb. Immediately over the lamp is a Chance infra-red filter with a small air gap above it. Cooling proved to be one of the more difficult problems to solve and requires the concentrated jet of air provided by a $1\frac{1}{2}$ in. d.c. Airmax axial flow fan directed at the bulb and air gap above it. Air is drawn from under the base of the instrument,which is raised from the bench by $\frac{1}{2}$ in. rubber feet, and escapes via a light trap louvre at the back. The fan is rubber-mounted to minimize noise. Optical condensing components were originally deliberately omitted in order to obtain uniform illumination, and have not subsequently been tried, but unless internal cooling could be dispensed with completely the present simple arrangement has much to recommend it. An optical filter is required to cut down the light intensity when positioning the film. This is brought into position in the air gap by means of a lever on the right-hand side of the box (P, Figs. 239 and 240).

The power supply for the lamp is a transistorized, highly stabilized d.c. power unit run from the mains. The regulation is at least as good as the best

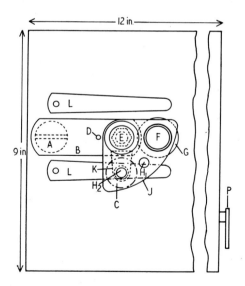

FIG. 240. Plan of the top of the photometer box.

448 APPENDIX X

that can be obtained from 'floating' accumulators and the unit is far smaller and more convenient to use.

The lead from the power unit goes to a special clamped variable resistance (Berco B140 3ω mica card resistance with tapping clip CL69) at the back of the photometer box and then to the lamp. Both this and the return lead to the unit are soldered to the bulb terminals, and all other contacts are either soldered or screwed.

Figure 240 shows the sloping top of the photometer box, made from $\frac{1}{4}$ in. high grade Paxolin, which is the working surface. The hole defining the light beam is far enough from the photocell support pillar A for all reflections on a normal Weissenberg film to be positioned over it. The separate circular Paxolin plate G containing this hole is a press fit in a recess in the main platform, so that it can be readily replaced if necessary. The top surface around the hole is coated with white cellulose paint to facilitate positioning the spot over the hole. It is very important that this coating should be smooth and uniform in the immediate neighbourhood of the hole, since even the slightest mark may make it difficult to position weak reflections over the hole. The circular hole normally used is 0·5 mm in diameter. The velvet-padded spring arms L hold the film down on either side of the hole.

The plate J containing the photocell housing C and the lens F (for positioning the film) can be swung on an accurate bearing E in the support bar B through 90° against stops. It is important that the photocell should be accurately located over the hole. K is the cover plate for the photocell to

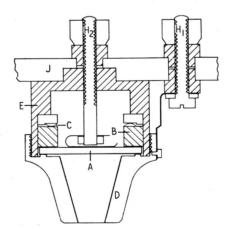

Fig. 241. Scale diagram of the photocell housing. Bakelite components hatched. All Bakelite bushes are threaded. Waved washer curvature shown conventionally. It is important that the compression of this washer should be within the manufacturer's tolerances, and that the contact spring on the back should be only just stiff enough to make a good contact.

prevent overloading (which would result in temporary loss of sensitivity) when the desk lamp is on.

Figure 241 shows details of the photocell housing. Strain can easily cause internal shorts in the selenium photocell A (1 in. diameter Megatron infrared) and it is essential that the housing should be spring-loaded. The ring contact on the front face of the photocell is clamped by a Bakelite disk B, spring-loaded to a predetermined tension by waved washer C (EMO-PLW. 4) against a raised ring on the cap D, which is screwed on to the Bakelite body E. D is connected to the outer terminal H_1. The conducting surface on the back of the photocell is in contact with a bent phosphor bronze strip on the end of the 2 BA threaded rod which forms the centre terminal H_2. This rod holds the assembly on to the rotating plate J (Figs 240 and 241) by means of a threaded Bakelite bush on top and the tapped hole in the body E below.

The contact surfaces on the housing can usefully be tin-plated, but must on no account be nickel-plated if contact potentials are to be avoided. The two terminals are connected directly to the galvanometer terminals with a resistance in series selected to give slightly under critical damping.

1.3 THE FILM MASK

A mask with hole punched for the reflections is made from thin, non-dusting card and tracing paper, both to protect the film during handling and to enable the spots to be easily identified by markings on the mask. A small foot-operated eyelet press fitted with punch and die enables the holes to be punched quickly and accurately.

In use, the thin tracing paper should be downwards so that the film is separated from the hole defining the light beam as little as possible.

2. Method of Operation

2.1 PRELIMINARY ADJUSTMENT

The galvanometer and scale are positioned so that the light spot can be obtained over the whole length of the 500 mm scale and the spot roughly zeroed. The lamp is switched on and allowed half an hour to warm up. The side of the box is removed and the cross slides of the lamp adjusted until the beam of light (seen by scatter from the surface of the lens) is central. The photocell is then swung into position and the deflection reduced by a filter to about 300 mm. Small adjustments are made to the cross slides to give maximum deflection. This should be a flat maximum. If the reading varies rapidly with movement in any direction and inspection shows that the beam is still fairly central, then the photocell should be changed. Finally, with the photocell over its cover, the lead is removed from one terminal. If contact potentials are absent there should be no appreciable movement of the zero. If appreciable movement (other than transient) does occur, the source of contact potential (or the light leakage) must be eliminated. A piece of processed unexposed film is then placed over the hole and the photocell swung into position just long enough to take a reading. By adjustment of the variable resistance (using a handled 4 BA box spanner) it should be possible to obtain a deflection of 450–500 mm. A piece of film with fairly high background (total density ~0·20) is now placed over the hole and the deflection again adjusted to 450–500 mm. If both adjustments cannot be made, the power unit voltage must be altered until they can, or a new photocell fitted.

2.2 NORMAL ADJUSTMENT

After the warming-up period the film to be used, adjusted in its envelope and carefully dusted with a 'blower-brush', is placed so that the background at about $2\theta = 90°$ (i.e. the lowest background of the film) is over the hole. The variable resistance is adjusted and a few minutes allowed for thermal equilibrium to be re-established. With the photocell swung over its cover, the zero is adjusted accurately by moving the scale. This adjustment should be checked before each reading (especially before measuring a dense reflection); but readjustment should only rarely be necessary, a drift of more than 0·5 mm h^{-1} indicating a faulty galvanometer or the existence of contact potentials.

A reflection to be measured, which should have a uniform central area of at least 0·6 mm square, is positioned over the illuminated hole. The desk lamp is switched off, the filter withdrawn, the photocell swung round against the stop just long enough to take the reading ϕ_R and swung back over its cover. The procedure is repeated with the background positioned over the hole (ϕ_B). If the background is varying rapidly two readings, one on either side of the reflection, should be taken and averaged. (For a discussion on background measurements see Section 18.1.2.2, p. 260). The density is given by $D = \log \phi_B/\phi_R$.

3. Accuracy

3.1 TESTS OF THE APPARATUS

(i) The linearity of the galvanometer deflection against voltage can be checked using a simple potentiometer method. The deflections should be linear to within the accuracy of reading the scale ($\pm 0·25$ mm).

(ii) The linearity of the photocell can be checked by a series of rotating sectors with known apertures placed in turn between the hole and the photocell. The graph should be again linear to within $\pm 0·001$.

(iii) Temperature variations make it necessary for the background measurement to be made immediately following that of the reflection. However, it is an incidental advantage of the integrating Weissenberg method over the positive film method that only short-term stability is required of the photometer.

In the warming-up period the rate of change of full-scale deflection should be less than 0·5 mm min^{-1}, and if the ambient temperature can be kept fairly constant it soon becomes negligible. Thus even in the worse case, if the background is measured within 30 sec of the reflection, the error due to temperature variation will be only just detectable, and under good conditions the variation over 5 or 10 min will be negligible.

3.2 Errors in D due to the Photometer Errors

If ϕ_B is taken as 500 mm and the independent reading errors of ϕ_B and ϕ_R are in the range 0–0·25 mm, the calculated values of ΔD and $\Delta D/D$ are given by Fig. 242. The r.m.s. error is below 0·5% for densities from 0·1 to 1·4 and

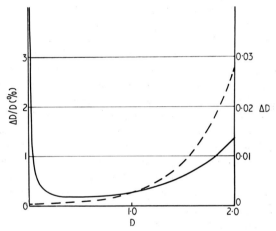

Fig. 242. Graph of relative and absolute errors (standard deviations) in optical density on the assumption that the scale can be read to the nearest 0·5 mm. In practice the errors may be somewhat larger, but still less than 1% over the range 0·1–1·4 in D.

is only 1·4% for density 2. For most films the range of linearity of density against X-ray intensity is below 1·4, so that measurements would not normally be made above this. At the low density end the absolute error tends to 0·0004 so that the percentage error increases rapidly below a density of 0·1. However, it is still below 5% at density 0·01, and for intensities less than this the spots are practically invisible. For densities less than 0·05 greater accuracy can be achieved by comparison with a standard scale than by using the photometer, and for very weak reflections it may be necessary to use the same method on a non-integrating Weissenberg photograph.

The results of using this instrument on test films show that the actual errors approximate to those given above, in contrast with the errors of 10–20% involved in eye estimation against a standard scale.

Answers to Examples

1. (i) $\sin \psi_n = \dfrac{n\lambda}{a}$ $\therefore n \leqslant \dfrac{0.85a}{\lambda} = 1.4$ $\therefore n = 1.$

(ii) $n = 1.$ $\therefore \lambda \leqslant 0.85a = 4200 \,\text{Å}.$ (Violet end of visible spectrum).

2. (i) To average any difference in the top and bottom measurements and reduce the error of measurement by the factor 2. (ii) $\cos v_2 = k\lambda/b$, since $\cos \mu_2 = 0$. (iii) No, since μ_2 remains 90°. (iv) 60·25 mm between ± 4 layer lines.

$$\therefore \frac{30.12}{30.0} = \cot v_2 = 1.005. \quad \therefore v_2 = 44.85°.$$

$$\text{(v)} \therefore b = \frac{4 \times 1.542}{\cos 44.85} = \frac{4 \times 1.542}{0.7090} = 8.70 \,\text{Å}.$$

Similarly, distance between ± 2 layer lines = 22·8 mm.

$$v_2 = \cot^{-1} \frac{11.4}{30.0} = 69.20°; \qquad b = \frac{2 \times 1.542}{0.3551} = 8.68 \,\text{Å}.$$

The first and third layer lines give a similar result, but the fourth layer result is the most accurate, both because the error in measurement is divided by a larger number, and also because the spacing of the layer lines is increasing rapidly and the relative error is therefore decreasing. This latter effect is only partly counteracted by the increasing width of the reflections, which makes measurement less accurate. (The accurate value for b is 8·747 Å.)

3. (i) $\phi_m = \dfrac{2\pi}{\lambda} mb \cos v_k = 2\pi mk$ (i.e. path difference for adjacent points,

$b \cos v_k = k\lambda$).

(ii) $\phi_m = \dfrac{2\pi}{\lambda} mb \cos (v_k + \Delta v) = \dfrac{2\pi mb}{\lambda} (\cos v_k - \sin v_k \, \Delta v)$

$$= 2\pi mk - \frac{2\pi mb \sin v_k}{\lambda} \Delta v \quad \text{(Fig. 243)}.$$

For $m = 1$ and $k = 1$, $\phi_1 = 2\pi - \dfrac{2\pi\ 8\cdot75\ \sin 79°50'\ \Delta v}{1\cdot542}$ rad

$$\equiv \frac{-360 \times 8\cdot75 \times 0\cdot9843\ \Delta v}{1\cdot542} = -\ 2010\ \Delta v°$$

since integral multiples of 2π can be ignored.

$$= -\ 20\cdot1° \text{ for } \Delta v = 0\cdot01 \text{ rad}$$

$$= -\ 40\cdot2° \text{ for } \Delta v = 0\cdot02 \text{ rad etc.}$$

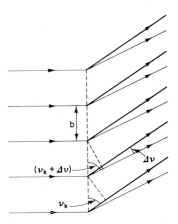

FIG. 243. Scattering from a row of equally spaced points. The heavy line shows the direction for constructive interference at the angle v_k with the row. In this direction the path difference between rays from adjacent points is $k\lambda$ and the scattered intensity is a maximum. In a direction making an angle Δv with that for maximum intensity, the path difference is no longer an integral number of wavelengths, and the intensity has to be found using the Argand diagram.

(iii) Figure 244 shows the Argand polygons up to $\phi_1 = -4 \times 20\cdot1°$, and the first zero clearly occurs for ϕ_1 somewhat less than $-80°$. For exact closure the phase angle is the exterior angle of a pentagon, i.e. $-360/5 = -72°$. (iv) For this value of ϕ_1, $\Delta v = 72/2010$ rad $\equiv 2\cdot05°$. This is the angle between the direction of maximum intensity and the direction of the first minimum on one side. (v) Only a difference of sign is involved; the angle would be the same. (vi) The 'spread' is therefore $4\cdot1°$. (vii) For 100 points ϕ_1 and Δv are $1/20$ of the previous values, i.e. the spread is $0\cdot205°$ and (viii) for 1000 points $0\cdot0205° = 1\cdot23'$. (ix) The size of the crystal in the c direction would be $10^3 \times 8\cdot75 \times 10^{-4} = 0\cdot875\ \mu m$.

Alternatively, a vector on the Argand diagram (App. I (17), p. 337) can be represented by a complex number $A \cos \phi + iA \sin \phi = A e^{i\phi}$. If ϕ is the phase angle for the first point, then the sum of all the N vectors on the diagram (the resultant vector, R) is given by

$$R = A + A e^{i\phi} + A e^{i2\phi} + A e^{i3\phi} + \dots A e^{i(N-1)\phi}$$

$$\therefore R = A(1 - e^{iN\phi})/(1 - e^{i\phi}) \quad \text{(sum of a geometrical series).}$$

$$\therefore \text{Intensity } I = RR^* = A^2(1 - e^{iN\phi})(1 - e^{-iN\phi})/(1 - e^{i\phi})(1 - e^{-i\phi})$$

$$= A^2(1 - e^{iN\phi} - e^{-iN\phi} + 1)/(1 - e^{i\phi} - e^{-i\phi} + 1)$$

$$I = A^2(1 - \cos N\phi)/(1 - \cos \phi) = A^2 \sin^2 \frac{N\phi}{2} \Big/ \sin^2 \frac{\phi}{2} .$$

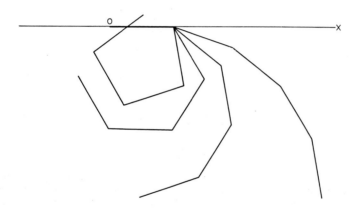

FIG. 244. The Argand diagram for 5 scattering points and various angles of deviation $\Delta\nu$ from the direction giving maximum intensity. In order to avoid complicating the diagram, the resultant, i.e. the closing side of the polygon, is not shown.

Substitute $\omega = \phi/2$.
Let $\omega = n\pi + \Delta$ and $\Delta \to 0$.

$$\frac{\sin N\omega}{\sin \omega} = \frac{\sin Nn\pi \cos N\Delta + \cos Nn\pi \sin N\Delta}{\sin n\pi \cos \Delta + \cos n\pi \sin \Delta} \to \frac{\pm N\Delta}{\pm \Delta} \to \pm N.$$

If $N\omega \to n\pi$, but $\omega \not\to m\pi$

$$\frac{\sin N\omega}{\sin \omega} \to 0$$

$$\therefore I = (NA)^2 \quad \text{for} \quad \omega = n\pi \quad \text{or} \quad \phi_n = 2n\pi.$$

456 ANSWERS TO EXAMPLES

For the 1st order $\phi_1 = 2\pi$ for maximum and $\omega = \pi$.

For the 1st zero on one side $N\omega = (N + 1)\pi$, $\phi_1 = \dfrac{N + 1}{N} \cdot 2\pi$

$$= 2\pi + \frac{2\pi}{N} \equiv \frac{2\pi}{N}, \text{ since integral multiples of } 2\pi \text{ can be ignored.}$$

For 5 points $\phi_1 = \frac{360}{5} = 72°$ and the calculation of Δv then proceeds as above. (In this case Δv is negative and the Argand diagram is the reflection in OX of that in Fig. 244).

4. See Fig. 247 (p. 458).

5. $CuK\alpha$, $d^*_{100} = \lambda/d_{100} = 1.54/5.0 = 0.308$ R.U.

Since $d_{200} = d_{100}/2$, $d^*_{200} = 2 \times d^*_{100} = 0.616$ and $d^*_{300} = 0.924$.

$MoK\alpha$, $d^*_{100} = 0.71/5.0 = 0.142$, $d^*_{200} = 0.284$, $d^*_{300} = 0.426$.

(i) and (ii) See Fig. 245.

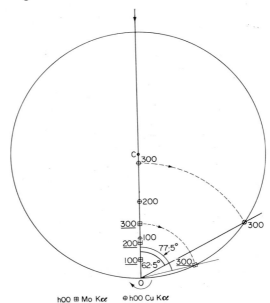

hOO ⊞ Mo Kα   ⊕ hOO Cu Kα

FIG. 245. The reciprocal points, 100, 200, 300 are marked along the normal to the set of planes (100), from the origin O, at the end of the diameter of the circle of reflection. Those for MoKα are surrounded by squares, with the indices underlined.

If the crystal is rotated about an axis perpendicular to the plane of the diagram, the normal with the reciprocal points on it rotates about O. When one of the points reaches the circle the crystal is in position to give the corresponding reflection.

From Bragg's Law $\sin \theta = \lambda/2d_{hkl}$.

For CuKα, $\sin \theta_{300} = 1{\cdot}54/2d_{300} = 1{\cdot}54 \Big/ \left(2 \times \dfrac{5{\cdot}0}{3}\right) = 0{\cdot}4625$; $\theta_{300} = 27°33'$.

For MoKα, $\sin \theta_{300} = 0{\cdot}71/3{\cdot}33 = 0{\cdot}2135$; $\theta_{300} = 12°20'$.

(iii) Rotation required for

$$\text{CuK}\alpha = 90° - 27°33' = 62°27' \ (62{\cdot}5° \text{ from Fig. 245});$$

$$\text{MoK}\alpha = 90° - 12°20' = 77°40' \ (77{\cdot}5° \text{ from Fig. 245}).$$

In marking off the points 100, 200 etc, where the distances from the origin are integral multiples of d^*_{100}, it is essential to calculate all the values of d^*_{100} and measure each off from the origin. Any other method, such as stepping off with dividers, introduces large errors in the positions of the far out points. This warning is of general application for many geometrical constructions in this subject.

l		h		
	0	1	2	3
0		14·08	7·05	4·68
		0·320	0·639	0·964
		32·0	63·9	96·4
1	9·42	6·82	4·90	3·72
	0·478	0·660	0·919	1·210
	47·8	66·0	91·9	121·0
2	4·68	4·00	3·40	2·85
	0·964	1·126	1·325	1·580
	96·4	112·6	132·5	158·0
3	3·05	2·80	2·52	2·25
	1·478	1·608	1·788	2·000
	147·8	160·8	178·8	200·0

FIG. 246. The numbers are: top, d_{h0l} in Å; middle, d^*_{h0l} R.U. for $\lambda = 4{\cdot}5$ Å; bottom, d^*_{h0l} mm to the scale of 1 R.U. $= 100$ mm.

6. Figure 246 gives the measurements of d_{h0l} and the calculated values of d^*_{h0l}.

In Fig. 247 the reciprocal points form a lattice with axes perpendicular to the crystal lattice axes. The coordinates of a point in the lattice are the same as its indices (i.e. the indices of the set of planes it represents) and the axial lengths are given by $a^* = d^*_{100} = \lambda/d_{100}$ and $c^* = d^*_{001} = \lambda/d_{001}$.

$$a^* = 0 \cdot 320, \qquad c^* = 0 \cdot 478 \text{ R.U.}$$

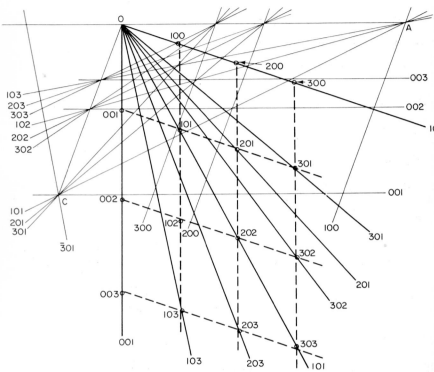

Fig. 247. The thin lines are the traces of the nearest plane to the origin of the set whose indices are given at the end of the line. The heavy full lines are the normals to these planes. The circles denote the position of the reciprocal points. Each reciprocal point is labelled with the indices of the set of planes from which it derives. The lattice of reciprocal points is outlined in heavy dashed lines.

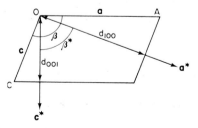

Fig. 248. The relation between the crystal lattice and RL axes for a monoclinic crystal

. From Fig. 248

$* = \lambda/d_{100} = 4\cdot5/(a \cos(\beta - 90))$

$= 4\cdot5/(15\cdot0 \cos 20) = 0\cdot319$ R.U.

$* = \lambda/d_{001} = 4\cdot5/(10\cdot0 \cos 20) = 0\cdot479$ R.U.

$* = \lambda/d_{010} = 4\cdot5/12\cdot0 = 0\cdot375$ R.U

$* = \beta - 2(\beta - 90) = 180 - \beta = 70°.$

. (i) $\mathbf{r} = 2\mathbf{i} - 2\mathbf{j} + 5\mathbf{k}; \quad r^2 = 4 + 4 + 25; \quad r = \sqrt{33}.$

.c.s: $2/\sqrt{33}, \quad -2/\sqrt{33}, \quad 5/\sqrt{33}.$

i) $\mathbf{d}^*_{101} = \mathbf{a}^* + \mathbf{c}^*; \quad \mathbf{d}^*_{111} = \mathbf{a}^* + \mathbf{b}^* + \mathbf{c}^*.$

$^*_{101} \cdot \mathbf{d}^*_{111} = d^*_{101}\, d^*_{111} \cos\psi.$

$\cos\psi = \mathbf{d}^*_{101} \cdot \mathbf{d}^*_{111}/d^*_{101}\, d^*_{111}$

$$= \frac{a^{*2} + c^{*2} + 2a^*c^* \cos\beta^*}{(a^{*2} + c^{*2} + 2a^*c^* \cos\beta^*)^{\frac{1}{2}}\,(a^{*2} + b^{*2} + c^{*2} + 2a^*c^* \cos\beta^*)^{\frac{1}{2}}}$$

$$= \frac{0\cdot09 + 0\cdot16 + 0\cdot24 \times 0\cdot5}{(0\cdot37 \times 0\cdot41)^{\frac{1}{2}}} = \frac{0\cdot37}{(0\cdot1515)^{\frac{1}{2}}} = \frac{0\cdot37}{0\cdot389} = 0\cdot951.$$

. $\psi = 18\cdot0°.$

ii) $[HKL] = H\mathbf{a} + K\mathbf{b} + L\mathbf{c}; \quad \mathbf{d}^*_{hkl} = h\mathbf{a}^* + k\mathbf{b}^* + l\mathbf{c}^*.$

$(H\mathbf{a} + K\mathbf{b} + L\mathbf{c}) \cdot (h\mathbf{a}^* + k\mathbf{b}^* + l\mathbf{c}^*) = 0 = (Hh + Kk + Ll)\lambda.$

. $Hh + Kk + Ll = 0.$

v) $\cos\psi = \dfrac{3a^{*2} + 4b^{*2} + 3c^{*2}}{(9a^{*2} + 4b^{*2} + c^{*2})^{\frac{1}{2}}\,(a^{*2} + 4b^{*2} + 9c^{*2})^{\frac{1}{2}}} = \dfrac{10}{14} = 0\cdot714.$

$\therefore \psi = 44°24'.$

) $\mathbf{r} \cdot \mathbf{d}^* = (x\mathbf{a} + y\mathbf{b} + z\mathbf{c}) \cdot (h\mathbf{a}^* + k\mathbf{b}^* + l\mathbf{c}^*)$

$= (hx + ky + lz)\lambda = (3 \times 0\cdot5 + 2 \times 0\cdot7 + 0\cdot2)2\cdot0$

$= 6\cdot2.$

(vi) The vectors determined by the third and fourth sides of the octagon are shown in Fig. 249(a). Figures 249(b) and (c) show how they can be derived. Sides 5 to 8 are merely the opposites of 1 to 4.

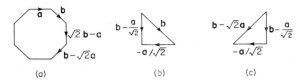

(a)   (b)   (c)

FIG. 249. The vectors determined by the sides of an octagon.

(vii) $\dfrac{\sqrt{3}}{2}\mathbf{i} + \dfrac{1}{2}\mathbf{j}; \mathbf{j}; -\dfrac{\sqrt{3}}{2}\mathbf{i} + \dfrac{1}{2}\mathbf{j}; \quad -\dfrac{\sqrt{3}}{2}\mathbf{i} - \dfrac{1}{2}\mathbf{j}; \quad \dfrac{1}{\sqrt{2}}\mathbf{i} + \dfrac{1}{\sqrt{2}}\mathbf{j};$

$-\dfrac{1}{\sqrt{2}}\mathbf{i} + \dfrac{1}{\sqrt{2}}\mathbf{j}.$

(viii) $\mathbf{a^*} \cdot \mathbf{a} = \mathbf{b^*} \cdot \mathbf{b} = \mathbf{c^*} \cdot \mathbf{c} = \lambda; \quad \mathbf{a^*} \cdot \mathbf{b} = \mathbf{a^*} \cdot \mathbf{c} = \mathbf{b^*} \cdot \mathbf{a} = \mathbf{b^*} \cdot \mathbf{c} = \mathbf{c^*} \cdot \mathbf{a}$

$$= \mathbf{c^*} \cdot \mathbf{b} = 0$$

The second set of equations means that an axis in one lattice (e.g. $\mathbf{a^*}$) is perpendicular to the other two axes ($\mathbf{b}$ and $\mathbf{c}$) in the other lattice. The first set defines the product of (1) the projection of a crystal lattice axis (e.g. $\mathbf{a}$) in the direction of the RL axis ($\mathbf{a^*}$), and (2) the length of the RL axis ($\mathbf{a^*}$). This product is equal to λ. Since $\mathbf{a^*}$ is normal to $\mathbf{b}$ and $\mathbf{c}$ and therefore to the set of planes 100, the projection of $\mathbf{a}$ in the direction of the normal is d_{100}.

$$\therefore d_{100} \times a^* = \lambda \quad \text{or} \quad a^* = \lambda/d_{100}.$$

Similarly $b^* = \lambda/d_{010}$ and $c^* = \lambda/d_{001}$.

(ix) $\overrightarrow{PQ} = \overrightarrow{OQ} - \overrightarrow{OP} = 5\mathbf{i} - 2\mathbf{j} + 4\mathbf{k} - (\mathbf{i} + 3\mathbf{j} - 7\mathbf{k}) = 4\mathbf{i} - 5\mathbf{j} + 11\mathbf{k}.$

$$\therefore PQ^2 = 16 + 25 + 121 = 162. \quad \therefore PQ = 9\sqrt{2}.$$

$$\text{d.c.s.} \quad \frac{4}{9\sqrt{2}}, \quad \frac{-5}{9\sqrt{2}}, \quad \frac{11}{9\sqrt{2}}.$$

(x) 243. Yes. The nth plane from the origin would cut off intercepts of $\dfrac{na}{h}, \dfrac{nb}{k}, \dfrac{nc}{l}$ and its indices would be $h/n, k/n, l/n$. At least one of these would

e a fraction since the indices of lattice planes cannot contain a common
ctor. (App. II, p. 341).

$$* = \frac{\lambda}{a \sin \beta} = \frac{1}{5 \times 0\cdot8} = \frac{1}{4}; \quad b* = \frac{\lambda}{b} = \frac{1}{6}; \quad c* = \frac{\lambda}{c \sin \beta} = \frac{1}{6};$$

$$\sin \beta* = 0\cdot8, \quad \cos \beta* = 0\cdot6.$$

$$*_{243} = 2\mathbf{a}* + 4\mathbf{b}* + 3\mathbf{c}*; \quad d*_{243} = (4a*^2 + 16b*^2 + 9c*^2 + 12a*c* \cos \beta*)^{\frac{1}{2}}$$

$$= (\tfrac{1}{4} + \tfrac{4}{9} + \tfrac{1}{4} + \tfrac{1}{2} \times 0\cdot6)^{\frac{1}{2}} = \sqrt{1\cdot245}$$

$$= 1\cdot116$$

$$d_{243} = \frac{\lambda}{d*_{243}} = \frac{1}{1\cdot116} = 0\cdot863 \text{ Å}.$$

$$\mathbf{d}*_{243} \cdot \mathbf{a}* = d*_{243} \, a* \cos \psi_1; \quad \cos \psi_1 = \frac{2a*^2 + 3a*c* \cos \beta*}{a* \times 1\cdot116}$$

$$= \frac{\tfrac{1}{2} + 0\cdot3}{1\cdot116} = 0\cdot716.$$

$$\therefore \psi_1 = 44°16'.$$

imilarly

$$\psi_2 = \frac{4b*^2}{b* \times 1\cdot116} = \frac{2/3}{1\cdot116} = 0\cdot599 \quad \therefore \psi_2 = 53°12'$$

$$\psi_3 = \frac{3c*^2 + 2a*c* \cos \beta*}{c* \times 1\cdot116} = \frac{\tfrac{1}{2} + 0\cdot3}{1\cdot116} = 0\cdot716 \quad \therefore \psi_3 = 44°16'.$$

he equality of ψ_1 and ψ_3 is an unfortunate mathematical coincidence due
o rounding off the values of the parameters to simplify the calculations.
otice that in this question λ was not specified and could be taken as $1\cdot0$ Å.

i) Velocity of the boat relative to the earth

$$= \mathbf{v} = 3\mathbf{i} + 4\mathbf{j} + \mathbf{i} - 3\mathbf{j} = 4\mathbf{i} + \mathbf{j}$$

$$\therefore v^2 = 16 + 1; \quad v = \sqrt{17}.$$

Speed of $\sqrt{17}$ mph in a direction $\tan^{-1} \tfrac{1}{4} = 14°2'$ N. of E. (Fig. 250).

FIG. 250. The resultant velocity of the boat relative to the earth.

(xii) $nd^*_{h_1k_1l_1} + md^*_{h_2k_2l_2} = nh_1\mathbf{a}^* + nk_1\mathbf{b}^* + nl_1\mathbf{c}^* + mh_2\mathbf{a}^* + mk_2\mathbf{b}^* + ml_2\mathbf{c}^*$

$\qquad = (nh_1 + mh_2)\mathbf{a}^* + (nk_1 + mk_2)\mathbf{b}^* + (nl_1 + ml_2)\mathbf{c}^*$

$\qquad = \mathbf{d}^*_{nh_1 + mh_2,\ nk_1 + mk_2,\ nl_1 + ml_2}.$

(xiii) Let $\mathbf{a}^*$, $\mathbf{b}^*$ define a primitive unit cell. Then $\mathbf{a}^{*\prime}$ and $\mathbf{b}^{*\prime}$ can be defined in terms of $\mathbf{a}^*$ and $\mathbf{b}^*$ as

$\mathbf{a}^{*\prime} = h_1\mathbf{a}^* + k_1\mathbf{b}^*$; $\quad \mathbf{b}^{*\prime} = h_2\mathbf{a}^* + k_2\mathbf{b}^*$ where h_1, k_1, h_2, k_2 are integers

(App. II, p. 340)

$\therefore \mathbf{a}^{*\prime} \times \mathbf{b}^{*\prime} = L\mathbf{a}^* \times \mathbf{b}^* = (h_1k_2 - h_2k_1)\mathbf{a}^* \times \mathbf{b}^*.$

$\therefore L = h_1k_2 - h_2k_1$ which is necessarily integral but may be positive or negative, depending on whether $\mathbf{a}^* \times \mathbf{b}^*$ and $\mathbf{a}^{*\prime} \times \mathbf{b}^{*\prime}$ are in the same or opposite directions.

Suppose

$$\mathbf{d}^* = p\frac{\mathbf{a}^{*\prime}}{L} + q\frac{\mathbf{b}^{*\prime}}{L} = r\mathbf{a}^* + s\mathbf{b}^* \qquad (1)$$

where p and q are any positive or negative integers. It is required to prove that r and s can have all possible positive and negative integral values.

Substituting in (1) and taking L to the right-hand side, we have

$$p(h_1\mathbf{a}^* + k_1\mathbf{b}^*) + q(h_2\mathbf{a}^* + k_2\mathbf{b}^*) = (h_1k_2 - h_2k_1)(r\mathbf{a}^* + s\mathbf{b}^*).$$

Equating coefficients of $\mathbf{a}^*$ and $\mathbf{b}^*$,

$$ph_1 + qh_2 = r(h_1k_2 - h_2k_1); \quad pk_1 + qk_2 = s(h_1k_2 - k_2h_1).$$

Solving for p and q gives

$$p(h_1k_2 - h_2k_1) = (h_1k_2 - h_2k_1)(rk_2 - sh_2).$$

$$\therefore p = rk_2 - sh_2 \quad \text{and} \quad q = -(rk_1 - sh_1).$$

Thus integral values of p and q to satisfy (1) can be found for any integral values of r and s. Q.E.D.

(xiv) Take the two given vectors as the directions of axes of a unit cell in the RL plane defined by the two vectors. Let the first be $h\mathbf{a}^{*\prime}$, where $\mathbf{a}^{*\prime}$ is the distance from the origin to the first relp in this direction. Similarly let the second vector be $k\mathbf{b}^{*\prime}$. h and k are positive integers. If $\mathbf{a}^{*\prime}$ and $\mathbf{b}^{*\prime}$ define an L-fold centred cell, choose another axis $\mathbf{b}^{*\prime\prime}$ which will define a primitive cell of the same 'hand', i.e. $\mathbf{a}^{*\prime} \times \mathbf{b}^{*\prime} = L\mathbf{a}^{*\prime} \times \mathbf{b}^{*\prime\prime}$, where L is positive.

$L\mathbf{b}^{*''} = p\mathbf{a}^{*'} + q\mathbf{b}^{*'}$ where p and q are integers (Ans. 8 (xiii), p. 462). Let the first relp in the direction of the third vector be defined by $u\mathbf{a}^{*'} + v\mathbf{b}^{*''}$, where u and v are positive or negative integers. The third vector can therefore be any RL vector in the plane (since $\mathbf{a}^{*'}$ and $\mathbf{b}^{*''}$ define a primitive unit cell). The direction is therefore given by $r'(u\mathbf{a}^{*'} + v\mathbf{b}^{*''})$ where r' is any positive number. It is required to prove that

$$r'(u\mathbf{a}^{*'} + v\mathbf{b}^{*''}) = nh\mathbf{a}^{*'} + mk\mathbf{b}^{*'}. \qquad (2)$$

Suppose this to be true for $r' = rL$.

Then $rLu\mathbf{a}^{*'} + rvL\mathbf{b}^{*''} = nh\mathbf{a}^{*'} + mk\mathbf{b}^{*'}$

$\therefore rLu\mathbf{a}^{*'} + rv(p\mathbf{a}^{*'} + q\mathbf{b}^{*'}) = nh\mathbf{a}^{*'} + mk\mathbf{b}^{*'}$

$\therefore \mathbf{a}^{*'}(rLu + rvp) + rvq\mathbf{b}^{*'} = nh\mathbf{a}^{*'} + mk\mathbf{b}^{*'}$

$\therefore nh = r(Lu + pv); \quad mk = rvq.$

$$\therefore r = \frac{mk}{vq} = \frac{nh}{Lu + pv}.$$

The simplest solution of these equations is:

$$n = (Lu + pv)k; \quad m = vqh; \quad r = hk \quad \text{and} \quad r' = Lhk.$$

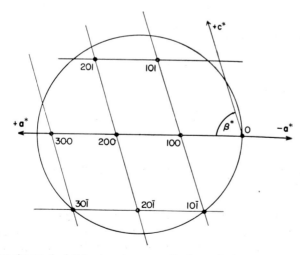

FIG. 251. Since the vertical RL plane is perpendicular to $\mathbf{b}$, it must be the $\mathbf{a}^*\mathbf{c}^*$ plane. Since $\mathbf{c}$ is along the vertical axis, $\mathbf{a}^*$ is horizontal. Only one of the two possible directions for $\mathbf{c}^*$ gives $\beta^* > 70(\beta < 110)$. $+\mathbf{c}^*$ is chosen upwards and $+\mathbf{a}^*$ is then to the left to give $\beta^* < 90(\beta > 90)$. The indexing follows, as shown.

Thus it is always possible to find values of m and n to satisfy (2), whatever the values of u, v, h and k, i.e. for any given pair of vectors and any RL vector in their plane. Q.E.D.

9. (i) See Figure 251.

(ii) $+\mathbf{b}^*$ is coming out of the plane towards the observer.

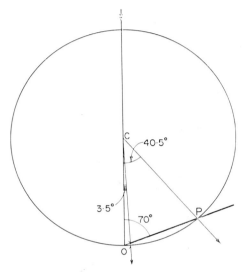

FIG. 252. The position of the weighted RL vector after a rotation of 70° from the initial position along the beam, which is shown in Figure 245 (p. 456). No points corresponding to a characteristic radiation are shown. The position of such points depends on the X-ray tube target material.

10. Refer to Answer 5 (p. 456)

(i) $\lambda_{\min} = \dfrac{12\cdot4}{50} = 0\cdot248$ Å.

Angle of Deviation OCP (Fig. 252) = 40·5° by measurement

$\qquad\qquad\qquad\qquad\qquad\qquad = 2(90 - 70)\ = 40\cdot0°$ by calculation.

$OP = 0\cdot688$ R.U. by measurement.

$d^*_{100} = 0\cdot308$ R.U., $\lambda = 1\cdot54$ Å.

$$\therefore \lambda_{100} = \frac{1 \cdot 54 \times 0 \cdot 688}{0 \cdot 308} = 3 \cdot 44 \text{ Å.}$$

$$d^*_{h00} = h \times 0 \cdot 308. \quad \therefore \lambda_{h00} = \frac{3 \cdot 44}{h}.$$

$$\lambda_{13,00} = 0 \cdot 264, \quad \lambda_{14,00} = 0 \cdot 246 \text{ Å.}$$

$\therefore$ 13 orders take part in the reflection with wavelengths $\lambda_{h00} = 3 \cdot 44/h$ Å for $h = 1, 2, \dots, 13$.

Alternatively $OP = d^*_{100} = \dfrac{\lambda_{100}}{d_{100}}$ R.U.

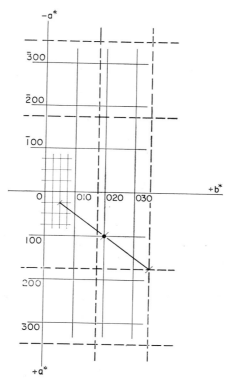

Fig. 253. The three RL nets superimposed. The medium lines give the net for the characteristic radiation, $\lambda = 1 \cdot 54 \text{Å}$. The small net of thin lines is for the 'cut-off' wavelength ($\frac{1}{4}$ the size). The large net ($1\frac{3}{4}$ the size) corresponds with the long wavelength 'fade-out'. The indices of the axial relps are given for the medium, characteristic wavelength net. Those for the other nets can be easily obtained by comparison. Reduced to a scale of 40 mm $\equiv$ 1 R.U.

$$\therefore \lambda_{100} = OP \cdot d_{100}$$
$$= 0.688 \times 5.0 = 3.44 \text{ Å.}$$
$$\lambda_{200} = OP \cdot d_{200}$$
$$\lambda_{h00} = OP \cdot d_{h00} = 0.688 \times \frac{5.0}{h} \text{ Å}$$

(ii) Since the Bragg angle, $\theta = 20° \ (90 - \widehat{COP} \text{ or } \widehat{OCP}/2)$ and

$$d_{100} = 5.0 \text{ Å,} \quad \lambda_{100} = 2 \times 5.0 \sin 20° = 10.0 \times 0.3420 = 3.42 \text{ Å.}$$
$$d_{h00} = 5.0/h, \quad \therefore \lambda_{h00} = 3.42/h \text{ Å.}$$

(iii) For $\lambda_{min} = 0.248 \text{ Å,} \ d^*_{100} = 0.248/5.0 = 0.049$ R.U.
The rod therefore becomes weightless at 0.049 R.U. from the origin; and as this point passes through the sphere, reflections cease at an angle of deviation (measured) of 3.5°. From Bragg's Law, $\sin \theta = 0.248/(2 \times 5.0) = 0.0248$; $\theta = 1°25'; \quad 2\theta = 2°50'$.

11. CuKα; $a^* = 1.54/5.13 = 0.30$; $b^* = 1.54/7.7 = 0.20$ R.U. (Fig. 253).

(i) $0.385 = 12.4/V \quad \therefore V = 12.4/0.385 = 32.3$ kV.

(ii) For a camera radius of 30 mm $2° \approx 1$ mm.

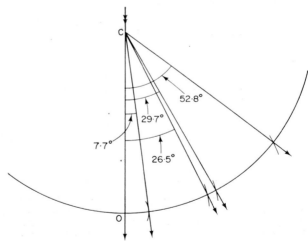

FIG. 254. The points at which the outer end, the Kα and Kβ relps and the inner end of the weighted 120 RL vector, cut the sphere of reflection. (The compass extensions for the arcs were taken direct from the original of Fig. 253 except for Kβ). The directions of the corresponding reflections are shown and the angles of deviation. Reduced to a scale of 50 mm ≡ 1 R.U.

∴ Cut-off occurs at 4 mm from direct beam. (Fig. 254).

CuKα at 15 mm from direct beam.

Long wavelength fade-out at 27 mm.

d_{120}^* for CuKα $= 0.50$ R.U.

∴ d_{120}^* for CuKβ $= 0.50 \times \dfrac{1.39}{1.54} = 0.45$ R.U.

∴ CuKβ at 13 mm.

Figure 255 shows approximately to scale the appearance of the resultant photograph.

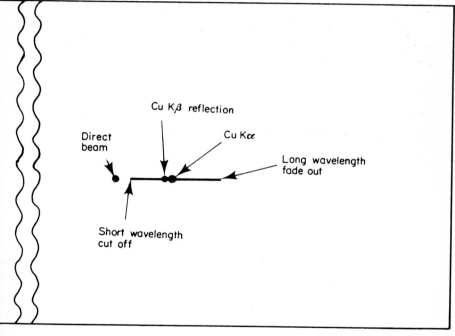

FIG. 255. The appearance on the unrolled cylindrical film of the continuous series of Laue reflections giving rise to the "Laue streak", with the Kα and Kβ reflections superimposed on it. The distances from the central spot in mm are approximately equal to $2\theta/2°$, for a film diameter of 60 mm (accurately so for a diameter of 57·3 mm).

(iii) The higher order reflections 240, 360 etc. and neighbouring weighted rods must be too weak to give appreciable intensity, otherwise there will be overlapping streaks obscuring the effect of 120.

12. (i) In Fig. 256 take the central incident ray and the ray reflected from the mean position of the lattice planes as reference directions. The extreme ray in the reflection plane (the plane of the diagram) has its angle of incidence changed by $\alpha - \varepsilon$. Therefore the extreme reflected ray makes an angle of $2(\alpha - \varepsilon)$ with the reference direction. The reflected beam hits the film at P', $D_1 \sec 2\theta$ from the crystal. $\therefore$ the spread of the beam perpendicular to CP' is $2r + 2D_1 \sec 2\theta \times 2(\alpha - \varepsilon)$ where r is the radius of the crystal. $\therefore$ length of spot equals

$$(2r + 4(\alpha - \varepsilon)D_1 \sec 2\theta) \sec 2\theta.$$

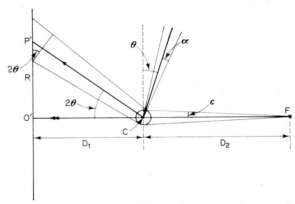

FIG. 256. Diagram showing reflection of X-rays from a point source F, by a spherical crystal C. C has an extreme range of orientation 2α, of the normals to the reflecting planes of its various mosaic blocks. The reflections are registered on a flat film perpendicular to the beam.

The breadth of the spot is obtained by rocking the plane of reflection about the central incident ray through $2\alpha \sec \theta$ to the two extreme positions of the normal, and adding the effect of oscillating the incident ray through 2ε in the horizontal plane.

$$\therefore b = O'P'\, 2\alpha \sec \theta + FO'\, 2\varepsilon = 2\alpha D_1 \tan 2\theta \sec \theta + 2(D_1 + D_2)\varepsilon. \quad (1)$$

$$l = 2r \sec 2\theta + 4(\alpha - \varepsilon)\, D_1 \sec^2 2\theta. \quad (2)$$

From (1)
$$\varepsilon = \frac{b - 2\alpha D_1 \tan 2\theta \sec \theta}{2(D_1 + D_2)}. \quad (3)$$

From (2)
$$\alpha - \varepsilon = \frac{l - 2r \sec 2\theta}{4D_1 \sec^2 2\theta}.$$

Substituting from 468(3),

$$\alpha - \frac{b}{2(D_1 + D_2)} + \frac{2D_1 \tan 2\theta \sec \theta}{2(D_1 + D_2)} \alpha = \frac{l - 2r \sec 2\theta}{4D_1 \sec^2 2\theta}$$

$$\therefore \alpha = \left(\frac{b}{2(D_1 + D_2)} + \frac{l}{4D_1 \sec^2 2\theta} - \frac{r}{2D_1 \sec 2\theta} \right) \Big/ \left(1 + \frac{D_1 \tan 2\theta \sec \theta}{(D_1 + D_2)} \right)$$

(1)

When one comes to measure the reflections on the photograph it immediately becomes clear that the mosaic spread is not uniform. Since the crystal had been ground spherical it is almost certain that most of the mosaic spread is in the outer layers. Only the stronger reflections register the effect of the outer layer. The weaker reflections give detectable diffraction only from the undisturbed inner core, which constitutes the greater part of the volume of the crystal. The weak reflections therefore show much less asterism than the stronger ones. Even in the outer shell the effect is not homogenous. Owing to the effects of absorption, the surface of the crystal around the positive direction of the normal will have more effect than that on the far side of the crystal. Asterism is more evident on the top right of the photograph than at the bottom left. This is almost certainly due to the differences in mosaic spread in different parts of the surface layer. All one can do is to get an estimate of the maximum spread by taking one of the longer spots (avoiding any which show signs of picking up K radiation).

The reflection at $x = 29$, $y = 16$ mm (top, right) has $l = 1 \cdot 25$, $b = 0 \cdot 25$ mm and $R = 32 \cdot 5$ mm. $D_1 = 25$, $D_2 = 100$ mm, $r = 0 \cdot 1$ mm. $\tan 2\theta = 1 \cdot 30$; $\sec \theta = 1 \cdot 13$; $\sec 2\theta = 1 \cdot 64$; $\sec^2 2\theta = 2 \cdot 69$ ($\theta = 26 \cdot 25°$).

Putting these values into (1) gives:

$$\alpha = \left(\frac{0 \cdot 25}{250} + \frac{1 \cdot 25}{100 \times 2 \cdot 69} - \frac{0 \cdot 1}{50 \times 1 \cdot 64} \right) \Big/ \left(1 + \frac{25 \times 1 \cdot 3 \times 1 \cdot 13}{125} \right)$$

$$= (0 \cdot 001 + 0 \cdot 0046 - 0 \cdot 0012)/(1 + 0 \cdot 30)$$

$$= 0 \cdot 0044/1 \cdot 30 = 0 \cdot 0034 \text{ rad} \equiv 0 \cdot 19°.$$

However, if one calculates ε from 468(3):

$$\varepsilon = \frac{0 \cdot 25 - 50 \times 1 \cdot 30 \times 1 \cdot 13 \times 0 \cdot 0034}{250} = \frac{0 \cdot 25 - 0 \cdot 25}{250} = 0 \cdot 0$$

whereas from the approximation of a point source

$$\varepsilon = r/D_2 = 0 \cdot 1/100 = + 0 \cdot 001.$$

It is clear that the source is, in fact, approximately the same size as the crystal, and the geometry more complicated than that for a point source.

However, the order of magnitude of the result is correct. The variability over the crystal does not justify a more elaborate analysis, which would have to take account of the fact that the effective focus is wider than it is high, as shown by the shape of the central spot.

In the case where the divergence (strictly the convergence from an extended source) of the beam is much greater than mosaic spread, so that its effect and that of the crystal size can be neglected, we have:

$$l = 4\varepsilon D_1 \sec^2 2\theta \quad \text{and} \quad b = 2\varepsilon D_1.$$

$$\therefore \, l/b = 2\sec^2 2\theta. \tag{1}$$

Since the conditions under which Fig. 43(j) was taken are far removed from those assumed in deducing (1), it is a coincidence that $2\sec^2 2\theta = 5\cdot4$ for the measured reflection which is nearly equal to the measured

$$l/b = 1\cdot25/0\cdot25 = 5\cdot0.$$

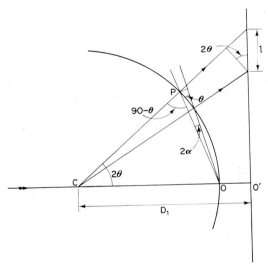

FIG. 257. The intersection with the sphere of reflection of the cone of RL vectors deriving from the misorientation of the mosaic blocks. A parallel beam of X-rays is assumed, and the limits of the reflection in the plane containing the X-ray beam and the RL vector are shown.

In terms of the sphere of reflection based on a parallel beam of X-rays, the weighted RL vector becomes a narrow cone (semi-angle α) of diffracting power, which intersects the sphere in an oval (Fig. 257). The projection of this on to the film gives an elongated spot. As a first approximation we can take the intersection with the tangent plane at P, where $OP = \mathbf{d}^*$. Then the angle subtended at C by the breadth of the oval $= d^* \times 2\alpha/1$, and the angle subtended by the length $= d^* \times 2\alpha \operatorname{cosec} \theta/1$. The length on the film,

$$l = 2d^*\alpha \operatorname{cosec} \theta \times D_1 \sec^2 2\theta$$

$$= 4 \sin \theta \times \alpha \operatorname{cosec} \theta \times D_1 \sec^2 2\theta$$

$$= 4D_1 \alpha \sec^2 2\theta.$$

$$b = 4 \sin \theta \times \alpha D_1 \sec 2\theta$$

$$= \frac{4 \sin \theta \cos \theta \sec 2\theta}{\cos \theta} D_1 \alpha$$

$$= 2D_1 \alpha \tan 2\theta \sec \theta.$$

These two expressions are equivalent to (1) and (2) if $r = \varepsilon = 0$. If θ is small,

$$\frac{b}{l} = \frac{\sin \theta \sec 2\theta}{\sec^2 2\theta} = \sin \theta (1 - 2 \sin^2 \theta) \simeq \theta.$$

In this case the cone axis is nearly perpendicular to the X-ray beam and the narrow cone intersects the sphere in a very elongated strip, similar in outline to the constant ρ curves for very small ρ (Fig. 94, p. 151). The corresponding low angle reflections become radiating streaks, and this accounts for the origin of the name 'asterism'.

(ii) The reflection just below the previous one, at $x = 30$, $y = 8 \cdot 5$ mm, $R = 30 \cdot 8$ mm, has an associated diffuse spot $1 \cdot 5$ mm nearer the direct beam than the centre of the reflection.

We have: $\tan 2\theta = 30 \cdot 8/25 = 1 \cdot 232$, $2\theta = 50 \cdot 9°$, $\theta = 25 \cdot 45$, $\sin \theta = 0 \cdot 4297$, $2 \sin \theta = 0 \cdot 8594$, $\cos \theta = 0 \cdot 9029$, $\cos 2\theta = 0 \cdot 6307$.

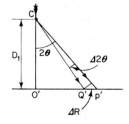

FIG. 258. The relation between $\Delta 2\theta$ and radial displacement ΔR on the flat film photograph.

From Fig. 258

$$\Delta 2\theta = \frac{\Delta R \cos 2\theta}{D_1 \sec 2\theta} = \frac{\Delta R}{D_1} \cos^2 2\theta.$$

$$d^* = 2 \sin \theta, \quad \therefore \ \Delta d^* = 2 \cos \theta \ \Delta \theta = \cos \theta \ \Delta 2\theta$$

$$\therefore \ \Delta d^* = \frac{\Delta R}{D_1} \cos^2 2\theta \cos \theta = \frac{1\cdot5}{25} \times 0\cdot6307^2 \times 0\cdot9029 = 0\cdot0216 \ \text{R.U.}$$

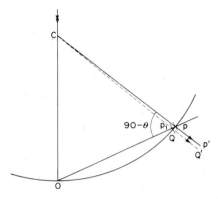

FIG. 259. The Laue reflection P', arising from the intersection of the weighted RL vector and the sphere at P, and the diffuse reflection Q', arising from the diffuse scattering volume centered at P_1.

Figure 259 shows the point P on the weighted RL vector which is in the sphere of reflection, giving rise to the Laue reflection. $OP = d^* = 0\cdot859$ R.U. The point Q, giving rise to the diffuse reflection associated with the Kα relp P_1, is in the sphere at the point $0\cdot0216$ R.U. nearer the origin than P. Figure 260 is an enlargement of the area around P. On the assumption that the volume of diffuse scattering power round P_1 is spherical, P_1Q is perpendicular

FIG. 260. An enlargement of the area around P and Q in Figure 259. Arcs are taken as straight lines, and CP and CQ as parallel.

to the sphere and to a first approximation parallel to the radius at P. The arc PQ is replaced by the tangent.

$$PQ_1 = \Delta d^*$$

$$\therefore PQ = \Delta d^* \sec \theta$$

$$P_1Q = PQ \tan \theta = \Delta d^* \frac{\sin \theta}{\cos^2 \theta}$$

$$= \frac{0 \cdot 0216 \times 0 \cdot 4297}{0 \cdot 9029^2} = 0 \cdot 0114 \text{ R.U.}$$

The distance between $K\alpha$ relps is of the order of $0 \cdot 2$ R.U. The effective radius is of the order of 1/20th of the distance between relps.

Thermal diffuse scattering is often referred to as phonon scattering, since the thermal vibrations for the crystal as a whole, which give rise to it, can be analysed into standing acoustic waves or phonons. The volume of phonon scattering power surrounding a $K\alpha$ relp must be carefully distinguished from the small volume (the shape transform) of Bragg scattering, due to the finite size of the crystal. The shape transform is the same for all relps, whereas the phonon scattering volumes are very varied and often very far from the spherical shape assumed in this example [19].

Phonon scattering takes place with a change of wavelength due to exchange of momentum between X-ray photon and thermal phonon. The change is so small that only with the Mossbauer effect can the difference be detected experimentally (Sec. **9.5.3**, p. 119).

13. (i) Successful indexing of a Laue photograph such as Fig. 174 depends on the careful use of an accurate scale. In Fig. 174, 2θ has a minimum of 16° and a maximum of 56°. The scale therefore need only be constructed for θ from 8 to 28°. Unfortunately the central spot does not contain a radiograph of the crystal, and an error of $0 \cdot 25$ mm in positioning the origin on the photograph can lead to a $2 \cdot 0$ mm error in the position of a relp, in the worst case. The origin can be found by drawing in the prominent zones containing **b***, i.e. straight lines of reflections passing through the central spot. These should intersect in the central spot and mark the origin. There are three such zones which are relatively prominent, i.e. the horizontal line and one on either side of the vertical direction. Tracing paper should be fastened with drafting tape over half the photograph, and these three directions drawn on the paper through the central spot. Their common point of intersection (which is not, in fact, at the centre of the direct beam spot) is the origin for the scale. Take the reflections a zone at a time and plot the relps on the paper

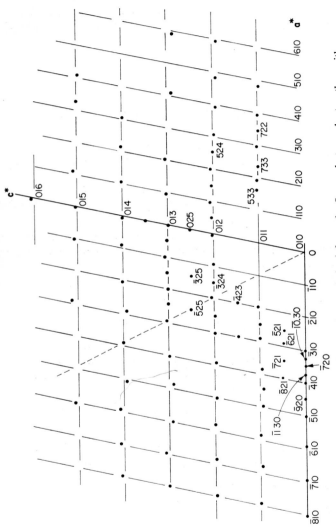

FIG. 261. The RL layer for $k = 1$, derived from the Laue photograph, together with additional points arising from higher layers.

by means of the scale. Figure 261 shows the plot for a sufficient proportion of the reflections to enable the net to be drawn, and examples of points from layers other than the first to be indexed.

Notice that the outer relps define the basic net, and that the points arising from higher layers only occur in the central portion of the net. The accuracy of plotting is not sufficient to determine with certainty which layer these additional points derive from, but the most probable indices for some of them are given in Fig. 261.

(ii)

$$b \geqslant \frac{\lambda_{min}}{2 \sin^2 \theta}; \qquad \lambda_{min} = \frac{12 \cdot 4}{50} = 0 \cdot 248 \text{ Å}; \qquad \tan 2\theta = \frac{14 \cdot 25}{50} = 0 \cdot 285;$$

$$\theta = 7 \cdot 95°; \qquad \therefore b \geqslant \frac{0 \cdot 248}{2 \times (0 \cdot 1383)^2} = 6 \cdot 5 \text{ Å} \qquad (cf. \ b = 8 \cdot 75 \text{ Å}).$$

(iii) For 43 (c), $\lambda_{min} = 0 \cdot 248$ Å. $\tan 2\theta = \dfrac{15 \cdot 25}{44} = 0 \cdot 3470.$ $\therefore \theta = 9°34'.$

$\therefore c = 4 \cdot 49$ Å. (cf. $c = 4 \cdot 38$ Å; the probable explanation of this result is that the peak voltage on the tube was slightly higher than the voltage registered on the meter.)

14. See Fig. 262

$$0 \cdot 5 \text{ R.U. for } \lambda = 1 \cdot 0 \text{ Å} \equiv 58 \cdot 8 \text{ mm.}$$

$$\text{CuK}\alpha. \quad 1 \text{ R.U. for } \lambda = 1 \cdot 54 \text{ Å} \equiv \frac{58 \cdot 8 \times 2}{1 \cdot 54} = \overset{\text{top (bottom)}}{76 \cdot 3 \ (25 \cdot 4) \text{ mm.}}$$

$$\text{MoK}\alpha. \quad 1 \text{ R.U.} \qquad \equiv \frac{58 \cdot 8 \times 2}{0 \cdot 71} = 165 \cdot 7 \ (55 \cdot 2) \text{ mm.}$$

(i) Zero intensity CuKα, 24, 51, 92·5° (possibly others off the diagram).

MoKα, 11, 27, 45° (certainly others off the diagram).

(ii) Scattering amplitude at 60° deviation = 2·9.

(iii) **D*** for this angle of scattering is 1 R.U. in the $-x$ direction.

(iv) With atoms numbered as shown, and putting

$$\frac{2\pi}{\lambda} \times \frac{180}{\pi} = 234,$$

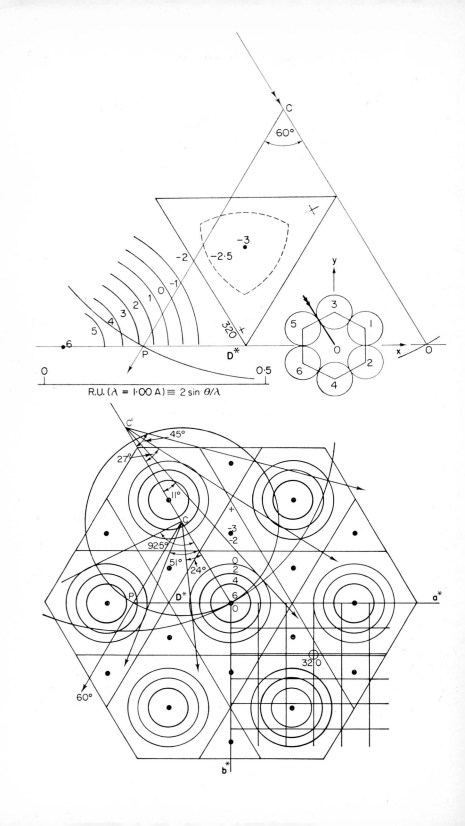

we have:

Atom No. (m)	$r_m \cdot D^*$	$\phi_m{}^\circ = 234 r_m \cdot D^*$	$\cos \phi_m$	$\sin \phi_m$
1	$-1 \cdot 21$	$-283 \,(\equiv 77)$	$0 \cdot 225$	$0 \cdot 974$
2	$-1 \cdot 21$	-283	$0 \cdot 225$	$0 \cdot 974$
3	0	0	$1 \cdot 0$	0
4	0	0	$1 \cdot 0$	0
5	$1 \cdot 21$	$283 \,(\equiv -77)$	$0 \cdot 225$	$-0 \cdot 974$
6	$1 \cdot 21$	283	$0 \cdot 225$	$-0 \cdot 974$
			$\overline{2 \cdot 90}$	$\overline{0 \cdot 0}$

There is no real need to calculate $\Sigma \sin \phi_m$. Since the distribution of atoms is centro-symmetric, for every ϕ_m there is a $-\phi_m$. The sines are therefore equal and opposite so that the sum is zero.

(v) Taking the incident X-ray beam 30° anti-clockwise to make it parallel to a side of the hexagon, and the scattered beam deviated at 60° on the right-hand side to make $D^* = 1$ R.U. in the y direction, we have:

m	$r_m \cdot D^*$	$\phi_m{}^\circ$	$\cos \phi_m$
1	$0 \cdot 70$	$163 \cdot 8$	$-0 \cdot 960$
2	$-0 \cdot 70$	$-163 \cdot 8$	$-0 \cdot 960$
3	$1 \cdot 40$	$327 \cdot 5$	$0 \cdot 843$
4	$-1 \cdot 40$	$-327 \cdot 5$	$0 \cdot 843$
5	$0 \cdot 70$	$163 \cdot 8$	$-0 \cdot 960$
6	$-0 \cdot 70$	$-163 \cdot 8$	$-0 \cdot 960$
			$\overline{-2 \cdot 154}$

The end of D^* is marked with a $+$ on the diagrams (taking the origin at the centre of the circles and rotating 60° in the top diagram), and the scattering amplitude $= -2 \cdot 25$.

15. (i) The scattering amplitude becomes zero except at the points of the reciprocal lattice, where, per unit cell, it has the same amplitude as before. The scattering diagram is multiplied by the RL with points of unit weight (and, of course, zero weight everywhere between points).

FIG. 262. The benzene transform with the incident beam and circles of reflection for CuKα and MoKα radiation. The various directions in which scattering is of zero amplitude are shown, and the scattering for an angle of deviation of 60°. The RL net for a hypothetical benzene crystal is shown (Answer 15) with the 320 relp marked. Some of the constructions have been carried out more accurately on the upper figure, to which the R.U. scale applies, and which is 3 times the scale of the lower figure.

This is a simple example of the transform of two convoluted functions, in this case the electron density of the unit cell and the crystal lattice with points of unit weight. Two 3-dimensional functions are said to be convoluted if every point in space is taken as the origin for the second function, which is then multiplied by the value of the first function at that point; and the value of the convoluted function at any point is the sum of the values of all the superimposed product functions at that point. In general this is a complicated integration; but if one of the functions is a point function, it has a simple interpretation. If we take the point lattice of the crystal as the first function, and place the origin of the electron density function (the unit cell contents) at a point in space and multiply the electron density by the value of the lattice function, the result will be zero except at lattice points. Since the value of the first function here is unity, the electron density distribution will be repeated round each of the lattice points. Nature, which does not allow interpenetrating atoms, makes sure that there are no mathematical problems from overlapping of neighbouring electron density functions, so the result of convoluting the unit cell contents with the point lattice is a crystal. In Fig. 29 (p. 38) the mask (c) is the convolution of (a) and (b).

There is a general theorem that the transform of two convoluted functions is the product of the transforms of the separate functions. The transform of the unit cell contents is the scattering diagram. The transform of the crystal lattice with points of unit weight is the reciprocal lattice, also with points of unit weight. The transform of a crystal (or its scattering diagram) is therefore the product of the scattering diagram for the unit cell contents and the reciprocal lattice with points of unit weight, as stated at the beginning. In Fig. 29 scattering diagram (c) is the product of (a) and (b). Since this result can be obtained without convolution theory (Section **9**.13 p. 97), it is not strictly necessary to try to understand this explanation. However, such understanding is useful because the analysis can be taken a stage further.

The scattering diagram (transform) produced by multiplying the transform of the cell contents by the RL with points of unit weight (i.e. the weighted RL) corresponds to an infinite crystal. The infinite crystal can be made finite by multiplying it by a shape function, which merely has the value unity inside the bounderies of the crystal and zero outside. The transform of the finite crystal is then obtained by convoluting the transform of the infinite crystal (the weighted RL) with the transform of the shape function (the shape transform). As before, the convolution with a point function simply means that the shape transform is placed at each lattice point and multiplied by the weight of each point. The relps become small volumes of scattering power, equal in extension, but each weighted with the Structure Factor of the relp.

The linear dimension of these volumes in any direction is approximately given by 379(1), i.e. $2\Delta d^* = 2\lambda/t$, where t is the linear dimension of the crystal in that direction. The relation is thus once again a reciprocal one. For a crystal mosaic block of 1500 Å, $\Delta d^* = 0.002$ R.U. for CuKα radiation. This is hardly detectable, but for electron diffraction by transmission through very thin crystals, which may be only 10 unit cells thick, Δd^* would be 1/10th of the RL dimension in that direction. Since the plate would be many hundreds of unit cells in extent in its own plane, Δd^* in any direction in this plane would be very small. The shape transform of a thin plate is thus a thin rod perpendicular to the plate, and in general the form of the shape transform is the reciprocal of that of the crystal (a very small perfect crystal in electron diffraction or a mosaic block for X-ray diffraction).

(ii) $a^* = \dfrac{1.54}{5.4} = 0.286$ $\qquad\qquad\qquad b^* = \dfrac{1.54}{5.8} = 0.266$ R.U.

$\equiv 0.286 \times 25.4 = 7.25$ mm. $\qquad\qquad \equiv 6.75$ mm

From the scattering diagram Fig. 262 (the $\oplus$ marks the corresponding point in the upper diagram):

$R'_{320} = -2.1$, $d^*_{320} = 25.7/25.4 = 1.011$ R.U. $\therefore \sin\theta = 0.506$.

$(\sin\theta)/\lambda = 0.506/1.54 = 0.328$. $\therefore f_c = 2.2 - 0.3 \times 0.28 = 2.12$.

$$\therefore F_{320} = -2.1 \times 2.12 = -4.45.$$

(iii) By calculation:

$$F_{320} = A_{320} = f_c \sum_m \cos 2\pi\,(3x_m + 2y_m).$$

$d^*_{320} = \sqrt{9 \times 0.286^2 + 4 \times 0.266^2} = \sqrt{0.735 + 0.283} = 1.009 = 2\sin\theta$.

$(\sin\theta)/\lambda = 1.009/(2 \times 1.54) = 0.328$. $\therefore f_c = 2.12$.

m	$x = X/a$	$y = Y/b$	$3x+2y$	$\phi_m{}^\circ$	$\cos\phi_m$
1	0.224	0.121	0.914	328.5	0.853
2	0.224	−0.121	0.430	155	−0.906
3	0	0.242	0.484	174	−0.994
4	0	−0.242	−0.484	−174	−0.994
5	−0.224	0.121	−0.430	−155	−0.906
6	−0.224	−0.121	−0.914	−328.5	0.853
					−2.094

$$\therefore F_{320} = 2.12 \times -2.094 = -4.43.$$

The figure from the scattering diagram is within 1% of the calculated figure, which is better than would be expected from such a small scale diagram.

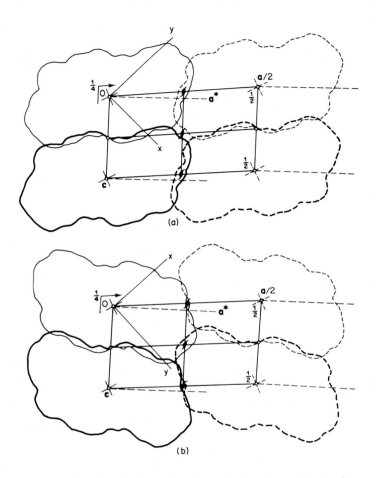

Fig. 263. The unit cell and symmetry elements, together with the best packing arrangements for flavanthrone molecules in the unit cell. The molecules outlined in full lines are equivalent; the difference in line thickness is only designed to avoid confusion at the overlap. The centres of these molecules lie on the centres of symmetry at the origin points. Similarly the other two molecules with dashed outlines are equivalent to each other, but related to the first pair by the screw axes and the glide plane at $\frac{1}{4}$. Their centres are at the centres of symmetry at $\frac{1}{2}$, $\frac{1}{2}$, 0. (a) has Oz upwards and (b) downwards. The fit for (b) is apparently slightly better, especially if the tilt axis is Ox, which would reduce the overlap near the central 2_1 axis (where the two molecules are in any case 1·9Å apart in the **b** direction). However it would be difficult to decide between the two possibilities on packing considerations alone, but the diffraction evidence turns out to be decisive in favour of (b) (Oz downwards), and the structure determination shows that the axis of tilt is almost exactly along Ox. (Whether the tilt is up or down at the central 2_1 axis depends on the choice of origin. If it is up with the origin as shown, it would be down if the origin were shifted to $\frac{1}{2}$, $\frac{1}{2}$, 0.).

16. (i) $Z = 4$ for the general positions in $P2_1/a$, so the molecules must lie on a pair of centres of symmetry. The molecule itself must therefore be centro-symmetric. If the origin is chosen to be at the centre of one molecule, the other will be at $\frac{1}{2}, \frac{1}{2}, 0$. (Fig. 263).

(ii) From Fig. 264:

$$\cos\psi = 3\cdot4/3\cdot8 = 0\cdot1895; \quad \psi = 26\cdot5°.$$

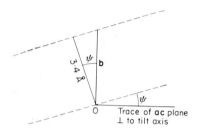

FIG. 264. The relation between the height of the unit cell b, the spacing between tilted planar molecules and the angle of tilt ψ.

(iii) Figure 263 (b) shows the best packing arrangement, with $\mathbf{a}^*$ making an angle of nearly 45° with Ox and Oy, and Oz in the direction of $-\mathbf{b}$. However if the molecules are turned upside down, with Oz upwards, an arrangement nearly as good can be found (Fig. 263 (a)) with the same direction in the molecule along $\mathbf{a}^*$. It would be difficult to decide, on packing considerations alone, which was the correct orientation of the molecule, bearing in mind that atoms related by the 2_1 axis have a height difference of $1\cdot9$ Å (in particular the ends of two adjacent molecules in the $\mathbf{a}$ direction have this height separation) and that further height differences and forshortening will arise from the unknown direction of tilt of the molecules.

(iv) Because of the a glide plane the projected unit cell repeats at $\mathbf{a}/2$.

(v) Since $\mathbf{a}$ is halved, $\mathbf{a}^*$ will be doubled, and the relps for h odd will be missing, i.e. $h0l$ reflections will only be present for h even.

(vi) Even though $\mathbf{a}^*$ has been doubled, it is still not much more than half the length of $\mathbf{c}^*$, so that $\mathbf{a}^*$ is the shorter RL axis. $\mathbf{b}^*$ is upwards.

(vii) Because the two molecules are related by the a glide, they are identical in projection, and the consequent halving of the projected cell leaves effectively only one molecule per cell. This would not be true for any other projection.

(viii) The transform is derived from a flat molecule. The length of **b** shows that the molecule from which the RL derives is tilted so that the projected shape will be slightly different. One would therefore not expect exact matching of the weighted RL and the transform.

(ix) With $-\mathbf{b}^*$ upwards, along OZ (i.e. the RL plane reversed from Fig. 176) a good match is obtained with $\mathbf{a}^*$ nearly at 45° to OX and OY. With $+\mathbf{b}^*$ upwards there is nothing approaching a match within 30° of this direction. Therefore Oz in the molecule is in the direction of $-\mathbf{b}$ (within the 25° angle of tilt).

(x) The angle between $\mathbf{a}^*$ and OX = 47; 46; 46°. Average 46·3°. The angle with OY is therefore 43·7°. (The angles from the structure determination = 45°).

(xi) The pseudo-hexagonal symmetry of the transform means that there are three pairs of directions (each pair related by 2-fold symmetry) which give a reasonable match with the weighted RL. Packing considerations are required to distinguish the correct pair.

(xii) The molecule is almost entirely composed of parallel benzene type hexagons, each of which gives the benzene transform. The sum of these benzene transforms is modified by two main wide fringe systems, perpendicular to the OY direction and to the direction at 60° to OY, due to the regular stacking of hexagons in the two directions. There is one set of narrow fringes which derives from fourteen pairs of atoms, all with the same separation and direction, leaving out only the two N and O atoms. If we take an atom from the left-hand vertical row of hexagons and the corresponding atom from the right-hand row, the line joining them makes an angle of 49° with Ox and its length is 6·42 Å, giving fringes of (1·542/6·42) R.U. ≡ 7·0 mm width. The very obvious narrow fringes giving the straight lines of zero amplitude which make an angle of 49° with OY, are 7·0 mm apart and must be due to these fourteen pairs of atoms.

The missing atoms caused by the sharing of edges produce other effects but in spite of this and the complicated pattern of fringe systems, the basic hexagonal benzene transform is very obvious. Comparison with Fig. 62 (p. 95) shows that the maxima occur in directions perpendicular to the sides of the hexagon at a distance of 1·285 R.U. for λ = 1·542 ≡ 37·5 mm on Fig 177. The six main maxima of the flavanthrone transform obviously occur in the same positions as those of the benzene transform, and give rise to the pseudo-hexagonal symmetry.

Because of the displacement of all the flavanthrone hexagons from the origin, the phases of the maxima in Fig. 177 are not the same as those of the benzene transform.

Although the sign almost certainly changes across the zero lines of the transform, it is not easy to determine the phases (0 or π) of the various maxima. This can be done by a further development of the optical transform [2(77)], but in fact in the original work the transform was computed, including, of course, the sign. The positive shaded areas on half the transform in Fig. 177 were obtained by comparison with the calculated diagram. The phases of the $h0l$ relps can be obtained by inspection from such a diagram, by superimposing the RL and seeing which relps lie in positive and which in negative areas. The phases for 67 $h0l$ reflections were obtained in this way, enabling the image of the molecule shown in Fig. 265 to be calculated.

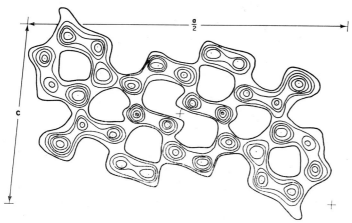

FIG. 265. The projected image of the molecule calculated by Fourier synthesis from 67 $h0l$ reflections, whose signs could be determined by comparison of the weighted RL layer with the transform of the molecule. The origin used for the calculation is at the centre of the molecule.

17. Figure 266 shows that the planes of the $(40\bar{2})$ set do, in fact, pass through the lines of intersection of the two sets (302) and $(10\bar{4})$.

18. (i) $\bar{2}02$.

(ii) $\mathbf{d}^*_{HKL} = \mathbf{d}^*_{h_1 k_1 l_1} - \mathbf{d}^*_{h_2 k_2 l_2}$.

$$\therefore H = h_1 - h_2, \qquad K = k_1 - k_2, \qquad L = l_1 - l_2.$$

(iii) $\mathbf{d}^*_{H'K'L'} = -\mathbf{d}^*_{HKL}$. $\therefore$ reflection is $\bar{H}, \bar{K}, \bar{L}$.

19. (i) Figure 267 shows the symmetry elements, with one end of the axes marked and the heavy great circles denoting the symmetry planes. (101 and the other 2-fold directions not in the primitive can be found from the net or by construction as shown for $0\bar{1}1$. The construction centres of the great circles

perpendicular to these directions are in the primitive). The intersections of the symmetry planes perpendicular to the 2-fold axes give the directions of the 3-fold axes. The {100}, {110} and {111} poles are labelled. (The indices of poles on the bottom half of the sphere are given in brackets).

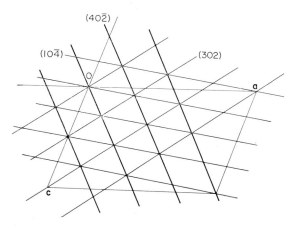

FIG. 266. Traces of the sets of planes (302) and (10$\bar{4}$) (thin lines) together with the set (3+1, 0+0, 2+$\bar{4}$) = (40$\bar{2}$) (thick lines). The 40$\bar{2}$ planes pass through the intersections of the other two sets.

(ii) 102 = 101 + 001 = 111 + 0$\bar{1}$1. The great circle passing through 111 and 0$\bar{1}$1 can be found from the net, and the two ends in the primitive marked. Because of the large radius the arc is most easily drawn with a variable radius curved ruler. Since 0$\bar{1}$1 is on the great circle, its opposite, 01$\bar{1}$, must also be on it. The dashed part of the arc shows the continuation of the great circle on the bottom half of the sphere. Since 111 + 01$\bar{1}$ = 120 = 110 + 010, the pole 120 is at the point where this great circle cuts the primitive. Similarly, by drawing the 111, $\bar{1}$01 great circle, 012 and 210 can be found. By drawing the circle through 1$\bar{1}$0, 111 and $\bar{1}$10 the remaining two of the {102} poles (201 and 021) are found. (To find the construction centre bisect one of the chords, and produce the perpendicular bisector to intersect the diameter, or the perpendicular bisector of a second chord.)

Since 101 + 110 = 211 = 111 + 100, the intersection of these two zones (already drawn) fixes 211, and the 011, 110 zone gives 121. 112 = 111 + 001 = 101 + 011. The 101, 011 zone is drawn in the same manner as the 1$\bar{1}$0, 111 zone; and this fixes the third of the {112} set, i.e. 112.

(iii) 123 = 111 + 012 = 112 + 011 and since both zones are already drawn, 123 can be marked in. Similarly, 213, 312, 321, 231 and 132 can also be marked in.

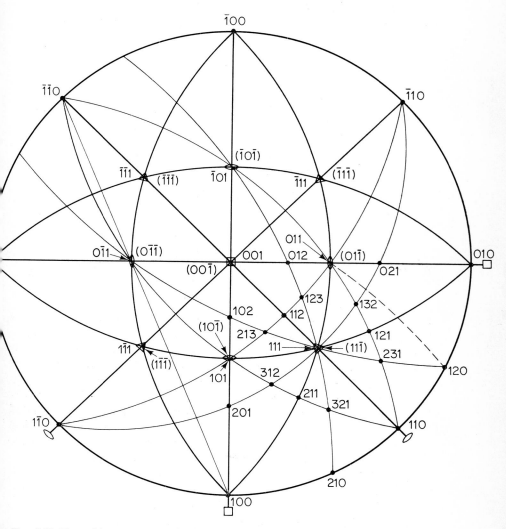

FIG. 267. The cubic stereogram, showing symmetry elements and the normals to various sets of planes. Only the 'positive quadrant' (actually 'octant') has been completed.

(iv)

hkl	100	110	111	102	112	123
p''	6	12	8	24	24	48
Sym.	$4\,mm$	$2\,m$	$3\,m$	m	m	1

(v) For the multiplicity relations below, $h \neq k \neq l$.

hkl. In any quadrant a diagonal symmetry plane produces a second equivalent pole, and the 3-fold axis makes this pair into 6. The 4-fold axes reproduce this 8 times, giving 48 equivalent poles or RL vectors.

hhl. A RL vector with two indices the same lies on a diagonal mirror plane. Two of the hkl poles coalesce to give one, and the number of RL vectors is reduced to 24.

$hk0$. One zero index puts the pole on a symmetry plane perpendicular to a 4-fold axis, and the number is again halved to 24.

$hh0$. This pole is on two symmetry planes, and the number of related vectors is therefore $48/(2 \times 2) = 12$.

hhh. This pole is on a 3-fold axis, and 6 of the hkl poles coalesce into 1, giving $p'' = 48/6 = 8$.

$h00$. 8 hkl poles coalesce on to a 4-fold axis to give $p'' = 48/8 = 6$.

(vi) The Laue symmetry determined from X-ray diffraction, assuming Friedel's law is obeyed, always contains a centre of symmetry. The Laue symmetry of point groups 23 and $(2/m)3$ is the same, i.e. $(2/m)3$. Only the axial mirror planes are present (no diagonal 2-fold axes or planes along the 3-fold axes) and reflections hkl and khl are no longer equal. However a geometrical diagonal mirror plane still exists in the RL, so that $d^*_{hkl} = d^*_{khl}$; the RL powder spheres of the two sets of relps coincide; and only the sum of I_{hkl} and I_{khl} can be found from powder photographs. In a sense the multiplicity is the same, but the result of powder diffraction intensity measurements is the average of $I_{hk'}$ and I_{khl}.

(vii) Although the regularity of the pattern shows that the crystal is cubic, the actual form is somewhat puzzling. If it were not for the first line and one or two others almost too faint to show up, it would appear to be the 2, 1, 2, 1, 2 ... pattern for an F-centred crystal ($N = 3, 4, 8, 11, 12, 16, 19, 20 ...$) (II (5.6), p. 364). However, the first line precludes this and the lattice must either be P or I.

The ξ value of a powder ring on the equatorial line is equal to d^*_{hkl} for the ring. By measuring the equatorial diameters of the rings on Fig. 113, adding 50% and transferring to a Bernal chart, we find that the ξ values of the first three rings are: 0·475, 0·575, 0·67 R.U. and $(10d^*)^2 = 22·56$, 33·06, 44·89. The N values are therefore either 2, 3, 4 for P or 4, 6, 8 for I. The indices and multiplicities of the first three lines are either:

	P			I		
hkl	110,	111,	200	200,	211,	220

or

p''	12	8	6	6	24	12

There are clearly more reflections in the first ring than the second, proving that the lattice is P.

The pattern of lines in Fig. 113 can be understood from the structure of of Cu_2O, which is such that the Cu atoms are on an F lattice and the O atoms on an I lattice. The combination of the two produces a P lattice. But since there are twice as many Cu as O and the scattering power of Cu is much greater than that of O, especially at high angles, the F arrangement of Cu dominates the pattern; except for the first line, 110, to which only the oxygens contribute. Very faint lines, also due only to O, can just be seen for $N = 6(211)$ and $10(310)$.

20. (i) The direction of the RL vector $\mathbf{d}^*_{hkl}$ is along the normal to the set of planes (hkl), and is not affected by the wavelength of the X-rays. The length, $d^*_{hkl} = \lambda/d_{hkl}$ is proportional to the wavelength. Along each RL vector there is therefore a continuous line of relps corresponding to the continuous range of wavelengths in the white radiation. The RL vectors become weighted radial rods. Whenever such a weighted rod cuts the sphere of reflection, Bragg's Law will be satisfied for the wavelengths which have relps at the point of intersection, and a Laue reflection will occur. The spectrum ceases abruptly at the short wavelength cut-off [1(52)] when $\lambda = (12·4/V)$ Å (V is the tube voltage in kV). Points near the origin on $\mathbf{d}^*_{hkl}$ which correspond to wavelengths shorter than this for a first order reflection (h, k, l no common factor), are therefore weightless, and no reflection can arise from an intersection of this part of the RL vector with the sphere. On the long wavelength side the intensity "fades out" around 3·0 Å. The nth order contribution starts n times as far from the origin as that of the first order, and the weight is spread n times as far along the vector. The main effect is therefore due to the low orders.

(ii) ρ is the angle ($\leqslant 90°$) between $\mathbf{d}^*_{hkl}$ and the rotation axis. ϕ is the angle ($\leqslant 180°$) between the plane containing the rotation axis and the direct beam, and the plane containing the rotation axis and $\mathbf{d}^*_{hkl}$.

(iii) Figure 268 shows that there is more of the weighted part of $\mathbf{d}^*_{120}$ inside the sphere at $\phi = 81 \cdot 5°$, so the streak must have ended at that point because the oscillation ended.

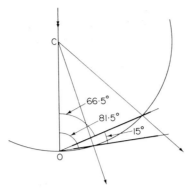

FIG. 268. The weighted 120 RL vector of Figure 253 (p. 465) is oscillated from $\phi = 66 \cdot 5$ to $\phi = 81 \cdot 5°$. The two extreme positions of the rod are shown, intersecting the sphere of reflection, and giving rise to the reflections at the outer and inner ends of the Laue streak. A further 10° clockwise rotation would have taken the weighted part of the rod out of the sphere. The inner end of the streak would then be due to the short wavelength cut-off.

(iv) From Fig. 268 $\phi = 81 \cdot 5 - 15° = 66 \cdot 5$.

(v) 90°.

(vi) The ϕ values of the inner and outer ends of the streaks defining the crescent are 82° and 77°, i.e. the oscillation is 5°. Therefore $\omega = 93 \cdot 5 + 5 = 98 \cdot 5°$ at the end of the oscillation, since ω increases with clockwise rotation.

(vii) At $\omega = 98 \cdot 5°$ the plane is 8° from being perpendicular to the X-ray beam. $\therefore$ at $\omega = 98 \cdot 5 + 8 = 106 \cdot 5°$ it will be perpendicular.

No.	$\theta°$	N	hkl	TB/LR	$\rho°$	$\phi°$
1	13	4	200	BR	83·5	67·5
2	15·5	6	211	BL	80·5	81
3	18	8	220	BR	41·5	47·5
4	15·5	6	211	TR	32	54

FIG. 269. The measurements of the four reflections shown on Figure 272. The position of the reflection on the film is listed; T = top or B = bottom, L = left or R = right. ϕ is measured at the anti-clockwise end of the oscillation, $\omega = 30°$, i.e. from the R.H. ends of the Laue streaks.

21. (i) Measurement and identification of low angle reflections.

Measure and list the θ, ρ and ϕ values for the four reflections on Fig. 178 with $\theta < 20°$ (Fig. 269). It is necessary to measure the angles to $\pm 0.25°$ if a stereogram of sufficient accuracy is to be obtained. It is useful to copy the reflections and streaks on to tracing paper, and mark the angles on as they are measured from the photograph (Fig. 270).

$$d^*_{hkl} = 2 \sin \theta_{hkl}$$

and

$$d^{*2}_{hkl} = a^{*2}(h^2 + k^2 + l^2) = \frac{\lambda^2}{a^2} N \text{ (for the cubic system).}$$

$$\therefore \quad \theta = \sin^{-1} \frac{\lambda}{2a} \sqrt{N}.$$

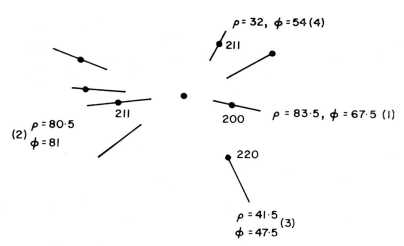

Fig. 270. A tracing of the measured Laue streaks and $K\alpha$ reflections of Fig. 178 (p. 298), annotated with the measurements.

Figure 271 gives values of N, *representative* indices hkl and θ up to $\theta = 20°$. (N is even because the cell is body-centred). By comparison, find the values of N and hkl and add to Fig. 269.

N	2	4	6	8	10
hkl	110	200	211	220	310
$\theta°$	8·92	12·68	15·58	18·08	20·26

Fig. 271. The θ-values of the possible reflections up to $\theta = 20°$.

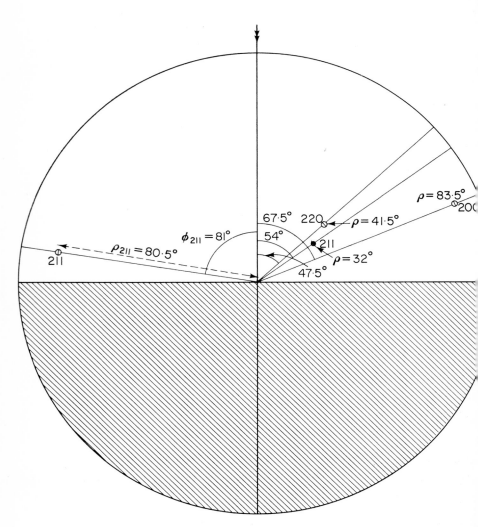

FIG. 272. The directions of the RL vectors at $\omega = 30°$, plotted on a stereogram with the rotation axis at the center and the X-ray beam coming towards the observer. The ρ and ϕ angles are shown. Where the RL vector is pointing downwards, the pole is circled. Since, in order to intersect the sphere of reflection, the RL vectors must make an acute angle with the X-ray beam, there can be no vectors plotted from the photograph in the shaded half of the stereogram.

(ii) Location of [111] directions. General method—applicable (with modifications) to all crystal systems.

(a) Plot the four poles on the stereogram with rotation axis in the centre and X-ray beam coming towards you. Use a 300 mm diameter full-circle Wulff net if possible, and draw the stereogram on tracing material, so that it can be rotated over the net about a pin through the centre. (Fig. 272).

(b) The angles between the normals (RL vectors) in Fig. 273 are calculated from the formula:

$$\cos\psi = \frac{\mathbf{d}^*_{h_1k_1l_1}\cdot\mathbf{d}^*_{h_2k_2l_2}}{d^*_{h_1k_1l_1}\,d^*_{h_2k_2l_2}} = \frac{h_1h_2 + k_1k_2 + l_1l_2}{\sqrt{(h_1^2 + k_1^2 + l_1^2)(h_2^2 + k_2^2 + l_2^2)}}$$

(for the cubic system).

$h_1k_1l_1 \ \ \widehat{\ \ } \ \ h_2k_2l_2$	$\cos\psi$	ψ
$111 \ \widehat{\ \ } \ 100$	$\pm 1/\sqrt{3}$	54°44′ or 125°16′
$111 \ \widehat{\ \ } \ 110$	$0,\ \pm 2/\sqrt{6}$	35°16′, 90°, 144°44′
$111 \ \widehat{\ \ } \ 211$	$(0,\ \pm 2,\ \pm 4)/\sqrt{18}$	19°28′, 61°52′, 90°, 118°8′, 160°32′
$111 \ \widehat{\ \ } \ 111$	$(\pm 1,\ \pm 3)/3$	70°32′, 109°28′
$100 \ \widehat{\ \ } \ 211$	$(\pm 1,\ \pm 2)/\sqrt{6}$	35°16′, 65°54′, 114°6′, 144°44′

FIG. 273. The possible angles between various RL vectors using *representative* indices.

(c) Plot the opposite poles of those in the bottom half of the stereogram, and draw small circles; of radius 35°12′ and 90° round 110 (220), and 54°44′ round 100 (200), as in Fig. 274. Points of intersection A and B are possible 111 poles.

Check the angular distances from 211. $211\ \widehat{\ \ }\ A = 62,\ 117\cdot5$; $211\ \widehat{\ \ }\ B = 90$, $89\cdot5$; $A\ \widehat{\ \ }\ B = 70\cdot25°$. This confirms A and B as 111 poles.

(d) Draw small circles round A and B, radius 70·5°. The intersection of these two at C is not a 111 pole since no three 111 normals make equal angles $< 90°$ with one another. The angles $211\ \widehat{\ \ }\ C$ are 25, 47°, which are not possible $111\ \widehat{\ \ }\ 211$ angles. The intersections D and E with the other circle about 110 are possible 111 poles. This identification is confirmed by the fact that they are 69° apart and 18·5, 60·5° and 19·25, 61° respectively from the 211 poles. (Fig. 274).

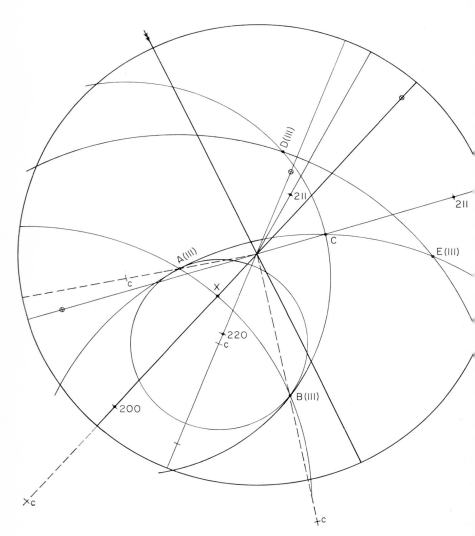

FIG. 274. The opposites of the poles in the bottom half of the stereogram (Figure 272) have been plotted for convenience, and small circles of 35°12′ and 90° drawn round 110(220). A small circle of 54°44′ round 100(200) gives points of intersection A and B which are possible 111 poles. Small circles of 70°33′ round A and B give intersection with the 90° circle about 110 at D and E, which are also possible 111 poles. The centres of the circles are marked '*c*'. The stereogram has been rotated anti-clockwise to accommodate the exterior centres.

(iii) Location of [111] directions. Method only applicable to the cubic system.

(a) Because of the high symmetry and invariant character of the cubic stereogram, it is usually much easier to bring the stereogram to the standard setting (with 001 at the centre) by conceptual rotations of the crystal. The 111 poles can then be put in and the crystal rotated back to its actual position, taking the 111 poles with it. In this case it would be possible by one large rotation to bring 200 to the centre; but it is probably easier to take it into the primitive and then rotate about it to bring 110 (220) into the vertical plane.

(b) Rotate 200 6·5° in the vertical great circle on to the primitive (to position 1, Fig. 275). 220 moves its position to 1 at the same time. Rotate about 200 in the primitive 18·5° to bring 220 on to the vertical great circle through 200 (position 2). This is now the standard setting (220 45° from centre), and the four 111 poles are put in on the vertical great circles at 45° to the one through 200 and at 54°44' from the centre. These (and 220) are rotated back through 18·5° to positions 1, and then through 6·5° to the positions A, B, D and E, found by the previous method.

The standard setting for the stereogram can be checked by moving one or both of the 211 poles at the same time as 220, and verifying that they lie in the correct positions. The one shown in Fig. 275 is 66° from the centre, and 35·5° from 200. The vertical great circle through the pole makes an angle of 26° with 200. (The angle 120 $\widehat{}$ 010 $= \cos^{-1} 2/\sqrt{5} = 26°10'$).

When the actual positions of the 111 poles have been found the arc and goniometer corrections can be obtained.

(iv) To move from goniometer reading 30° to 6°15' (bottom arc parallel to beam) requires an anti-clockwise rotation of 23°45'. Rotate the stereogram through this angle anti-clockwise, keeping the direction of the X-ray beam fixed, and draw in the direction parallel to the beam. This represents the *plane* of the bottom arc, and therefore the two poles in the primitive represent the *axis* of the top arc (since the bottom arc is at zero). The axis of the bottom arc is also in the primitive, at right angles to that of the top arc. To avoid mistakes, rotate the stereogram back to the position at $\omega = 30°$. (Fig. 276).

Select the nearest 111 pole to the primitive (i.e. E) and try each arc axis in turn, to see what rotation is required to bring E into the primitive. In this case using the bottom arc requires an impossible rotation of 78°, whereas the rotation required on the top arc is 17°. This cannot be significantly reduced by a preliminary rotation on the bottom arc (even a 25° rotation would only reduce it to 14°), so the top arc is moved through 17° (clockwise,

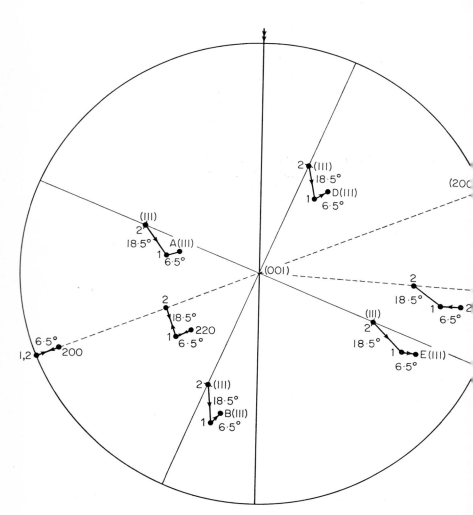

Fig. 275. The 100 and 110 poles on the top half of the stereogram are rotated into the standard setting and the four 111 poles marked in. Then they, as well as the other poles are rotated back to the starting position, bringing the 111 poles to positions *ABD* and *E* as in Figure 274. One of the 211 poles has also been rotated into the standard position for checking purposes.

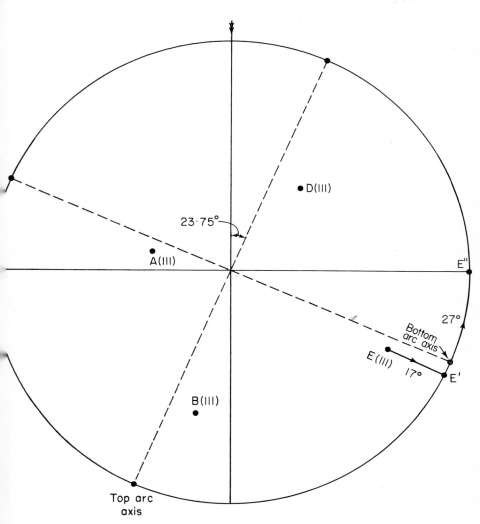

FIG. 276. The four 111 poles are plotted and the stereogram rotated anti-clockwise by 23°45′ to bring the bottom arc parallel to the beam. The axes of the arcs are marked and the stereogram rotated back to the $\omega = 30°$ position, with the top arc axis making an angle of 23°45′ with the beam. E, the nearest 111 pole to the primitive, is brought down to the primitive, i.e. horizontal, at E' and rotated to E'' perpendicular to the beam.

facing the scale, to 17°L) bringing E to E'. (If the vertical plane containing E had been more nearly at 45° to the arc axes, it would have been preferable to share the adjustment approximately equally between the two arcs. In the present example, if the axes had been at 45° to the positions given, it would require undesirably large rotations of 22 or 24° on the arcs separately to bring E into the primitive, but only 11 and 12° if both arcs are used together.)

(v) An anti-clockwise rotation of 27° will bring E' to E'' at right angles to the X-ray beam. This corresponds to a goniometer reading of $\omega = 3°$. (Fig. 276).

Notes

(a) *Representative indices*

All poles of the same form are labelled with the same representative indices, since there is no point in attempting to distinguish the actual arrangement and sign of the indices for the present purposes. Also, since only directions are involved on the stereogram, hkl and nh, nk, nl can be used interchangeably.

(b) *Drawing circles which cut the primitive*

The standard method for drawing a circle about a pole is to find the two ends of the diameter on the vertical great circle through the pole, and hence the centre as the mid-point and the radius as half the distance between the ends. (Except for circles about the centre point of the stereogram, the centre of the circle on the stereogram is not the same point as the pole.) This method of finding the centre is simple, using a Wulff net, if both ends of the diameter

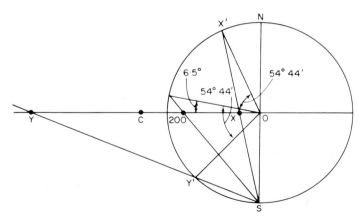

FIG. 277. Diagram showing the method of calculating the position of the centre of a small circle on the stereogram, when the outer end of the diameter is too far away for easy construction.

lie within the primitive; and if one end lies not too far outside, it can be easily found by construction. But if one end is a long way outside, the position of the centre must be found partly by calculation. Taking as an example the 54°44 circle about 200, the inner end of the diameter, X, is marked from the Wulff net. Figure 274 shows that the outer end of the diameter, Y, is too far away for easy construction. However, from Fig. 277, $\overset{\frown}{NSY'} = \frac{1}{2}\overset{\frown}{NOY'}$ $= \frac{1}{2}(90 + 54 \cdot 75 - 6 \cdot 5) = 69 \cdot 1°$.

$$\therefore OY = OS \tan \overset{\frown}{OSY} = 150 \times 2 \cdot 621 = 393 \text{ mm}.$$

By measurement on the stereogram $OX = 38$ mm.

$$\therefore XY = 393 - 38 = 355 \text{ mm}.$$

$\therefore$ Radius, $Xc = 177 \cdot 5$ mm and the centre of the circle about 220 can be marked on the line through OC, $177 \cdot 5$ mm *from* X.

If X is on the far side of O, as in the case of the 90° circle about 200, OX is added to OY instead of being subtracted; but the method is otherwise the same, the radius being measured from X and *not* O, in both cases.

The various centres on the stereogram are labelled c.

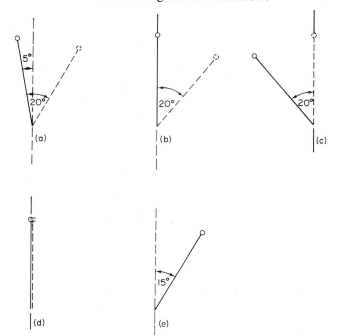

FIG. 278. Diagrams illustrating the steps required for reasonably accurate adjustment of the supporting fibre, by pushing it over in the plasticene mount.

22. (i) The tangent to the strong Laue streak at the centre makes an angle of 24·5° with the horizontal, and requires an anti-clockwise correction of this amount. The bottom arc after correction therefore reads 24·5° R.

(ii) The steps are illustrated by Figs 278(a–e). Vertical cross-wire—long dashes; desired position of fibre—short dashes. View in the microscope:

(a) with the bottom arc perpendicular to the microscope axis;

(b) after moving the arcs *B*.5°L; *T*.7°R;

(c) after moving the bottom arc 20° anti-clockwise to 15°R, so that the desired position is along the vertical cross-wire;

(d) after pushing the fibre upright to the desired position (check that is it also upright viewed perpendicular to the top arc.)

(e) after moving both arcs back to zero so that the fibre is in the required position.

Since the fibre may have been slightly rotated in the plasticene there may be an error of more than a degree, although all the operations can be carried out to better than 0·25°. The error should however be less than 5°. At least rough recentring must be carried out after, or in conjunction with, each arc movement.

The microscope will invert the diagrams, but the direction of rotation will not be affected.

23. (i) The positions of the arcs for ω equals (a) 150·5°; (b) 60·5°; (c) 105·5°; (d) 285·5° are shown in Figs 279(a–d).

(ii) $x_L = 0·0$ mm; $x_R = 2·4$ mm; measured at $2\theta = 90$ (47 mm from the radiograph of the crystal) or as near to that position as possible. (If there are no reflections within a mm or so of this position measure two pairs, one on either side, and use linear interpolation.)

$$\alpha_L = 0·0°. \qquad \alpha_R = 2·4 \times 0·68 = 1·65°.$$

(iii) The RL vector in the plane of the bottom arc gives rise to the reflections on the right-hand side of the photograph, and is therefore corrected from

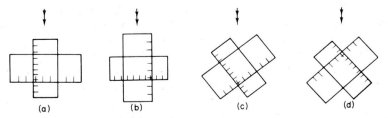

FIG. 279. The positions of the arcs for the four setting oscillations.

the right-hand measurement. Since for the longer exposure (position (c)) the RL vector is too high on the right-hand side, a clockwise rotation is required to bring it down into the equatorial plane.

(iv) The arc readings after correction are:

$0.0°$ for the top arc and $1.65°$ left for the bottom arc.

24. (i) Draw in the directions of the X-ray beam, the *axes BB'*, TT' of the arcs, and the RL axes $b*$ and $c*$ on the stereogram, for $\omega = 97°$. Use the stereographic net to draw small circles of $80°$ about $b*$ and $70°$ about $c*$. The intersection of these circles gives the direction of $a*$. (Fig. 280).

(a) The direct method is to bring $a*$ into the plane perpendicular to BB' by rotation about TT', and then to the centre by rotation about BB'. Since this method provides examples of a number of stereographic methods it will be demonstrated first, before going to the more indirect but simpler process. Bring T and $a*$ on to the same great circle on the net and measure the angle between them. $\widehat{T\ a*} = 82°$. Find the points on the primitive L, M, which are $82°$ from T. The circle through $La*M$ is the small circle about T which $a*$ will move on when rotated about T. Use compasses to draw in this circle. Rotation about T will bring $a*$ to A_1, but since this is on a small circle it cannot be measured directly. Draw the great circles through $Ta*T'$ and TA_1T' (T' can be drawn by construction outside the primitive to get the third point on the circle) and the great circle at $90°$ to T, intersecting the others at E and F. E and F are the points on the equator for T as N-pole, and the distance along the equator measures the angle between two planes of longitude, i.e. the rotation about the N–S axis to go from one longitudinal position to another. The rotation about the axis T to go from $a*$ to A_1 is therefore given by the great circle distance from E to F. This can be measured on the chart, or, by a general construction for angles along inclined great circles [4(29)], by drawing straight lines from T through E and F to the primitive at E' and F' and measuring the angle along the primitive from E' to F'. Both methods give $13°$ as the rotation to bring $a*$ to A_1.

A_1 is then brought to the centre at R (the camera rotation axis) by a rotation of the bottom arc of $17°$. $b*$ is likewise rotated about T through $13°$ to B_1. The position of B_1 on the small circle B_1b*K about T (all at $148°$ from T) is obtained by tracing the great circles $TGb*T'$ and THB_1T', where G and H are $13°$ apart on the equator $EFGH$. The angular distance of B_1, outside the primitive, is measured by construction as $5°$, and the corresponding point in the bottom hemisphere, $\bar{B}_1$, is plotted inside the primitive. $\bar{B}_1$ is then rotated about BB' through $5.5°$ to the primitive and a further $11.5°$ in the top hemisphere to B_2, making $17°$ rotation about BB'. B_2 is thus the position of

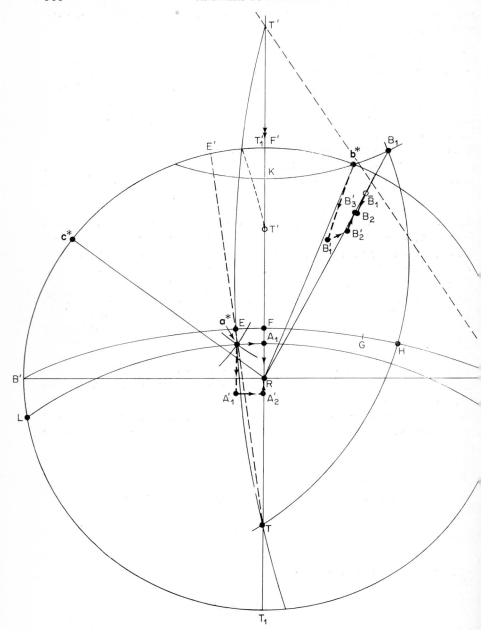

FIG. 280. A stereogram showing two methods of finding the arc corrections required to align a triclinic RL axis along the rotation axis.

b* when **a*** is at R. The plane **a*b*** then makes an angle of 27·5° clockwise from the plane of the bottom arc.

(b) In the more indirect method the bottom arc is first considered to be rotated (about BB') to the zero setting, bringing TY' into the primitive at T_1T_1'. **a*** is then rotated about T_1T_1' to come into the vertical plane TRT', and the bottom arc given a further rotation about BB' to bring **a*** to R. All small circles are then about points in the primitive, and the net can be used to plot all the movements directly. The dashed lines show the path of this triple process. The net movements of the arcs and the final positions of **b***(B_2 and B_3') are, of course, the same by the two methods (differences of 0·5° may arise from drawing errors) and the final readings of the arcs are T. 0·3°L, B. 7·3°L.

(ii) The arc corrections required if both axes are horizontal can be read directly from the net, and are 13° and 16°, with an error of only 1° in the bottom arc correction. Since an error of about 0·5° is normally to be expected from this method, it is hardly worthwhile to take the inclination of the top arc axis into account. However, the angle which the **a*b*** plane makes with the plane of the bottom arc after this simplified procedure is 22°, giving an error of 5·5°. On this account it is probably worth using the second process above (which only requires the use of the net) if the bottom arc is off zero by more than 5°.

(iii) If the arc readings had been on the opposite side of zero the corrections could not have been put on the arcs, and the fibre would have to be pushed over on the plasticine, using the technique of Example 22. This may also require to be done in order to keep the arc readings near zero, to avoid mechanical fouling which may occur with arcs which have been offset through considerable angles.

25. (i) The corresponding reflections to the left and right of the central spot are at 21·0 and 23·4 mm from the centre respectively. Using 163(1), we have,

$$\delta = 14{\cdot}3\cos^2 2\theta\,\frac{2{\cdot}4}{25},$$

where θ is the average for the two sides.

$$\therefore\, 2\theta = \tan^{-1}\frac{22{\cdot}2}{25} = 41°36'; \qquad \cos 2\theta = 0{\cdot}748.$$

$$\therefore\, \delta = 14{\cdot}3 \times 0{\cdot}748^2 \times 0{\cdot}096 = 0{\cdot}765°.$$

This is rather large for the approximate formula, so the exact equation 162(1) should be used as a check.

$$\tan^{-1}\frac{21\cdot0}{25} = 40°2'; \quad \tan^{-1}\frac{23\cdot4}{25} = 43°6'; \quad \therefore \delta = \frac{43°6' - 40°2'}{4} = 0\cdot765°.$$

(ii) Clockwise correction required. $\therefore \omega = 215\cdot6 + 0\cdot75 = 216\cdot35°.$

(iii) Since the mis-setting is about the vertical axis, the corresponding reflections will be in the equatorial plane and therefore on the horizontal line through the central spot.

The obvious reflections to choose are the two on the prominent elliptical zones. Although the intensity pattern is different on the two sides, the geometry of the RL includes a vertical symmetry plane, and the third, almost undistorted elliptical zone is an aid in following the geometrical pattern, from the equivalent reflections at the intersections of the three main zones, to the equatorial reflections. These are the sixth reflection on each side, if one ignores the first reflection on the left-hand side whose counterpart on the right-hand side is too weak to be visible.

(iv) On the left-hand, set photograph the spots are smaller and rounder than on the right-hand photograph; and although the effective exposure is some 50% greater, as judged by the densities of the weaker reflections, the central background is little, if any, greater.

The spot shapes and background fogging indicate a more divergent beam of large cross-section in the case of the right-hand photograph, and this is confirmed by the size of the direct beam spots.

In fact the left-hand photograph was taken using a Hilger fine focus tube, while the right-hand was taken with a normal sealed tube with an extended focus. The reason for this difference in conditions was that it proved impossible, using the fine focus tube, to repeat an earlier observation that missetting by about $0\cdot75°$ produced a considerable number of reflections which included a K component and were therefore very intense. Since the symmetry related planes in general did not reflect a K component, this effect apparently destroyed (or disguised) the trigonal symmetry. Not until the camera was set up against a normal large focus tube could the effect be reproduced.

The small divergence of the beam from a fine focus tube is thus very advantageous in taking Laue photographs for symmetry determination.

26. (i) Only 111, since the indices must all be even or all odd for an F lattice.

(ii) It can loosely be said that the three zone axes appear as reflections on a Laue photograph, although strictly the reflections arise from the weighted

RL vectors, $\mathbf{d}^*_{100}$, $\mathbf{d}^*_{110}$ and $\mathbf{d}^*_{111}$. However, in the cubic case ($Fm3m$ is a cubic space group) the zone axes and RL vectors with the same indices are parallel to one another, so that as long as only directions are involved they are equivalent. The reflections corresponding to [100] and [110] are present because weighted RL vectors in these directions arise from the 200 and 220 relps (in general $2h,00$ and $2h,2h0$), even although the 100 and 110 reflections are absent.

(iii) See Answer 21 (p. 489).

hkl	100	110	111
100	90	45 90	54·75
110	—	60 90	35·25 90
111	—	—	70·5

(iv) Since these directions are the same as those of the corresponding weighted RL vectors, they are given by the reflections which lie at the intersection of a number of important zones on the Laue photograph.

Refl.	TB/LR	ε†	γ	ψ‡	
A	TR	$+20\cdot5$	19·5	13R	} 35°
B	TL	,,	,,	22L	
A	TR	$-\overline{35\cdot5}$	0·5	23·5L	} 54·5°
C	BL	,,	,,	31R	
F	BL	,,	,,	15·5R	
B	TL	$-\overline{78}$	20·5	21·5R	} 45°
C	BL	,,	,,	23·5L	
D	TL	,,	,,	3R	
A	TR	$+\overline{85}$	20·5	11L	
E	BR	,,	,,	10R	
G	BR	$-\overline{35}$	20	0·5L	
E	BR	,,	,,	12L	
H	BR	,,	,,	10R	

† Measured with the chart protractor from the downward direction of the vertical; + clockwise; − anti-clockwise; $-\overline{20}$ means that the zero of the chart protractor is 20° anticlockwise from the *upward* direction of the vertical, i.e. measured with the photograph upside down.

‡ L or R with the chart protractor towards the observer.

FIG. 281. The Greninger chart measurements of the various zones, and marked reflections on Figure 101 (p. 165). The chart is kept with the protractor towards the observer and the photograph turned to align the various zones with the constant γ hyperbolic curves on the chart. In plotting the poles, the stereogram and Wulff net are turned to the same orientation, which accounts for the variation in direction of the lettering in Figure 282.

(v) A, B, C, possibly D and E.

(vi) Figure 281 gives the Greninger chart measurements as defined in Section 13.5.3 (p. 164).

(vii) The angles between the RL vectors corresponding to A, B and C are consistent with A being [111]; B, [110]; and C, [100].

(viii) Figure 282 shows the stereographic plot of the poles (a pole on several zones is plotted for each zone, the slight differences being due to the errors in reading the chart). The round dots with undashed letters are the original plot, and the square dots with dashed letters give the positions of the poles after C [100] has been brought to the centre of the stereogram by a rotation of 31° about an axis perpendicular to CC'. Multiple positions have been averaged for this operation. The main zone axes are as expected in the standard cubic positions, and the stereogram has been oriented to bring the poles into the positive quadrant as shown.

(ix) D', being on the zone containing 010 and 001, must have indices $0kl$. $D'C' = 26\cdot5°$ and comparison with the angle table identifies D' as 012. Similarly F' is hhl and $F'C' = 16°$, so that F' is 115. E' is on the hkh zone and $A'E' = 22°$. E' is therefore 313. The $D'G'$ zone cuts $A'C'$ in a pole hhl, which the angle table shows is 113, and passes through 101. The $E'G'$ zone cuts the 101, 001 zone at a point $h0l$, which is identified in the same way as 102. As a check the other reflection on that zone has been plotted and rotated to the standard position at H', i.e. this is the 102 reflection. G' is thus between 101, 113 and also 102, 313. Simple combinations of these pairs of indices do not produce a common set for G', and it is probably quicker to employ the systematic method of Appendix II (5.8), (p. 366) to find the zone axes of the two zones, and then, by the same method, find the RL vector which is perpendicular to both zone axes, i.e. lies in both zones.

$$
\begin{array}{c|cccc|c}
1 & 0 & 1 & 1 & 0 & 1 \\
  & \times & \times & \times & & \\
1 & 1 & 3 & 1 & 1 & 3 \\
\end{array}
\qquad
\bar{1}\ \ \bar{2}\ \ 1
$$

$$
\begin{array}{c|cccc|c}
1 & 0 & 2 & 1 & 0 & 2 \\
  & \times & \times & \times & & \\
3 & 1 & 3 & 3 & 1 & 3 \\
\end{array}
\qquad
\bar{2}\ \ 3\ \ 1
$$

$$
\begin{array}{c|cccc|c}
\bar{2} & 3 & 1 & \bar{2} & 3 & 1 \\
  & \times & \times & \times & & \\
\bar{1} & \bar{2} & 1 & \bar{1} & \bar{2} & 1 \\
\end{array}
\qquad
5\ \ 1\ \ 7
$$

$$H_1\,K_1\,L_1 \qquad\qquad H_2\,K_2\,L_2 \qquad\qquad h\ \ k\ \ l$$

The indices of G' are thus $517 = 4(101) + 113 = 2(102) + 313$.

(x) In Fig. 45 [$11\bar{1}$] is along the beam, with the hhl zone horizontal, $11l$ to the left and $hh1$ to the right. The problem is therefore to bring A (Fig. 282) to the centre of the stereogram and AC horizontal. The operation is best done in a number of stages, using a sketch of the arcs, remembering that we are looking at the photograph from the direction of the X-ray tube, with the incident beam going *down* through the film.

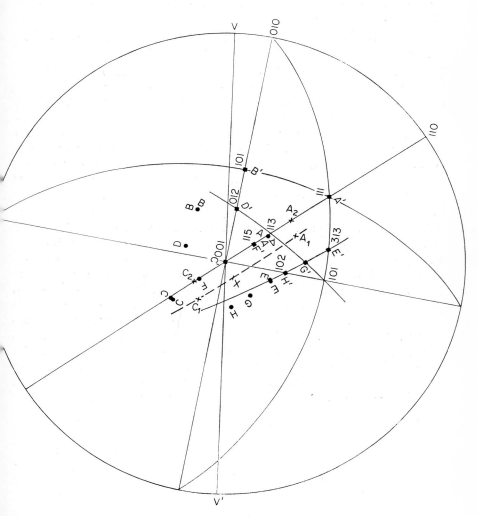

FIG. 282. The round dots *A–H* are the stereographic plot of the measurements in Figure 281; the square dots *A′–H′* are the corresponding positions after *C* has been brought to the standard position at *C′*. C_1, C_2 and A_1, A_2 show the movements of *C* and *A* in the derivation of arc adjustments required to produce Figure 45 (p. 64). The orientations of the letters *A–H* and the indices correspond to the positions of the stereogram when the poles were plotted. The X-ray beam is going *downwards* at the centre of the stereogram.

The stages, in outline, are as follows.

(a) Bring the axis of the bottom arc along the beam by rotation about the vertical axis. (b) Bring the AC zone through the centre of the stereogram, using the top arc, whose axis is now perpendicular to the beam. (c) Rotate the AC zone horizontal using the bottom arc. (d) Rotate A into the centre of the stereogram using the vertical rotation axis. The details of each stage are given below.

(a) A $12 \cdot 5°$ anti-clockwise rotation about V brings A to A_1 and C to C_1. The position T_0, of the top arc axis when the bottom arc is at zero, is in the primitive, $90°$ from V.

(b) The $A_1 C_1$ zone is shown by the dashed curve. Since the bottom arc reads $20 \cdot 7°$L (i.e. clockwise looking at the scale), the top arc axis T is $20 \cdot 7°$ anti-clockwise from T_0 (looking in the opposite direction). The point X on the $A_1 C_1$ zone and perpendicular to T is brought to the centre of the stereogram by an $8 \cdot 5°$ anti-clockwise rotation (looking at the scale) of the top arc. The scale reading is therefore $3 \cdot 2°$R. A_1 and C_1 go to A_2 and C_2.

(c) A $35 \cdot 5°$ clockwise rotation (looking from the tube) about the bottom arc axis brings $A_2 C_2$ horizontal. The bottom arc scale now reads $14 \cdot 8°$R. (Since the rotation is anti-clockwise looking at the scale).

(d) A_2 is brought to the centre of the stereogram by a $36 \cdot 0°$ clockwise rotation, giving a net clockwise rotation of $23 \cdot 5°$ and a scale reading of $\omega = 283°$.

(xi) $r = s \tan 2\gamma$.

$$\text{For} \quad \gamma = 30°, s = \frac{r}{\tan 60} = \frac{r}{\sqrt{3}}.$$

From measurement on Fig. 103, $r = 29 \cdot 3$ mm. $\therefore s = 29 \cdot 3/1 \cdot 732 = 16 \cdot 9$ mm.

27. (i) The Bernal chart is used, with the outer edge over the hole, to identify the $10,0\bar{8}$ reflections. The following readings in mm (averages of three settings) were obtained on the original film using a travelling microscope:

	L.H.S.		R.H.S.	
	α_1	α_2	α_2	α_1
Upper	82·10	82·64	114·40	114·94
Lower	82·10	82·68	114·46	114·98

Let T_1 and T_2 be the average separation of the α_1 and α_2 pairs and r the effective radius of the camera i.e. the radius of the undistorted central section of the film.

If $\phi = 90 - \theta$, $T/r = 4\phi$. $r = 30\cdot10$ mm

	T†	$4\phi°$	$\theta°$	$\sin\theta$	$d^*_{10,0\overline{8}}$
α_1	$32\cdot86 \pm 3$	$62\cdot549$	$74\cdot363$	$0\cdot96298$	$1\cdot92596$
α_2	$31\cdot77 \pm 3$	$60\cdot476$	$74\cdot881$	$0\cdot96540$	$1\cdot93080$

† These figures were obtained from the original film. The reproduction is slightly smaller.

(ii) $d^* = 2\sin\theta$; $\quad \dfrac{\Delta d^*}{d^*} = \cot\theta\,\Delta\theta$.

$\Delta\theta = 0\cdot03/(30 \times 4) = 0\cdot00025$ $\quad\therefore\quad$ $\Delta d^* = 1\cdot9 \times 0\cdot28 \times 0\cdot00025 = 0\cdot00013$.

(iii) $d^{*2}_{10,0\overline{8}} = 1\cdot92596^2 = 100a^{*2} + 64c^{*2} - 160a^*c^*\cos\beta^* = 3\cdot7093$.

$\qquad a^* = d^*_{100} = 1\cdot8924/13 = 0\cdot14557$; $\quad a^{*2} = 0\cdot021193$.

$\qquad c^* = d^*_{001} = 1\cdot8336/9 = 0\cdot20373$; $\quad c^{*2} = 0\cdot041504$.

$\therefore\quad \cos\beta^* = \dfrac{-3\cdot7093 + 2\cdot1193 + 2\cdot6563}{160 \times 0\cdot14557 \times 0\cdot20373} = 0\cdot22420$.

$$\therefore\quad \beta^* = 77\cdot05°.$$

$$\cos\beta^* = \frac{d^{*2} - h^2a^{*2} - l^2c^{*2}}{2hla^*c^*} = \frac{X}{Y}.$$

Take logs and differentiate

$$-\tan\beta^*\Delta\beta^* = \frac{1}{X}(2d^*\Delta d^* - 2h^2a^*\Delta a^* - 2l^2c^*\Delta c^*) - \frac{2hl}{Y}(a^*\Delta c^* + c^*\Delta a^*).$$

Putting in the values and treating the errors as independent:

$4\cdot3\Delta\beta^*$

$$= \left[\frac{1}{1\cdot2}(0\cdot0005^2 + 0\cdot0009^2 + 0\cdot0015^2) + \left(\frac{1\cdot60}{4\cdot7}\right)^2(0\cdot00081^2 + 0\cdot00063^2)\right]^{\frac12}$$

$$= \frac{1}{100}\left[\frac{1}{1\cdot2}(0\cdot033) + 0\cdot12(0\cdot010)\right]^{\frac12} = \frac{1}{100}[0\cdot027 + 0\cdot001]^{\frac12} = 0\cdot0017$$

$\therefore\quad \Delta\beta^* = 0\cdot00039$ rad $= 0\cdot02°$.

28. (i) From 177(1) and 177(2)

$$\zeta = \frac{25}{(25^2 + 30^2)^{\frac{1}{2}}} = \frac{25}{39\cdot05} = 0\cdot640.$$

$$\xi = (1 + \frac{900}{1525} - 60 \times 0\cdot786/39\cdot05)^{\frac{1}{2}}$$

$$= 0\cdot618.$$

(ii) From $\triangle$ *CNO* Fig. 105 (p. 176):

$$CN^2 = CO^2 + NO^2 - 2\,CO.NO\cos\phi.$$

$$\therefore \quad \cos\phi = \frac{-\sin^2\psi + 1 + \xi^2}{2\xi} = \frac{\cos^2\psi + \xi^2}{2\xi} = \frac{\zeta^2 + \xi^2}{2\xi}$$

$$= \frac{0\cdot618^2 + 0\cdot640^2}{1\cdot236} = \frac{0\cdot791}{1\cdot236} = 0\cdot641.$$

$$\therefore \quad \phi = 50\cdot1°.$$

(iii) From Fig. 105, $\rho = \tan^{-1} \xi/\zeta = 44\cdot0°$.

$$d^* = (\zeta^2 + \xi^2)^{\frac{1}{2}} = (0\cdot791)^{\frac{1}{2}} = 0\cdot889.$$

(iv) $\therefore \sin\theta = 0\cdot4445; \quad \theta = 26\cdot4°.$

(v) The values measured with the charts should be within $0\cdot005$ for ξ and ζ and $0\cdot25°$ for the angles, of the values calculated.

29. (i) An inspection of both photographs shows that there is a horizontal mirror plane in the Laue symmetry of the crystal. Since there is only one such plane in the monoclinic system, perpendicular to the 2-fold axis along **b**, the **b**-axis must be along the rotation axis.

(ii) The Bernal chart reading of $\pm$4th layers of Fig. 109 gives, $\zeta_4 = 0\cdot705$ R.U.

The spacing between crystal lattice points along the rotation axis $= \lambda/\zeta_1 = 1\cdot542/0\cdot1762 = 8\cdot76$ Å. Therefore the rotation axis is **b**.

(iii) From the measurement on Fig. 52 of the ϕ-values of the two prominent equatorial reflections which are at the intersection of many important zones (45°L and 57°R), it seems almost certain that these are $+\mathbf{a}^*$ and $-\mathbf{c}^*$ (or, of course, $-\mathbf{a}^*$ and $+\mathbf{c}^*$ which is equivalent—we have to make the choice) with an angular separation of 103° (see Sec. **6.5.5.2**, p. 75). From the Laue photograph it is not possible to tell which is which, but at $\omega = 78°$ either **a*** or $-$**c*** makes an angle of 45° on the left-hand side of the beam. The oscillation photograph must be used to decide between the two possibilities, and also which is the direction of positive **b***. Since this latter is not yet known, the net should be drawn on tracing paper so that it can be reversed if necessary.

(iv) The construction of the **a*c*** RL net (Scale 100 mm = 1 R.U.).

$$a^* = \lambda/d_{100} = \frac{1\cdot542}{10\cdot87\cos 13°} = 0\cdot1458.$$

$$c^* = \lambda/d_{001} = \frac{1\cdot542}{7\cdot763\cos 13°} = 0\cdot2040.$$

The values of ha^* and lc^* out to 2 R.U. ($\equiv 200$ mm) *must be calculated* to the nearest 0·1 mm, and the points marked off from the table. (Fig. 283). On no account must the points be stepped off using dividers, or the accumulation of errors towards the outer edges of the net will lead to incorrect indexing. For problems such as this it is useful to have the circle of reflection drawn on a clear plastic sheet, so that it can be placed in various positions on the net and the results compared with the ξ values of the reflections on the photograph. For the zero layer these are given in Fig. 284.

				mm			
h,l	1	2	3	4	5	6	7
ha^*	14·6	29·2	43·8	58·4	73·0	87·5	102·1
lc^*	20·4	40·8	61·2	81·6	102·0	122·4	142·8

			mm				
h,l	8	9	10	11	12	13	14
ha^*	116·7	131·2	145·8	160·4	175·0	189·5	204·1
lc^*	163·2	183·6	204·0				

FIG.283. The values of ha^* and lc^* used for drawing the RL net.

Since one axis is at 45° on the left-hand side of the X-ray beam at $\omega = 78°$, the angle will be 50° for $\omega = 73°$, and 35° at the other end of the oscillation. If we first try **a*** in this position, the nearest ξ value of a relp to 0·555 on the left-hand side is 0·525, and this is not even quite within the oscillation lune. On the right-hand side, there is no relp in or near the lune corresponding to $\zeta = 0.53$. If we turn the net over and take $-\mathbf{c}^*$ on the left-hand side, the ζ values fit almost exactly over the whole range as shown in Fig. 110 (p. 182). This establishes the correctness of our orientation, and positive $\mathbf{b}^*$ is therefore upwards for right-hand axes.

(v) The indices of the reflections can be obtained by inspection of Fig. 110 and are given in Fig. 284.

L.H.S.			R.H.S.		
ζ	I	$h0l$	ζ	I	$h0l$
0·555	m	$\bar{2}0\bar{2}$	0·53	s	301
0·745	s	$\bar{2}0\bar{3}$	0·665	vs	401
0·94	s	$\bar{2}0\bar{4}$	0·805	s	501
1·07	s	$\bar{1}0\bar{5}$	0·89	s	600
1·135	s	$\bar{2}0\bar{5}$	1·15	m	$80\bar{1}$
1·275	s	$\bar{1}0\bar{6}$	1·18	m	800
1·445	m	$00\bar{7}$	1·295	s	$90\bar{1}$
1·48	s	$\bar{1}0\bar{7}$	1·445	s	10, $0\bar{1}$(10, $0\bar{2}$?)
1·62	vw	$10\bar{8}(20\bar{8}?)$	1·575	vw	11, $0\bar{2}$
1·81	vw	$40\bar{9}(30\bar{9}?)$	1·595	s	11, $0\bar{3}$
1·83	vw	$50\bar{9}$	1·645	vw	11, $0\bar{4}$
1·86	m	$60\bar{9}$	1·765	w	12, $0\bar{4}$
1·95	s	$80\bar{9}$	1·90	w	12, $0\bar{6}$(11, $0\bar{7}$?)

FIG. 284. ζ and I values measured from the zero layer of the photograph, and indices obtained from the interpretation. (The ζ values measured from the reproduction are slightly smaller.)

Points to notice: (a) The net was drawn from measurements made on original photographs, but the interpretation measurements are from a paper reproduction. In consequence the discrepancies between the net and the measured ζ values are greater than would normally be expected. However because the measured ζ values are consistently larger than the net values, it is possible to index unambiguously a number of close pairs of reflections which would have been ambiguous for each considered separately. One of the reasons for marking the measured arcs on the lunes is to enable such relationships to be appreciated at a glance. (b) Relp 501 is just outside the right-hand lune, but the reflection is present. The measurements on the Laue photograph using the ρ, ϕ, chart are only accurate to the nearest 0·5°, and were also made from a reproduction (they gave a β^* angle of 78° instead of 77°, which is a further indication of the inaccuracies involved). Therefore the ϕ angles of $-\mathbf{c}^*$ for the two ends of the oscillation can easily be out by 0·5°. A clockwise rotation of the circles of reflection by this amount would bring 501 into the lune. It would also bring 301; 10,0$\bar{1}$; $\bar{1}0\bar{5}$; $60\bar{9}$ and $80\bar{9}$ definitely into the lune, instead of in the edge, as at present; and $\overline{3}0\overline{3}$; $\overline{3}0\overline{4}$; $00\overline{8}$; $20\overline{9}$; $11,0\overline{5}$; $10,0\overline{3}$; $90\overline{2}$, would be definitely outside the lunes. Such slight re-adjustments of the positions of the lunes can often be made during an interpretation, and used in the interpretation of other layer lines. If a number of sequential oscillation photographs are being taken, the improved accuracy of orientation can be used in the interpretation of subsequent photographs. (c) When everything possible has been done there will still be some am-biguities in the indexing, as indicated in Fig. 284. These can only be resolved by taking photographs, normally with a smaller oscillation angle, designed to bring only one relp at a time into the lune. (d) It is important to list the absent (and possibly absent) reflections. In this case, on the R.H.S.: 700 (1·28); $12,0\overline{5}$ (1·82); and on the L.H.S.: $\overline{2}0\overline{6}$ (1·33). The photograph can then be re-scrutinised in the neighbourhood of the ζ values for these absent re-flections, to see if any are in fact visible. One of any ambiguous pair can be an absent reflection, and this must be taken into account. (e) Instead of turning the RL net upside down to get the correct orientation, it would be possible to turn the photographs upside down. In relation to the apparatus this can be confusing, since it corresponds to adjusting the crystal standing on one's head, or holding the camera upside down! The additional possibility of error involved in dealing with this situation makes it worthwhile getting the net in the right orientation. This is particularly true when dealing with a triclinic crystal or a monoclinic crystal about **a** or **c**. (In the latter case, since the **c*b*** (or **a*b***) net is orthogonal, it is only necessary to get the positive directions of the axes correct. This cannot be found from the zero layer, but only by interpretation of a non-zero layer (see Example 30). (f) Since in monoclinic Laue symmetry there is always a mirror plane perpendicular to **b***, the $h\overline{k}l$ reflection must be equal to hkl. That is to say, it does not matter which direction we choose as positive **b***, or, in other words, whether we have right-hand or left-hand axes. However it does matter in the triclinic system and in the case of any space group lacking a centre of symmetry, when anomalous scattering (i.e. the breakdown of Friedel's Law) is being used to find the absolute configuration of a molecule. It is thus worthwhile to adopt right-hand axes as a routine.

(vi) Interpretation of the fourth layer above zero.

Since $+\mathbf{b^*}$ is upwards the indices will be $h4l$.

$$\zeta_4 = 4\lambda/b = 4 \times 1·542/8·747 = 0·706.$$

$\therefore$ Radius of circle of reflection $= R_4 = \sqrt{1 - 0·706^2}$

$$= \sqrt{1 - 0·499} = 0·708 \text{ R.U.}$$

Correct the directions of the X-ray beams by 0·5°. Draw the circles of reflection for the 4th layer (Fig. 110, p. 182).

Measurements and indices are given in Fig. 285.

	L.H.S.			R.H.S.	
ξ	I	h4l	ξ	I	h4l
0·615	s	04$\bar{3}$	0·315	vs	24$\bar{1}$
0·82	w	04$\bar{4}$	0·445	vs	34$\bar{1}$
1·005	s	14$\bar{5}$	0·58	s	44$\bar{1}$
1·03	m	04$\bar{5}$	0·715	m	54$\bar{1}$
1·205	s	14$\bar{6}$(24$\bar{6}$?)	0·855	w	64$\bar{1}$
1·405	vw	34$\bar{7}$(24$\bar{7}$?)	1·005	s	14$\bar{2}$(74$\bar{1}$?)
1·615	w	44$\bar{8}$	1·155	m	84$\bar{2}$
			1·20	s	84$\bar{3}$
			1·335	vw	94$\bar{3}$
			1·60	vvw	10,4$\bar{5}$

FIG. 285. The measurements and indices of the fourth layer. (The ξ values measured from the reproduction are slightly smaller.)

Absent reflections on the L.H.S.: 04$\bar{2}$? (0·408); 54$\bar{8}$ (1·63); 64$\bar{8}$ (1·67); and on the R.H.S.: 94$\bar{2}$? (1·28); 94$\bar{4}$ (1·38); 10,4$\bar{4}$ (1·51); 10,4$\bar{6}$ (1·58). The last reflection however would be off the edge of the film, and is therefore not necessarily absent.

30. (i) (a) The crystal is either orthorhombic, monoclinic or triclinic from the optical evidence. (b) In Fig. 180, since there is no horizontal plane of symmetry (even geometric), the crystal lattice vector along the rotation axis cannot be an orthorhombic axis or monoclinic **b**. (c) There is a vertical zone on the left-hand side of the photograph, i.e. the shape of the curve of reflections is symmetrical about the equatorial line. This zone contains the reflection labelled *A*, arising from a major horizontal RL vector which is at

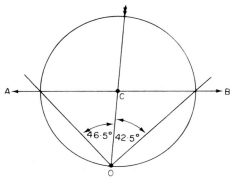

FIG. 286. The positions of the two major horizontal RL vectors and the corresponding reflections A and B.

90° (within the combined accuracies of the ρ, ϕ chart and the reproduction) to the corresponding vector giving rise to the reflection labelled B on the right-hand side. (d) To investigate this vertical zone we bring its zone axis along the X-ray beam. The ϕ angle of the major horizontal RL direction is 46·5° (Fig. 286) so that to bring it perpendicular to the beam requires an anti-clockwise rotation of $90 - 46·5 = 43·5°$. (e) Figure 50 shows that the zone axis direction has 2-fold symmetry (with due allowance for the effect of Kα radiation, Sec. **6.5.2**. p. 63) It is therefore either an orthorhombic axial direction or monoclinic **b**. Since it contains no planes of symmetry it is not an orthorhombic axis and must therefore be monoclinic **b**. The crystal system is thus established as monoclinic, and the vertical direction can be provisionally chosen as +**c**.

(ii) A photograph at 90° to Fig. 50 would have shown a vertical plane of symmetry, but would not have distinguished between orthorhombic and monoclinic, since the other axes and planes in an orthorhombic crystal could be at an angle to the direct beam and would not then show up in the photograph.

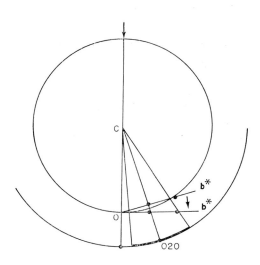

FIG. 287. The position of **b*** for the second oscillation.

(iii) (a) Figure 183 shows an almost undistorted picture of the **a*****c*** vertical RL plane in the circle outlined by the Laue streaks to the left of the direct beam (Sec. **4.2** p. 34), and confirms the choice of axes (with +**a*** to the right to give $\beta^* < 90°$, and therefore $+\mathbf{b}^*$ back along the X-ray beam for right-hand axes). By measurement with the Bernal chart, $3a^* = 0.445$ R.U. and $4\zeta_c = 0.795$ (where ζ_c is the RL interplanar spacing in the $\mathbf{c}$ direction). By direct measurement of the angle on the film, $\beta^* = 77°$, and this is confirmed with increased accuracy from Fig. 50 (Sec. 6.5.5.1 p. 73). (b) In Fig. 184 the $\mathbf{b}^*$ axis finishes up the oscillation nearly perpendicular (87.5°) to the direct beam. The $\mathbf{b}^*$ axis therefore comes out on the right-hand side of the direct beam (Fig. 287). The first reflection from the centre has a ζ value of 0.35, but the distance apart of reflections on the layer lines above and below it indicates that this reflection is 020, and that 010 and 030 are therefore absent. These measurements complete the determination of the unit cell, as follows:

$$c = \frac{\lambda}{\zeta_c} = \frac{1.542 \times 4}{0.795} = 7.77 \text{ Å}.$$

$$a^* = 0.445/3; \quad a = \frac{\lambda}{a^* \sin \beta^*} = \frac{1.542 \times 3}{0.445 \times 0.974} = 10.69 \text{ Å}.$$

$$b^* = 0.35/2; \quad b = \frac{\lambda}{b^*} = \frac{1.542 \times 2}{0.35} = 8.80 \text{ Å}.$$

$$\beta = 180 - \beta^* = 103°.$$

$$a^* = 0.148; \quad b^* = 0.175; \quad c^* = \zeta_c/\sin \beta^* = 0.199/0.974 = 0.205 \text{ Å}.$$

(iv) The interpretation of the zero layer of Fig. 183 is carried out as in Example 29. The results are shown in Figs 288 and 289.

	L.H.S.			R.H.S.	
ξ	I	$hk0$	ξ	I	$hk0$
0.15	m	$\bar{1}00$	0.47	s	310
0.30	m	$\bar{2}00$	0.68	s	420
0.445	vs	$\bar{3}00$	0.815	s	520
0.62	m	$\bar{4}10$	0.905	w	530
0.765	mw	$\bar{5}10$	1.02	m	540 (or 630)
0.90	w	$\bar{6}10$	1.135	mw	640 (or 550)
0.955	ms	$\bar{6}20$	1.255	w	650
1.095	m	$\bar{7}20$	1.385	w	660
1.17	ms	$\bar{7}30$	1.45	mw	570
1.38	ms	$\bar{8}40$	1.53	mw	670
1.49	w	$\bar{8}50$ (or $\bar{7}60$)	1.605	mw	580
1.62	mw	$\bar{7}70$	1.675	m	680
1.715	m	$\bar{8}70$	1.705	vw	490
1.76	w	$\bar{7}80$			
1.90	w	$\bar{7}90$ (or $\bar{5}10,0$)			

FIG. 288. The measurements and interpretation of the zero layer of Figure 183.

The measured ξ values are systematically greater than those on the net, and the difference appears to be proportional to k. b^* is therefore probably slightly greater than 0·175, but the net is sufficiently accurate for interpretation if this point is borne in mind.

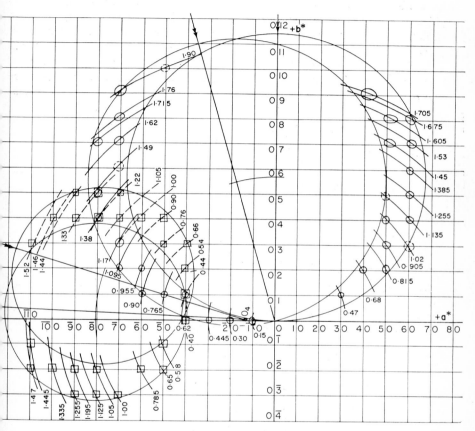

FIG. 289. The $\mathbf{a^*b^*}$ net, circles of reflection and ξ arcs for the interpretation of the zero layer of Figure 183 and for the $+4$ layer of Figure 184. Where the two lunes overlap, the ξ arcs for the 4th layer are broken, and the identified relps are surrounded by squares for the 4th layer and circled in the case of the zero layer. Alternative identifications are shown by dotted squares or circles. Reduced to 1 R.U. $\equiv$ 39·5 mm in reproduction.

(v) For the 4th layer above zero we have to find the position of the origin of construction (Sec. **14.1.3** p. 181). Since $+\mathbf{c}$ has been chosen upwards, O_4 will be in the $-\mathbf{a^*}$ direction from the origin of the net at a distance

$$s = 4c^* \cos \beta^* = 4 \times 0.205 \times 0.225 = 0.184 \text{ R.U.}$$

The radius of the circle of reflection, $R_4 = \sqrt{1-\zeta_4^2} = \sqrt{1-0\cdot795^2} = 0\cdot606$. The directions of the X-ray beam at the two ends of the oscillation are drawn to intersect at O_4, and the rest of the construction follows as for Example 29, always using O_4 as the origin for all constructions. The interpretation is shown in Figs. 289 and 290. The systematic discrepancies indicate that the value of a^* used to construct the net is slightly too large; and for the purpose of designing oscillations to produce high angle reflections for accurate parameter determination, the net should be redrawn with $b^* = 0\cdot177$, $a^* = 0\cdot147$ R.U.

	L.H.S.			R.H.S.	
ξ	I	$hk4$	ξ	I	$hk4$
0·40	vs	$\bar{4}04$	0·44	s	$\bar{4}14$
0·58	s	$\bar{5}14$	0·54	ms	$\bar{4}24$
0·65	vw	$\bar{5}24$	0·66	mw	$\bar{4}34$
0·785	vw	$\bar{6}24$	0·76	ms	$\bar{5}34$
1·00	m	$\bar{7}34$	0·90	m	$\bar{5}44$
1·05	m	$\bar{8}24$	1·00	w	$\bar{6}44$
1·125	vw	$\bar{8}34$	1·105	vw	$\bar{7}44$
1·195	w	$\bar{9}24$	1·22	m	$\bar{8}44$ (or $\bar{7}54$)
1·255	vw	$\bar{9}34$	1·33	vw	$\bar{9}44$ (or $\bar{8}54$)
1·335	vw	$\overline{10,2}4$	1·44	vw	$\bar{9}54$
1·445	vw	$\overline{11,1}4$	1·46	vw	$\overline{10,4}4$
1·47	m	$\overline{11,2}4$	1·52	vw	$\overline{11,3}4$

FIG. 290. The measurements and interpretation of the fourth layer of Figure 184. (Some of the vw reflections are not visible on the reproduction.)

(vi) Figure 184 has some β reflections coming through and spots at the ends of the Laue streaks. (Spots at the outer ends on the right-hand side and the inner on the left-hand side show that the crystal was stationary for a short time at the anti-clockwise end of the oscillation).

31. (i) The reflections lie on constant ξ lines and an important RL vector is therefore parallel to the rotation axis. The constant ζ layer lines show that an important crystal lattice vector is also parallel to the rotation axis. The crystal probably has at least monoclinic symmetry. We therefore take the axes in the zero layer to be $\mathbf{a}^*$ and $\mathbf{c}^*$.

The measured values of ξ and the values of $d^{*2} = \xi^2$ (for the zero layer) calculated from them are given in Fig. 291. (Actually $(10\xi)^2$ is listed for convenience).

Set out the values of d^{*2} to scale along a line as in Fig. 292. We start by assuming that $d_1^* = d_{100}^* = a^*$ and $d_2^* = d_{001}^* = c^*$. These measurements must be made with the greatest possible accuracy. Line 15 is measured as a check since ξ is only just over $0\cdot55$ R.U.